MODELING FOR CASTING AND SOLIDIFICATION PROCESSING

MATERIALS ENGINEERING

1. Modern Ceramic Engineering: Properties, Processing, and Use in Design: Second Edition, Revised and Expanded, *David W. Richerson*
2. Introduction to Engineering Materials: Behavior, Properties, and Selection, *G. T. Murray*
3. Rapidly Solidified Alloys: Processes • Structures • Applications, *edited by Howard H. Liebermann*
4. Fiber and Whisker Reinforced Ceramics for Structural Applications, *David Belitskus*
5. Thermal Analysis of Materials, *Robert F. Speyer*
6. Friction and Wear of Ceramics, *edited by Said Jahanmir*
7. Mechanical Properties of Metallic Composites, *edited by Shojiro Ochiai*
8. Chemical Processing of Ceramics, *edited by Burtrand I. Lee and Edward J. A. Pope*
9. Handbook of Advanced Materials Testing, *edited by Nicholas P. Cheremisinoff and Paul N. Cheremisinoff*
10. Ceramic Processing and Sintering, *M. N. Rahaman*
11. Composites Engineering Handbook, *edited by P. K. Mallick*
12. Porosity of Ceramics, *Roy W. Rice*
13. Intermetallic and Ceramic Coatings, *edited by Narendra B. Dahotre and T. S. Sudarshan*
14. Adhesion Promotion Techniques: Technological Applications, *edited by K. L. Mittal and A. Pizzi*
15. Impurities in Engineering Materials: Impact, Reliability, and Control, *edited by Clyde L. Briant*
16. Ferroelectric Devices, *Kenji Uchino*
17. Mechanical Properties of Ceramics and Composites: Grain and Particle Effects, *Roy W. Rice*
18. Solid Lubrication Fundamentals and Applications, *Kazuhisa Miyoshi*
19. Modeling for Casting and Solidification Processing, *edited by Kuang-O (Oscar) Yu*

Additional Volumes in Preparation

MODELING FOR CASTING AND SOLIDIFICATION PROCESSING

EDITED BY
KUANG-O (OSCAR) YU
RMI Titanium Company
Niles, Ohio

CRC Press
Taylor & Francis Group
Boca Raton London New York

CRC Press is an imprint of the
Taylor & Francis Group, an **informa** business

CRC Press
Taylor & Francis Group
6000 Broken Sound Parkway NW, Suite 300
Boca Raton, FL 33487-2742

First issued in paperback 2019

© 2002 by Taylor & Francis Group, LLC
CRC Press is an imprint of Taylor & Francis Group, an Informa business

No claim to original U.S. Government works

ISBN-13: 978-0-367-39684-8

**Visit the Taylor & Francis Web site at
http://www.taylorandfrancis.com**

**and the CRC Press Web site at
http://www.crcpress.com**

Foreword

Modeling of casting and solidification processes, as we think of it today, can be traced back to the precomputer days of the first half of the twentieth century. To pick just a few examples, the first application of the error function solution to the solidification of ingots was in 1930. An elegant mathematical solution to the microsegregation problem was available in the 1940s. Simple fluid flow analyses were widely employed by foundrymen in the 1950s and thereafter. Of course, because of analytical and computational limitations, these models were necessarily highly simplified and therefore of only limited practical value.

With the advent of the computer and its development in the 1960s, the situation began to change quickly. The 1960s saw the development of quite detailed models for flow and solidification in complex sand castings and in continuous castings. From that point on, the steady and rapid advance of speed and power of computers has changed the world of design and production of cast metal parts. Today computation is an essential tool in modern foundries and cast shops for mold and process design and process control.

In a book I coauthored 40 years ago, we wrote that "metal casting has traditionally been an art and a craft, with secrets of the trade passed jealously from father to son. Only in the last century have science and engineering made noticeable in-roads on materials and processes of the foundrymen. But casting will always be one of the most economical routes from raw material to finished metal products, and it was inevitable the art of the founder would yield to the economy and precision of the engineering approach."

Those words were correct then, but this book shows how much truer they are today. The engineering "rules of thumb" of which we were so proud at mid-century have yielded to the precision of modern modeling. The guesses and the trials and errors we made in reaching suitable gating and risering processes have yielded to modern computational packages—with final results far better optimized than those we achieved with our mid-century "combination of art and science." Of course, in one sense the metal casting "art" of the future will survive; in using modeling in new and innovative ways to produce better components more inexpensively and quickly.

The combination of theory and application presented in this book represents the "new engineering" of casting processes. It is recommended reading for the experienced as well as for the newcomer to the metal casting field, to provide tools for the present as well as an understanding of the direction and power of this new engineering.

Merton C. Flemings
Toyota Professor of Materials Processing
Department of Materials Science and Engineering
Massachusetts Institute of Technology
Cambridge, Massachusetts

Preface

Modeling is a method that uses mathematical equations and computer algorithms to represent certain physical phenomena. The application of modeling techniques to solve engineering problems provides many advantages over conventional trial-and-error methods. With the rapid advances in computer hardware and software, casting process modeling is being increasingly accepted by foundries and molten metal processing plants as a viable engineering tool to solve routine production problems. In order to effectively utilize modeling, process engineers need to have a thorough understanding of the principles of both casting/solidification processing and computer/numerical analysis.

Although numerous technical papers regarding casting process modeling are being published in technical journals and conference proceedings each year, very few books have attempted to provide a systematic introduction to the casting process modeling technology. The objective of this book is therefore to provide a comprehensive technical background as well as practical application examples regarding the technology. The ultimate goal is to increase the application of casting process modeling in production by enhancing the process engineer's understanding of this technology. It can be used as a reference book for process engineers in industry as well as casting/solidification researchers in academia and research institutes. In addition, it can also be used as a textbook for graduate and undergraduate students, the source of future process engineers in casting foundries and molten metal processing plants. This book includes three parts: Theoretical Background, Application to Shape Castings, and Application to Ingot Castings and Spray Forming. Presenting such an extensive amount of information constitutes a tremendous task. The approach that has been taken in the preparation of the book was to involve many experts with different backgrounds. Indeed, 26 dedicated experts from industry, research institutes, and academia contributed to this book.

I am grateful to the contributing authors for the time and effort they devoted to their respective chapters. I also appreciate the contributions of Dr. Francois Mollard, Mr. Patrick A. Russo, and Mr. Dun-Wei Yu, who assisted by proofreading and commenting on the text. Last but not least, I would like to acknowledge my secretary, Ms. Carol Muszik, who patiently prepared the manuscript.

Kuang-O (Oscar) Yu

Contents

Foreword   Merton C. Flemings   *iii*

Preface   *v*

Contributors   *xi*

1   Introduction   1
    Kuang-O (Oscar) Yu

PART I   Theoretical Background

2   Fundamentals of Casting Process Modeling   17
    Daniel L. Winterscheidt and Gene X. Huang

3   Stress Analysis   55
    Umesh Chandra and Alauddin Ahmed

4   Defects Formation   95
    Vijay Suri and Kuang-O (Oscar) Yu

5   Microstructure Evolution   123
    Doru M. Stefanescu

6   Thermophysical Properties   189
    Juan J. Valencia and Kuang-O (Oscar) Yu

7   Quick Analysis   239
    Chungqing Cheng

8   Electronic Data Interchange   263
    Gerald M. Radack

PART II   Application to Shape Castings

9   Sand Casting   291
    Michael L. Tims and Qizhong Diao

10 Lost Foam Casting   317
    Chengming Wang

11 Investment Casting   333
    Dilip K. Banerjee and Kuang-O (Oscar) Yu

12 Permanent Mold Casting   373
    Chung-Whee Kim

13 Die Casting   391
    Horacio Ahuett-Garza, R. Allen Miller, and Carroll E. Mobley

14 Semi-Solid Metalworking   417
    Michael L. Tims

PART III   Application to Ingot Castings and Spray Forming

15 Continuous Casting   499
    Brian G. Thomas

16 Direct Chill Casting   541
    Hallvard G. Fjær and Dag Mortensen

17 Vacuum Arc Remelting and Electroslag Remelting   565
    Lee A. Bertram, Ramesh S. Minisandram, and Kuang-O (Oscar) Yu

18 Electron Beam Melting and Plasma Arc Melting   613
    Yuan Pang, Shesh Srivatsa, and Kuang-O (Oscar) Yu

Contents

19 Spray Forming 655
Huimin Liu

Index *695*

Contributors

Alauddin Ahmed   Pratt & Whitney, East Hartford, Connecticut

Horacio Ahuett-Garza*   Industrial, Welding, and Systems Engineering Department, The Ohio State University, Columbus Ohio

Dilip K. Banerjee   Global Product Engineering, GE Global Exchange Services, Gaithersburg, Maryland

Lee A. Bertram   Department of Chemical and Materials Process Modeling, Sandia National Laboratories, Livermore, California

Umesh Chanda   Modern Computational Technologies, Inc., Cincinnati, Ohio

Chunqing Cheng   NetScreen Technologies, Inc., Sunnyvale, California

Qizhong Diao   ASAT Inc., Fremont, California

Hallvard G. Fjær   Materials and Corrosion Technology Department, Institute for Energy Technology, Kjeller, Norway

Gene X. Huang   The Company Procter & Gamble, Cincinnati, Ohio

Chung-Whee Kim   EKK Inc., Walled Lake, Michigan

Huimin Liu   Ford Motor Company, Dearborn, Michigan

R. Allen Miller   Industrial, Welding, and Systems Engineering Department, The Ohio State University, Columbus, Ohio

Ramesh S. Minisandram   Research & Development, Allvac, an Allegheny Technologies Company, Monroe, North Carolina

Carroll E. Mobley   Materials Science and Engineering Department, The Ohio State University, Columbus, Ohio

---

Current affiliation: Center for Product Design and Innovation, ITESM Campus Monterrey, Monterrey, Mexico

Dag Mortensen   Materials and Corrosion Technology Department, Institute for Energy Technology, Kjeller, Norway

Yuan Pang   Concurrent Technologies Corporation, Pittsburgh, Pennsylvania

Gerald M. Radack   Concurrent Technologies Corporation, Johnstown, Pennsylvania

Shesh Srivatsa   Materials and Process Engineering Department, GE Aircraft Engines, Cincinnati, Ohio

Doru M. Stefanescu   Department of Metallurgical and Materials Engineering, The University of Alabama, Tuscaloosa, Alabama

Vijay Suri   ALCOA CSI, Crawfordsville, Indiana

Brian G. Thomas   Department of Mechanical and Industrial Engineering, University of Illinois at Urbana-Champaign, Urbana, Illinois

Michael L. Tims   Concurrent Technologies Corporation, Johnstown, Pennsylvania

Chengming Wang   Concurrent Technologies Corporation, Johnstown, Pennsylvania

Daniel L. Winterscheidt   Concurrent Technologies Corporation, Johnstown, Pennsylvania

Juan J. Valencia   Concurrent Technologies Corporation, Johnstown, Pennsylvania

Kuang-O (Oscar) Yu   RMI Titanium Company, Niles, Ohio

1
Introduction

Kuang-O (Oscar) Yu
RMI Titanium Company, Niles, Ohio

I. WHY MODELING?

The production of almost all metallic components involves melting and solidification processes. When a molten metal is poured into a mold to make a product with a specific shape, the process is called casting. However, in processes such as water, gas, vacuum, and centrifugal atomization, the molten metal is first disintegrated into small molten droplets which then solidify as powder.

Casting and solidification processing involves many physical phenomena such as fluid flow, heat transfer, electromagnetic force, thermal stress, defect formation, and microstructure evolution. The quality of the final product depends on the mechanisms of defect formation and microstructure evolution, which are controlled by heat transfer, fluid flow, and thermal stress. How to control processing parameters, such as metal superheat, pouring/casting speed, and mold preheat temperature/cooling condition, to provide proper solidification conditions and satisfactory quality castings, has been the subject of intensive investigations.

In production environments, process engineers typically use the trial-and-error method based on empirical relationships between processing parameters and the quality of the resultant castings. This method usually leads to long process development times and high production costs. Process modeling, on the other hand, enables process engineers to make virtual castings using computer techniques. As a result, the effects of processing parameters on the quality of the resultant castings can be evaluated without incurring the cost of actually .making castings. The processing parameters can easily be modified until a set of processing parameters that will result in castings with satisfactory quality is found. By applying casting process modeling, the time and cost of developing new, and enhancing existing, processes can be significantly reduced.

As mentioned earlier, casting and solidification processing involve many complex physical phenomena. Developing a comprehensive model to represent all these phenomena is a very challenging task. From an engineering point of view, it is not always possible or even necessary to have a comprehensive model which simulates all the involved physical phenomena. In general, each casting and solidification processing process has its own unique characteristics which have a dominant effect on the quality of the resultant product. Thus, developing a model that provides an effective way to simulate these unique characteristics is often not only technically sufficient, but also cost effective. In this introduction chapter, a general description of the various metal manufacturing processes and their applications is first presented. The modeling approach used to effectively simulate the unique characteristics of each process is then discussed.

II. UNDERSTANDING METAL MANUFACTURING PROCESSES

The knowledge necessary to establish a model that effectively simulates a casting process is rooted in a general understanding of the various metal manufacturing processes relying on casting as one of their key steps.

There are three basic types of metallic products: cast, wrought, and those made by powder metallurgy (PM). Cast products are used in their as-cast form with little or no machining. The most important feature of the shape casting processes is the capability to produce near net shape components, resulting in significant savings in machining cost. In addition, castings also permit design simplification and parts count reduction. However, since no mechanical work is applied to the final casting to refine its microstructure, mechanical properties and microstructural uniformity are usually inferior to those of wrought products. Mechanical work such as forging, rolling, and extrusion is used to change the shape and refine the microstructures of cast products, resulting in wrought products which typically have finer microstructures and better mechanical properties than the original cast products. The major disadvantages of wrought products are the high machining cost and low material yield typical of the conversion of the input stock into the final products. PM products are made by consolidating metal powder into near net shape components. Conventional PM components usually have lower densities and mechanical properties than wrought products. On the other hand, advanced PM processes can produce fully densed materials which have mechanical properties that are equivalent to or better than those of wrought products.

A. Cast Products

Many different casting processes are used to make shaped components; the following sections briefly describe some of the most common.

1. Sand Casting

Sand casting is the most widely used shape casting process. It uses bonded sand as the mold material and can produce castings that weigh from only a few grams to more than a hundred tons. The sand casting process is applicable to a wide range of metals including aluminum, steel, cast iron, etc. Sand cast products are used by almost all industries, from the high volume, cost sensitive automobile industry to the high unit cost and top quality aerospace industry. Sand casting is always performed in air atmosphere, with the sand mold at room temperature. As a result, sand casting usually results in a relatively rough surface product; it also has only a limited capability to make thin wall components.

2. Investment Casting

The investment casting process uses ceramic molds and can be carried out in vacuum as well as in air. The ceramic molds may be preheated to very high temperatures (e.g., up to 1550°C for nickel-base superalloys), allowing for the producing of thin wall castings. Because of the high ceramic mold preheat temperatures, radiation heat loss from the mold surfaces strongly affects solidification conditions. The use of the vacuum environment enables the investment casting of superalloys and titanium alloys, which have a chemical composition otherwise difficult to control in air. On the other hand, aluminum alloys, steels, and cobalt alloys are typically cast in air. Investment castings are mostly used for aerospace and medical implant applications, which tend to have a relative low production volume but high unit cost. Recently, investment cast golf club heads have become an important nonaerospace application for titanium alloys.

3. Lost Foam Casting

The lost foam casting process has features of both investment casting and sand casting. It uses a coated polystyrene foam pattern imbedded in traditional unbonded sand. During mold filling, the foam pattern is decomposed by the heat of the molten metal. The metal replaces the foam pattern and duplicates all the features of the pattern. The permeable refractory coating on the pattern allows the gases from the decomposing foam to escape rapidly from the mold, yielding castings with a smooth surface. The major advantage of the lost foam casting process is that it can produce castings with a quality similar to that of investment casting, but at a cost close to that of sand casting. Lost foam castings (aluminum alloys and cast iron) are mostly used in the automobile industry.

4. Permanent Mold Casting

Permanent mold casting uses metallic molds; cooling and/or heating channels are sometimes imbedded in critical locations of the mold to facilitate the control of the solidification process. Permanent mold casting process is particularly suitable for high volume production of castings with fairly uniform wall thickness and limited undercuts or intricate internal coring. Compared to sand casting, permanent mold casting can produce castings with more uniform wall thickness, closer dimensional tolerances, superior surface finish, and improved mechanical properties. Alloys that can be cast by the permanent mold casting process include aluminum, magnesium, zinc, copper, and hypereutectic gray iron. Because of the high cost of the metallic tooling, the permanent mold casting process is primarily used for making high-volume components such as those intended for automobile applications.

5. Die and Squeeze Casting

Die casting is another casting process using metallic dies/molds to make high-volume components that are particularly suitable for the automobile industry. However, instead of gravity mold filling as in the permanent mold casting process, die casting relies on pressure to provide very rapid filling of the metallic die. The jetting associated with the extremely rapid mold filling process can cause the entrapment of air in the resultant castings. The entrapped air will then expand to form bubbles during subsequent heat treatment. Because of this, die castings are typically not heat treatable and are limited to applications that do not require high mechanical strength. Recently, squeeze casting, one special form of die casting, has been developed to overcome this shortcoming by employing a slower and more controllable mold filling process to avoid the entrapment of air in the casting. Aluminum, zinc, magnesium, and copper alloys are most commonly made by either die or squeeze casting process.

6. Semi-Solid Metalworking

Semi-solid metalworking (SSM) also relies on a metallic die/mold; it bears some similarities to the die casting process. The two most important features of the SSM process are its reliance on input materials with a unique fine grain microstructure, and an operating temperature between the melt liquidus and solidus temperatures, i.e., in the mushy region. The unique fine grain microstructure of the input material is largely maintained in the final casting, resulting in mechanical properties superior to those of die castings. The major advantage of the SSM process is that it produces components with complex geometries similar to die casting, and yet with mechanical properties comparable to those of wrought products. SSM is a fairly new process and its products are primarily used in automobile and other high volume–high mechanical strength applications. Alloys that have been cast by the SSM process include aluminum, magnesium, and copper.

B. Wrought Products

The majority of metal components are made by making wrought products. For a long time, the input material for wrought processing was made by the conventional ingot casting process. Since the 1960s, continuous and semicontinuous casting processes have been gradually introduced into production. Now, with very few exceptions, most of input materials for wrought products are cast by either continuous or semicontinuous casting processes.

1. Conventional Ingot Casting

Ingots made by the conventional metal (mostly cast iron) mold casting process were the primary source for wrought processing before the 1960s. The productivity of the conventional ingot casting process is inherently low. Large ingots are usually octagon shaped and individually cast, whereas small ingots may have a square cross section and are cast in clusters. A thermally insulated or heated molten metal reservoir at the top of the ingot (called the hot top) is used to feed the solidification shrinkage and reduce the size of the shrinkage pipe or void. The hot top, and the part of the ingot with the shrinkage pipe, are cut off before subsequent forming (forging, rolling, and extrusion) operations, resulting in a significant material loss. Severe macrosegregation may also happen, which has a detrimental effect on ingot quality and hence limits the size of the ingot that can be cast. Today, the conventional ingot casting process is used primarily for small quantity production of certain specialty alloys.

2. Continuous Casting

The advantages of continuous casting in primary metals production have been recognized for more than a century. The dramatic growth of this technology, however, has only been realized since the 1960s. The principal advantages of the continuous casting process are high productivity, high material yield, good product quality, and low energy consumption. The primary purpose of continuous casting is to bypass conventional ingot casting and to cast a form that is directly rollable on finishing mills. The cross-sectional shapes of continuously cast blooms/billets/slabs can be round, square, or rectangular. The principle of the continuous casting process is to form the cast bloom/billet/slab in a continuously withdrawn water-cooled copper mold. To prevent sticking of the frozen casting surface to the copper mold, the mold is normally oscillating during the casting operation and a lubricant is added to the mold metal interface, resulting in a smooth as-cast surface. Beyond the mold, water spray is used to speed up the heat removal; this results in a fast cooling rate and a reduced degree of macrosegregation in the casting. The length of the casting can, in theory, be infinite. A vertical continuous casting machine is most commonly used. The solidifying casting is first curved from a vertical to a horizontal position. The completely solidified casting is then cut to length and subjected to subsequent rolling operations. Very little material is lost due to the hot top, and thus the material yield is high. In addition, the production rate of continuous casting is very high, typically in the hundreds of thousands or even millions of tons per year for steel. Continuous casting is primarily used for ferrous alloys, especially low carbon steels. Currently, most of the world's near 800 million tons of steel produced each year is made by the continuous casting process.

3. Direct Chill Casting

Compared to steel, aluminum and copper have a significantly higher thermal conductivity and thus solidify much faster. As a result, the curved continuous casting process cannot be used for aluminum and copper alloys. In addition, the quantity of metal to be produced for all nonferrous alloys is significantly lower than for steel. Thus, there is no economical incentive to use expensive continuous casting machines to cast nonferrous metals. Consequently, the principal continuous casting process for nonferrous alloys is the direct chill casting (DC casting) process. The vertical DC casting process is a semicontinuous process widely used for making aluminum and copper alloy billets and slabs. DC casting is similar to the ferrous continuous casting process, except that the resultant billets/slabs have a finite length, typically around 8–10 m. As a result, the slab/billet curving and cutting operations, which are important components of the ferrous continuous casting process, are not necessary in the DC casting process. Consequently, the capital cost for DC casting machines is significantly lower than for continuous casting equipment. Recently, horizontal DC casting process has been developed to cast large aluminum alloy slabs for rolling to plate and strip. In general, the withdrawal speed for DC casting is up to 0.2 m/min, significantly lower than that used for the continuous casting of steel (typically 1 m/min). For both DC and continuous casting processes, metal melting and billet/slab casting are uncoupled. Consequently, the molten metal superheat and the billet/slab casting speed can be controlled independently.

4. Vacuum Arc Remelting and Electroslag Remelting

Vacuum arc remelting (VAR) and electroslag remelting (ESR) are two second-ary remelting processes widely used for producing ingots of high performance alloys such as titanium (VAR) and nickel-base superalloys and specialty steels (VAR and ESR). These two processes are semicontinuous and bear some similarities with the DC casting process. The major characteristic of VAR and ESR is the use of a precast or prefabricated electrode as the input material. This electrode is then melted by vacuum arc (VAR) or slag joule heating (ESR). Molten metal droplets falling from the electrode tip accumulate in the water-cooled copper mold or crucible to form an ingot. High power input results in high electrode melting rate (i.e., high ingot casting rate) and high molten metal superheat. This coupled relationship between electrode melt-ing and ingot casting results in a limited processing window yielding ingots with a desirable structure. Contrary to continuous casting and DC casting processes, the principal driving force for using VAR and ESR is ingot quality enhancement, rather than productivity improvement. In fact, the productivity of VAR and ESR is quite low. VAR and ESR ingots have a faster cooling rate, lower inclusion content, lower degree of macrosegregation, better grain struc-ture, and improved forgeability than ingots made by conventional ingot casting process. Because of this, many segregation-prone alloys, which could not be made by conventional ingot casting processes, can now be produced routinely by VAR and ESR processes. Currently, some nickel-base superalloys and titanium alloys used for aerospace applications can only be produced by the VAR process.

5. Electron Beam Melting and Plasma Arc Melting

Electron beam melting (EBM) and plasma arc melting (PAM) are two rela-tively new secondary remelting processes used to improve the quality of tita-nium alloys by removing detrimental inclusions. EBM also produces nickel-base superalloy remelt stock. The heating source for EBM is an electron beam, and PAM is heated by a plasma arc generated by the ionization of helium and/or argon gases. Both processes use a water-cooled copper hearth to hold the molten metal. High density inclusions (HDI) such as tungsten carbide bits are removed since they sink to the bottom of the molten pool in the hearth. Refined clean molten metal then flows into an open mold to form a continu-ously cast ingot. EBM can cast both cylindrical ingots as well as rectangular slabs, whereas PAM currently can only cast cylindrical ingots. The application of EBM and PAM processed titanium alloys is primarily focused on the aero-space industry. EBM is also widely used to recycle commercially pure (CP) titanium.

C. Powder Metallurgy Products

Conventional and advanced powder metallurgy (PM) products have very different processing routes as well as properties. In addition, their intended markets are also different. Spray forming is a relatively new process derived from the advanced PM process.

1. Conventional Powder Metallurgy

The conventional PM process uses sintering to consolidate powders to form complex shaped components. Because sintering is not a melting and solidification process, the density of the resultant PM components is lower than the alloy theoretical density; these components contain porosity and are not fully dense. As a result, the mechanical properties of conventional PM products are lower than those of wrought products made from the same alloys. Sometimes a close die forging operation is used to forge PM preforms to produce components with improved mechanical properties. Powders are made by melt atomization as well as by hydrometallurgy processes. In general, conventional PM products are mostly used for high volume, complex shape, and relatively low mechanical property components. Both ferrous and nonferrous alloys are processed by conventional powder metallurgy.

2. Advanced Powder Metallurgy

The need for high mechanical properties is the primary driving force for using advanced PM products. For jet engine turbine disk applications, conventional superalloy disk alloys such as Inconel 718 and Waspaloy are produced by the VAR process. For alloys like IN100, MERL76, René 95, and René 88, a higher content of strengthening elements (aluminum and titanium) is used to develop superior mechanical strength and temperature capability. However, a higher strengthening element content also results in a stronger segregation tendency during solidification. The resulting ingots typically have an unacceptable forgeability for subsequent open die forging operations. As a result, these alloys cannot be produced by conventional secondary remelting processes (VAR and ESR); they have to be produced by advanced PM processes. In the advanced PM processing route, gas atomized or vacuum atomized powders are consolidated by hot compaction, hot isostatic pressing (HIP), or extrusion to produce fully dense billet material or disk preform. These billets or disk preforms have a forgeability that is significantly better than those of the wrought billets and can be easily close die forged to make the final disks. These disks have a uniform fine grain structure and no macrosegregation. Consequently, their mechanical properties are typically better than or equivalent to those of wrought products. The advanced PM process is primarily used for producing jet engine superalloy turbine disk materials, although some tool steels, high strength aluminum alloys, and titanium alloys are also amenable to this process.

3. Spray Forming

Although the advanced PM process can produce satisfactory products, its processing steps are complex and its production cost is high. Spray forming has the potential to make products with mechanical properties that are equivalent to those of advanced PM products, but at lower cost. The principle of spray forming is to use a mandrel or drum to catch molten metal droplets, produced by gas atomization, before they are completely solidified. The metal droplets hit the surface of the mandrel or drum, are flattened, and accumulate layer by layer to form billets or hollow cylindrical tubes/preforms. The billets can be used as input material for close die forging to make jet engine turbine disks. The tubes can be used in the as-sprayed condition whereas hollow cylindrical preforms can be ring-rolled to form engine frame components. Although superalloy components are the primary applications for spray forming, other high performance alloy components have also been produced by this process.

III. APPLICATION OF CASTING PROCESS MODELING

The application of casting process modeling in a production environment is not just a scientific exercise; it is an important technical step which can have a significant impact on the quality, yield, and hence, cost of the final products. To be successful in applying casting process modeling, process engineers need to have a good understanding of currently available technologies, and their capabilities and limitations, as well as their relevance to practical production issues. The following sections present general instructions on how to successfully apply casting process modeling in a production environment

A. Understanding the Role of Modeling

Modeling is a tool for helping engineers do a better job. As an engineering tool, modeling provides engineers with a way to understand the process dynamics and evaluate the quantitative effects of various process variables on the quality of the resulting products. Furthermore, casting modeling allows process engineers to make virtual castings and to optimize their casting process in terms of quality and yield without actually making castings. These capabilities make modeling more powerful than any other tools previously available to process engineers.

Because of its powerful capabilities, modeling is increasingly accepted as a technology which can improve quality and decrease cost in foundries and molten metal processing plants. For shape casting foundries, the combination of process modeling and rapid prototyping technology makes concurrent product and process development technologies a reality. The widespread use of casting process modeling has already made some positive impact on the production floors of foundries and molten metal processing plants. However, the powerful capabilities of process modeling also sometimes create a false understanding of the essence of modeling technology.

As powerful as it is, modeling is still just a tool. It is up to process engineers, not computers, to make the final decision. Modeling can be used to help process engineers develop new processes as well as optimize current production processes. However, the true power of modeling is in enhancing the engineer's understanding of process dynamics and ability to make a more intelligent judgment. The other important benefit of modeling is that it requires engineers to follow a strict discipline to define, as well as control, the process variables. This situation then results in lower variabilities in process control and product quality. Modeling should not be seen as providing a magic box where one can just push some buttons, and good results will automatically come out. Modeling should also not be treated as a panacea; not all the metal casting problems can be solved by modeling. The best way to apply casting process modeling is first to have a good understanding of the problem, and then to decide whether modeling can help. If the answer is yes, then the next step is to develop a suitable model to address that particular problem.

The justification for applying casting process modeling in foundries and molten metal processing plants is to provide process engineers with a better way to solve the complex technical problems they face in production. Thus, the usefulness of casting process modeling must be justified by its success in solving practical production problems. Having the capability to understand the heat transfer and fluid flow phenomena and to predict the mold filling sequence and molten metal pool profile is just a first step toward that goal. To be able to understand why, and predict when, defects will form is a further step in that direction, but it is still not enough. As one foundry manager once said: "We do not sell defects; we sell good castings." Process engineers have to demonstrate that they can use modeling results to develop a strategy for eliminating defects and producing good products quickly and cost effectively.

B. Possessing the Appropriate Technical Background

Casting and solidification processing involve many physical phenomena, such as fluid flow (mold filling, natural and forced convection), heat transfer, electromagnetic field, solidification, defect formation, and microstructure evolution. It is obvious that one who wants to perform casting process modeling needs a good understanding of the physical meaning as well as the mathematical representation of all these phenomena. In addition, one needs some background in the numerical analysis of differential equations and computer programming. However, it should be emphasized that, from an application point of view, process engineers should concentrate their efforts on understanding the problem, establishing an appropriate model to represent that particular problem, making sense out of the model prediction, and developing a strategy to solve the problem. Thus, process engineers should first have a very good understanding of the production process they are working on. They need just enough background in mathematical equations, numerical analysis, and computer programming techniques to allow them to effectively perform their own tasks. It is not necessary for process engineers to have a deep technical background in differential equation solving and computer code writing in order to perform casting process modeling.

C. Understanding Each Process's Unique Characteristics

Each casting and solidification process has its own unique characteristics that have a major influence on the quality of the resultant products. Understanding the unique characteristics of the particular process that one is using is the first step in establishing a proper model for successfully modeling that process.

1. Differences Between Shape and Ingot Castings

Shape casting processes involve complex shape components and require three-dimensional models to perform process simulation effectively. Model building activities, such as accurately, quickly, and cost effectively inputting the complex casting geometry into the model, as well as establishing a suitable finite element mesh, have a critical impact on the successful application of casting process modeling on the foundry floor. Electronic data interchange (EDI) and automated meshing technologies play major roles in these areas. On the other hand, ingot castings typically have simple geometries such as round, square, and rectangular cross-sectional shapes. In many instances, ingot casting processes can be effectively simulated by using two-dimensional models.

From the technical point of view, the mold filling events have a significant effect on the solidification conditions and structural integrity of shape castings. Once the mold is full, however, natural convection in the remaining molten metal plays only a minimal role in affecting casting quality. Conversely, the mold filling sequence generally has little effect on the structure of continuously or semicontinuously cast ingots, whereas the effect of natural convection can be significant, controlling the macrosegregation severity, especially in large size ingots. In addition, critical defects are also quite different for shape and ingot castings. For example, porosity (macroshrinkage and microporosity) is the most important defect in shape castings, but macrosegregation, which impacts ingot chemistry uniformity and formability in subsequent forging and rolling operations, is of primary concern for most ingots. Finally, shape castings are cast one by one, and always in a transient condition. As a result, true three-dimensional transient models are required. On the other hand, ingots are cast either in truly steady state conditions (continuous casting) or in quasi-steady state conditions (semicontinuous casting processes). In most cases, two-dimensional steady state models are adequate.

2. Differences Among Shape Casting Processes

The characteristics of the various shape casting processes are quite different; thus, different models are needed to effectively simulate the unique characteristics of each process. For example, the sand casting mold can be treated as having a semi-infinite thickness, and the temperature of the mold outer surface can be considered to be a constant. Thus, the boundary condition for the sand casting mold can be simply a constant temperature. Conversely, preheat temperatures for investment casting molds are quite high and the radiation heat transfer rate at the mold surface has a significant effect on the casting solidification conditions. Consequently, radiation view factor calculation is a very critical step on modeling the investment casting process. In addition, because the mold and the molten metal temperatures are quite close to each other, or even identical in directional solidification and single crystal casting processes, mold filling analysis is not needed for investment cast columnar grains and single crystal superalloy turbine airfoils. For metallic mold/die casting processes (die/squeeze casting, permanent mold casting, and SSM), since the mold/die is used repeatedly, the quasi-steady state temperature distribution in the mold/die has important effects on the solidification condition and quality of the resultant castings. Consequently, knowing how to establish an accurate quasi-steady state mold/die temperature distribution is crucial for accurately modeling metallic mold/die casting processes.

3. Differences Among Ingot Casting Processes

Conventional ingot casting is a discrete process; thus, a true transient model is needed. On the other hand, the continuous casting process takes place under truly steady state conditions and hence, a steady state model is commonly used. For semicontinuous casting processes (DC casting, VAR, ESR, EBM, and PAM), both steady state and transient phenomena are important. At the top of the ingot, it is critical to know how to establish an effective hot top procedure to reduce the size of the shrinkage pipe and increase the material yield.

Thus, a transient model is needed to simulate the hot top procedure. However, in the middle portion of the ingot, where a quasi-steady state condition is reached, the shapes of the liquid metal pool and the mushy zone are relatively constant. Thus, a steady state model can be used to predict the liquid pool and mushy zone profiles, as well as their impact on the macrosegregation pattern in the resultant ingot.

In the continuous casting and DC casting processes, the molten metal superheat and ingot casting speed are not related and can be specified independently. On the other hand, in the VAR and ESR processes, electrode melting and ingot casting rates are coupled and usually cannot be controlled independently. This situation leads to narrow processing windows to produce ingots with a desirable structure. In addition, the strong electromagnetic field in VAR, ESR, and PAM processes has important effects on fluid flow behavior and macrosegregation formation tendency in these ingots. Because VAR and ESR processes are primarily used to melt high performance and high segregation tendency alloys, such as superalloys, titanium alloys, and tool steels, macrosegregation has a major effect on the quality of the resultant ingots. Thus, developing a model which can provide an accurate way to evaluate the ingot macrosegregation formation tendency has very practical benefits for the VAR and ESR processes.

4. Differences Between Spray Forming and Casting Processes

Spray forming is a free form deposition process and does not produce products with highly precise geometrical dimensions. Many physical phenomena and defects (e.g., mold filling, shrinkage pipe and macrosegregation) associated with regular casting processes are not present in the spray formed products. Thus, the modeling approach for the spray forming process is quite different from those of the regular casting processes. The major technical challenge for modeling the spray forming process is to predict the molten metal droplet size distribution during gas atomization, the individual droplet cooling/solidification rate before it hits the mandrel/drum, and the consolidation condition of the metal droplets during the deposition process. Porosity formation is primarily due to the entrapment of gas during deposition, not the volumetric change during solidification.

D. Developing a Suitable Model

A "suitable model" has two different meanings. First, as discussed in the above sections, a suitable model should include all the technical features necessary to simulate the unique characteristics of a particular process. Second, a suitable model must simulate the dominant effects that impact current product quality.

Shape castings typically exhibit many different types of defects. However, in practice, process engineers can develop a process based on their experience to produce a particular casting with only one or two defects which are difficult to eliminate. Thus, a suitable model need only be established to eliminate those defects without causing any other defects to form. For example, superalloy single crystal turbine airfoils exhibit many defects such as microporosity, hot tears, cold cracking, dimensional distortion, equiaxed grains, freckles, and recrystallized grains. When making turbine airfoils with one particular alloy, René N6, freckles tend to form. The best approach to solve this problem is to develop a casting process, based on past experience, that can produce castings without any other types of defects, except freckles. Then a suitable model must be developed to eliminate the freckle problem without causing any other defects to form.

E. Ensuring the Model's Accuracy

Model predictions should always be compared with experimental results to ensure the accuracy of the model. Two methods can be used. In the first one, melt temperatures recorded by thermocouples as a function of time are compared with model-predicted cooling curves to verify the casting thermal history. This method is time consuming, very costly, and hence is not always feasible. The second method compares model-predicted defect formation tendency with foundry inspection results. This method is easier to perform and most commonly adopted by foundries. In practice, an approach commonly employed by production foundries is to first establish a baseline simulation condition by comparing the model predictions with foundry experimental results, including both thermocouple data and defect inspection records. Once an appropriate baseline model is established, various process parameters are modified to evaluate their effects on the defect formation tendency. These process parameters are then optimized to develop a process that produces defect-free castings. Finally, actual castings are poured and their inspection results are compared to the model predictions.

It should be noted that only "relatively good" accuracy is needed to verify the model. It is not necessary to have model predicted cooling curves perfectly match experimental thermocouple data. The other point worth mentioning is that modeling is best used to compare the relative differences between different sets of processing parameters and to indicate in which direction they should be modified. Eventually, process engineers need to evaluate all model-predicted results, make sense out of them, and settle on the final casting process.

F. Using Models to Solve Practical Problems

As mentioned earlier, casting process modeling is not just a scientific exercise; it can have an important impact on a company's production performance. Thus, the success of casting process modeling should be judged by its ability to solve practical problems. In general, casting process modeling can be used by process engineers to develop new processes, as well as optimize existing ones. In practice, however, there are some differences in this regard between shape casting processes and ingot casting processes.

For shape casting foundries, each component has its own unique geometry configuration and quality requirement, and hence each requires a unique process to obtain good castings. It is common for a foundry to have 30–50 different parts in production at the same time. As time goes on, some of the old parts drop out and new parts come in. As a result, developing a new casting process for a new part happens all the time. Consequently, the application of casting process modeling in shape casting foundries puts more emphasis on developing new processes than on enhancing existing ones.

In ingot casting, due to the simple geometry and large volume production rate, enhancing existing processes happens more frequently than developing new processes. For example, Inconel 718 and Waspaloy are two nickel-base superalloys which are widely used for jet engine turbine disk applications. For the last 30 years, these two alloys have been produced by the VAR process. The industry standard process is to melt a 432 mm (17 in.) diameter electrode in a VAR furnace to make a 508 mm (20 in.) diameter ingot. It can be seen that the need for developing a new process for this product is minimal. However, there are two separate requirements for enhancing the existing process. First, since the electrode is produced by the conventional ingot casting process, longitudinal shrinking pipe and horizontal cracks are sometimes present. During VAR processing, the shrinkage pipe and the cracks can cause arc instability and result in variability in ingot quality. The combination of process modeling and experimentation can provide an insight into understanding and developing a way to minimize this variability. Second, the current process needs to be modified to make ingots that are larger than the current 508 mm diameter. However, as the ingot diameter becomes larger, so does the tendency to form freckle-type defects. Process modeling is an effective way to simulate ingot solidification conditions and provide information for developing a modified process. One possible approach is to develop a modified process for the large diameter ingot, which has a similar Local Solidification Time (LST) as the currently produced 508 mm diameter ingot. Because the formation of freckles is related to the ingot LST, similar LST for both ingot sizes will increase the probability to produce freckle-free, large diameter ingots.

IV. SUMMARY

A general overview regarding the application of various casting and solidification processing processes in metal manufacturing, as well as instructions on how to effectively model these processes, has been presented. The following chapters will first introduce the general background on casting process modeling. The application of modeling to each specific process will then follow.

2
Fundamentals of Casting Process Modeling

Daniel L. Winterscheidt
Concurrent Technologies Corporation, Johnstown, Pennsylvania

Gene X. Huang
The Procter & Gamble Company, Cincinnati, Ohio

I. INTRODUCTION

Casting process modeling involves the simulation of mold filling and solidification of the cast metal. At the macroscopic scale, these processes are governed by basic equations which describe the conservation of mass, momentum, and energy. This chapter focuses on fundamentals of casting process modeling. Special emphasis will be placed on heat transfer (which is of obvious importance in solidification simulation), fluid dynamics (which is necessary to model mold filling and the natural convection which may occur during solidification), and general procedures for performing casting process modeling. Thermal-mechanical modeling (stress analysis) is described in Chapter 3. The modeling of microstructural evolution (micromodeling) is discussed in Chapter 5.

The chapter is outlined as follows. The fundamentals of heat transfer are first described in Sec. II and then applied to solidification heat transfer modeling in Sec. III. Next, the principles of fluid dynamics are presented in Sec. IV and applied to mold filling simulation in Sec. V. General numerical methods and special techniques used to solve the governing equations are then discussed in Secs. VI and VII. The types of commercially available software are described in Sec. VIII, and step-by-step modeling procedures are explained in the last section of the chapter.

II. HEAT TRANSFER

Heat transfer is perhaps the single most important discipline in casting simulation. The solidification process depends on heat transfer from the part to the
mold and from the mold to the environment. There are three possible modes of
heat transfer: (1) *conduction*, (2) *convection*, and (3) *radiation*. Conduction
refers to the heat transfer that occurs as a result of molecular interaction.
Conduction is important in modeling heat transfer in the cast part (both liquid
and solid states) and is the primary mode of transfer through the mold (solid
state only). Convection refers to heat transfer that results from the movement
of a fluid. Convection heat transfer is important to the liquid metal both during
mold filling (forced convection) and after the mold is filled (natural convection)
as well as cooling of the mold exterior to the atmosphere. Radiation heat
transfer refers to the transfer of electromagnetic energy between surfaces, a
process which does not require an intervening medium. Radiation heat transfer
is most important in investment casting processes. The different modes of heat
transfer are described in the following subsections. The interested reader may
consult one of several textbooks, such as Ref. 1, for additional details.

A. Conduction

The temperature at any point in a medium is associated with the energy of the
molecules in the vicinity of the point. When molecules collide, energy is transferred from the more energetic (higher temperature) molecules to the less energetic (lower temperature) molecules. Thus conduction heat transfer must occur
in the direction of decreasing temperature. The rate of heat transfer by conduction is given by Fourier's law:

$$\mathbf{q} = -k\nabla T \tag{1a}$$

In Cartesian coordinates,

$$q_x = -k\frac{\partial T}{\partial x} \qquad q_y = -k\frac{\partial T}{\partial y} \qquad q_z = -k\frac{\partial T}{\partial z} \tag{1b}$$

The above expression simply states that *heat flux* (heat transfer rate per unit
area) is proportional to the temperature gradient. The proportionality constant
k is the *thermal conductivity* of the material. The minus sign indicates that heat
is transferred in the direction of decreasing temperature.

   The differential equation describing heat conduction is given by

$$\rho c\frac{\partial T}{\partial t} + \nabla \cdot \mathbf{q} = \dot{Q} \tag{2a}$$

where $\dot{Q}$ is a heat generation term, ρ is the *density*, c is the *specific heat*, and t represents time. In Cartesian coordinates, the heat conduction equation is

$$\rho c \frac{\partial T}{\partial t} - \frac{\partial}{\partial x}\left(k\frac{\partial T}{\partial x}\right) - \frac{\partial}{\partial y}\left(k\frac{\partial T}{\partial y}\right) - \frac{\partial}{\partial z}\left(k\frac{\partial T}{\partial z}\right) = \dot{Q} \qquad (2b)$$

The solution of the above equation in a given domain (region of space) requires knowledge of initial and boundary conditions. The *initial conditions* define the temperature distribution throughout the domain at some initial point in time. The *boundary conditions* describe the conditions that must be satisfied on the boundary of the domain. For equations such as Eq. (2), boundary conditions must be specified on the entire boundary. These conditions may describe (1) the value of the dependent (unknown) variable, (2) the value of the spatial derivative of the variable in a direction normal (perpendicular) to the boundary surface, or (3) a combination of these conditions. When applied to the heat conduction equation, these conditions become (1) prescribed temperature, (2) prescribed normal heat flux (which is a condition on the spatial derivative), and (3) a convection and/or radiation condition discussed in the following sections.

B. Convection

In addition to the energy transfer due to random molecular motion, energy is transferred by the bulk motion of a fluid. Convection heat transfer can be categorized as either forced or natural (free). The term *forced convection* is used when the flow is caused by some external means, such as a fan. The term *natural convection* is used when the flow is caused by buoyancy forces in the fluid.

Of particular interest is the heat transfer that occurs between a fluid (i.e., liquid metal) in motion and a stationary surface, such as the mold surface. In casting, liquid metal is in contact with the mold's interior surface and air (or possibly water) is in contact with the mold's exterior surface. In addition to a velocity boundary layer, a thermal boundary layer exists where the temperature varies from the surface temperature to the fluid freestream temperature. Heat transfer at the mold exterior could be determined by modeling the fluid dynamics and heat transfer in the fluid (air or water) in conjunction with the heat transfer in the solid (mold). Such an approach is fundamentally sound but requires a great deal of computational effort. The most common approach in this case is to approximate the heat flux at the solid surface using Newton's law of cooling:

$$q = h(T_s - T_\infty) \qquad (3)$$

Here the *heat flux* (normal to the surface) is taken to be proportional to the difference between the surface and freestream fluid temperature. The proportionality constant h is the *convection heat transfer coefficient*. The value of this parameter depends on the nature of the fluid flow as well as the properties of the fluid and the solid.

Empirical relationships for the convection coefficient are frequently expressed in terms of certain dimensionless groups. For forced convection, the *Nusselt* number (Nu) is typically a function of the *Reynolds* (Re) and the *Prandtl* (Pr) numbers, defined as

$$\text{Nu} = \frac{hL}{k} \qquad \text{Re} = \frac{LV\rho}{\mu} \qquad \text{Pr} = \frac{c\mu}{k} \tag{4}$$

where L is a characteristic length of the problem and V is the freestream velocity. The conductivity k, density ρ, specific heat c, and the viscosity μ are fluid properties. The convection coefficient for natural convection is generally given in terms of the *Grashof* number:

$$\text{Gr} = \frac{g\beta\rho^2(T_s - T_\infty)L^3}{\mu^2} \tag{5}$$

where g is the gravitational acceleration and β is the coefficient of thermal expansion:

$$\beta = -\frac{1}{\rho}\left(\frac{\partial \rho}{\partial T}\right)_p \tag{6}$$

C. Radiation

Thermal radiation is energy emitted by all matter at temperatures above absolute zero. The energy of the radiation field is transported by electromagnetic waves and therefore does not require a material medium. The radiative heat flux emitted from a surface is given by

$$q = \varepsilon\sigma T_s^4 \tag{7}$$

where σ is the Stefan-Boltzmann constant ($\sigma = 5.67 \times 10^{-8}\,\text{W/m}^2\,\text{K}^4$) and the *emissivity* ε is a radiative property of the surface. A perfect emitter (blackbody) has an emissivity value of unity. In many cases heat transfer due to radiation is negligible compared to convection. In other cases, such as vacuum investment casting processes, radiation is the dominant (perhaps the only) mode of heat transfer.

The above expression describes the radiation emitted by a surface. Determination of the net rate at which radiation is exchanged between surfaces is much more complicated. A rigorous approach involves determining a *radiosity* value for each surface element. The radiosity includes direct emission from the surface as well as the reflected portion of irradiation received by the surface. This method requires the simultaneous solution of radiosity values for each surface element. Since each element interacts with many other elements, this approach can be very computationally demanding for three-dimensional problems.

A frequently used alternative approach is to approximate the radiative heat flux from a surface element to the surrounding environment as

$$q = F\varepsilon\sigma(T_s^4 - T_\infty^4) \tag{8a}$$

where F (radiation view factor) is the fraction of radiation leaving the surface which reaches the surroundings. The view factor F is necessary in order to account for obstructions in the view path. The above expression may be cast in the following convective form:

$$q_{rad} = h_{rad}(T_s - T_\infty) \tag{8b}$$

by defining a radiation coefficient as

$$h_{rad} = F\varepsilon\sigma(T_s^2 + T_\infty^2)(T_s + T_\infty) \tag{9}$$

III. SOLIDIFICATION

Solidification modeling involves the application of the heat transfer concepts described in the previous section along with techniques to account for the release of latent heat during solidification. The mold and any other solid materials (chill, insulation, etc.) are modeled using the standard heat conduction equation (2). For the solidifying metal, special procedures are required to accurately model the latent heat release.

A. Fraction of Solid

The extent of solidification at any location within the casting is represented by the *fraction of solid* f_s. At temperatures greater than or equal to the *liquidus* temperature, the cast metal is in a completely liquid state with a solid fraction value of zero. As the latent heat is removed, the fraction of solid increases and reaches a value of unity when the metal is in a completely solid state. The temperature at this point is called the *solidus*. The region where the solid fraction is between zero and unity is referred to as the *mushy zone*.

One of two approaches may be used to determine the solid fraction value.

1. Solid Fraction–Temperature Equilibrium

With this widely used approach the solid fraction is assumed to be a known function of temperature, essentially a temperature dependent "property" of the metal. There are several ways to describe the solid fraction variation between the liquidus and solidus temperatures. The simplest approach is to assume that the solid fraction varies linearly in the mushy zone. Alternatively, an analytical expression such as the Scheil equation [2] may be used. Perhaps the best approach is to determine the solid fraction–temperature relationship using experimental measurements.

2. Solidification Kinetics

The previously described method is simple but only approximate. The reality of the solidification process is that the solid fraction evolves in time in a manner that depends on several parameters. The solidification kinetics approach involves the time integration of a solid fraction evolution equation [3]. This approach permits the accurate prediction of phenomenon such as undercooling. However, the solidification kinetics approach requires detailed metallurgical data which may not be known. Readers interested in the modeling of solidification kinetics can refer to Chapter 5 and Ref. 3.

B. Latent Heat

The release of latent heat during solidification can be accounted for by a heat "generation" term:

$$\dot{Q} = \rho \, \Delta H_f \, \frac{\partial f_s}{\partial t} \tag{10}$$

where ΔH_f represents the latent heat of solidification. The above expression assumes that the latent heat varies in proportion to the solid fraction, which is a reasonable approximation for casting process modeling. There are three common methods of incorporating the latent heat term given by Eq. (10) into the heat conduction equation (2) [4–7].

1. Latent Heat Source Term

Here the latent heat release is treated as a source term and is determined from known parameters (not dependent on the unknown temperature). Substituting Eq. (10) into Eq. (2b) yields:

$$\rho c \frac{\partial T}{\partial t} - \frac{\partial}{\partial x}\left(k\frac{\partial T}{\partial x}\right) - \frac{\partial}{\partial y}\left(k\frac{\partial T}{\partial y}\right) - \frac{\partial}{\partial z}\left(k\frac{\partial T}{\partial z}\right) = \rho \Delta H_f \frac{\partial f_s}{\partial t} \qquad (11)$$

This method is commonly used to couple the solidification kinetics approach described in the previous section with a general heat transfer solver.

2. Apparent Specific Heat

The latent heat release rate described by Eq. (10) depends on the change in solid fraction which depends on the temperature change. It is therefore necessary to include this temperature dependency on the left-hand side of Eq. (11). The latent heat release rate can be expressed as

$$\rho \Delta H_f \frac{\partial f_s}{\partial t} = \rho \Delta H_f \frac{\partial f_s}{\partial T}\frac{\partial T}{\partial t} \qquad (12)$$

The above term can be included in the left-hand side of Eq. (11) by defining an apparent specific heat as

$$c_{app} = c - \Delta H_f \frac{\partial f_s}{\partial T} \qquad (13)$$

For the case where the solid fraction is assumed to vary linearly between liquidus and solidus, the above expression becomes

$$\begin{aligned} c_{app} &= c + \frac{\Delta H_f}{\Delta T} \qquad T_s < T < T_l \\ c_{app} &= c \qquad\qquad T \le T_s \text{ or } T \ge T_l \end{aligned} \qquad (14)$$

where ΔT is the difference between the liquidus and solidus temperatures.

The apparent specific heat method is easily implemented as a modification to the temperature dependent specific heat values. Although simple to implement, the method may develop problems when the solidification range ΔT is small. The first problem is one of iterative convergence. A frequently encountered condition is one where some of the temperature values "jump" back and forth (around the mushy zone) and fail to converge to a particular value. The second problem is that it is possible for a temperature value to prematurely jump below the solidus temperature and miss some of the latent heat release. Modifications that improve the performance of the basic method are described in Ref. 4.

3. Enthalpy Method

The enthalpy method is currently the most common approach used to model latent heat release during solidification. With this method the latent heat is included in the definition of enthalpy:

$$H(T) = \int_{T_{ref}}^{T} c(T)\, dT + (1 - f_s)\, \Delta H_f \tag{15}$$

and the heat conduction equation is expressed as

$$\rho \frac{\partial H}{\partial t} - \frac{\partial}{\partial x}\left(k \frac{\partial T}{\partial x}\right) - \frac{\partial}{\partial y}\left(k \frac{\partial T}{\partial y}\right) - \frac{\partial}{\partial z}\left(k \frac{\partial T}{\partial z}\right) = 0 \tag{16}$$

The method involves iteration between temperature and enthalpy values until convergence is achieved. The enthalpy method ensures that all of the latent heat release is accounted for, even when fairly large time steps are used in the calculation.

C. Initial and Boundary Conditions

1. Initial Conditions

The most accurate method of determining the initial conditions is to perform a mold filling simulation (described in Sec. V). The temperature distribution at the end of the mold filling simulation will then serve as the initial temperature distribution for the solidification simulation. However, because of the computational demand, the mold filling simulation is frequently omitted and some reasonable initial temperature values are used. For the cast metal, the initial temperature usually is set somewhere between the liquidus temperature and pouring temperature, depending on the estimated loss of superheat during mold filling. The initial mold temperature depends on the type of casting. For sand castings the initial mold temperature will most likely correspond to the ambient temperature. For investment castings, the mold is preheated to a specified temperature which depends on the metal to be cast. Permanent mold and die castings typically have high production rates and therefore the mold achieves a quasi-steady thermal condition after several castings have been poured. It may be necessary to perform several simulations to accurately determine the thermal condition of the mold. Interested readers can refer to the appropriate chapters of this book.

2. Boundary Conditions

a. *Mold Exterior.* Although the temperature or heat flux can be prescribed at the mold exterior, the convective condition given by Eq. (3) is most frequently employed. The data shown in Table 1, taken from Ref. 6, should serve as a guide in determining the convection coefficient.

In some casting processes, such as investment casting, heat loss due to radiation is more significant than convection. For these cases, the radiation

Table 1   Convection Heat Transfer Coefficients

Mode	h $(W/m^2 {}^\circ C)$	h $(Btu/hr\, ft^2 {}^\circ F)$
Free convection, $\Delta T = 30^\circ C$		
Vertical plate 0.3 m (1 ft) high in air	4.5	0.79
Horizontal cylinder, 5 cm diameter, in air	6.5	1.14
Forced convection		
Airflow at 2 m/s over 0.2 m^2 plate	12	2.11
Airflow at 35 m/s over 0.75 m^2 plate	75	13.17
Air at 2 atm flowing in 2.5 cm diameter tube at 10 m/s	65	11.41
Airflow across 5 cm diameter cylinder at 50 m/s	180	31.60

Source: Ref. 6.

condition given by Eq. (8) may be used to replace or perhaps supplement the convection condition.

b. Metal/Mold Interface.   The metal and mold are generally not in perfect contact and therefore a temperature discontinuity exists at the metal/mold interface or gap. If the characteristics of the gap such as the thickness variation with time and the properties of the gas within the gap were known, then the heat transfer across the gap could be computed directly. A much simpler and more common approach is to express the interfacial heat flux as

$$q = h_{gap}(T_c - T_m) \tag{17}$$

where T_c and T_m are the temperatures of the casting surface and mold surface, respectively. The gap heat transfer coefficient h_{gap} must be determined empirically. Typical values, taken from Ref. 6, are given in Table 2.

The actual gap heat transfer coefficient depends strongly on the state of the solidifying metal. Although there is generally no gap at temperatures above the liquidus temperature, thermal contact between the casting and the mold is imperfect because of surface tension effects, oxide layers, and mold coatings. The gap coefficient is therefore fairly large, but not infinite. In the mushy region, there is partial contact between the surfaces resulting in a lower gap heat transfer coefficient. Below the solidus temperature, it can be assumed that a gap has formed causing heat loss by convection and radiation. All of this behavior can be modeled using a gap heat transfer coefficient which is dependent on metal surface temperature, as shown in Fig. 1. Such temperature dependent coefficients have been determined using an inverse heat transfer approach [7].

Table 2   Metal/Mold Interface Heat Transfer Coefficients h_{gap}

Casting situation	h_{gap} $(W/m^2\,°C)$	h_{gap} $(Btu/hr\,ft^2\,°F)$
Ductile iron in cast iron mold coated with amorphous carbon	1709	300
Steel in cast iron mold	1025	180
Aluminum alloy in small copper mold	1709–2563	300–450
Steel chilled by steel mold		
Before gap forms	399–1025	70–180
After gap forms	399	70
Aluminum die castings		
Before gap forms	2506–5012	440–880
After gap forms	399	70

Source: Ref. 6.

c. Cast Metal Surface.   At the inflow surface and open risers, the cast metal is exposed to the ambient, typically air. The heat transfer at these surfaces may be considered negligibly small, or can be modeled using a convection boundary condition.

D.   Approximate Analysis

With certain assumptions, it is possible to obtain an analytical solution to the heat transfer problem. The results of such analysis may be used to develop quick and approximate methods which are discussed in Chapter 7. These approximate solutions are also sometimes used to avoid the expense of actual

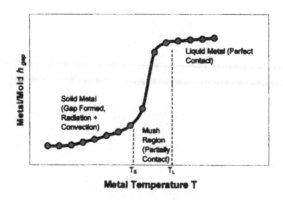

Figure 1   Metal/mold interfacial heat transfer coefficient as a function of temperature.

modeling the mold region. In these cases, only the part region is modeled and the heat flux at the part surface is approximated using analytical expressions. Such methods are most commonly used to approximate solidification in thick molds, which can be approximated as infinitely thick. Assuming constant values for the metal temperature (melting point) and mold thermal properties, analytical solutions can be obtained for castings of simple geometry. For example, the heat flux from a spherical casting surface into a concave, spherical mold wall is given by [8]

$$q = k(T_s - T_i)\left(\frac{1}{R} + \frac{1}{\sqrt{\pi \alpha t}}\right) \tag{18}$$

where T_s is the temperature of the mold surface, T_i is the initial mold temperature, R is the radius of the spherical mold surface, and α is the thermal diffusivity of the mold. Equation (18) can be employed to approximate the heat flux on surfaces of arbitrary geometry by substituting the surface radius of curvature for R in the above expression.

IV. FLUID DYNAMICS

Both the mold filling process and the natural convection that may occur during casting are governed by general fluid dynamics principles [9, 10]. In this section the governing equations of fluid dynamics are presented and several important concepts are discussed.

A. Fluid Dynamics Equations

In general, the equations which describe fluid dynamics are more complicated and require greater computational effort to solve than the heat conduction equation (2). In conservative form, the governing equations can be expressed as [11]

1. Conservation of mass:

$$\frac{\partial \rho}{\partial t} + \nabla \cdot (\rho \mathbf{V}) = 0 \tag{19}$$

where $\mathbf{V}$ is the velocity vector. Equation (19) is known as the continuity equation.

2. Conservation of momentum:

$$\frac{\partial}{\partial t}(\rho \mathbf{V}) + \nabla \cdot (\rho \mathbf{V}\mathbf{V} - \mu \nabla \mathbf{V}) + \nabla P = \mathbf{B} + \mathbf{S}_v \tag{20}$$

where P is the pressure, $\mathbf{B}$ is the body force vector, and $\mathbf{S}_v$ consists of viscous terms other than those expressed by $\nabla \cdot (\mu \nabla \mathbf{V})$. The body force vector typically includes the force of gravity $\mathbf{B} = \rho\mathbf{g}$, but may also include other terms such as Coriolis and centrifugal forces in the case of a rotating frame of reference or forces induced by electromagnetic fields. Equation (20) is a vector equation, describing the conservation of momentum in the three coordinate directions. These equations are known as the Navier-Stokes equations.

3. Conservation of energy:

$$\frac{\partial}{\partial t}(\rho H) + \nabla \cdot (\rho \mathbf{V} H - k\nabla T) = S_e \tag{21}$$

where S_e is a source term which may include internal heat generation, viscous dissipation, and other effects.

Equations (20) and (21) may be expressed in the following general form:

$$\frac{\partial}{\partial t}(\rho\phi) + \nabla \cdot \mathbf{J} = S \tag{22}$$

where ϕ is a general dependent variable and $\mathbf{J}$ is a flux vector which consists of convection and diffusion fluxes. The above equations are written in (flux) conservative form. The equations may also be expressed in nonconservative form. For example, using the continuity equation (19), the energy equation (21) may be expressed as

$$\rho\frac{\partial H}{\partial t} + \rho\mathbf{V} \cdot \nabla H - \nabla \cdot (k\,\nabla T) = S_e \tag{23}$$

Numerical formulations based on finite/control volume concepts (discussed in Sec. VI) are always applied to equations expressed in conservative form. Other numerical methods may be applied to the equations in conservative or nonconservative form.

Appropriate initial and boundary conditions are required to obtain a solution to the fluid dynamics problem. The conditions required to solve the energy equation are the same as described for the heat conduction equation (2). The initial velocity of the fluid is generally taken to be zero. For an internal flow problem, boundary conditions must be prescribed along the walls and at the inflow and outflow regions. For an inviscid fluid, the velocity normal to stationary solid walls must be zero, but the tangential components are not constrained (slip is permitted). For a viscous fluid, all velocity components should be zero (no-slip condition). In practice, however, some type of slip condition is sometimes permitted for high Reynolds number cases. At porous walls, the normal component of velocity is prescribed. At the inflow and outflow boundaries either the velocity or the pressure is required.

B. Types of Fluid Flow

Fluid flow can be described as either compressible or incompressible. The fluids in compressible flows are frequently gases that obey an equation of state which relates the density to the pressure and temperature. For incompressible flows, the fluid density is considered to be independent of pressure. Most liquid flows are considered incompressible because the fluid density is essentially independent of pressure. In casting processes, the metal flow is treated as incompressible, although the density is generally temperature dependent.

Fluids are commonly categorized according to their viscous behavior. A fluid is referred to as a *Newtonian fluid* if the viscous stress varies linearly with the shear rate. Simple fluids such as water, oils, gases, and most liquid cast metals are considered to behave in this manner. Other materials whose viscous stress varies nonlinearly with the shear rate are called *non-Newtonian fluids*. The interested reader may refer to a fluid dynamics text such as Ref. 12 for additional details.

Fluid flow may be classified as either laminar or turbulent. Laminar flow is characterized by smooth and predictable movement of fluid particles. In contrast, turbulent flow is erratic and chaotic with the velocity of the fluid fluctuating in an apparently random manner. The transition from laminar to turbulent flow is associated with some critical value of the Reynolds number. For example, for flow in a pipe, the critical Reynolds number is found to be between 2000 and 13,000, depending upon the smoothness of the entrance conditions and wall roughness [13].

It is generally believed that the governing equations (19) to (21) describe turbulent flows completely. A direct numerical approach is impractical, however, because turbulent motion involves both large and extremely small length and time scales. The current practice is to describe turbulent motion in terms of time-averaged quantities rather than instantaneous ones. The instantaneous quantities are expressed as the sum of the time-averaged value and a fluctuation. Due to the nonlinearity of the governing equations, this process leads to a number of unknown correlations between the fluctuating components of velocity and temperature. Turbulence modeling involves approximating these correlations by expressing them in terms of mean-flow quantities [13].

One of the best known turbulence models is known as the k-ε model [9]. In this model, the turbulence field is characterized in terms of two variables, the turbulent kinetic energy k and the viscous dissipation rate of turbulent kinetic energy ε. A turbulent viscosity μ_t can be directly related to the turbulent quantities k and ε. It is necessary to solve equations for the turbulent quantities in addition to the basic conservation equations. These equations involve a number of turbulence model "constants" which can only be determined empirically. Fortunately, the fluid flow during casting is generally laminar. However, turbulence is experienced in certain cases such as a mold filling simulation for die casting.

V. MOLD FILLING SIMULATION

The simulation of mold filling involves the application of the fluid dynamics principles described in the previous section along with methods to model free surface dynamics and account for factors unique to the casting process.

A. Free Surface Modeling

Modeling the free surface dynamics during mold filling is one of the most challenging aspects of casting process simulation. Although many methods have been proposed to model free surface flow problems, the volume of fluid (VOF) method developed by Hirt [14] is most commonly used. With this method, a fractional VOF function F is employed, where fluid (liquid metal) regions are assigned values of $F = 1$ and gas (typically air) regions are assigned values of $F = 0$. A region, such as a computational cell, with $0 < F < 1$ is considered to be partially filled with liquid. The VOF equation is obtained by applying the continuity equation (19) to the fluid region:

$$\frac{\partial(\rho_l F)}{\partial t} + \nabla \cdot (\rho_l V F) = 0 \tag{24}$$

where ρ_l is the density of the liquid (metal). The free surface is tracked by solving Eq. (24) for the VOF function F along with the basic governing equations. Solution of the above equation requires special techniques to ensure that there are no overshoots ($F > 1$) or undershoots ($F < 0$) and that there is little "smearing" in the solution. At the free surface, the VOF function should sharply change from $F = 1$ on the liquid side to $F = 0$ on the gas side. Standard solution methods which avoid overshoots and undershoots typically result in significant "smearing" (spreading of fluid) over several mesh cells. Accurate free surface modeling requires that special methods be employed to eliminate or greatly reduce the level of smearing

B. Back Pressure Modeling

During mold filling, the gas (typically air) present in the mold escapes through vents and through the mold itself, if porous. In sand and lost foam casting processes the mold is porous and generally does not require venting. Investment, die, and permanent mold castings require vent holes to permit the gas to escape. If not adequately vented, gas pressure will build up, preventing normal mold filling. Furthermore, the entrapped gases result in high levels of porosity. Gas flow through a porous mold can be approximated using D'Arcy's law [15]:

$$V = \frac{p \, \Delta P}{\mu \, L} \tag{25}$$

where p is the specific permeability of the mold, ΔP is the pressure drop across the thickness of the mold L, and μ is the viscosity of the gas. A similar equation can be used to describe gas escape through vent holes. The gas pressure could be modeled directly using the fluid dynamics equations along with a gas equation of state. Equation (25) would serve as a boundary condition to apply at the part/mold interface. This approach is generally not taken, however, because of the large computational cost involved. Instead, the fluid dynamics equations are typically solved only in the liquid metal region. The gas pressure can be estimated from knowledge of the volume of gas and the change in gas mass as a result of venting.

C. Effect of Solidification on Fluid Flow

The fact that the liquid metal is in the process of solidification must be accounted for in the fluid dynamics computations. This effect may be accounted for by using an apparent viscosity model where the viscosity would increase rapidly as the solid fraction increased, becoming very large at some critical solid fraction value. A more common approach is to use an interdendritic fluid flow model [16]. Here it is assumed that interdendritic flow behaves like flow through a porous material. This assumption results in a friction drag term to be added to S_v in Eq. (20).

D. Initial and Boundary Conditions

The heat transfer conditions were described in Sec. III.C. The fluid flow and free surface modeling conditions are described below.

1. Initial Conditions

Generally the liquid metal is assumed to be initially at rest. The VOF function is initialized depending on the actual conditions. Frequently, a region such as a pouring basin will be assumed to be filled with liquid metal at the start of the simulation.

2. Boundary Conditions

a. Inflow.   Inflow conditions can involve a known pressure or fluid velocity, which may be determined from the average pouring rate. In some cases, a relationship between the pressure and entrance velocity is determined based on approximate analysis [17].

b. Part/Mold Interface.   For the liquid metal, a no-slip condition can be imposed or a partial slip condition may be employed. Gas escape through the mold may be modeled using Eq. (25).

c. Gas Escape Vents.   Gas escape through vents may be modeled by using an equation similar to Eq. (25).

d. Liquid Metal Free Surface.   The total stress must be continuous across the free surface. This condition establishes a relationship between the liquid viscous stress, liquid pressure, gas pressure, and surface curvature if surface tension is taken into account. The gas pressure may be computed as described in Sec. V.B, or may be assumed equal to the ambient pressure if the mold is well vented.

VI.   DISCRETIZATION OF THE GOVERNING EQUATIONS

Discretization involves representing the continuous variables with a number of discrete values associated with the cells (elements) or vertices (nodes) of a computational grid (mesh). Through this process, the differential equations are approximated by a set of algebraic equations, which may be solved for the unknown discrete values.

A.   Grid Structure

Computational grids may be classified as structured or unstructured [18]. In structured grids, all the mesh points lie on the intersection of curvilinear coordinate lines. One of two approaches may be used to generate a structured grid for a domain of arbitrary geometry. A two-dimensional example of the first approach is illustrated in Fig. 2. Here a regular grid is overlaid on the problem geometry. Curved boundaries are approximated by a "stair step" type of pattern. This meshing approach is fast and easy to implement, but the representation of the boundary surface is only approximate. With the second approach, a grid is constructed which conforms to the problem geometry. In this case, the boundary representation is very accurate, but the mesh generation procedure is more complex. For most practical problems, it is necessary to divide the domain up into a number of "blocks" and mesh each block separately.

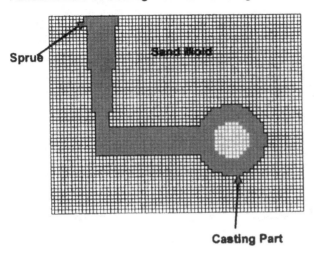

Figure 2 A structured two-dimensional mesh of a cast part and its sand mold.

When applied to three-dimensional problems, this approach can be extremely time consuming.

An unstructured (finite element type) grid for the same geometry is illustrated in Fig. 3. Here the domain may be approximated by triangular and/or quadrilateral cells. A three-dimensional domain may be represented by a combination of hexahedron (brick), triangular prism (wedge), and tetrahedron elements. A typical three-dimensional finite element mesh is shown in Fig. 4. An unstructured mesh permits an accurate representation of geometry, but can be time consuming to generate. In two dimensions, automatic mesh generators generally perform very well for quadrilateral as well as triangular elements. In three dimensions, fully automatic mesh generation is possible only with tetrahedral elements. It should be pointed out that the results obtained from properly constructed quadrilateral (2-D) and brick (3-D) meshes are generally superior to those obtained using triangular and tetrahedral elements, respectively. The superior accuracy stems from both the regularity of the mesh (formal accuracy) and the ability to grade or refine the mesh in regions where the variables are expected to change rapidly.

The vast majority of casting simulation software employ either (1) a regular overlaid structured grid or (2) an unstructured (finite element type) grid. Both approaches have certain advantages and disadvantages. A regular grid is easily generated, but may require an extremely large number of cells to adequately represent castings of complex geometry. A high quality unstructured grid generally requires quite a bit of time to generate, but permits very

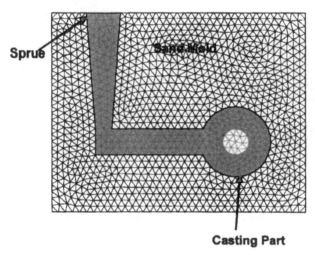

Figure 3 An unstructured two-dimensional mesh of the cast part and mold shown in Fig. 2.

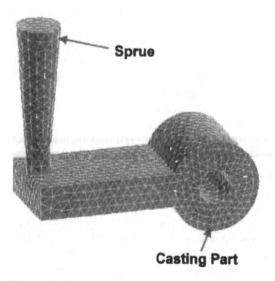

Figure 4 An unstructured three-dimensional mesh of a casting part.

accurate representation of the domain. Unstructured codes require greater memory and CPU time per element (cell) than structured codes, but generally use a mesh of variable density which requires fewer total elements.

B. Spatial Discretization

The three most common numerical methods for fluid flow and heat transfer analysis are the *finite difference method (FDM)*, *finite (control) volume method (FVM)*, and *finite element method (FEM)* [18]. The various methods will be illustrated using the following one-dimensional model equation:

$$\frac{\partial \phi}{\partial t} = \alpha \frac{\partial^2 \phi}{\partial x^2} \tag{26}$$

The finite difference method uses a structured grid and is perhaps the simplest method to understand and implement on problems of simple geometry. In basic terms, the method consists of replacing the derivatives in the differential equations by finite difference approximations. For example, the second derivative can be approximated using the following central difference formula:

$$\frac{\partial^2 \phi_i}{\partial x^2} \cong \frac{\phi_{i+1} - 2\phi_i + \phi_{i-1}}{\Delta x^2} \tag{27}$$

where a value of the dependent variable ϕ is associated with each cell or grid point and Δx is the mesh spacing. The formal accuracy of the above expression is said to be of second order. An expression is accurate to order n provided $|TE| \le K|\Delta x|^n$ as $\Delta x \to 0$ (sufficiently small Δx) where K is a positive constant, independent of x. The *truncation error (TE)* is the difference between the partial derivative and its finite difference representation. Substituting the finite difference expression into Eq. (26) results in an algebraic equation:

$$\dot{\phi}_i - \alpha \frac{\phi_{i+1} - 2\phi_i + \phi_{i-1}}{\Delta x^2} = 0 \tag{28}$$

where $\dot{\phi}_i$ represents the time derivative at point i. A discretized equation is written for each cell or grid point in the mesh. The resulting set of algebraic equations are then solved for the unknown ϕ_i values.

The finite (or control) volume method is a generalization of the finite difference method that may be used with either structured or unstructured grids. The unknowns are typically associated with the cells and are considered to represent the cell average value rather than the value at a particular point. Integrating Eq. (26) over the control volume,

$$\frac{\Delta x}{\alpha}\dot{\phi}_i = \int_{x-\Delta x/2}^{x+\Delta x/2}\frac{d^2\phi}{dx^2}\,dx = \frac{d\phi}{dx}\bigg|_{i+1/2} - \frac{d\phi}{dx}\bigg|_{i-1/2}$$

$$\cong \frac{\phi_{i+1}-\phi_i}{\Delta x} - \frac{\phi_i-\phi_{i-1}}{\Delta x} \tag{29}$$

$$\dot{\phi}_i - \alpha\frac{\phi_{i+1}-2\phi_i+\phi_{i-1}}{\Delta x^2} = 0 \tag{30}$$

Note that for this case the finite volume result, Eq. (30), is identical to the finite difference result, Eq. (28). The key feature of the approach is that the finite volume method is based on flux integration over the control volume surfaces. The method is implemented in a manner that ensures local flux conservation, regardless of the grid structure.

The finite element method is an inherently unstructured method that generally employs some type of weighted integral solution:

$$\int_\Omega W_i\left(\frac{\partial\phi}{\partial t} - \alpha\frac{\partial^2\phi}{\partial x^2}\right)d\Omega = 0 \tag{31}$$

where Ω represents the computational domain and W_i is a weight function. When the weight functions are chosen to be the same as the nodal interpolation functions, the method is referred to as the Galerkin finite element method [18]. Applying this method with linear interpolation between nodes yields the following algebraic equation:

$$\frac{1}{6}\dot{\phi}_{i-1} + \frac{2}{3}\dot{\phi}_i + \frac{1}{6}\dot{\phi}_{i+1} - \alpha\frac{\phi_{i+1}-2\phi_i+\phi_{i-1}}{\Delta x^2} = 0 \tag{32}$$

Note that the standard finite element method expresses the time derivative term as a kind of weighted average. This is referred to as a "consistent mass" formulation. Frequently a "lumped mass" approximation is used, which for this case produces results identical to the finite difference and finite volume formulations.

When applied to a simple one-dimensional diffusion equation with constant properties on a regular grid, the finite difference, finite volume, and finite element (with mass lumping) methods produce identical discretized equations. However, when applied to multidimensional problems with variable properties and grids, the methods in general result in different algebraic equations.

C. Convection Modeling

In this section the finite difference and finite volume methods are applied to a model convection problem in two dimensions which can be expressed as

$$\frac{\partial \phi}{\partial t} + u \frac{\partial \phi}{\partial x} + v \frac{\partial \phi}{\partial y} = 0 \tag{33a}$$

in nonconservative form, and

$$\frac{\partial \phi}{\partial t} + \frac{\partial (u\phi)}{\partial x} + \frac{\partial (v\phi)}{\partial y} = 0 \tag{33b}$$

in conservative form. For this example the velocity is assumed to be a known function of space and time. The numerical methods will be applied using a regular mesh with cells of size $\Delta x \cdot \Delta y$. The index i will be used to number cells in the x direction, and the index j will be used to number cells in the y direction.

Consider the finite difference approximation in nonconservative form using Eq. (33a). Assuming that both components of velocity are positive, an "upwind" difference equation is

$$\frac{\partial \phi_{i,j}}{\partial t} + u_{i,j} \frac{\phi_{i,j} - \phi_{i-1,j}}{\Delta x} + v_{i,j} \frac{\phi_{i,j} - \phi_{i,j-1}}{\Delta y} = 0 \tag{34}$$

The one-sided differencing of the first derivatives is referred to as "upwinding." In this case, "full upwinding" is used, which means that downstream values of ϕ (such as $\phi_{i+1,j+1}$) are not permitted to have any influence on the solution for $\phi_{i,j}$. This treatment of the convective terms results in a monotone (oscillation free) solution, but is formally only first-order accurate.

A finite volume expression analogous to Eq. (34) is obtained by considering the integral form of Eq. (33b):

$$\int_{\Omega} \frac{\partial \phi}{\partial t} + \int_{\Gamma} \phi_{uw} \mathbf{V} \cdot d\Gamma = 0 \tag{35}$$

where Ω represents the finite volume and Γ represents the surface of the volume. The expression ϕ_{uw} indicates that an upwind value of the dependent variable is to be used. Application of Eq. (35) yields:

$$\Delta x \, \Delta y \frac{\partial \phi_{i,j}}{\partial t} + \Delta y \, u_{i+1/2,j} \phi_{i,j} + \Delta x \, v_{i,j+1/2} \phi_{i,j} - \Delta y \, u_{i-1/2,j} \phi_{i-1,j}$$
$$- \Delta x \, v_{i,j-1/2} \phi_{i,j-1} = 0 \tag{36}$$

Upon dividing by the volume $\Delta x \, \Delta y$, we obtain

$$\frac{\partial \phi_{i,j}}{\partial t} + \frac{u_{i+1/2,j} \phi_{i,j} - u_{i-1/2,j} \phi_{i-1,j}}{\Delta x} + \frac{v_{i,j+1/2} \phi_{i,j} - v_{i,j-1/2} \phi_{i,j-1}}{\Delta y} = 0 \tag{37}$$

Several mold-filling codes have been developed based on the SOLA-VOF approach [13], which uses finite difference expressions very similar to Eq. (34). More recently developed codes are based on conservative finite volume equations similar to Eq. (37). Note that when applied to a regular grid under conditions of constant velocity, the two methods produce identical results.

The above expressions for the convective terms are only of first-order accuracy, which has the effect of adding "false viscosity" to the equations. To correct this deficiency, upwinding schemes of second-order accuracy [19] may be found in general computational fluid dynamics (CFD) codes. To the authors' knowledge, such schemes have not been used in commercial casting simulation software.

D. Time Integration

The equations which describe the casting process are partial differential equations of the following form:

$$\frac{\partial \phi}{\partial t} = f\left(\phi, \frac{\partial \phi}{\partial x}, \frac{\partial \phi}{\partial y}, \frac{\partial^2 \phi}{\partial x^2}, \ldots\right) \tag{38}$$

The above expression states that the time rate of change of the variable is some function f of the variable itself and the spatial derivatives of the variable. Discretization of the spatial derivatives by finite difference, finite volume, and finite element methods were described in the previous sections. We now consider the treatment of the transient aspect of the equations. The time derivative can be approximated as

$$\frac{\partial \phi}{\partial t} \cong \frac{\phi^{n+1} - \phi^n}{\Delta t} \tag{39}$$

where the superscript n refers to the time level and the time step $\Delta t = t^{n+1} - t^n$. The function f in Eq. (38) can be evaluated using either *explicit* or *implicit* time integration.

When explicit time integration is used, the function f is evaluated at time level n:

$$\frac{\phi^{n+1} - \phi^n}{\Delta t} = f^n \tag{40}$$

Explicit time integration permits the direct advancement of the solution using only values at time level n. A disadvantage of this approach is that the time step must be limited in order to obtain stable results. For example, consider the limits for (1) convection and (2) conduction in one dimension. For convection (with full upwinding), the following condition must be satisfied:

$$\frac{u \, \Delta t}{\Delta x} < 1 \tag{41}$$

which is known as the CFL condition (named after Courant, Friedrichs, and Lewy) [20]. For conduction, the time step restriction is given by

$$\frac{\alpha \Delta t}{\Delta x^2} < \frac{1}{2} \tag{42}$$

where α is the thermal diffusivity $\alpha = k/\rho c$.

With implicit time integration, the function f is generally evaluated using a weighted average of the values at time level n and $n + 1$:

$$\frac{\phi^{n+1} - \phi^n}{\Delta t} = \theta f^{n+1} + (1 - \theta) f^n \tag{43}$$

Selecting the implicit factor θ to be unity results in a "fully implicit" scheme, whereas selecting θ to be zero reduces to the explicit scheme. The scheme is unconditionally stable for $\frac{1}{2} \leq \theta \leq 1$, although accuracy considerations may place limits on the permissible time step size. The time integration is first-order accurate for all values of θ except $\theta = \frac{1}{2}$, which is second-order accurate. Although formally less accurate, values of θ greater than $\frac{1}{2}$ are often used to reduce or eliminate numerical oscillation in the initial transient solution. The cost to be paid for the unconditional stability of the implicit approach is the requirement to solve a system of simultaneous equations for the unknown ϕ^{n+1} quantities at each time step.

Mold filling is generally characterized by convection dominated free surface flow. For this type of simulation, explicit time integration may be used to accurately model the free surface movement and correctly predict the liquid metal temperature distribution. For conduction-dominated solidification, implicit time integration is often used in order to permit time steps much larger than the explicit limit.

VII. SOLUTION OF SIMULTANEOUS EQUATIONS

The numerical formulations presented in the previous sections result in large sets of simultaneous equations that must be solved at least once and possibly several times during each time step of the simulation. It is therefore essential to employ solution methods which minimize computing time and memory requirements. The methods used to solve the discretized equations depend on the characteristics of the matrix equation. All of the numerical methods which have been discussed in this chapter (FDM, FVM, and FEM) result in a large, sparse matrix equation. In other words, only a relatively small number of the matrix elements are nonzero. The precise pattern of sparsity depends on the grid structure and the method of discretization. Some standard sparse matrix solution methods will be mentioned briefly. Recently, very efficient solution methods, particularly multigrid methods, have been employed in general purpose commercial software.

A. Methods for Structured Grids

With structured grids, the simultaneous algebraic equations may be solved efficiently with minimum memory requirements by taking advantage of the inherent order of the grid. For a one-dimensional problem, an implicit algorithm results in a tridiagonal system of equations. A tridiagonal matrix is easy to solve because nonzero elements are found only on the diagonal and one element on each side of the diagonal. For multidimensional problems, alternating direction implicit (ADI) or fractional step methods [20] are generally employed which involve splitting the problem into separate steps for each dimension of the problem. Each step involves implicit operations in only one coordinate direction. As a result of this "splitting," only tridiagonal systems of equations need to be solved.

B. Methods for Unstructured Grids

A finite element type of grid has no inherent structure and therefore the previously described methods cannot be employed. For unstructured grid methods, it is usually more economical to compute and store the matrix elements rather than compute them "on the fly" each time they are needed. Currently a common approach is to use an iterative method such as the conjugate gradient method [20] for symmetric matrices or a generalization of the method for nonsymmetric matrices. With a symmetric matrix, the lower triangular half of the matrix is a mirror image of the upper half. This means that only the diagonal and either the lower or upper triangular half of the matrix need be stored. In either case, special sparse matrix storage schemes are used because the matrix is extremely sparse with no particular sparsity pattern.

VIII. TYPES OF MODELING SOFTWARE

Many casting process modeling software packages are commercially available. Although all of this software have been developed for the purpose of providing assistance in optimizing the casting process, selecting the right package to satisfy a particular foundry's specific needs can be a difficult task. Perhaps the most obvious difference between the various software packages is the structure of the grid used in the computations. As described in Sec. VI.A, computational grids may be classified as either structured or unstructured. Essentially all of the simulation software is based on either (1) a regular, structured grid using finite difference and/or finite volume numerical methods or (2) an unstructured grid using finite element methods. In addition, many software packages offer some type of quick (and approximate) analysis capability. The choice of modeling software depends on the expertise, preferences, and computational resources of the user as well as the process modeling requirements. The characteristics of each type of software is summarized in the following paragraphs.

A. Structured Grid Software

The first commercially available casting modeling software used a structured grid approach. The original codes used FDM; more recently developed codes may employ FVM. The most important advantage of structured grid software is that it is fairly easy to use. Because the mesh consists of simple rectangular cells, the meshing process is a straightforward task. However, the resulting "zigzag" mesh does not represent the geometry very accurately, especially for complex thin walled castings. Generally, a uniform cell size is used which can result in a large number of cells if the casting has both thin and thick sections. FDM/FVM software has been successfully used to model the heat transfer, fluid flow, and solidification during casting. At present, this type of software cannot calculate casting stress distribution directly. In order to calculate thermal stress, a separate FEM stress analysis is needed. If stress analysis is desired, the computed temperature history must be written out and mapped onto an FEM mesh of the same geometry. The thermal stress analysis is then performed by the FEM software using the temperature data obtained from the previous FDM/FVM analysis.

B. Finite Element Software

This type of software uses a completely unstructured FEM type of mesh which represents the model geometry very accurately. However, finite element meshing can be a complicated task. Automatic tetrahedral meshing is commonly employed to reduce the time demand on the user. This type of software generally requires greater memory and CPU time per element (cell), but may use elements of variable size to reduce the required number of elements. The most important advantage of this type of software is that it can perform heat transfer, fluid flow, and stress analysis with a single model. The stress analysis can be performed separately or simultaneously with the heat transfer calculations. Because stress analysis is a time consuming task, a separate calculation (only one stress analysis is performed after several heat transfer calculation steps) is most widely used.

C. Quick Analysis

This type of software typically uses a modulus (surface to volume ratio) approach to calculate the casting solidification sequence and identify hot spots (the last places to solidify) which tend to form macroshrinkage type defects. This type of approach usually does not properly account for the geometry of the different solid materials (insulation, chills, etc.) in the model. Recently, however, methods have been developed to better account for the geometry of the various materials and also generate the thermal history (temperature distribution as a function of time) [21]. Quick and approximate methods have also been developed to model the mold filling sequence. The most important advantage of quick analysis software is that the computation may be performed in minutes instead of hours required by traditional numerical methods. The disadvantage of this type of method is that the results are generally less accurate than traditional computations and may be unsuitable for prediction of microstructure evolution and defect formation (e.g., amount of porosity) during casting.

IX. GENERAL PROCEDURES OF CASTING MODELING

The previous sections laid out the fundamentals of fluid flow and heat transfer phenomena involved in modeling casting processes. These fundamentals not only serve as the basis for casting software development, but are also helpful to a software user in building casting models and running simulations. A software user, such as a process engineer in a foundry plant, might not have to understand the details of how the software is written, but needs to know how to efficiently employ the software to conduct computer modeling for his own casting process. This section will discuss the general procedures of casting process modeling using simulation software, either in-house developed software or a commercial casting simulation package.

Casting computer modeling usually consists of three stages: (1) preprocessing, (2) running the simulation, and (3) postprocessing. Preprocessing is most critical to the success of the modeling effort since this is the stage where the user actually builds the computer model for his particular casting process. The preprocessing stage consists of several steps: building geometry, meshing, assigning material properties, specifying initial and boundary conditions, selecting simulation control parameters, etc. Running the simulation involves performing the primary calculations and monitoring the results. Postprocessing is the stage to retrieve, process, and most importantly, make sense out of the simulation results. Postprocessing can be the most exciting when a nice, smooth, converged solution is obtained from the simulation, but can also be the most disappointing when meaningless temperature contours appear on the computer screen.

A. Building Geometry and Meshing

1. Building Geometry

The geometry for the casting process model includes part geometry itself and process-related geometry such as runners, gates, and risers. There are two options to build the geometry of the part. The first option is to create the geometry from the *blueprints* of the part. Some of the commercial casting simulation packages have a built-in capability to allow the user to create a geometry model consisting of complex shapes and dimensions. It is convenient to use this built-in capability if the geometry of the part is simple. The second option is to use an *electronic data interface*. At the part development and design stage, an electronic data file of a solid model of the part is created by a design engineer with CAD (computer aided design) or CAM (computer aided manufacturing) software. When casting process modeling is needed, the electronic data file can be retrieved and translated with electronic interface software to a format that can be read by the casting simulation package. All of the commercial casting simulation packages have the capability to directly read some type of electronic files created by CAD/CAM packages. The interested user can refer to Chapter 8 for details of the electronic data interface process. The most convenient way to build the complete model geometry is to import cast part geometry data file and then use the built-in geometry creation capability to add runners, gates, risers, etc.

2. Meshing

The part to be cast is usually of complex geometry. Meshing such a part is not a trivial exercise. It takes quite a bit practice for a user to develop the skilled experience necessary to generate a high quality mesh. The quality of the mesh affects not only the accuracy of the results, but also the stability and convergence of the solution. In addition, mesh density directly determines the computational efficiency, i.e., the computer CPU time and memory as well as the hard storage space required to run the simulation. In general, the higher the mesh density (meaning more elements and nodes in a specified volume), the higher the simulation accuracy. However, higher mesh density also significantly increases the computer CPU time, memory, and storage space. Thus, it is important to establish minimum mesh density required to ensure a reliable and stable numerical solution. For example, generally a minimum three elements is required to obtain a reliable solution for fluid flow calculation. Most commercial casting simulation packages have internal meshing capability regardless of whether the package uses a structured or an unstructured mesh. Some of the packages also allow the user to import mesh files generated from other software.

B. Assigning Material Properties

Material properties can be assigned to individual cells (or elements) based on the material type occupying the cells (or elements). The material properties related to fluid flow and heat transfer include thermophysical properties, such as density, specific heat, thermal conductivity, radiative emissivity, viscosity, thermal expansion coefficient, surface tension coefficient, solidus and liquidus temperatures, and latent heat of fusion. These properties are usually temperature and composition dependent. Most of the casting simulation packages have *built-in material databases* that can be retrieved through the pre-processor. The databases store the thermophysical properties for the materials that are most commonly used in casting processes, such as cast irons, ductile irons, carbon and stainless steels, aluminum and its alloys, copper and its alloys, superalloys, titanium alloys, various sands, and some ceramics. Once a particular material is assigned to a certain region of the computational domain, the property values will be retrieved from the database and assigned to the computational cells in the region. The user can modify and edit the property data through the pre-processor to meet his particular needs.

C. Specifying Initial Conditions

The user is generally required to provide initial conditions, such as initial metal positions and initial temperature distributions. Sections III.C and V.D briefly described basic concepts of initial conditions for heat transfer and mold filling.

1. Heat Transfer

When only heat transfer is calculated, the mold cavity is assumed to be initially filled with metal. In this situation, the initial temperature has to be provided or specified. Most of the commercial casting simulation packages allow the user, through the pre-processor, to specify *constant temperatures* for metal and mold, respectively. The metal can be given a constant value of temperature, such as the pouring temperature, and the mold can be given another one, such as the mold preheated temperature. This is suitable for situations involving quick filling of a thick wall casting. If the mold filling takes place slowly or the mold is preheated nonuniformly, an initial *temperature distribution* must be provided. This can be implemented through a user subroutine similar to the one described in the previous section. The temperature distribution can be specified manually based on the user's knowledge, or obtained through a mold filling simulation.

2. Mold Filling

When mold filling is involved, an additional initial condition must be provided for the metal position. This can easily be implemented through the pre-processor of the casting simulation package.

D. Specifying Boundary Conditions

The boundary conditions are usually much more complicated than the initial conditions described in the previous section. Generally there are four types of boundaries that are referred to here: mold exterior walls, metal flow ingates, metal/mold interface, and metal free surfaces. The latter two types are sometimes called "interior boundaries" since they are not usually located at the outside boundaries of the computational domain. Sections III.C and V.D included a short discussion on the boundary conditions for different types of boundaries.

1. Mold Exterior Walls

The mold exterior walls can be given one or more of the following conditions: a constant temperature, a constant heat flux, a convection condition that could be natural or forced convection, and a radiation condition. All these conditions are available in most of the commercial casting simulation packages and can be specified during pre-processing. Which type of condition(s) should be applied really depends on the process that is modeled. For example, generally a constant temperature or convection condition should be applied to sand casting mold exterior walls and a radiation condition should be employed to investment casting exterior ceramic shell surfaces. The reader should refer to Secs. III.C and V.D as well as Chapter 11 in the latter part of this book for the details.

2. Metal Flow Ingates

The conditions at the metal flow ingates, sometimes called inflow conditions, must be specified when fluid flow (mold filling) is involved. This is fairly easy to implement, as described in Sec. V.D. For example, constant temperature and velocity may be specified at the ingates. This can be implemented through the pre-processor of any simulation package. Some packages allow the user to specify time-dependent ingate conditions.

3. Metal/Mold Interface and Metal Free Surfaces

The conditions at the boundaries, such as metal/mold interface and metal free surfaces, are discussed in Secs. III.C and V.D. These conditions can be specified through the preprocessor. Most of the simulation packages also set some default conditions, such as a perfect contact and no-slip condition at the metal/mold interface and a zero-heat-flux and zero-stress condition at the metal free surfaces.

4. Other

Some processes may have particular boundaries, either exterior or interior. For example, a permanent mold or die casting is usually equipped with gas escape vents. Special models might be needed to specify conditions for these particular boundaries. In this situation, the user should check the manual.

It should be pointed out that the above discussion on boundary conditions is quite general. These conditions change from case to case for the different casting processes. The reader can refer to a corresponding chapter of this book when dealing with a particular process.

E. Selecting Simulation Control Parameters

Simulation control parameters include total simulation time (the operation time in the casting process, not the computer CPU time), time step size control variables, convergence criteria, computational relaxation factors, and frequency of result output, etc. Although from a theoretical point of view these parameters look trivial, in practice they significantly affect the solution accuracy, stability, convergence, and computational efficiency. One of the most frustrating experiences in casting modeling is to find that a long battle for a good simulation was a result of a bad choice of simulation control parameters. Although some tips, given in the following paragraphs, can be used to select a workable combination of these parameters, there are unfortunately no general rules that guarantee an accurate, smooth, and converged solution. It may take several trial-and-error simulations for the user to obtain a good feeling about what will work and what will not. The user is recommended to read the manual provided by the simulation package that he/she chooses and consult experienced modeling personnel instead of sitting in front of a computer and wasting a huge amount of time on trial and error.

1. Total Simulation Time

Total simulation time should be determined by the process to be simulated. This time in turn directly determines the computer CPU time. It is suggested that the total simulation time for the first trials be set to such a number that fewer time steps are needed. This will significantly shorten the trial-and-error time, especially for a simulation that is expected to take more than several hours of CPU time to complete. After smooth and converged solutions are obtained for the first several time steps, a complete job covering the entire process can be submitted. In practice, two methods are commonly used to decide the total simulation time. One method is to stop the simulation when the highest temperature in the casting reaches a specified value, usually several degrees lower than the metal solidus temperature. The other method is to stop the simulation when the total simulation time reaches a specified value, based on an estimation how long the casting will take to completely solidify.

2. Time Step Size

Most commercial casting simulation packages have an option of automatic adjustment of time step sizes (or time increments). If the solution converges at the current time step and the temperature change falls in a predetermined value, the package will enlarge the time step size for the next time step. When using this option, the user is required to provide an allowed minimum and maximum time step size. Some packages stop the simulation if the allowed minimum time step size is reached. The maximum time step size should be selected based on two criteria: stability requirement and time truncation error tolerance. For a solution method using explicit time integration, the CFL criterion, Eq. (41), must be satisfied to assure the solution stability. This will usually result in a satisfactory time truncation error. For a solution using implicit time integration which is unconditionally stable, a relatively large time step size can be selected as long as the time truncation error is kept below a tolerable level. However, there are no measures to directly calculate the time truncation error. It can only be estimated by examining the results or comparing them with experimental data. Frequently, the maximum time truncation error is specified in terms of CFL number.

3. Convergence Criteria

Convergence criteria are used only in the solution schemes with an implicit time integration where an iteration sequence is needed to solve a nonlinear system of equations. There are typically two types of convergence criteria: the residual of the discretized equation and the deviation error of the variable between two consecutive iterations. It is best if both criteria are satisfied. For example, the residual of the discretized energy equation at the current iteration falls to a level 2 orders of magnitude smaller than that at the first iteration. At the same time, the temperature difference between the current and last iteration becomes less than 1°C. Appropriate values should be selected for the convergence criteria. Too large a value will result in inaccurate solutions and too small a value will consume unnecessarily a large amount of computer CPU time. Most of the simulation packages have default values set for the convergence criteria. The user is recommended to use these default values.

4. Computational Relaxation Factors

Computational relaxation factors usually refer to the under-relaxation factors used in nonlinear iteration sequences. Similar to the convergence criteria discussed above, the computation relaxation factors are only applied in the solution schemes using implicit time integration. The simulation packages that use the implicit solution schemes may require the user to specify the underrelaxation factors through the preprocessor. The values of the underrelaxation factors can be different for each variable, such as temperature, solid (or liquid) fraction, pressure, and velocity components. Again, appropriate values must be selected for the underrelaxation factors. Too large values may cause convergence problems and too small values will unnecessarily slow down the simulation significantly.

5. Data Output Frequency

A complete simulation of a casting process may need several hundred or thousand time steps. Because of the restriction of computer hard storage space, it is impossible and unnecessary to store the results at every time step. Most of the casting simulation packages allow the user to specify an output frequency for writing out the simulation results. Some packages even allow the user to use a variable output frequency. This gives the user the flexibility to limit computer storage space and still be able to catch the most important characteristics of the results in the time sequence.

F. Selecting a Computational Model

Most of the casting simulation packages include several computational models. The user can select a specific model or a combination of some models to meet his own simulation needs.

1. Quick Analysis Model

Some casting simulation packages provide the quick analysis model for heat transfer known as "modulus model," such as the one described in Chapter 7. The model does not solve the partial differential equations but calculates heat balance based on an integrated effect of boundaries. This simple analysis model has its unique advantage: superior computational efficiency. The model takes little CPU time to complete a simulation (less than an hour for a mesh with a million nodes). The disadvantage of the model is its accuracy. It may be used as a tool for rough estimates or prestudy. It can also be used to conduct parametric studies if the model can be validated through comparisons with experiments or with a more accurate model, such as the mold filling and heat transfer model described in Sec. V.

2. Heat Conduction Model

The heat conduction model is a good choice if the mold filling takes place quickly and its effect on metal and mold temperatures is small. This model needs more CPU time (maybe a few to several hours) than the quick analysis model but is still much faster than a coupled mold filling and heat transfer model.

3. Mold Filling and Heat Transfer Model

This is the most accurate but least computationally efficient model. It may take several days to complete a simulation. The model is used only when it is necessary, for example, when mold filling is very slow, or an incomplete filling due to solidification is suspected, or a high degree of accuracy is desired.

4. Thermal Stress Model

When stress-related defects are of concern, thermal stress calculation can provide some insight to better understand the problem. Stress-related defects include hot tearing, cold cracking, dimensional distortion, and recrystallized grains in single crystal and directionally solidified turbine airfoils.

G. Running the Simulation

After the simulation job is submitted to the computer, it should be monitored as frequently as possible. This will help the user save modeling cycle time which refers to the time needed from building the geometry to the point when a good solution is achieved. The user should use the postprocessor to exam the results that are written out for the existing output time steps. If serious problems are found in the results, he should stop the simulation and go back to tune the model. If the problem is minor, he might restart the simulation and save the results for future reference. At the same time, he may start another simulation with better tuned model parameters.

H. Postprocessing

The postprocessing includes two aspects: visualization of the simulation results and analysis of these results to provide a better understanding of the casting process. The simulation results include direct output quantities of the FDM/ FEM analyses and indirect output quantities which are calculated from those direct output quantities. The direct output quantities are raw data and provide only limited information regarding the details of the casting process. To have a better understanding of the casting process details and to find a solution for eliminating potential casting defects, indirect output quantities are more useful. Most of the casting simulation packages have a built-in visualization software which can be used to visualize simulation results. Some of the visualization software allow the user to animate the results such as the mold filling and solidification sequence.

The direct output quantities of the FDM/FEM analyses are:

Heat transfer analysis: Temperature distribution at various time steps
Fluid flow analysis: Free surface locations, velocity vectors, and pressure distribution at various time steps
Stress analysis: Stress and strain distribution at various time steps

Although direct output quantities provide some understanding of the casting process, other indirect quantities would be more useful to predicting defect formation and microstructure evolution during casting. Some of the most commonly used indirect output quantities and their physical meanings are briefly described in the following part of this section. The reader should refer to Chapters 4 and 5 for details on how to use these quantities to predict defect formation and microstructure evolution.

1.   Mold Filling

Metal fluid flow velocities, pressure, and free surface location can be plotted in a time sequence to visualize mold-filling process. This mold-filling sequence will reveal how the metal flows from the pouring sprue to the end of casting, and consequently reveals potential problems associated with the mold filling. These problems include the potential danger of no filling, formation and locations of weld lines, and residual gas bubbles.

2.   Isotherms

Temperature spatial distributions can be plotted out as isotherms at different times. A series of isotherms can be generated to visualize a whole picture of thermal evolution process during mold filling and cooling. These isotherms can show solidification sequences and hot spots, which in turn indicate the location and size of potential macroshrinkage.

3. Isochrons

Another more effective way to show hot spots and predict potential macroshrinkage is through the use of isochrons. An isochron plot shows the time to reach a specific temperature, such as liquidus or solidus temperature. One isochron plot is equivalent to a series of isotherm plots in terms of the information concerning hot spots and macroshrinkage locations.

4. Cooling Curves

Temperature history at each nodal point can be plotted out as cooling curves. These cooling curves can be used for model validation purposes by comparing them with experimental thermocouple (TC) measurements. Figure 5 shows cooling curves at four spatial locations in the metal and mold during a permanent mold casting of Al 6010 alloy. Thermocouple (TC) data was used to verify the accuracy of the model results.

5. Cooling Rate

Cooling rate has the unit of °C/s. There are two types of cooling rates which can usually be calculated and used for defect and microstructure prediction.

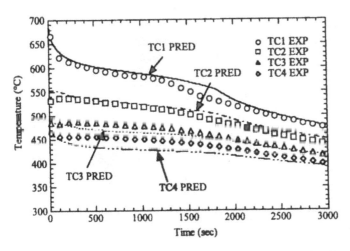

Figure 5 Cooling curves at four spatial locations in metal and mold during a permanent mold casting of Al 6010 alloy. Also shown are thermocouple data represented by symbols.

Instantaneous cooling rate is calculated at a specified temperature and *average cooling rate* is calculated over a temperature interval such as the interval between solidus and liquidus temperatures. These cooling rates can be used to develop empirical relations to predict dendrite arm spacing (DAS) and grain sizes.

6. Local Solidification Time (LST)

Local solidification time (LST) is the time needed for a specific spatial location to cool down from liquidus temperature to a eutectic or solidus temperature. LST can be calculated based on the temperature history and is often used in the prediction of DAS.

7. Temperature Gradient G

Temperature gradient G is defined as the rate of change in temperature over the spatial coordinates. It has the unit of °C/m and is actually a vector which can be decomposed to three components, e.g., G_x, G_y, and G_z in a Cartesian coordinate system. These components can easily be calculated from the simulation results and the vector sum can then be deduced.

8. Solidification Rate R

Solidification rate R is defined as the moving speed of solidification front over the spatial coordinates. It has the same unit as velocity, i.e., m/s. It can be calculated at a specific time and a specific spatial location.

9. Combination of G and R

A combination of solidification rate R with temperature gradient G can be very useful in prediction of solidification structure, feeding capability, and microporosity formation tendency. For instance, G/R is usually used to predict solidification structure. As shown in Fig. 6, the values of G/R determine whether a particular region of the casting part will have an equiaxed dendritic structure, or columnar dendritic structure, or plane front structure. Another use of the combination of G and R is the prediction of the microporosity formation tendency of the casting part, such as Niyama criterion for microporosity formation in steels and LCC criterion in aluminum alloys.

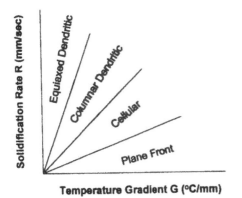

Figure 6   Effect of cooling rate and solidification rate on solidification structure.

REFERENCES

1. FP Incropera, DP Dewitt. Fundamentals of Heat and Mass Transfer. 2nd ed. New York: John Wiley & Sons, 1985.
2. MC Flemings. Solidification Processing. New York: McGraw-Hill, 1974.
3. DM Stefanescu. Methodologies for modeling of solidification microstructure and their capabilities. ISIJ International 35(6):637–650, 1995.
4. VR Voller, CR Swaminathan, BG Thomas. Fixed grid techniques for phase change problems: a review. Int J Numer Methods Eng 30:875–898, 1990.
5. CR Swaminathan, VR Voller. A general enthalpy method for modeling solidification processes. Metallurgical Transactions 23B:651–664, 1992.
6. DR Poirier, EJ Poirier. Heat transfer fundamentals for metal casting. The Minerals, Metals & Materials Society, 1992.
7. G Li, BG Thomas. Transient thermal model of the continuous single-wheel thin-strip casting process. Metall Trans 27B:509–525, 1996.
8. HS Carslaw, JC Jaeger. Conduction of Heat in Solids. 2nd ed. London, UK: Oxford University, 1959, p. 248.
9. J Szekely. Fluid Flow Phenomena in Metals Processing. New York: Academic Press, 1979.
10. DR Poirier, GH Geiger. Transport Phenomena in Material Processing. Warrendale, PA: TMS, 1994.
11. SV Patankar. Numerical heat transfer and fluid flow. Washington, DC: Hemisphere, 1980.
12. RB Bird, WE Stewart, EN Lightfoot. Transport Phenomena. New York: John Wiley & Sons, 1960.
13. H Schlichting. Boundary-Layer Theory. 7th ed. New York: McGraw-Hill, 1979.
14. W Hirt, BD Nichols. Volume of fluid (VOF) method for the dynamics of free boundaries. J of Computational Physics 39:201–225, 1981.

15. J Bear. Dynamics of Fluids in Porous Media. New York: American Elsevier Publishing Co., 1972.
16. VR Voller, C Prakash. A fixed grid numerical modeling methodology for convection-diffusion mushy region phase-change problems. Int J Heat Mass Transfer 30(8):1709–1719, 1987.
17. GH Geiger, DR Poirier. Transport Phenomena in Metallurgy. Reading, MA: Addison-Wesley, 1973.
18. C Hirsch. Numerical Computation of Internal and External Flows. Volume 1. Chichester, UK: John Wiley & Sons, 1988.
19. C Hirsch. Numerical Computation of Internal and External Flows. Volume 2. Chichester, UK: John Wiley & Sons, 1990.
20. DA Anderson, JC Tannehill, RH Pletcher. Computational Fluid Mechanics and Heat Transfer. New York: Hemisphere Publishing Corporation, 1984.
21. C Cheng, KO Yu. Innovative approach for modeling the heat transfer during casting solidification. In: SA Argyropoulos, F Mucciardi, ed. Computational Fluid Dynamics and Heat/Mass Transfer Modeling in the Metallurgical Industry. Metallurgical Society of CIM, 1996, pp. 56–67.

3
Stress Analysis

Umesh Chandra
Modern Computational Technologies, Inc., Cincinnati, Ohio

Alauddin Ahmed
Pratt & Whitney, East Hartford, Connecticut

I. INTRODUCTION

This chapter deals with the development of stresses and distortions in shaped castings and their prediction using numerical methods, specifically the finite element method. It also discusses the topics of hot tears, hot cracks, and patternmakers' allowance which are related to the more general subject of stresses and distortions.

Casting of a part involves a complex interaction between several metallurgical and mechanical phenomena, e.g., solidification and subsequent cooling of the molten material; grain nucleation and growth; transfer of heat from the melt to the mold and, finally, to the environment; development of stresses and distortions; and so forth. In order to make the task of computer simulation manageable, Chandra has proposed a scheme shown in Fig. 1 [1]. According to this scheme, after a finite element or finite difference mesh of the cavity, mold, and the feeding system is prepared, three separate simulations can be performed in sequence. The first of these, that is, the mold-fill simulation, is based on the mathematics of fluid mechanics and convective heat transfer. It is carried out until the mold is completely filled and the convective effects have become negligible. Such simulation is capable of predicting loss of superheat, turbulence levels, mold erosion, and misrun (nonfill) and cold shut type defects. The output of this simulation in terms of temperatures at various locations or nodes at the end of mold filling is passed on to the next step, that is, the solidification simulation, to serve as the initial conditions.

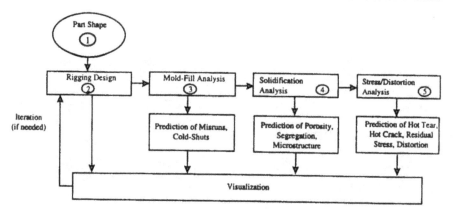

Figure 1 Schematic of the casting process simulation procedure.

The solidification simulation is based on the mathematics of conduction heat transfer in the melt and the mold, kinetics of solidification, and a convective or radiative heat loss from the mold surface. Convection within the melt may be neglected during this period. This simulation step assists in predicting porosity, macroshrinkage, and segregation type defects, cast microstructure and, at least conceptually, the mechanical properties of the cast part. The output of this simulation in terms of temperatures and grain density and size at various locations in the assembly of the cast part, mold, and feeding system can be passed on to the last step, that is, the stress/displacement analysis, which could predict hot tears, hot cracks, residual stresses, and distortions in the cast part, mold cracks, and shrinkage allowance for pattern making.

The scheme shown in Fig. 1, involving three consecutive simulations, was appropriate a few years ago when the primary interest in casting process simulation lied in the prediction of porosity. At that time, the efforts in the areas of mold-fill simulation and microstructure prediction were in their early stages, and the subject of coupled thermal-mechanical analysis was not yet fully developed. Due to recent advances in all of these fields, it is now possible to reduce the simulation steps from three to two. The first step is again of mold-fill simulation based on the mathematics of fluid mechanics and convective heat transfer. Thermal output of the first step is fed to the second step which now combines solidification, subsequent cooling, and stress/displacement analyses together. Simulation in this step may also include the mathematics for microstructure and/or hot tear prediction. The original scheme of three-step simulation (Fig. 1) is the basis of a sequential thermomechanical analysis, whereas the two-step simulation scheme forms the basis of a coupled thermomechanical analysis.

In this chapter, it is assumed that a mold-fill simulation has already been performed up to a point that the mold is completely filled, and the convective effects have become negligible (also see Chapter 2). The temperatures at the end of mold-filling are available at the various locations or nodes in the assembly of the cast part, mold, and feeding system which serve as the initial condition for a subsequent coupled thermomechanical analysis.

In the discussion so far, the casting was assumed to be confined within its mold. But, the stresses and distortions in a finished cast part are also influenced by subsequent operations in its processing history, e.g., mold removal, finish machining, heat treatment and/or straightening. Therefore, in order to accurately predict stresses and distortions in a casting, these operations should also be analyzed. In this chapter, the stresses and distortions due to solidification, cooling, mold removal, and finish machining will be discussed, but the subjects of heat treatment and straightening will not be considered.

In Sec. II, some basic concepts related to stress/displacement analysis of casting processes are reviewed. Then, in Sec. III, a brief review of the transient, nonlinear thermomechanical analysis is presented; both sequential and coupled forms are discussed. In Secs. IV through IX, the treatment of several special topics such as the release of latent heat and formation of hot tears in the mushy region, thermal and mechanical interaction between the mold and the metal, and others, is discussed. It is the treatment of these topics which separates the simulation of the casting processes from a conventional thermomechanical analysis employed in product evaluation. However, a detailed discussion of these topics would require an entire book and is beyond the scope of this chapter. Finally, in Sec. X, some results of thermomechanical analyses of simple test castings are presented to highlight the formation of gap and contact at the mold-metal interface, prediction of hot tears, residual stresses, and distortions.

II. BASIC CONCEPTS

Before going further, it is instructive to clarify certain definitions and concepts related to hot tears, hot cracks, distortions, and transient as well as residual stresses in shaped castings. A clear understanding of these will help the reader appreciate the full extent of the task at hand. This discussion is divided into two parts: before mold removal and after mold removal.

A. Before Mold Removal

Consider a simple I-section casting shown in Fig. 2a. The cooling curve of an arbitrary point within this casting is shown in Fig. 2b. Temperatures T_P, T_L,

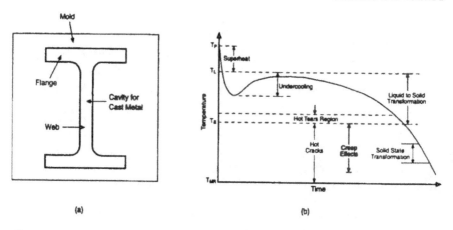

Figure 2 (a) A hypothetical *I* section casting and (b) cooling curve of an arbitrary point.

T_S, and T_{MR} refer to metal pour, liquidus, solidus, and mold removal, respectively; the mold removal temperature may be higher than the room temperature. Also, the initial or pour temperature of the cast metal is much higher than that of the mold. Figure 2b identifies several regions of interest, e.g., superheat, liquid to solid transformation (mushy region), hot tears, hot cracks, creep effect, and solid state phase transformation. The *I* section of Fig. 2a is reconsidered in Fig. 3 at an intermediate stage of its cooling. Figure 3a identifies several important features, e.g., a hot spot, the gap between the mold and the

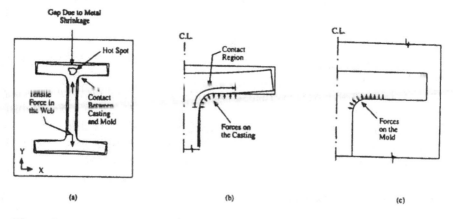

Figure 3 (a) Thermomechanical response of the *I* section, (b) forces on the casting, and (c) forces on the mold.

metal, the contact between the mold and the metal, a tensile force in the web of the *I* section, and bending (distortion) of its flanges.

The hot spot shown in Fig. 3a is a potential site for hot tears in addition to, of course, porosity. A hot tear is a fracture in the liquid film surrounding the solid grain [2]. It occurs in a hot spot at a temperature slightly higher than the solidus (Fig. 2b), when the strain in the liquid film surrounding solid grains exceeds a certain critical value. Hot cracking, on the other hand, refers to the formation of an intergranular crack during cooling subsequent to solidification. A hot crack occurs when the transient stress at a point during cooling exceeds the tensile strength of the material at the instantaneous temperature. It is, thus, similar to a quench crack.

When a casting is still in its mold, the stresses and distortions in it are caused by (1) mechanical constraint imposed by the mold during shrinkage of the cast metal, (2) temperature gradient between different parts of the casting, and (3) volumetric change and transformation plasticity associated with the solid state phase transformations.

The effect of mechanical constraint imposed by the mold on the casting can be explained with the help of Fig. 3. As the cast metal undergoes solidification and subsequent cooling, it tends to shrink. This shrinkage causes the web to contract along the *Y* axis, which is resisted by the relatively rigid mold, resulting in tensile stresses in the web. The constraint due to the rigidity of the mold may also result in forces acting on the flanges of the *I* section. If, at a certain location in the casting, the transient stresses due to mechanical constraints of the mold exceed the elastic limit of the cast material at any time during the cooling history, nonuniform plastic deformation takes place, which, when the cooling is complete, results in residual stresses in the cast part [3–6].

The stresses due to temperature gradients between different parts of the casting are very similar to those encountered in quenching and welding processes. The difference in temperature in two adjacent parts of the casting causes differential shrinkage. Then, to maintain compatibility, the two parts develop equal and opposite forces. Again, if the stresses in any of these two parts exceed the yield strength at any instant during solidification and cooling, permanent deformation takes place, which on final cooling results in residual stresses [4, 5].

During cooling of a casting the effect of creep at elevated temperatures (near solidus) may be significant even though the stresses are small [7]. Further deliberations revealed that some other factors combine to make the problem extremely complex [8]. For example, (1) a large casting made by the sand mold process may stay in its mold for 2–3 days which implies that both primary and secondary creep may be significant, (2) this creep occurs under the condition of varying stress (e.g., in the *I* section of Fig. 3, the axial force in the web varies with time), and finally, (3) the body forces due to the weight of the metal are significant and should be considered.

Many metals, during cooling subsequent to solidification, may undergo one or more solid state transformations. In addition to the release of a small amount of latent heat, these solid state transformations are accompanied by a volumetric change and the so-called transformation plasticity. Both of these effects contribute further to the development of transient and residual stresses in the cast part.

B. After Mold Removal

Figure 3b,c shows a system of equal and opposite forces acting on the casting and the mold at the contact surface prior to mold removal. When the mold is removed, the equilibrium of forces at the contact surface is satisfied if the forces previously acting on the mold are transferred to the casting. This causes a redistribution of stresses in the interior of the casting and some additional distortions [3–6]. The final size and shape of the cast part after removal of the mold will not be the same as those of the cavity in which the molten metal was poured. The difference in the two determines the patternmakers' allowance [6, 8].

In complex castings with abrupt changes in thickness or sharp corners, the as-cast fillet radii and some part dimensions are often larger than those in the finished products. The excess material is removed by finish machining which causes additional distortions in the casting and a redistribution of the residual stresses [3–5].

It is apparent from the foregoing discussion that a proper thermome-chanical analysis of the casting process should be able to account for a number of phenomena, such as (1) solidification (including the release of latent heat) and subsequent cooling of the molten metal, (2) formation of hot tears, (3) loss of heat from the outer surface of the mold, (4) creep under variable stress, (5) hot cracks, (6) transformation plasticity, (7) thermal and mechanical interaction between the metal and the mold, and (8) effect of mold removal and finish machining on stresses and distortions in the cast part. Simulation of the investment casting process should also account for enclosure radiation. Furthermore, simulation of the directional solidification process (e.g., the Bridgman process) should account for mold withdrawal. Obviously, if a finite element code has to qualify for a stress/displacement analysis of the casting processes, it should be able to account for all these phenomena. No such code is presently available. A commercial thermome-chanical analysis code [9] has been used and enhanced by some researchers [3–6].

III. FUNDAMENTALS OF TRANSIENT THERMOMECHANICAL ANALYSIS

In its simplest form, the transient thermomechanical analysis of a casting metal and mold system can be divided into two sequential steps, as shown in boxes 4 and 5 of Fig. 1: (1) a nonlinear thermal analysis and (2) an elastic-plastic-creep stress/displacement analysis. The finite element mesh consists of solid elements to represent the metal and mold, and special contact elements to represent the mold-metal interface. Thermal analysis is first performed at a number of time steps between the pour and room temperatures. The results of this analysis are saved and used as input for the subsequent stress/displacement analysis.

During thermal analysis, the following heat transfer equation is solved for the metal-mold system:

$$\nabla \cdot (k \, \nabla T) + \dot{Q}_c = \rho C_p \frac{\partial T}{\partial t} \tag{1}$$

where T is the temperature at an arbitrary location in the system at time t, k is the thermal conductivity of the material, $\dot{Q}_c$ is the rate of heat generated per unit volume, ρ is the density, C_P is the specific heat, and ∇ is the differential operator; all material properties are assumed to be temperature dependent.

The initial conditions are either obtained from a mold-fill analysis or assumed to be the same as the pour temperature. The thermal boundary conditions can be of several possible types, e.g., (1) a chill, (2) insulation, (3) exothermic hot top, and (4) surface convection and radiation. As discussed later in this section, the treatment of heat transfer at the mold-metal interface requires special attention. For brevity, the details of solving Eq. (1) are not discussed here. A general discussion of the subject can be found in Refs. 10–12.

Mechanical analysis entails the determination of transient stresses and displacements at various locations in the metal-mold system due to thermal loads. The underlying theory is extremely complex and can not be adequately discussed in this short chapter. It begins with the decomposition of the total strain term, ε_{ij}, at time t, in accordance with the small strain theory:

$$\varepsilon_{ij} = \varepsilon_{ij}^{el} + \varepsilon_{ij}^{th} + \varepsilon_{ij}^{p} + \varepsilon_{ij}^{cr} + \varepsilon_{ij}^{v} + \varepsilon_{ij}^{tr} \tag{2}$$

where superscripts el , th , p, cr, v, and tr refer to elastic, thermal, plastic, creep, volumetric change, and transformation plasticity components, respectively. The treatment of the first three strain terms is well documented in the literature [13, 14]. These terms are considered in practically all nonlinear thermomechanical finite element analysis codes. A specialized treatment of these terms for prediction of hot tears is given in Sec. IV. Although the treatment of the creep strain term for a general case is discussed in Refs. 13 and 14, its further customization to the casting process is discussed in Sec. V. Finally, the treatment of the strain terms associated with volumetric change and transformation plasticity during solid state phase transformation is briefly discussed in Sec. VIII.

A major problem with the traditional form of sequential thermomechanical analysis, discussed above, is in its treatment of the heat transfer coefficient at the mold-metal interface. As discussed further in Sec. VI, this coefficient is an extremely complex function of several variables, including the gap thickness, temperature dependent properties of the medium in the interfacial gap, and surface roughness. As the solidification and subsequent cooling progresses, the thickness of the interfacial gap increases. It also varies spatially. In the absence of a mechanical analysis, the thickness of the gap at a given time and location is not known. The heat transfer coefficient is then generally assumed to be a function of the temperatures at the two sides of the interface at the previous time step. This method is approximate since it does not account for the gap thickness.

In the fully coupled thermomechanical analysis the governing equations of temperature and stress/displacement are solved simultaneously at each time step. The three displacement components and temperature are the four nodal degrees of freedom for the solid (metal and mold) as well as the interface elements. Thus, updated values of nodal temperature and displacements can be found at each time step. This enables the determination of accurate values of gap thickness and, hence, a more reliable treatment of the interface heat transfer. Additionally, from these basic quantities of nodal temperature and displacements, other quantities of interest such as grain size and morphology, liquid film thickness, nodal stresses, and contact stresses can be determined at each time step.

Another alternative is to use a compromise between the sequential and the fully coupled thermomechanical analysis scheme [4]. In this approach, the complete simulation between metal pour and mold removal temperatures can be divided into several large time intervals. For the first time interval, a perfect contact between the metal and the mold can be assumed and a transient thermal analysis followed by a transient mechanical analysis can be performed. The gap thickness at the end of the first time interval can be used in the second time interval and so forth. However, this approach is tedious since it requires interrupting and restarting the thermal and mechanical analyses several times during a complete simulation. It may be used if the available code does not have a robust coupled thermomechanical analysis capability.

IV. MUSHY STATE

In this chapter, two important phenomena taking place in the mushy state: (1) the release of latent heat and (2) the formation of hot tears are discussed. A proper treatment of the former is essential for accurate determination of the temperature histories at various points in the casting during solidification and subsequent cooling. This subject has been discussed extensively in the literature, including this book, and will be addressed in this section only in an implied manner. The primary focus of this section will be on the prediction of hot tears in case of equiaxed dendritic solidification.

A. Prediction of Hot Tears

The prediction of hot tears is one of the most difficult and least developed topics in casting simulation. In 1952, Pellini [15] presented his strain theory of hot tear formation in castings and weldments, which is now widely accepted as the most appropriate explanation [16–18]. However, no attempt was made during the following 40 years to utilize Pellini's theory in developing a predictive methodology until Chandra proposed a scheme for this purpose in 1995 [4].

Earlier efforts for the prediction of hot tears can be divided into two categories: (1) crack susceptibility coefficient or CSC type and (2) continuum mechanics type. The CSC approach was first suggested by Clyne and Davies [19], and modified by Campbell [18]. It is empirical in nature and is suited for a qualitative assessment of relative hot tear tendencies of different alloys when cast in a one-dimensional (bar) shape encased in an infinitely rigid mold, which completely suppresses the thermal strain in the casting. The continuum mechanics approach by Decultieux et al. [20] treated the cast metal in its mushy state as a homogeneous single-phase viscoplastic material and failed to capture the peak strains in the liquid film surrounding the solid grains which (as will be seen later) can be as high as 150 times the average strain. As clearly demonstrated by Pellini in 1952 [15], this peak strain in the liquid film is actually responsible for the formation of hot tears. Therefore, to be meaningful, a casting process simulation must be able to predict this peak film strain.

The approach proposed by Chandra [4] was based upon the following observations: (1) a hot tear occurs during the later stages of solidification, e.g., when the solid fraction is between 0.8 and 0.98 [15]; (2) it occurs when the strain in the liquid film surrounding the solid grains exceeds a certain critical value; and (3) thermal strain does not cause hot tears—on the contrary, it is the part of thermal strain which is suppressed by the mold and/or by the already solidified metal that causes hot tears. This part of suppressed thermal strain was termed as the "mechanical strain" by Cheng et al. [21]. Thus, the objective of the finite element analysis is to compute the mechanical strain in the liquid film.

To compute the mechanical strain in the liquid film, the cast metal is first treated as a homogeneous single-phase material. Ignoring the last two terms, Eq. (2) can be rewritten in the following form [21]:

$$\varepsilon_{ij} = \varepsilon_{ij}^{m} + \varepsilon_{ij}^{th} \tag{3}$$

where ε_{ij}^{m} is the mechanical strain in the homogenous single-phase material. It consists of the elastic ε_{ij}^{el}, plastic ε_{ij}^{p}, and creep ε_{ij}^{cr}, components. For a short freezing range alloy, such as Ni-Al-bronze, the creep component can be ignored because the duration of the film stage is small. This component can, however, become important when dealing with long freezing range alloys.

The mechanical strain tensor ε_{ij}^{m} can be resolved into three principal strains, ε_{1}^{m}, ε_{2}^{m}, and ε_{3}^{m}, with ε_{1}^{m} being the largest and ε_{3}^{m} being the smallest. Although all these principal strains can contribute to a hot tear, at this stage it is assumed that only the maximum principal mechanical strain ε_{1}^{m} is responsible for such failure [18].

In Ref. 21, Cheng et al. presented a scheme to convert the maximum principal strain in the homogeneous single-phase material ε_{1}^{m} into strain in the liquid film surrounding equiaxed dendritic grains. As shown in Fig. 4, utilizing Campbell's suggestion, the grains were assumed to grow as spheres of uniform size. These spheres were then replaced by cubes of equivalent

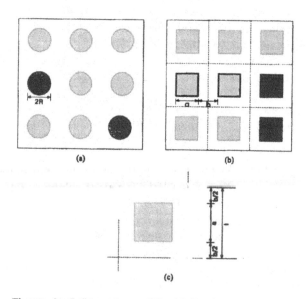

Figure 4 Solid grains and liquid film in the mushy region: (a) spherical grains, (b) equivalent cubic grains, (c) a unit cell containing a cubic grain and liquid film.

volume. After some further mathematical manipulation, the strain in the liquid film ε_f^m was found to be

$$\varepsilon_f^m = \frac{\varepsilon_1^m}{1 - f_s^{1/3}} \tag{4}$$

where f_s is the fraction solid.

Thus, the quantity $1/(1 - f_s^{1/3})$ may be viewed as a magnification factor which converts the average strain in a homogeneous single-phase material ε_1^m into the strain in the liquid film ε_f^m, which is eventually responsible for hot tears. The significance of the magnification factor becomes apparent by noting that when the solid fraction is 0.95, the magnification factor is 59. But, with a small increase in solid fraction to 0.98, the magnification factor becomes 149. Thus, any attempt to predict hot tears by treating the cast metal as a homogeneous single-phase material (e.g., Ref. 20) will fail to recognize a drastic increase in the film strain as the material passes through the tearing range.

Comparison of the computed principal mechanical strain in the liquid film, ε_f^m, with the experimentally determined value of critical fracture strain ε_{fr} provides an indication of hot tearing. A hot tear is assumed to occur if

$$\varepsilon_f^m \geq \varepsilon_{fr} \tag{5}$$

B. Determination of Solid Fraction

Equation (4) represents the strain in the liquid film that can be used to characterize the formation of hot tears in castings. In order to compute this strain, one must compute the solid fraction accurately because of the strong dependence of the "magnification factor" on solid fraction. Solid fraction can be calculated by one of the several ways including the Scheil's equation, the lever rule, and solidification kinetics. The solidification kinetics approach is known to be more comprehensive and accurate. Its application is discussed below. In addition, a new, semiempirical approach is also described.

1. Solidification Kinetics Model

Solidification kinetics is based on the concept of grain nucleation and growth. Under the assumption that the equiaxed grains grow spherically, the solid fraction f_s at time t is given by [22]:

$$f_s = \tfrac{4}{3}\pi R(t)^3 N(\dot{T}, t) \tag{6}$$

where R is the average grain radius and is a function of time, and N is the number of grains per unit volume or grain density and is a function of cooling rate and time. This cooling rate refers to the one just prior to the liquidus, and should not be confused with the average cooling rate of a casting.

Equation (6) is coupled with Eq. (1) through the source term $\dot{Q}_c$, as follows:

$$\dot{Q}_c = L\left(\frac{\partial f_s}{\partial t}\right) \tag{7}$$

where L is the total latent heat.

The rate of solidification can be obtained by differentiating Eq. (6):

$$\left(\frac{\partial f_s}{\partial t}\right) = \left[4\pi R^2\left(\frac{\partial R}{\partial t}\right)N(t) + \tfrac{4}{3}\pi R^3\left(\frac{\partial N}{\partial t}\right)\right] \tag{8}$$

According to Ref. 22, the effective area of the interface between the solid grain and liquid is weighted by the factor $1 - f_s$ and $\partial N/\partial t$ may be assumed to be zero; i.e., N remains constant with time. Then, Eq. (8) becomes

$$\left(\frac{\partial f_s}{\partial t}\right) = 4\pi R^2\left(\frac{\partial R}{\partial t}\right)N(1 - f_s) \tag{9}$$

The grain growth rate $\partial R/\partial t$ is given by

$$V = \left(\frac{\partial R}{\partial t}\right) = \mu\,(\Delta T)^2 \tag{10}$$

where ΔT is the undercooling and μ is a morphological constant. In a strict sense, ΔT should include all components of undercooling, i.e., thermal, constitutional, and curvature. But, in order to minimize computational time, it is a common practice to treat ΔT at a particular location as the difference between the liquidus temperature and the actual temperature at that location, at any particular instant in time. As discussed in Ref. 21, the determination of the morphological constant μ requires the knowledge of several material constants such as partition coefficient, mass diffusivity, and Gibbs-Thomson coefficient.

Following Ref. 22, a parabolic relationship between the cooling rate $\dot{T}$ and the grain density N may be assumed:

$$N(\dot{T}) = a_1 + a_2\dot{T} + a_3\dot{T}^2 \tag{11}$$

where a_1, a_2, and a_3 are material constants which have to be determined experimentally.

Equations (6) through (11) can be used to compute the change in solid fraction and thickness of liquid film as solidification progresses, provided the assumptions involved in the mathematics are justified and the various material constants are readily available.

The experience with Ni-Al-bronze revealed the following limitations of the solidification kinetics approach [21, 23]: (1) experimental determination of reliable values for grain density and size, especially as a function of preliquidus cooling rates, is extremely difficult; (2) the determination of a reliable value of the morphological constant μ requires a whole new set of experiments; and (3) when the theory of solidification kinetics was implemented in a general purpose finite element software [9], it required a large amount of CPU time for the determination of solid fraction vs. time relationship, especially when the grain density was high.

2. A Semiempirical Model

A much simpler and computationally efficient alternative to account for the latent heat and to determine solid fraction is based upon the use of differential scanning calorimetry (DSC) data [24]. A typical DSC plot is shown in Fig. 5. It provides information on liquidus and solidus, and the variation of specific heat (sensible heat plus latent heat) between the two temperatures. The combined sensible and latent heat can be considered as an effective or apparent specific heat.

The solid fraction at any temperature T between the liquidus T_L and solidus T_S can be found by integrating the curve in Fig. 5, resulting in the following expression:

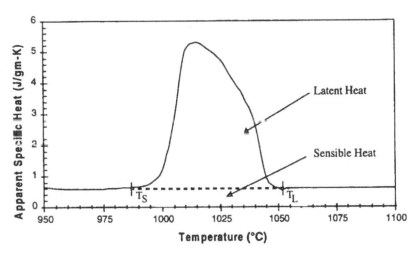

Figure 5 Apparent specific heat vs. temperature for Ni-Al-bronze.

$$f_s = \frac{1}{L} \int_T^{T_L} \left[C_P^{\text{app}}(\tau) - C_P \right] d\tau \tag{12}$$

where $T_S < T < T_L$; $T < \tau < T_L$; $C_P^{\text{app}}(\tau)$ is the apparent specific heat at temperature τ, C_P is the average specific heat outside the solidification range or the sensible heat, and L is the latent heat of solidification. The latent heat L can be given by the integral:

$$L = \int_{T_S}^{T_L} \left[C_P^{\text{app}}(\tau) - C_P \right] d\tau \tag{13}$$

The resulting plot of solid fraction as a function of temperature is shown in Fig. 6. For the purpose of this discussion, it is termed as the "experimental" plot. Now, a mathematical function, which most closely fits the experimental plot of Fig. 6, especially in the hot tear region is needed. Complete details of the curve fitting procedure are given in Ref. 24. One possible function is

$$f_s = \frac{1 - ax \sin(\pi x)}{\cosh(bx^2)}; \qquad 0 \leq x \leq 2 \tag{14}$$

where $x = 2(T - T_S)/T_L - T_S$, and a and b are adjustable parameters that must be determined from the experimental data. It is found that when $a = 0.04$ and $b = 1.28$, the above equation fits the experimental data extremely well, as shown by the solid line in Fig. 6. It needs to be emphasized here that the above equation is not the only function that can be used to mathematically

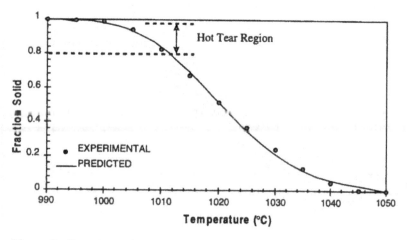

Figure 6 Experimental and predicted solid fraction vs. temperature.

describe the experimental solid fraction data; other functions, just as good, may be found to fit the data.

C. Correction to Mechanical Strain

The procedure for the prediction of hot tears, outlined above, solves the equations of solid mechanics and heat transfer, assuming that the solidifying casting consists of a homogeneous, single-phase material. Therefore, most general purpose finite element codes will predict certain values of stress and mechanical strain even when the cast metal is in the liquid state or at a very early stage of solidification. It is obvious that in the liquid stage, the molten metal is free to flow and should not develop stress and mechanical strain. Metallurgical evidence also indicates that alloys begin to develop strength when a certain level (coherency point) of solid fraction is reached [25–27]. Thus any mechanical strain and corresponding stress, computed prior to the coherency point, are fictitious. If these fictitious stresses and strains are not accounted for, they will cause an overprediction of film strain and transient as well as residual stresses.

There are two possible approaches to account for such fictitious stresses and strains. These are (1) to use low values of modulus up to the coherency point, or (2) to subtract the fictitious stresses and strains from the computed values after the coherency point is reached. The first approach should be used with caution as it may cause numerical difficulties or ill conditioning in the solution if the value of the modulus is very low. The second approach does not affect the numerical stability and convergence of the solution procedure and has been used by Ahmed and Chandra [8].

V. THERMAL-ELASTIC-PLASTIC-CREEP MATERIAL MODEL

At high temperatures, all alloys exhibit creep even if the stresses are small. This effect of creep on strains (distortions) may become especially pronounced in large sand castings which remain confined in the mold for several days. Additional complexity may arise due to the fact that in such cases, the stress varies continuously as the casting cools, and plasticity and creep occur simultaneously. Many commercial codes are not able to handle such situations and require enhancement.

A special thermal-elastic-plastic-creep constitutive model has been derived by Ahmed and Chandra [8]. This constitutive model is based upon the published works by Snyder and Bathe [14], Hult [28], and

Manson [29], and suitably modified to account for the variable stress condition. All material properties involved in the derivation of this constitutive model (e.g., Young's modulus, yield stress, thermal expansion, etc.) are assumed to be functions of temperature. This is an extremely lengthy work and its full details are given in Ref. 8. However, a summary is presented in the following paragraphs.

Equation (2), minus the last two terms, is the basis for the derivation of the thermal-elastic-plastic-creep constitutive model. Due to the nonlinear nature of plasticity and creep, the following time-rate form of stress-strain relationship applies:

$$\dot{\sigma}_{ij} = E_{ijkl}\left(\dot{\varepsilon}_{kl} - \dot{\varepsilon}^{p}_{kl} - \dot{\varepsilon}^{cr}_{kl} - \dot{\varepsilon}^{th}_{kl}\right) + \dot{E}_{ijmn}\varepsilon^{el}_{mn} \qquad (15)$$

where ε_{kl} are the various strain terms defined earlier and E_{ijkl} are the components of the modulus tensor.

For the plastic strain term, von Mises yield criterion, isotropic hardening law, and Pradtl-Reuss flow rule are used. For the creep strain component, a power-law relationship between the equivalent strain, equivalent stress, and time is used [28, 29]. It is then augmented by a special form of time hardening rule shown conceptually in Fig. 7. After some mathematical manipulations, the following general form of the constitutive model was obtained by Ahmed and Chandra [8]:

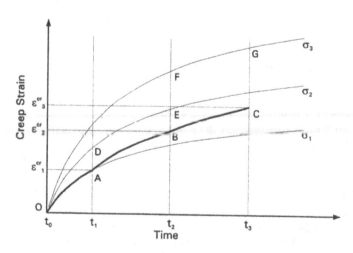

Figure 7 Creep strain vs. time for constant and variable stresses.

$$\dot{\sigma}_{ij} = \left\{ E_{ijkl} - \frac{E_{ijmn} S_{mn} S_{rs} E_{rskl}}{\frac{4}{9} h \sigma_y^2 + S_{pq} E_{pqrs} S_{rs}} \right\} (\dot{\varepsilon}_{kl} - \varepsilon_{kl}^{cr})$$

$$+ \left(E_{ijkl} S_{kl} F - E_{ijkl} \frac{\partial \varepsilon_{kl}^{th}}{\partial T} + \frac{\partial E_{ijmn}}{\partial T} \varepsilon_{mn}^{el} \right) \dot{T} \qquad (16)$$

where

$$F = \frac{S_{rs} \left(E_{rsuv} \frac{\partial \varepsilon_{uv}^{th}}{\partial T} - \frac{\partial E_{rsuv}}{\partial T} \varepsilon_{uv}^{el} \right) + \frac{2}{3} \sigma_y \frac{\partial \sigma_y}{\partial T}}{\frac{4}{9} h \sigma_y^2 + S_{pq} E_{pqrs} S_{rs}} \qquad (17)$$

For the meaning of the various terms in Eqs. (16) and (17), Ref. 8 should be consulted. Also, following Fig. 7, the accumulated creep strain can be given by

$$\varepsilon_k^{cr} = \sum_{i=1}^{k} A_i \sigma_i^{m_i} \left(t_i^{n_i} - t_{i-1}^{n_{i-1}} \right) \qquad (18)$$

The preceding constitutive law was implemented into a commercial finite element analysis code. Full details of this work, including its validation and application to a simple test casting, are discussed in Ref. 8. The application part is also summarized in Sec. X of this chapter.

VI. INTERACTION BETWEEN THE MOLD AND THE CASTING

As the liquid metal solidifies and further cools, it shrinks away from the mold, creating a gap at the mold-metal interface. On the other hand, at certain locations, the solidified metal rests against the mold. This formation of intermittent gap and contact affects the thermal as well as the mechanical response of the mold-metal system.

From the thermal point of view, the interfacial gap provides resistance to heat flow from the metal to the mold. The interface may consist of intermittent contacts at the asperities with gaps in between. Conductive heat transfer takes place in the region of solid to solid contact with conduction and/or radiation across the gas film between the asperities. Due to the small thickness of these gas films, convective heat transfer may not be significant [5]. Since it is impractical to model every asperity, it is a common practice to assume that the outer surface of the casting and the inner surface of the mold are smooth, although the gap thickness may still vary along the mold-metal interface. Then the interface heat transfer can be represented either as equivalent convection or as equivalent conduction [30]. The two forms are

$$q = h_{\text{eff}} A (T_{\text{metal}} - T_{\text{mold}}) \tag{19}$$

$$q = \frac{k_{\text{eff}} A (T_{\text{metal}} - T_{\text{mold}})}{\delta} \tag{20}$$

where q is the rate of heat transfer, A is the area of the surface (or surface element), and δ is the thickness of the gap between the metal and mold.

If one intends to perform only a thermal analysis of the casting process (box 4 of Fig. 1), it may be adequate to treat the interface heat transfer as a time or temperature dependent equivalent convective coefficient without explicitly tracking the thickness of the interface gap. This involves two steps: (1) an experimental determination of the interface heat transfer coefficient using castings of simple shapes instrumented with thermocouples, followed by the application of inverse heat transfer technique [31–35], and (2) the implementation of a suitable algorithm in the computer software which provides a heat flux continuity between the metal and the mold utilizing the values obtained in step 1 [36, 37].

When the objective of the casting process simulation is to perform a thermomechanical simulation, i.e., boxes 4 and 5 of Fig. 1, it is more appropriate to track the change in thickness of the interfacial gap and to use the equivalent gap conductance approach.

A proper treatment of the mechanical aspect of interaction between two solid bodies is extremely complex. It involves tracking the progression of relative normal and tangential motions, i.e., contact/gap and sliding, and the stresses or forces transmitted between them at their interface. This introduces additional complexities in the finite element formulation since the area of contact between the mold and the casting is continuously changing and has to be found as part of the solution. Also, the stresses at the contact surface are not uniform.

The treatment of the mechanical interaction between two solid bodies can be traced back to the classical Hertzian elastic contact problem [38]. Its finite element treatment has been discussed extensively in recent years, with special emphasis on the simulation of forging and other large deformation forming processes [39–44]. The present capabilities include treatment of linear and nonlinear contact with small or finite sliding and choice of friction models, with or without heat transfer across the interface. An extended discussion of this subject is beyond the scope of this work. Instead, in Sec. X, mechanical interaction between some simple castings and their molds and the implication of such interaction when the molds are removed is briefly discussed.

VII. HEAT LOSS FROM THE MOLD SURFACE

The heat content of the molten metal is ultimately lost to the ambient from the mold surface via convection and/or radiation. Thus, convection and/or radiation can be viewed as the thermal boundary conditions at the mold surface. Sand molds are poor conductors of heat, and it takes a relatively long time for the heat to reach the outside surface of the mold. Heat is released slowly to the ambient, and the process is more likely to be dominated by convection. The convective heat transfer coefficient, or the surface film coefficient, which must be calculated theoretically or determined experimentally, dictates the rate of heat transfer through the mold surface.

In the investment casting process, on the other hand, the loss of heat through the outer surface of the mold is primarily by radiation. A proper mathematical treatment of radiative heat transfer at the mold surface can be very complex. In order to put the subject in proper perspective, some of the basic concepts are summarized in the following paragraphs.

Consider a casting allowed to cool in a large room with ambient temperature T_∞, as shown in Fig. 8. Radiation takes place from surfaces A through G, which are assumed to be at different temperatures at any given moment. However, for simplicity, each surface is assumed to have one uniform temperature. Surfaces A, B, C, and G radiate heat to the ambient. While the mold surfaces are assumed to be gray bodies, the ambient is assumed to act as a black body. The radiant heat transfer rate from surface A, B, C, or G is given by

$$q = \sigma \varepsilon A \left(T_s^4 - T_\infty^4 \right) \tag{21}$$

where σ is the Stefan-Boltzmann constant, ε is the emissivity of the radiating surface, A is the surface area, T_s is the absolute surface temperature, and T_∞ is the absolute ambient temperature.

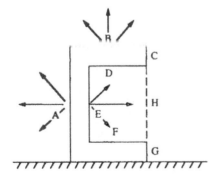

Figure 8   Radiation from a casting cooling in air.

Now consider surfaces D, E, and F. In a strict sense, these surfaces exchange radiant energy with each other and with the ambient, which may be represented by a hypothetical surface H. This type of radiation is termed "enclosure" or "cavity" radiation. Two approaches are available to treat enclosure radiation. The so-called net-radiation approach involves the use of a rigorous definition of the term view factor and consideration of irradiation and radiosity. The view factor F_{ij} is defined as the fraction of the diffusely radiant energy leaving a surface (or surface element) i that arrives at another surface (or surface element) j. Mathematically, the view factor is defined as

$$F_{ij} = \frac{1}{A_i} \int_{A_i} \int_{A_j} \frac{\cos \theta_i \cos \theta_j \, dA_i \, dA_j}{\pi r^2} \tag{22}$$

The terms used in Eq. (22) are defined in Fig. 9 [45]. Several schemes for computing view factors F_{ij}, in addition to the specialized topics of blocking or shadowing, axisymmetry, cyclic symmetry, etc., are discussed in the literature [46–48].

Irradiation is the total radiation incident upon a surface per unit time and per unit area, and radiosity is the sum of the energy emitted and the energy reflected (when no energy is transmitted) per unit time and unit surface area. The net energy leaving a surface is the difference between the radiosity and the irradiation. This net-radiation approach is computationally intensive, especially for large and complex castings that may involve several thousand finite elements.

A simplified approach takes advantage of the fact that the difference in temperature between surfaces D, E, and F is usually much smaller than that between any surface and the ambient. Thus, the heat exchange between these surfaces can be neglected in comparison to the exchange between a surface and the ambient (represented by the hypothetical surface H in Fig. 8). The radiant flux at D, E, or F is then treated in the same way as at surfaces A, B, C, or G with the exception that a simplified form of view factor F_i is introduced. F_i is

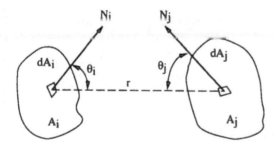

Figure 9 Explanation of terms for view factor calculation. (From Ref. 45.)

defined as the fraction of the diffusely radiant thermal energy leaving the surface D, E, or F that arrives at the hypothetical surface H, which is assumed to behave as a black body. The value of this view factor is less than unity. Equation (21) is modified by multiplying its right hand side by the view factor F_i. In simple castings, this treatment of radiation boundary condition at the mold surface is often adequate. A simple scheme for computing the view factors F_i was recently developed by Upadhya et al. [49].

In Fig. 10, the casting of Fig. 8 is shown to cool in a furnace instead of air. In this case, the furnace walls also exchange radiant energy with various surfaces of the casting. They have to be treated as gray bodies, and the exchange of energy between various mold surfaces and furnace walls has to be accounted for through the application of the net-radiation approach outlined above. An even more complex treatment of radiative heat loss at the outer mold surface is required for the simulation of the directional solidification process, e.g., the Bridgman process, where the mold is withdrawn from the furnace, as shown in Fig. 11. In this case, the view factors F_{ij} for all surface elements change as the mold is withdrawn and have to be recalculated periodically. Schemes for this purpose have been discussed by Thomas et al. [50] and by Hediger and Hofmann [51].

VIII. SOLID STATE PHASE TRANSFORMATIONS

Many metals such as steels undergo one or more solid phase transformations during cooling subsequent to solidification. These solid state transformations are accompanied by a release of latent heat, a change in volume, and a pseudo-

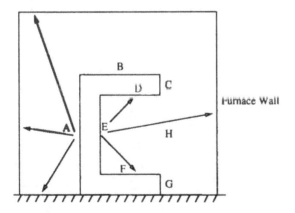

Figure 10 Radiation from a casting cooling in a furnace.

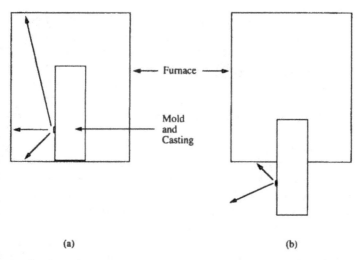

Figure 11 Radiation from an element at the mold surface during Bridgman process: (a) mold completely within the furnace, (b) mold partially withdrawn from the furnace.

plasticity effect commonly known as the transformation plasticity. The evolution of latent heat during the solid state phase transformation is very similar to that during liquid to solid transformation, although of a much smaller amount. The volume change and transformation plasticity may, however, contribute significantly to the residual stresses and distortions in a casting. It is therefore, important to include these effects in the numerical modeling of stresses and distortions in castings if the alloy under consideration is known to exhibit this behavior.

Figure 12 shows a simple example of the transformation plasticity effect in steel based upon the results of a constrained dilatometry experiments [52]. This figure shows that, during cooling in the phase transformation regime, the presence of even a very low stress can result in residual plastic strains in the test piece. Two widely accepted mechanisms for transformation plasticity are due to Greenwood and Johnson [53] and Magee [54]. According to the Greenwood-Johnson mechanism, the difference in volume between two coexisting phases in the presence of an external load generates a microscopic plasticity in the weaker phase. This in turn leads to macroscopic plastic flow even though the external load is not sufficient to cause plasticity on its own. According to the Magee mechanism, if a martensitic transformation takes place under an external load, martensite plates are formed with a preferred orientation affecting the overall shape of the body. As indicated earlier, an external load or stress is present in castings during cooling because of the thermal gradients in different parts and/or due to the mechanical constraints imposed by the mold.

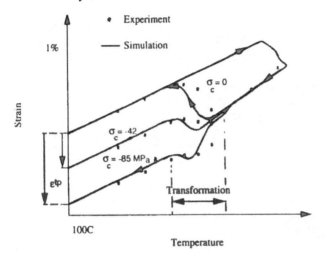

Figure 12 Concept of transformation plasticity. (From Ref. 52.)

The modeling of the effects of volumetric change and transformation plasticity during solid state transformation were first addressed in the context of quenching and welding processes. Their significance is now being recognized in the simulation of casting processes as well [4, 7, 55]. Early attempt by Rammerstorfer et al. [56] to model transformation plasticity in steel was by simply lowering the yield strength of the material within the transformation range. Since there is no evidence of exceptionally low yield stress during austenite to martensite transformation, the use of this approach in now considered unreliable.

The current approach for the treatment of volumetric change and transformation plasticity effects during solid state phase change ignores the creep strain term ε_{ij}^{cr} in Eq. (2). The fractions of the various decomposed phases are determined using the theories of transformation kinetics, and suitable expressions are derived for the strain terms ε_{ij}^{v} and ε_{ij}^{tr} in Eq. (2). For mathematical details related to the theory of transformation kinetics and the two strain terms, Refs. 52, 57–62 may be consulted.

IX. MOLD REMOVAL, FINISH MACHINING, AND PATTERNMAKER'S ALLOWANCE

A. Mold Removal

As discussed earlier, when a casting cools in the mold during and after solidification, the mold tends to constrain the thermal shrinkage of the casting. This results in distributed equal and opposite forces in the casting and mold at the points (rather surfaces) of contact. The forces in the casting-mold system are in complete equilibrium under this condition. When the casting is removed, the constraints imposed by the mold are removed and the casting adjusts to a new equilibrium state through a redistribution of stresses and additional distortions. The final as-cast dimensions and the as-cast residual stresses in the casting depend on what happens during the mold removal process. Therefore, analysis of mold removal forms an important part of a complete procedure for thermomechanical analysis of castings.

For modeling purposes, the process of mold removal is very similar to that of material removal by machining discussed in the following section. In fact, the former is simpler since it does not involve the complexities of thermal and mechanical effects of the cutting tool. Chandra and coworkers utilized this observation in studying the redistribution of stresses and distortions due to mold removal in simple castings [6, 8]. The highlights of this work are reported in Sec. X.

B. Finish Machining

Castings need to be machined for several reasons. All the gates, runners, risers, and other rigging systems, added to the part for the metal flow and for providing feed metal, must be removed by machining. Sometimes extra material is added in certain locations in the part in order to make the part more castable. This additional material must be removed by subsequent machining. Features such as holes, which are difficult to cast are often produced by finish machining. Finish machining may also be necessary to meet the surface finish requirements.

When material is removed during finish machining, the residual stresses in the material being removed are transferred to the remaining material causing a redistribution of stresses and additional distortion in the part. Thermal and mechanical effects of the cutting tool cause additional stress redistribution and distortions. As a result of these, it is often difficult to maintain the dimensions of a cast part within specifications if shape is complex or if its walls are thin.

Ignoring the effects of the cutting tool, Chandra [3] explained the concept of stress redistribution and distortions during machining with the help of several simple models. He further tested the accuracy of the finite element code for this purpose. As discussed in the previous section, he further extended this concept to modeling the mold removal process [4]. Wenyu [63], on the other hand, has presented a finite element treatment of the subject using a continuum approach. A proper theory for modeling the effects of cutting tools is more complex and yet to be developed.

C. Patternmakers' Allowance

The dimensions of a cast part are influenced by its entire processing history. The final dimensions are not the same as the dimensions of the mold cavity or the pattern. The patternmaker uses his experience in choosing the dimensions of the pattern that will eventually result in the desired dimensions of the cast part. It requires a lot of trial and error on the foundry floor. A thermomechanical simulation of the casting process, as discussed in this chapter, yields the final dimensions of the part after it is taken out of the mold. Obviously, the difference in the final dimensions of the part and the initial dimensions of the pattern is indicative of the patternmakers' allowance. Thus, an upfront computer simulation of the process can be used to a considerable advantage in deciding the patternmaker's allowance and in saving a lot of trials on the floor.

X. EXAMPLES

Simulation of three simple hypothetical castings is presented in this section. These examples serve to numerically demonstrate some of the concepts discussed in the preceding paragraphs, such as the formation of gap and contact between the surfaces of the mold and the metal, formation of hot tears, effect of creep on transient and residual stresses, effect of mold rigidity and mold removal on residual stresses and distortions, and patternmakers' allowance. Two of these castings are shown in Fig. 13 and the third is shown in Fig. 14.

The casting in Fig. 13a is essentially the same I section that was used earlier in Figs. 2 and 3 to explain the basic concepts. It is symmetrical about the x and y axes, and is of the plane strain type. The channel section in Fig. 13b is symmetrical about the x axis only and is also of the plane strain type. The casting in Fig. 14 is an axisymmetric version of the I section of Fig. 13a.

Due to symmetry, only one quarter of the I section of Fig. 13a was analyzed. Creep effect was not considered and an approximate constitutive model was used to represent the sand mold. Some interesting results of this analysis are shown in Fig. 15, which were obtained after cooling of the metal-mold assembly to room temperature. The figure shows (1) the gap between the casting and the mold, (2) contact between the flange of the cast part and the mold, (3) bending of the flange, (4) stress concentration at the fillet, and (5) uniform stress distribution in the web. Also, notice that underneath the flange, the stresses in the mold are compressive, but in the outer region of the mold they are tensile and small. In all probability, the sand mold will be able to withstand such small tensile stresses.

The effect of mold removal is shown in Figs. 16 and 17, where the distortions are magnified by a factor of 3 for ease of visualization. The former

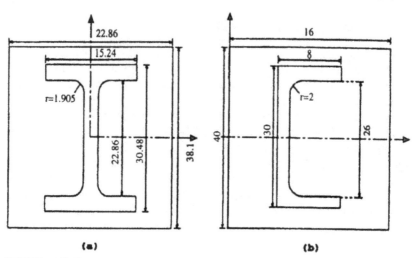

NOTE: All dimensions are in cm.

Figure 13 Test castings: (a) an *I* section and (b) a channel section.

shows the shape of the part before and after mold removal. It may be noted that the length of the cast part after mold removal is less than the original cavity which was filled by the molten metal and the flange remains bent after mold removal. These two factors determine the patternmaker's shrinkage and distortion allowance. Figure 17 shows the stresses in the cast part before and after mold removal. It may be noted that before mold removal, the stresses in the mold were high, but after mold removal they became practically zero, except near the fillet which is an area of stress concentration. The stresses after mold removal may be referred to as the residual stresses.

Since the channel section of Fig. 13b is not symmetrical about the *y* axis, its web (in addition to the flange) distorts during cooling and remains so after the mold removal, as seen in Fig. 18.

The axisymmetric casting shown in Fig. 14 was analyzed very thoroughly using an enhanced version of a commercial finite element analysis code [8] that could account for hot tears prediction and thermal-elastic-plastic-creep. The simulation also utilized measured values of thermophysical and mechanical properties of the cast metal, including the creep parameters; more realistic properties of the molding sand and a concrete type constitutive model that permitted use of different yield strengths in tension and compression. The simulation further accounted for the ficticious strains and stresses prior to the coherency point in the alloy. Full details of the simulation are available in Ref. 8; some of the important results are discussed below.

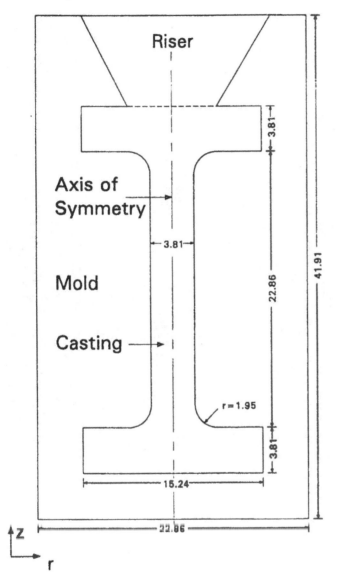

Figure 14 An axisymmetric testing casting.

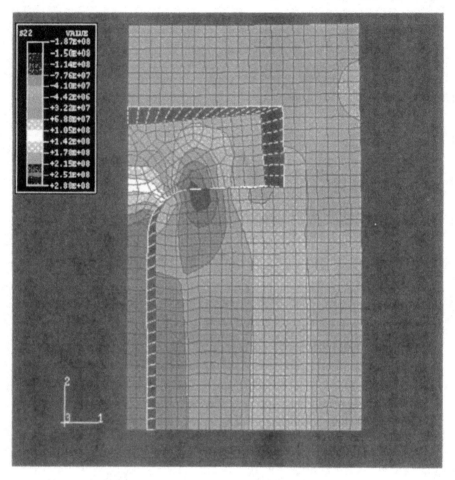

Figure 15 *I* section casting and mold after cooling to room temperature.

Figure 19 shows the contour plots of solid fraction in the casting at different times during cooling. These contours show the presence of a hot spot at the junction of the web and the lower flange. In order to check for the possibility of hot tear in this region, film strain at several nodes were examined. It was found that, in case of a rigid mold, film strain at one of the nodes could be as high as 43%, which could lead to hot tears. But, for a flexible mold the maximum film strain was about 13%, which may not be high enough to cause any hot tear in the cast metal. Thus, the use of a flexible mold reduces the tendency for hot tears.

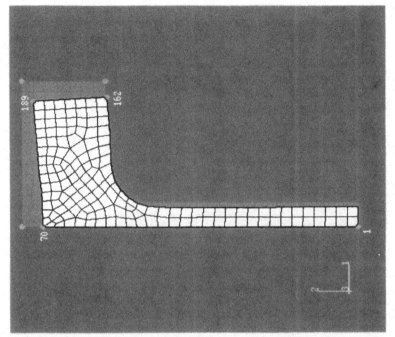

(a)

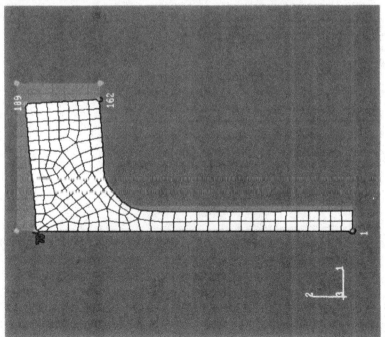

(b)

Figure 16   Distortions of the *I* section casting: (a) before and (b) after mold removal.

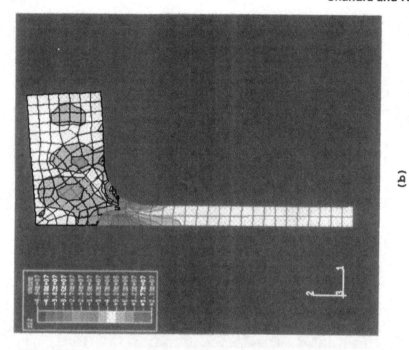

(b)

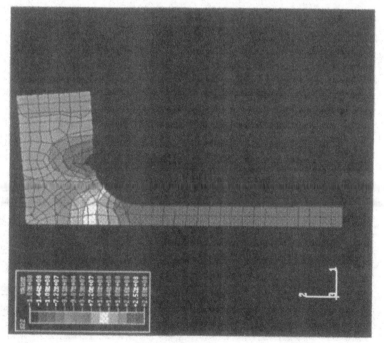

(a)

Figure 17 Stresses in the *I* section: (a) before and (b) after mold removal.

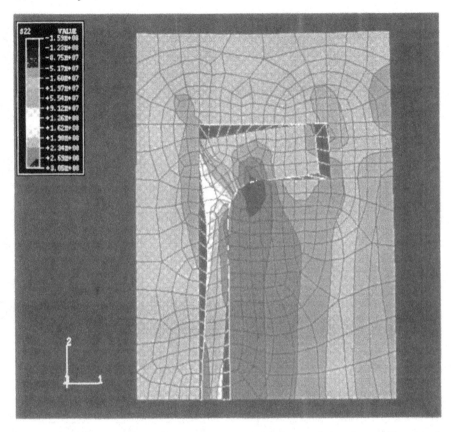

Figure 18   Distortion of the channel section: (a) before and (b) after mold removal.

Figure 20 shows the importance of considering creep effects in the simulation. The figure shows axial stresses over a section A-A in the web far removed from the two flanges. When creep is not included in the simulation, axial stresses are high. Including the creep effect reduces the stresses and makes them more evenly distributed over the section.

The effects of mold rigidity on axial stresses along section A-A before and after mold removal are shown in Figs. 21 and 22. For a valid comparison, the effect of creep is included in all cases. It may be seen that, prior to mold removal, a rigid mold will result in higher and more evenly distributed stresses than a flexible mold. Also, in case of a rigid mold, mold removal can significantly reduce the stresses, whereas in case of a flexible mold, the reduction in stresses is not so pronounced. Finally, after the mold removal, the stresses are higher and less uniform if the mold used was flexible.

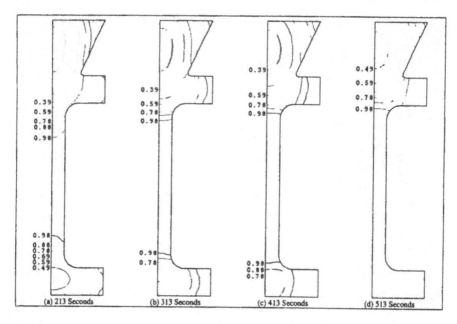

Figure 19   Solid fraction evolution in the axisymmetric casting.

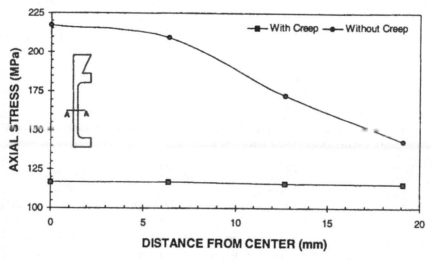

Figure 20   Effect of creep on axial stress along section A-A before mold removal.

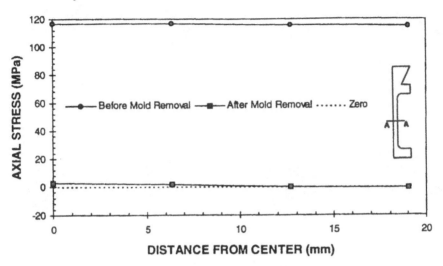

Figure 21 Axial stress along A-A due to a rigid mold.

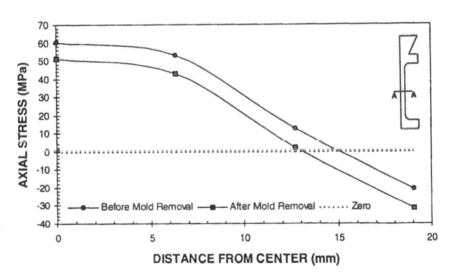

Figure 22 Axial stress along A-A due to a flexible mold.

The foregoing observations can be explained as follows. A very flexible mold offers practically no resistance to the shrinkage of the casting during its cooling. In that case, the stresses in the casting are similar to those developed during quenching, characterized by steep stress gradients. Also, the contact forces at the mold-metal interface are small. When the mold is removed, these forces do not appreciably alter the existing stress distribution in the casting. On the other hand, a rigid mold provides significant constraint to shrinkage during cooling and, as a result, large contact forces are developed at the mold-metal interface. When the mold is removed, these forces are capable of altering the existing stress distribution significantly.

The simulation also showed that the stiffness of the mold has a considerable effect on the patternmakers' allowance—a flexible mold requiring a larger allowance than a rigid mold.

XI. SUMMARY AND CONCLUDING REMARKS

Historically, the simulation of the casting processes began with solidification analysis. It then progressed to the analysis of mold filling. After the successful development of technology for the analysis of these two phenomena, in recent years the attention has shifted to the analysis of stress and distortion. In this chapter, recent advances in this field have been summarized.

The determination of stresses and distortions in a cast part requires the use of a very sophisticated coupled thermomechanical finite element code. This code should be able to account for a number of important phenomena, such as (1) solidification and subsequent cooling of the molten metal, (2) formation of hot tears, (3) loss of heat from the outer surface of the mold, (4) creep under variable stress, (5) constitutive models for the cast metal and the molding sand, (6) thermal and mechanical interaction between the metal and the mold, (7) transformation plasticity, and (8) effect of mold removal and finish machining on stresses and distortions in the cast part. If the code has to perform simulation of the investment casting process, it should also have a good constitutive model for the ceramic mold and an algorithm for enclosure radiation. Similarly, simulation of the directional solidification process would require an algorithm for mold withdrawal.

At this point, no code with all of the aforementioned capabilities is available. Ahmed and Chandra used a commercial finite element code [9] in their work and enhanced it for several features listed above. Using this enhanced version, some simple castings were analyzed and many important conclusions were made. These conclusions are discussed in the chapter.

Based upon the work presented in this chapter, it is concluded that the underlying technology has progressed to a point that it is now possible to develop a customized finite element code fully dedicated to the determination of stresses and distortions in complex parts made of the sand mold, permanent mold or investment mold casting processes.

REFERENCES

1. U Chandra. Finite element simulation of the investment casting process for manufacture of aircraft engine parts. Proceedings of the Engineering Foundation Conference on the Modeling of Casting, Welding and Advanced Solidification Processes V, M Rappaz, MR Ozgu, KW Mahin (eds.), Warrendale, PA: TMS, 1991, pp. 629–634.
2. Metals Handbook: Casting, 9th ed., vol. 15, DM Stefanescu (volume chair), Metals Park, OH: ASM International, 1988.
3. U Chandra. Validation of finite element codes for prediction of machining distortions in forgings. Communications in Numerical Methods in Engineering 9:463–473, 1993.
4. U Chandra. Computer prediction of hot tears, hot cracks, residual stresses and distortions in precision castings: basic concepts and approach. Light Metals 1995, Proceedings of the 1995 TMS Annual Meeting, J. Evans (ed.), Warrendale, PA: TMS, 1995, pp. 1107–1117.
5. U Chandra. Computer simulation of manufacturing processes—casting and welding. Computer Modeling and Simulation in Engineering 1:127–174, 1996.
6. U Chandra, R Thomas, SC Cheng. Shrinkage, residual stresses and distortions in castings. Computer Modeling and Simulation in Engineering 1:369–383, 1996.
7. BG Thomas. Stress modeling of casting process: an overview. Proceedings of the Engineering Foundation Conference on the Modeling of Casting, Welding and Advanced Solidification Processes VI, TS Piwonka, V Voller, L Katgerman (eds.), Warrendale, PA: TMS, 1993, pp. 519–533.
8. A Ahmed, U Chandra. Prediction of hot tears, residual stresses, and distortions in castings including the effects of creep. Computer Modeling and Simulation in Engineering 2:419–448, 1997.
9. ABAQUS, Version 5.4, Pawtucket, RI: Hibbitt, Karlsson and Sorensen, Inc., 1994.
10. G Comini, S Del Giudice, RW Lewis, OC Zienkiewicz. Finite element solution of non-linear heat conduction problems with special reference to phase change. International Journal for Numerical Methods in Engineering 8:613–624, 1974.
11. RW Lewis, K Morgan, RH Gallagher. Finite element analysis of solidification and welding processes. Numerical Modeling of Manufacturing Processes, RF Jones, Jr., H Armen, JT Fong (eds.), ASME, PVP-PB-025, 1977, pp. 67–80.
12. JN Reddy, DK Gartling. The finite element method in heat transfer and fluid dynamics. Boca Raton, FL: CRC Press, Inc., 1994.
13. A Levy, AB Pifko. On computational strategies for problems involving plasticity and creep. International Journal for Numerical Methods in Engineering 17:747–771, 1981.

14. MD Snyder, K-J Bathe. A solution procedure for thermo-elastic-plastic and creep problems. Nuclear Engineering and Design 64:49–80, 1981.
15. WS Pellini. Strain theory of hot tearing. Foundry 80:124–133, 1952.
16. RA Dodd. Hot tearing of castings: a review of the literature. Foundry Trade Journal 321–331, 1956.
17. SA Metz, MC Flemings. Hot tearing in cast metals. AFS Transactions 77:329–334, 1969.
18. J Campbell. Castings. Oxford, UK: Butterworth-Heinmann, 1991.
19. TW Clyne, and GJ Davies. Comparison between experimental data and theoretical predictions relating to dependence of solidification cracking on composition. Solidification and Casting of Metals, London: Metals Society, 1979, pp. 275–278.
20. F Decultieux, P Vincent-Hernandez, C Levaillant. Hot tearing test: experimental and FEM modeling. Proceedings of the Engineering Foundation Conference on the Modeling of Casting, Welding, and Advanced Solidification Processes VI, TS Piwonka, V Voller, L Katgerman (eds.), Warrendale, PA: TMS, 1993, pp. 617–624.
21. S Cheng, S Sundarraj, J Jo, U Chandra. Computer prediction of hot tears in castings. Proceedings of the National Heat Transfer Conference, vol. 1, V Prasad et al. (eds.), ASME-HTD vol. 323, 1996, pp. 59–68.
22. DM Stefanescu, G Upadhya, D Bandyopadhyay. Heat transfer-solidification kinetics modeling of solidification in castings. Metallurgical Transactions 21A:997–1005, 1990.
23. JJ Jo, ED Peretin, U Chandra. Experimental study of grain nucleation and growth in Ni-Al-Bronze. Report No. TR 97-050, Johnstown, PA: Concurrent Technologies Corporation, 1997.
24. A Ahmed, U Chandra. A solidification model for use in the prediction of hot tears in castings. Proceedings of the Engineering Foundation Conference on the Modeling of Casting, Welding and Advanced Solidification Processes VIII, BG Thomas, C Beckermann (eds.), Warrendale, PA: TMS, 1998, pp. 891–898.
25. ARE Singer, SA Cottrell. Properties of the aluminum-silicon 1035 alloys at temperatures in the region of the solidus. Journal of the Institute of Metals 73:33–54, 1946.
26. JA Williams, ARE Singer. Deformation, strength, and fracture above the solidus temperature. Journal of the Institute of Metals 96:5–12, 1968.
27. AK Dahle. Mushy zone properties and castability of aluminum foundry alloys. PhD dissertation, The Norwegian University of Science and Technology, 1996.
28. JAH Hult. Creep in Engineering Structures. Waltham, MA: Blaisdell, 1966.
29. SS Manson. Thermal Stress and Low-Cycle Fatigue. New York: McGraw-Hill, 1966.
30. U Chandra. Benchmark problems and testing of a finite element code for solidification in investment castings. International Journal for Numerical Methods in Engineering 30:1301–1320, 1990.
31. K Ho, RD Pehlke. Metal-mold interfacial heat transfer. Metallurgical Transactions B, 16B:585–594, 1985.

32. B Dorri, U Chandra. Determination of thermal contact resistance using inverse heat conduction procedure. Proceedings of the Seventh International Conference on Numerical Methods in Thermal Problems, RW Lewis, JW Chin, GM Homsy (eds.), Swansea UK: Pineridge Press, 1991, pp. 213–223.

33. JV Beck, B Blackwell, CR StClair. Inverse Heat Conduction: Ill-Posed Problems. New York: John Wiley & Sons, 1985.

34. D O'Mahoney, DJ Browne. A study of the variation of heat transfer coefficients in aluminum investment castings. Modeling of Casting, Welding and Advanced Solidification Processes VIII, BG Thomas, C Beckermann (eds.), Warrendale, PA: TMS, 1998, pp. 1031–1038.

35. SA Argyropolous, NJ Goudie, M Trovant. The estimation of thermal resistance at various interfaces. Fluid Flow Phenomena in Metals Processing, Proceedings of the 1999 TMS Annual Meeting, N. El-Kaddah, DGC Robertson, ST Johanesen, VR Voller (eds.), Warrendale, PA: TMS, 1999, pp. 535–542.

36. A Shapiro. TOPAZ — a finite element heat conduction code for analyzing 2-D solids. Livermore, CA: Lawrence Livermore National Laboratory, 1984.

37. MT Sammonds. Finite element simulation of solidification in sand mould and gravity die castings. PhD dissertation, University of Wales, Swansea, UK, 1985.

38. SP Timoshenko, JN Goodier. Theory of Elasticity. New York: McGraw-Hill, 1970.

39. SK Chan, IS Tuba. A finite element method for contact problems of solid bodies. Part I: Theory and validation. International Journal of Mechanical Science 13:615–625, 1971.

40. SI Oh. Finite element analysis of metal forming processes with arbitrary shaped dies. International Journal of Mechanical Science 24:479–493, 1982.

41. JT Oden, EB Pires. Algorithms and numerical results for finite element approximations of contact problems with non-classical friction laws. Computers and Structures 19:137–147, 1984.

42. JW Joo, BM Kwak. Analysis and application of elasto-plastic contact problems considering large deformation. Computers and Structures 24:953–961, 1986.

43. AB Chaudhary, K-J Bathe. A solution method for static and dynamic analysis of three-dimensional contact problems with friction. Computers and Structures 24:855–873, 1986.

44. WW Tworzydlo, W Cecot, JT Oden, CH Yew. New asperity-based models of contact and friction. Contact Problems and Surface Interaction in Manufacturing and Tribological Systems, MH Attia, R Komanduri (eds.), ASME, 1993, pp. 87–104.

45. ANSYS–Engineering Analysis System, Version 5.1, Houston, PA: ANSYS, Inc., 1993.

46. R Siegel, JR Howell. Thermal radiation heat transfer. New York: Hemisphere Publishing Corporation, 1981.

47. AB Shapiro. FACET—A radiation view factor computer code for axisymmetric, 2D planar, and 3D geometries with shadowing. Livermore, CA: Lawrence Livermore National Laboratory, 1983.

48. JP Holmann. Heat Transfer, New York: McGraw-Hill, 1990.

49. GK Upadhya, S Das, U Chandra, AJ Paul. Modeling the investment casting process: a novel approach for view factor calculations and defect prediction. Applied Mathematical Modelling 19:354–362, 1995.

50. BG Thomas, DD Goettsch, KO Yu, MJ Beffel, M Robinson, D Pinella, RG Carlson. Modeling of directional solidification process. Proceedings of the Engineering Foundation Conference on the Modeling of Casting, Welding and Advanced Solidification Processes V, M Rappaz, MR Ozgu, KW Mahin (eds.), Warrendale, PA: TMS, 1991, pp. 603–610.

51. F Hediger, N Hofmann. Process simulation for directionally solidified turbine blades of complex shapes. Proceedings of the Engineering Foundation Conference on the Modeling of Casting, Welding and Advanced Solidification Processes V, M Rappaz, MR Ozgu, KW Mahin (eds.), Warrendale, PA: TMS, 1991, pp. 611–619.

52. JM Bergheau, JB Leblond. Coupling between heat flow, metallurgy and stress-strain computations in steels: the approach developed in the computer code SYSWELD for welding or quenching. Proceedings of the Engineering Conference on the Modeling of Casting, Welding and Advanced Solidification Processes V, M Rappaz, MR Ozgu, KW Mahin (eds.), Warrendale, PA: TMS, 1991, pp. 203–210.

53. GW Greenwood, RH Johnson. The deformation of metals under small stresses during phase transformations. Proceedings of the Royal Society, London, 1965, pp. 403–421.

54. CL Magee. Transformation kinetics, microplasticity and aging of martensite in FE31Ni. PhD dissertation, Carnegie Institute of Technology, Pittsburgh, PA, 1966.

55. R Song, G Dhatt, A Ben Cheikh. Thermo-mechanical finite element model of casting systems. International Journal for Numerical Methods in Engineering 30:579–599, 1990.

56. FG Rammerstorfer, DF Fisher, W Mitter, K-J Bathe, MD Snyder. On thermo-elastic-plastic analysis of heat-treatment processes including creep and phase changes. Computers and Structures 13:771–779, 1981.

57. DF Watt, L Coon, M Bibby, J Goldak, C Henwood. An algorithm for modelling microstructural development in weld heat-affected zones (part A) reaction kinetics. Acta Metallurgica 36:3029–3035, 1988.

58. C Henwood, M Bibby, J Goldak, D Witt. Coupled transient heat transfer-micro-structure weld computations (part B). Acta Metallurgica 36:3037–3046, 1988.

59. JB Leblond, J Devaux, JC Devaux. Mathematical modeling of transformation plasticity in steels I: case of ideal-plastic phases. International Journal of Plasticity 5:551–572, 1989.

60. JB Leblond. Mathematical modeling of transformation plasticity in steels II: coupling with strain hardening phenomena. International Journal of Plasticity 5:573–591, 1989.

61. S Das, G Upadhya, U Chandra. Prediction of macro-residual stresses in quenching using phase transformation kinetics. Proceedings of First International Conference on Quenching and Control of Distortions, G Totten (ed.), Metals Park, OH: ASM International, 1992, pp. 229–234.

62. DJ Bammann, VC Prantil, JF Lathrop. A model of phase transformation plasti-
     city. Proceedings of the Engineering Foundation Conference on Modeling of
     Casting, Welding and Advanced Solidification Processes VII, M Cross, J
     Campbell (eds.), Warrendale, PA: TMS, 1995, pp. 275–285.
63. Z Wenyu. Determination of residual stresses in components of complex shapes—
     correction of measurement using finite element method. Material Science and
     Technology 4:1030–1033, 1988.

4
Defects Formation

Vijay Suri
ALCOA CSI, Crawfordsville, Indiana

Kuang-O (Oscar) Yu
RMI Titanium Company, Niles, Ohio

I. INTRODUCTION

Defects reduce the yield and increase the cost of castings. As a result, understanding the mechanism of why defects form, developing tools to predict when defects occur, and establishing a processing strategy to prevent the formation of defects are crucial to casting process engineers. However, it should be understood that the definition of defects is both alloy and application dependent. Defects such as shrinkage, porosity, distortion, and cracking are considered as "universal" defects to all types of castings. On the other hand, microstructure related defects are relevant in castings of superalloys, titanium alloys, and aluminum alloys [1, 2]. Currently, a number of postcasting repairing techniques are used to improve the casting integrity. These include welding repair for shrinkage defects, HIP (hot isostatic press) for internal pores (Fig. 1) [3], and mechanical straightening for geometrical distortions. By using these repairing procedures, the requirements for the casting as-cast quality can sometimes be relaxed to a certain extent.

The formation of casting defects is related to casting processing conditions such as heat transfer, fluid flow, and thermal stress as well as casting materials, i.e., metal, mold, and core. Not all the defect formation mechanisms are well understood, and currently only a few are predicted by modeling. In general, mechanisms for processing related defects have been studied extensively and are better understood. In addition, the types of defects as well as their formation mechanisms are quite different for shape and ingot castings. In

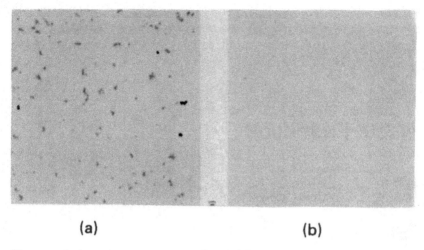

(a)   (b)

Figure 1 Effect of hot isostatic pressing (HIP) on porosity elimination of aluminum Alloy A356 (a) before HIP, (b) after HIP. (From Ref. 3.)

this chapter, discussion is focused on defects that occur in shape castings. The formation mechanisms for all types of casting defects will first be reviewed, followed by detail discussion on the prediction of common casting defects such as shrinkage and porosity.

II. TYPES OF DEFECTS

Details of the international classification and inspection procedures of common casting defects can be found in Ref. 4. Although there are many types of casting defects, they can be classified into the following categories based on the mechanism of their formation: mold filling, segregation, shrinkage, stress, micro/grain structure, and mold/core related defects.

A. Mold Filling Related Defects

During mold filling, the combination of heat transfer condition, mold filling sequence, and the molten metal free surface profile can result in no-fill, entrapped gas, and weld line defects. These defects can be predicted by using isotherm and isochron (time to reach a specified temperature) plots which can be obtained from postprocessing results of heat transfer analysis.

1. No-Fill

No-fill (Fig. 2) is a defect where the liquid metal temperature is below the alloy liquidus temperature so that the fraction of solid is too high to complete filling [5]. It usually occurs when liquid metal superheat or mold preheat temperature is not adequate to maintain a fully liquid flow front during mold filling process. Thin wall castings with intricate geometrics have a higher tendency to form no-fill defects than thicker castings. No-fill can be predicted by isotherm or iso-chron plots. When the temperature of a location of the casting falls below a reference temperature without a good feeding of liquid metal, no-fill is assumed to form in that location.

2. Entrapped Gas

For castings poured in air or under inert gas atmosphere, gas can be entrapped inside castings due to a turbulent mold filling pattern. During postcasting solution heat treatment, the entrapped gas begins to expand resulting in

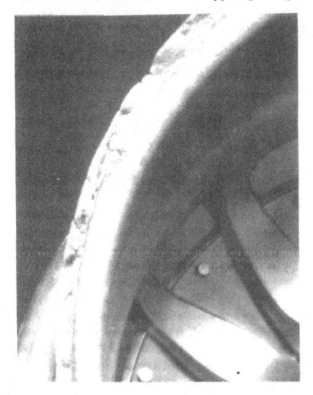

Figure 2   No-fill defect. (From Ref. 5.)

build up of internal pressure. When the heat treatment temperature is high, internal gas pressure may be higher than the yield strength of the cast metal resulting in the formation of a gas bubble. For a nonturbulent mold filling pattern, the location of entrapped gas defects can be predicted by isotherm and isochron plots. On the other hand, in high mold filling velocity casting processes such as die casting, jetting, instead of gradual filling, is common. Jetting causes splashing of liquid metal when the liquid hits the mold/die wall, increasing the potential for the formation of entrapped gas defects. Currently, there is no reliable way to model the gas entrapment process due to jetting and splashing during mold filling.

3. Weld Line

When two free surface flow fronts meet during mold filling (Fig. 3), a weld line defect will form if the flow fronts do not join intimately upon contact. Causes for the formation of weld lines include inadequate temperature of the flow fronts, oxides and impurities at the flow fronts, and presence of air/gas that can prevent intimate contact of flow fronts [6]. Unlike entrapped gas defects,

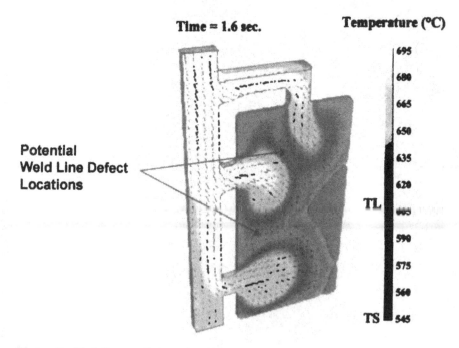

Figure 3   Modeling prediction of weld line defect. (From Ref. 6.)

weld line defects can be formed for components that are cast under air/inert gas atmosphere as well as vacuum environment. For chemically active metal such as aluminum, the oxide film at the molten metal free surface flow fronts presents an additional resistance for these metals to completely weld together and, hence, increase the propensity to form weld line defects. From a modeling point of view, the prediction of weld line defect is quite similar to that of the entrapped gas defects. When the mold filling turbulence is relatively moderate, the formation of weld line defects can be modeled by isotherm and isochron plots (Fig. 3). For high mold filling velocity casting process, e.g., die casting, the weld line defects due to jetting and splashing phenomena are extremely difficult to model.

B. Shrinkage and Porosity

The formation of shrinkage and porosity related defects have been studied extensively. The details of the modeling approach to predict these defects will be discussed in Secs. III and IV. Here, only a brief description of the formation mechanism for these defects will be presented.

1. Macroshrinkage

Most of alloys contract on solidifying. During casting, macroshrinkage forms in the region which is isolated and last to solidify. Equiaxed grain castings usually have a higher tendency to form macroshrinkage than directionally solidified and single-crystal castings. The formation of macroshrinkage is controlled by casting bulk heat transfer conditions and can be modeled by isotherm and isochron plots. Recently, advances have been made to predict not only the location but also the actual size of the macroshrinkage. This is particularly useful for those large diameter ingot castings in which the formation of shrinkage pipe can significantly reduce the material yield of the resultant ingot.

2. Microporosity

Microporosity forms near the end of solidification in the interdendritic regions when capillary feeding becomes insufficient. The propensity of the microporosity formation is related to the casting feeding ability during the last stage of solidification, which is related to the casting macro heat transfer parameters (temperature gradient G, solidification rate R, and combination functions of G and R) and dendrite structure (primary and secondary dendrite arm spacing) in the mushy zone. Microporosity can show up or beneath the surface of the casting. It should be noted that even a macro heat transfer condition that results in no isolated last solidifying spots, microporosity can still form in the regions where feeding capability is low.

3. Gas Porosity

Aluminum has a high hydrogen gas solubility. During molten metal processing, if the liquid aluminum is not properly treated, a large amount of hydrogen will be dissolved in the molten aluminum. During solidification, temperature decreases and so does the hydrogen solubility in the molten aluminum, excess amount of hydrogen will be expelled from the metal and form gas porosity. The formation of this type of defect is primarily related to the molten metal processing and not related to the casting processing. Unlike microporosity, which is irregular shape (Fig. 1), gas porosity (Fig. 4) generally has a round shape [5]. It should be noted that although aluminum castings have the highest tendency to form gas porosity defects, castings of other alloy systems can also have gas porosity defects.

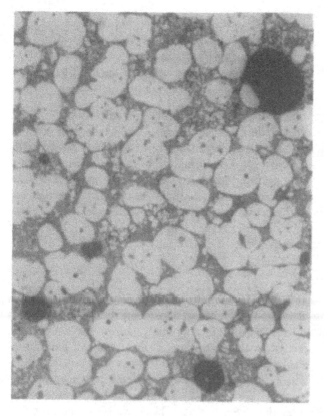

Figure 4 Micrograph showing spherical gas porosities. (From Ref. 5.)

C. Segregation Related Defects

Segregation is a natural phenomenon associated with the solidification process. Under normal solidification conditions, most of the engineering alloys result in equiaxed and/or columnar dendritic structure. The degree of segregation and the magnitude of the dendrite arm spacing (DAS) are related to the casting cooling rate and local solidification time (LST). High cooling rate and low LST result in a fine DAS and the associated segregation. Because the magnitude of DAS is in the order of micrometer, this segregation phenomenon is commonly referred as microsegregation. Microsegregation can have a detrimental effect on casting final mechanical properties and can usually be reduced to a certain degree by a postcasting solution heat treatment. For ingot/slab/billet castings that are used to make wrought products, microsegregation can decrease the casting incipient melting temperature and hot workability during the ingot/slab/billet breakdown step. These ingot/slab/billet castings are sometimes given a homogenization treatment before they break down to reduce the degree of microsegregation and increase the incipient melting temperature and hot workability. The modeling approach for simulating the microsegregation is discussed in detail in Chapter 5.

When casting size gets bigger, the intensities of the natural convection in bulk liquid, as well as the interdendritic fluid flow in mushy zone, become stronger. The macro scale segregation, i.e., macrosegregation, forms. For conventional steel mold ingot castings (Fig. 5) and sand mold roller castings, natural convection in bulk liquid is the primary driving force for macrosegregation such as A-type and V-type defects [7]. For continuous casting and semicontinuous casting processes such as direct chill casting (DC casting), vacuum arc remelting (VAR), electroslag remelting (ESR), plasma arc melting (PAM), and electron beam melting (EBM), the interdendritic fluid flow in the mushy zone is responsible for channel-type defects like freckles for nickel-base superalloys and beta flecks for titanium alloys (see Chapter 17). For VAR and ESR processed materials, the formation of freckles and beta flecks is related to the ingot solidification and the freckles and beta flecks are referred to as indigenous defects. The formation of exogenous defects, e.g., white spots in superalloys and hard alpha particles in titanium alloys, is due to the exogenous materials which drop in the molten pool and remain unmelted. The details of the formation mechanism for both indigenous and exogenous defects, as well as the modeling approach for predicting the formation of indigenous defects in VAR and ESR, are presented in Chapter 17.

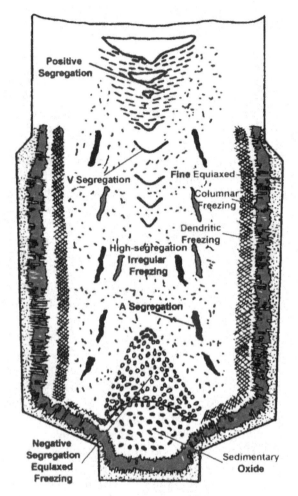

Figure 5 Macrosegregation defect in conventional large steel ingots. (From Ref. 7.)

D. Stress Related Defects

Most metals contract during solidification. Nonuniform cooling and/or hindered shrinkage will result in the formation of thermal stress in the casting. If the stress level remains in the elastic range of the cast metal, no defects will form. When the thermal stress is higher than the yield strength of the cast metal, plastic deformation occurs. Plastic deformation results in localized yielding and subsequently nonuniform contraction which then results in casting geometric distortion. Another type of defect that can result from the plastic deformation is recrystallized grains in single crystal and directionally solidified turbine airfoils (see Chapter 11). However, it should be noted that recrystallized grains actually form during postcasting solution heat treatment and not during casting processing. When the thermal stress is higher than the ultimate tensile stress of the cast metal, cracks may form in the casting. Cracks (Fig. 6) that form during solidification, i.e., in the mushy region, are called hot tearing [5]. Cold cracking is referred to the defects that form after the solidification is completed, i.e., in the solid region. The modeling approach for the calculation of thermal stress is presented in Chapter 3.

E.   Micro/Grain Structure Related Defects

For most alloy systems, casting microstructure is typically not a part of the quality specification. Only for aluminum alloys, superalloys, and titanium alloys, casting microstructures are strictly specified [1, 2]. The microstructure

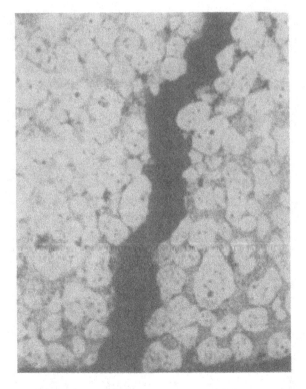

Figure 6   Micrograph showing hot tears defect. (From Ref. 5.)

requirement for aluminum alloy castings is DAS, whereas the requirements for superalloy and titanium alloy castings are grain structures. Aluminum castings can be made by die casting/squeeze casting/semisolid metalworking, sand casting, and investment casting processes. Superalloy and titanium alloy castings are mostly produced by the investment casting process.

The formation of as-cast microstructure can be predicted by the macro thermal parameters as well as by more advanced micro modeling approaches. For macro thermal parameter prediction, casting microstructure morphology is related to the ratio of temperature gradient G and solidification rate R. Figure 7 shows the casting microstructure morphology changes from planar, cellular, columnar dendritic to equiaxed dendritic as the G/R values decrease [8]. For the dendritic solidification, the width of DAS (Figs. 8 and 9) is decreased with an increase of casting cooling rate [9]. Details of the micro modeling, are presented in Chapter 5. Details of applying the macroscopic thermal parameter modeling, as well as the micro-modeling to predict the casting microstructure related defects for superalloys and titanium alloys, are discussed in Chapter 11.

F. Mold/Core Related Defects

During casting, chemical reaction between mold/core and molten metal may take place and result in the increase of oxygen content in molten titanium and

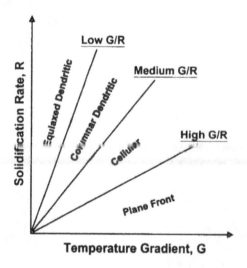

Figure 7 As-cast grain morphology as functions of temperature gradient G and solidification rate R. (From Ref. 8.)

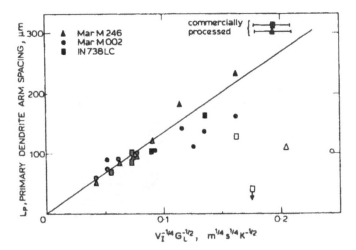

Figure 8 The primary dendrite arm spacing L_p plotted as a function of $G_L^{-1/2}V_I^{-1/4}$ for three directionally solidified nickel-base superalloys. (From Ref. 9.)

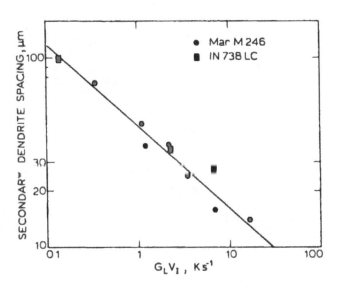

Figure 9 Secondary dendrite arm spacings λ_s of directionally solidified nickel-base superalloys as a function of the cooling rate $G_L V_I$. (From Ref. 9.)

the loss of active element such as Y in Y-containing nickel-base single-crystal alloys [10]. The increase of oxygen content in molten titanium results in the so-called alpha case at the surface of the titanium casting. On the other hand, the loss of active element Y from the molten superalloy is usually accompanied by the increase of casting inclusion content. However, in most of the cases, casting inclusion content is originated from the entrapped slag during molten metal processing. For sand casting, the mold/core may crack during casting and molten metal may penetrate into sand, resulting in a mixture of sand and metal adhering to the surface of a casting. Another type of defect is core shift during casting, resulting in a casting with unacceptable wall thickness. In general, mold/core related defects are very difficult to predict by process modeling. The approaches for modeling the formation of alpha case in titanium castings and the loss of Y in nickel-base superalloys are discussed in Chapter 11.

III. PREDICTION OF MACROSHRINKAGE

The first step in prediction of macroshrinkage is to establish the thermal history of the casting as it solidifies. Since the temperature of the metal begins to change from the moment mold filling starts, both the filling process and temperature changes during solidification must be modeled comprehensively to predict defects accurately.

Occurrence of macroshrinkage is associated with adequacy of feeding and, hence, is related to features such as risers, gates, vents, and chills that are incorporated into the casting design. Reviews of various factors that determine the size and location of these features are available in Refs. 11 and 12.

A. Prediction by Temperature Field

Since the formation of macroshrinkage or bulk shrinkage cavities is closely related to the temperature and solid fraction distribution in the casting, tracking isotherms and isochrons in the solidifying casting offers a first estimate of the size and location of shrinkage defects.

By mapping the isotherms from a thermal model of a casting at different stages during solidification, sections of the casting which freeze last can be identified as illustrated in Fig. 10 [13]. Shrinkage due to phase change occurs most often in these areas, and can result in shrinkage cavities if they occur in the bulk of the casting (hot spots). Depending on the alloy system, the isotherm that is of interest is the eutectic or solidus or any temperature between solidus and liquidus below which fluidity is expected to diminish severely. Similar

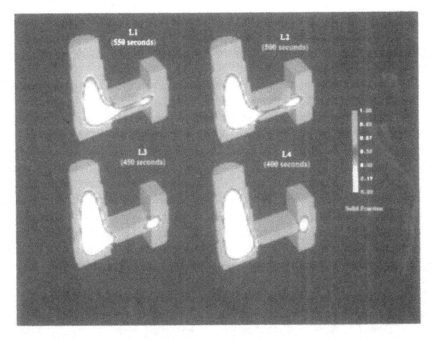

Figure 10 Example of a hot spot formation in a T-plate casting. (From Ref. 13.)

information can also be obtained by using isochrons or maps of solid fraction distribution.

Obvious advantage of these methods is their simplicity. They help the casting engineer in designing the layout of gates, runners, and risers, so that hot spots can be driven from the casting into the risers. Significant disadvantages of the temperature based maps are that (1) they typically do not account for mold filling and this may result in inaccurate temperature distribution and (2) quantitative prediction of shrinkage cavities is difficult (rejected gas in the metal and gravity affect the size, shape, and location of the cavities). The following section describes a model to calculate the size and location of cavities more rigorously than those predicted by thermal maps [14, 15].

B. Calculation of Shrinkage Pipe Location and Volume

When the mold cavity is first filled by the liquid metal, the entire casting is made up of one single liquid pool, which is usually connected to the riser. As solidification proceeds, this liquid pool will grow smaller and, in an ideal casting, it will remain connected to gates/risers till the end of casting solidification.

With inadequate rigging, it is possible that one or more smaller liquid pools bounded by solid metal may evolve in different sections of the casting. In each of these pools, the potential for the formation of a shrinkage cavity is evaluated as described below.

For a given volume of molten pool of metal in the casting, say in the riser section of a casting, volume depletion in the casting due to shrinkage will result in a drop of the liquid metal level in the riser, if it is still in a fluid state. As the fraction of solid in the casting increases with the progress of solidification, the fluidity of the liquid metal decreases. The critical temperature or solid fraction at which fluidity becomes negligible is difficult to estimate as it is dependent on the alloy as well as the size and shape of dendrite structure that evolves during solidification. Assuming spherical dendrites, the critical solid fraction f_{scr} can be assigned a value of 0.74, which represents close packing factor in a face-centered cubic lattice. The corresponding critical liquid fraction $f_{lcr} = 0.26$ is the value below which liquid can not feed shrinkage. Figure 11 illustrates various domains in a casting used to analyze shrinkage and to predict of cavities. The liquid fraction in region A is above f_{lcr}. This includes 100% liquid regions as well as solid-liquid mushy regions. Region B represents the highly viscous mushy region where the liquid fraction is below f_{lcr}. Region C represents the solidified area. It must be noted that shrinkage due to volume contraction takes place continuously in both liquid and solid phases until the entire casting is cooled to ambient temperature. However, in this model, cavity formation due to shrinkage is limited to areas above f_{lcr}.

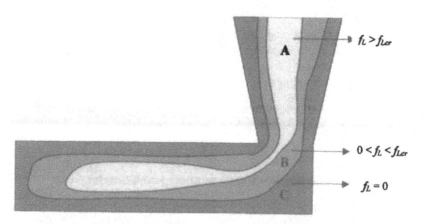

Figure 11   Representation of various domains in a casting during solidification used for shrinkage void prediction.

When a unit volume of the casting changes phase from a completely liquid state to a solid state, the resulting volume fraction change β is given by

$$\beta = \frac{\rho_s - \rho_l}{\rho_l} \tag{1}$$

where ρ_s and ρ_l are densities of solid and liquid, respectively. Volume fraction change is positive for metals that shrink upon solidification. To compute the shrinkage volume and location of voids, the volumes containing liquid ($f_L \geq f_{Lcr}$, region A), highly mushy zone ($0 < f_L < f_{Lcr}$, region B) and solid zone (region C) must be continuously tracked in the various sections of the casting during solidification. Net change in volume due to shrinkage over a time interval during solidification in a zone with an initial volume V and $f_L > f_{lcr}$ is then calculated by

$$\Delta V = \beta \cdot V \cdot \Delta f_L \tag{2}$$

Depending on the location of the molten metal pool, this volume is either compensated with a lowering of liquid metal level (e.g., at the top of the riser), or by introducing a void. In either case, the shape of the void is primarily governed by the sequence of solidification, as well as by the value chosen for the critical liquid fraction f_{Lcr}. Controlled casting experiments can be designed to obtain the value of f_{Lcr} for a given alloy system to fine-tune void prediction. The shrinkage ratio β of eutectic ductile iron varies as a function of temperature, as are shown in Fig. 12 [16]. Taking this into account, the computed bulk shrinkage at the end of solidification in a hypothetical ductile cast iron automotive piston [15] is shown in Fig. 13a. The model predictions indicate that

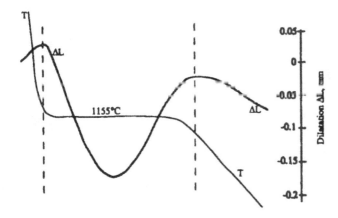

Figure 12   Volume change during solidification in ductile irons. (From Ref. 16.)

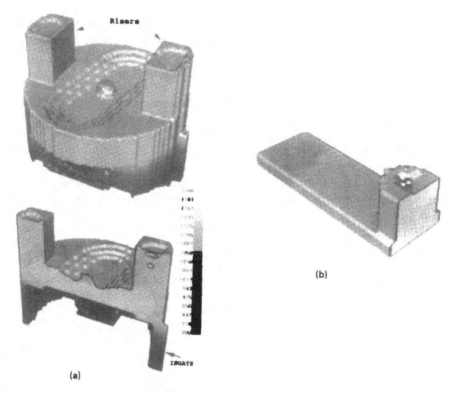

Figure 13 Examples of shrinkage void predictions in (a) ductile iron automotive piston and (b) test casting plate. (From Ref. 15.)

shrinkage cavities occur inside the riser closer to the ingate, whereas some extent of piping is seen in both risers of the casting. Figure 13b illustrates computed volume shrinkage at the end of solidification in a test plate casting. In this Al-7wt%Si casting made in a sand mold, there is a clear indication of piping in the riser. The results are in good qualitative agreement with experimental observations [15].

IV. PREDICTION OF MICROPOROSITY

Microscopic cavities can occur in castings either due to incorrect feeding of liquid metal which experiences a localized contraction on solidification, or due to the rejection of dissolved gas upon melt solidification. Most often it is a combination of these two factors that result in cavities. These cavities which may be present in a variety of forms and locations are often referred to as dispersed porosity or microporosity. Mathematically, porosity may be defined as

$$\text{Porosity}\% = \frac{(\rho_{max} - \rho_s)100}{\rho_{max}} \tag{3}$$

where ρ_{max} is the theoretical density of the solid phase and ρ_s is the actual density of the solid. There is a broad range of work relating to pore formation in shaped castings. This includes the developments of simple heat transfer based guidelines to minimize porosity formation, as well as more complex approaches to study the evolution of shrinkage and gas porosity. The thermal conditions prevalent during solidification have an important effect on the size and distribution of porosity in a casting. Historically, the level of porosity was controlled by manipulating the solidification time, thermal gradient, and other thermal parameters in the casting. Most of these methods were based on simple principles of heat transfer in the metal and mold media. An outline of these methods is given below.

A. Total Solidification Time

The basis for using total solidification time as a measure of casting soundness is that the riser must solidify later than the casting so that there will be adequate feeding for shrinkage in the casting. Chvorinov's rule [17] relates the total solidification time t_f to the volume V to surface area A ratio, called the modulus M based on simple one-dimensional heat transfer across metal mold interface.

$$M = \frac{V}{A} = \frac{2}{\sqrt{\pi}} \left(\frac{T_m - T_0}{\rho_s H} \right) \sqrt{K_m \rho_m C_m} \sqrt{t_f} \tag{4}$$

$$t_f = C \left(\frac{V}{A} \right)^2 \tag{5}$$

where
T_m = melting temperature of the metal
T_0 = initial temperature of the metal
ρ_s = density of solid metal
H = heat of fusion of the metal
K_m = thermal conductivity of the liquid metal
ρ_m = density of the liquid metal
C_m = specific heat of the liquid metal
C = a constant for a given metal-mold combination

The above relationship can be used to estimate the sizes of risers and other rigging features for a given casting in order to ensure liquid feeding till the end of solidification. In a good design, the modulus value of the riser M_R is greater than that of the casting M_C. Since the ratio M_R/M_C determines the material yield of the casting, careful choice of riser shape and size is required to maintain high yield levels. In complex castings, modulus estimation can be cumbersome. In these cases, a numerical discretization scheme (meshing) can be used effectively to compute modulus in the casting and riser volumes. This method has been incorporated into several commercial casting simulation models and it is often used as a first approximation of casting and rigging design or shrinkage defect prediction [18–20].

Though the total solidification time or modulus method has been used to produce porosity free castings, its success is limited since porosity formation is essentially a localized phenomena. Simulation methods referred above attempt to localize the computation of solidification time, but their accuracy is limited as they typically use one-dimensional methods to estimate the temperature field.

B. Minimum Temperature Gradient

In addition to estimating suitable sizes for risers using Chvorinov's rule, feeding characteristics in the casting are improved by incorporating chills. A chill constitutes a local heat sink incorporated into the casting assembly to increase the rate of solidification and the thermal gradient within the casting. It has been observed that the occurrence of porosity can be minimized by maintaining a *minimum* temperature gradient in the casting [11, 21, 22]. Most of this work has been studied with short freezing range alloys. Figure 14 illustrates a map of the computed final solidification time for low carbon steel castings of various sizes, against empirical values of critical gradients [22].

Temperature gradient at any location in the casting (local temperature gradient) can be calculated by dividing the difference in temperature between two neighboring points, with the distance between the two points (Fig. 15):

$$G_0 = \frac{1}{n}\sum_1^n \frac{\Delta T_i}{\Delta x_i} \qquad i = 1, 2, 3, \ldots \tag{6}$$

$$G_0 = \max\left(\frac{\Delta T_i}{\Delta x_i}\right) \qquad i = 1, 2, 3, \ldots \tag{7}$$

It is obvious that the value of G at a given location can vary depending on the direction of neighboring points chosen. In simulation models, an average value of gradient [Eq. (6)] is suitable when a fine mesh is used to descretize the casting volume. Maximum value of gradient [Eq. (7)] may be used when a

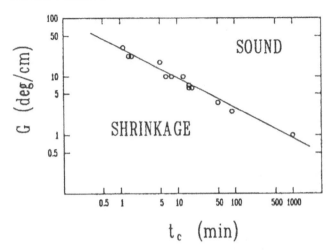

Figure 14 Thermal gradient criterion suggested by Niyama [22].

coarse mesh is employed since it would give a correct representation of the driving force for feeding over a given length in the solidifying region. A map of the thermal gradient of the entire casting will reveal areas that need additional chills or risers. It can also be used to control the direction of the solidification front, which is beneficial in designing the ejection sequence in permanent mold or die casting process.

A major limitation of using thermal gradients to design feeding in castings is that the minimum gradient may vary with alloy composition and possibly with casting shape and size.

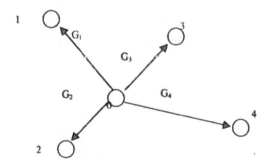

Figure 15 Schematic for temperature gradient calculation between neighboring points Eqs. (6) and (7).

C. Comprehensive Analysis of Microporosity

Pore formation, though affected by the local thermal conditions, is also dependent upon the local flow field existing in the mushy region of the solidifying casting. Knowledge of the flow and pressure fields also allows one to evaluate gas evolution and stability of a gas pore.

Analysis of transport phenomena at a localized or microscale is necessary for understanding the driving forces that form a stable pore. An important issue in this analysis is where a pore initially occurs before it grows. It is generally agreed that pore nucleation is a heterogeneous phenomena and occurs within interdendritic networks and in locations near mold walls, inclusions, and other irregularities. In addition, pore nucleation is often a result of the combined effects of shrinkage and release of dissolved gas during solidification.

A number of analytical methods based on microscale transport in the mushy region have been published [23–25]. Typically, the analysis involves a balance of the resistance to flow in the interdendritic region, the presence of dissolved gas content and the gas-liquid surface tension. According to Piwonka and Flemings [23], a critical initial dissolved gas content, which is expected to be considerably higher than the gas content required for the formation of porosity due to the combined effects listed above, is required to form "gas-caused" porosity. Thus the basic equation for a stable pore to exist is

$$P_g = P + \frac{2\sigma_{Lg}}{r} \tag{8}$$

where P is the local liquid pressure, P_g is the equilibrium dissolved gas pressure in the liquid, σ_{Lg} is the gas-liquid surface tension, and r is the characteristic dendrite arm spacing.

Of these terms, P is calculated from a balance of mass, momentum, and energy within the mushy region [24]. The biggest hurdle in this approach is the estimation of the resistance to flow in the mushy region and solution rests on our ability to predict the evolving mushy region structure accurately. Even when the structure is known at any given instant, estimating the friction drag for flow within this structure is another important challenge. To circumvent these difficulties, a common approximation that is used to describe the resistance to flow in the mushy region is to use a modified form of the Darcy's law. Used extensively to describe flow in a porous medium, Darcy's law, in its basic form, is given by

$$\Delta P = \frac{\mu}{K} f_L v \tag{9}$$

where μ is the viscosity of the fluid, K is the permeability of the porous medium, f_L is the void fraction (liquid fraction), and v is the flow velocity.

In solidification, the value of K changes with time as a function of the liquid fraction. Various expressions for K have been used for describing mushy region permeability, but the accuracy and generality of any of these remains questionable. Some of these expressions are listed below [22, 24]:

$$K = \alpha \cdot f_L \tag{10}$$

where α is a constant.

$$K = pf_L^q \tag{11}$$

where p and q are constants.

$$K = \frac{f_L^3 d^2}{180(1 - f_L)^2} \tag{12}$$

where d is the primary dendrite arm spacing.

To eliminate singularity in computing K at fractions of liquid $f_L = 0$ and 1, the limits of computing permeability are chosen as $0.05 < f_L < 0.95$ beyond which the value of K is assumed to remain constant. The primary dendrite arm spacing d is inversely proportional to the local temperature gradient G and the solidification growth velocity V_s and can be expressed generally as [26],

$$d = aG^{-b}V_s^{-c} \tag{13}$$

where b and c are constants. It must be noted here that d can also be expressed as a function of the local solidification time [27] if there is pertinent material data.

The second term in the porosity equation [Eq. (8)], gas pressure P_g, is obtained from the thermodynamics of gas-liquid equilibrium, as follows for hydrogen dissolution and by Sievert's law:

$$[H_0] = (1 - f_L)[H_s] + f_L[H_L] + \alpha \frac{P_g f_p}{T} \tag{14}$$

$$[H_s] = K_{SH} P_g^{1/2} \tag{15}$$

$$[H_L] = K_{LH} P_g^{1/2} \tag{16}$$

where f_p = fraction of porosity
H_0 = initial concentration of hydrogen
H_s = hydrogen concentration in solid
H_L = hydrogen concentration in liquid
α = a constant relating to hydrogen solubility
K_{SH} = equilibrium constant for solubility of hydrogen in solid
K_{LH} = equilibrium constant for solubility of hydrogen in liquid

The surface energy term in Eq. (8) consists of σ_{Lg}, the surface tension at the liquid-gas interface, and r, the radius of the pore. Assuming spontaneous nucleation, and that porosity occurs at the roots of primary dendrite arms, r is obtained as half the dendritic arm spacing d given by Eq. (13).

After solving for the pressure balance of the porosity equation, a distribution of pore size can be predicted in various regions of the casting. Kubo and Pehlke [25], and Suri and Paul [15] demonstrated the utility of this method for aluminum alloys and steels.

In a different approach, Suri et al. [28] also developed a comprehensive method to predict porosity in which they attempt to overcome some of the limitations of the above mentioned methods. An important aspect of Suri's method is the generalized definition of criterion for pore formation. The extent of feeding in the mushy region depends on the pressure drop, both static and dynamic, along the length of the channel. A dynamic pressure field results due to effects of friction drag, shrinkage, and thermal and solutal convection. The static forces acting along the boundaries of this flow field are the gas pressure, surface tension of a gas bubble, and external pressure on the system. Feeding of liquid toward a solidifying front will be physically impossible if the net pressure drop ΔP_{net} along the length of the mushy region is greater than the sum of all static forces P_{static} acting on the system. Thus porosity will occur if

$$\Delta P_{net} > P_{static} \tag{17}$$

$$P_{static} = P_0 + \Gamma_s - P_g - \rho_L g h \tag{18}$$

$$\Delta P_{net} = P_{static} - P_L \tag{19}$$

where P_0 = external pressure on the system
        Γ_s = surface tension of a gas bubble
        P_g = gas pressure
        ρ_L = density of liquid
        g = gravity constant
        h = height of liquid
        P_L = local liquid pressure

Thus the above condition in Eqs. (17) to (19) represents choking of flow with incomplete feeding. In other words, porosity occurs when P_L falls to a very low value and the driving force for feeding vanishes. An equation of this form is convenient to use for porosity prediction in a wide range of casting conditions. For example, both degassed and nondegassed melts can be studied by assigning suitable values to P_g. Also, casting soundness can be analyzed under various processing parameters such as external pressures, gravity (sand and permanent mold), with or without vacuum (die casting), etc.

The gas pressure component of total static pressure may be computed using thermodynamic relations of gas dissolution as described earlier. Suri et al. [28] used a unique approach to compute the local liquid pressure P_L. The method recognizes that resistance to feeding in the mushy region depends on the specific surface area of the solid phase particles within the mushy region. That is, the magnitude of friction drag and hence local liquid pressure depends not only on the size of solid particles but also on their morphology. Hence separate expressions are developed for resistance to feeding in primarily columnar dendritic mushy regions and equiaxed/eutectic mushy regions. Details of the analysis are available in Ref. 28. Once the local liquid pressure P_L is known, Eqs. (17) to (19) can be used to assess the feeding and potential for pore formation.

D. Criteria Functions to Predict Porosity

Although the models discussed above are reasonably comprehensive and provide a good approximation of porosity distribution, they tend to be computationally intensive. Also, unlike the shrinkage-only models, they fail to provide a convenient tool to aid in the design of casting/rigging to minimize occurrence of porosity in terms of controllable casting variables. This may well be the reason that rigorous methods to predict porosity defects have only limited use in the casting industry. Lumped parameters, generally called porosity criteria functions, fill this crucial gap for the casting designer.

Criteria functions attempt to capture the primary forces that facilitate feeding or limit feeding causing porosity. Thus they are often extracted from transport equations describing flow in the mushy region during solidification.

A notable porosity criterion was developed by Niyama et al. [22], who expressed the pressure drop in the mushy region as an inverse function of $G/R^{1/2}$, where G and R are the thermal gradient and the solidification rate, respectively, at the end of solidification. Assuming that porosity occurs when the pressure drop in the mushy region is excessively high, a minimum critical value can be assigned to $G/R^{1/2}$ below which shrinkage pores may be expected. Niyama et al. showed the applicability of this criterion for low-carbon steels, which are primarily short freezing range alloys.

To apply the Niyama criterion to long freezing range alloys, care should be given to the calculation of the thermal parameters G and R. Figure 16 illustrates two points in the solidification regime where these values can be calculated for use in criteria functions. If a large amount of primary phase precipitates during solidification followed by a eutectic phase, it is reasonable to calculate the thermal parameters near the end of the primary phase. If, on the other hand, eutectic phases dominate the solidification, G and R may be

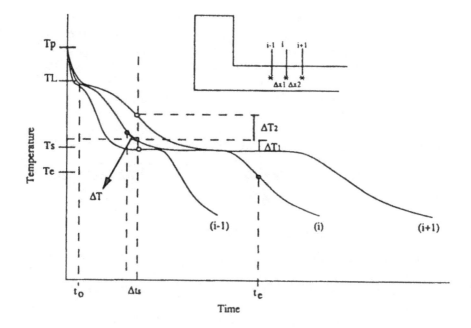

$$\text{Cooling Rate, } R = \frac{\Delta T}{\Delta t_s}$$

$$\text{Thermal Gradient, } G = (\Delta T_1/\Delta x_1 + \Delta T_2/\Delta x_2)/2$$

$$\text{Solidus Velocity, } V_s = R/G$$

Figure 16 Definition of thermal parameters used for the computation of the porosity criterion functions.

calculated at the end of solidification. A composite approach may be developed for casting conditions where both the primary and eutectic phases evolve in significant amounts.

Following the Niyama criterion, Lee et al. [29] developed a criteria function for long freezing range aluminum alloys, sometimes referred to as LCC after the authors. The most important difference between LCC and the Niyama criterion lies in the way they correlate permeability to liquid fraction in the mushy region. The criterion is given by

$$LCC = \frac{G t_s^{2/3}}{V_s} \tag{20}$$

where V_s is the sold front velocity and t_s is the local solidification time. As described earlier, these thermal parameters are calculated either at the end of solidification or at the end of a primary phase precipitation. Like the Niyama criterion, LCC takes a minimum critical value when feeding becomes difficult and conditions for pore formation are favorable.

Feeding Resistance Number (FRN) is another criterion of interest to predict the occurrence of porosity in castings [30]. It is based on the basis that the resistance to feeding in the mushy region is dependent on the specific surface area of the evolving solid as well as the prevailing thermal conditions.

$$FRN = \frac{N \mu \, \Delta T}{\rho_L \beta G V_s D^2} \tag{21}$$

where N is a constant, μ is the viscosity of the melt, ΔT is the freezing range of the alloy, ρ_L is the density of liquid, β is the shrinkage ratio, and D is the characteristic size of solid particles. The values of N and D depend on casting macrostructure. $N = 16\pi$ for columnar dendrites, and 216 for equiaxed dendrites and eutectic phases. D represents the characteristic length scale of the solid phase, i.e., either the primary dendrite arm spacing in columnar dendrites [Eq. (13)], or the equiaxed dendrite or eutectic grain size D_0. Otherwise, the value of D may be approximated to an estimated value of dendrite/grain size.

A high value of FRN indicates higher resistance to feeding and hence higher potential for pore formation. FRN is useful in casting simulations which address solidification kinetics during freezing.

E. Correlation Between Criteria Functions and Porosity

The criteria functions described above, depending on their magnitude in different locations of the casting, are only indicative of the relative propensity of pore formation. In a given region of the casting, higher values of Niyama and LCC criteria and lower values of FRN compared to other regions of the same casting, indicate better feeding conditions. It must be noted that the range of these values vary with casting conditions and alloy systems.

It is easy to establish quantitative correlations between the criteria functions and porosity and reduce the ambiguity that exists in using them in casting simulation models:

1. Make an experimental casting with the desired alloy. The casting should be well risered and preferably have a one-dimensional feeding, such as a long plate or bar with a chill at one end.

2. Measure the actual porosity distribution along the length of the casting.
3. From a casting simulation, estimate the values of a porosity criteria function along the length of the casting.
4. Plot the criteria function against the measured porosity, and obtain a correlation between the two.

The correlation thus obtained can be directly employed in a computer simulation to predict potential porosity in castings poured with the same alloy as the experimental casting produced under similar conditions. An illustration of this procedure applied to sand castings of aluminum alloy A356 is available in Ref. 30.

V. SUMMARY

Casting defects have been broadly classified into mold fill related defects (nonfill, entrapped gases, and weld line); shrinkage and porosity related defects (macroshrinkage and microporosity); segregation related defects (microsegregation and macrosegregation); micro/grain structure related defects; mold/core related defects; and stress related defects. Prediction methods to identify the potential of occurrence of some of these defects have been discussed in this chapter. It is evident from the discussion that the accuracy of defect prediction is dependent upon the accuracy of (1) the phenomenological description of the defect formation or conditions for occurrence, and (2) prediction of the conditions that affect the defect, such as temperature and flow fields, distribution of solid, liquid, and gas phases in the solidifying casting. Accuracy of calculation of shrinkage and porosity related defects is also affected by (1) thermal property data, e.g., solid and liquid densities, solid fraction, and permeability variation with temperature, (2) the extent of mesh refinement, and (3) intervals of time steps used in the model calculations. While using criterion functions are convenient for predicting porosity, preliminary experimental work with test casting shapes is necessary to establish the range of criterion function values that are applicable to a given casting system and casting conditions.

In prediction of defects, there is a clear compromise that the casting engineer has to exercise in balancing speed of computing and accuracy. For example, predictions of hot spots and shrinkage based on the modulus method described above provide excellent "first approximations" of casting soundness and rigging design. These designs can then be fine-tuned with further analyses which require more comprehensive approaches.

REFERENCES

1. KO Yu, JJ Nichols, M Robinson. Finite-element thermal modeling of casting microstructures and defects. JOM 6:21–25, 1992.
2. SAE ARP 1947. Determination and acceptance of dendrite arm spacing of structural aircraft quality D357 aluminum alloy castings. Warrendale, PA: Society of Automotive Engineers, 1996.
3. JM Eridon. Hot isostatic pressing of castings. Metal Handbook, 9th ed., vol. 15, Casting, ASM International, 1988, pp. 538–544.
4. ASM Committee on Nondestructive Inspection of Castings. Testing and inspection of casting defects. Metal Handbook, 9th ed., vol. 15, Casting, ASM International, 1988, pp. 544–561.
5. MP Kenney, JA Courtois, RD Evans, GM Farrior, CP Kyonka, AA Koch, KP Young. Semisolid metal casting and forging. Metal Handbook, 9th ed., vol. 15, Casting, ASM International, 1988, pp. 327–338.
6. CM Wang, AJ Paul, WW Fincher, OJ Huey. Computational analysis of fluid flow and heat transfer during the EPC process. AFS Trans 897–904, 1993.
7. WH Bailey. Refining and casting of large forging ingots. Ironmaking and Steelmaking 4(2):72, 1977.
8. KO Yu, MJ Beffel, M Robinson, DD Goettsch, BG Thomas, RG Carlson. Solidification modeling of single crystal investment castings. AFS Trans 417–428, 1990.
9. M McLean. Directionally Solidified Materials for High Temperature Service. London: The Metals Society, 1983.
10. KO Yu, JA Oti, M Robinson, RG Carlson. Solidification modeling of complex-shaped single crystal turbine airfoils. Superalloys 1992. SD Antolovich et al. (eds.), Warrendale, PA: TMS, 1992, pp. 135–144.
11. John Campbell, Casting, Oxford, UK: Butterworth-Heinemann Ltd.
12. WS Pellini. Factors which determine riser adequacy and feeding range. AFS Trans 61:61–80, 1953.
13. Rapid cast takes the guesswork out of casting design. Concurrent Technologies Corporation, 1992.
14. I Imafuku, K Chijiiwa. A mathematical model for shrinkage cavity prediction in steel castings. AFS Trans 10:527–540, 1983.
15. VK Suri, AJ Paul. Modeling and prediction of micro/macro-scale defects in castings. AFS Trans 144:949–954, 1993.
16. R Hummer. Relationship between cooling and dilation curves of ductile iron melts and their shrinkage tendency. Cast Metals 1(2):62–68, 1988.
17. N Chvorinov. Giesserei 27:201–208, 1940.
18. GK Upadhya, AJ Paul. A comprehensive casting analysis model using a geometry based technique followed by a fully coupled, 3D fluid flow, heat transfer and solidification kinetics calculations. AFS Trans 100:925–933, 1992.
19. G Upadhya, AJ Paul, JL Hill. Optimal design of gating and risering in castings: an integrated approach using empirical heuristics and geometric analysis. In: TS

Piwonka, et al. (eds.). Modeling of Casting, Welding and Advanced Solidification Processes VI. Warendale, PA: TMS, 1993, pp. 135–142.

20. G Upadhya, AJ Paul. Rational design of gating and risering for castings: a new approach using knowledge base and geometric analysis. AFS Trans 919–925, 1993.

21. DR Irani, V Kondic. Casting and mold design effects on shrinkage porosity of light alloys. AFS Trans 77:208–211, 1969.

22. E Niyama, T Uchida, M Morikawa, S Siato. Method of shrinkage prediction and its application to steel casting practice. AFS Int Cast Metals J 7:52–63, 1982.

23. TS Piwonka, MC Flemings. Pore formation in solidification. Trans Met Soc of AIME, Aug/236:1157–1165, 1966.

24. DR Poirier, K Yeum, AL Maples. A thermodynamic prediction for microporosity formation in aluminum-rich Al-Cu alloys. Met Trans Nov/18A:1979-1987, 1987.

25. K Kubo, RD Pehlke. Mathematical modeling of porosity formation in solidification. Met Trans, June/16B:359–366, 1985.

26. W Kurz, DJ Fisher. Fundamentals of solidification. Trans Tech Publ, Aedermannsdorf, Switzerland, 1986, pp. 85–87.

27. DR Poirier. Permeability for flow of interdendritic liquid in columnar-dendritic alloy. Met Trans 18B:245–255, 1987.

28. VK Suri, N El-Kaddah, JT Berry. Theoretical and experimental studies on pore formation during casting solidification. PhD dissertation, The University of Alabama, 1993.

29. YW Lee, E Chang, et al. Modeling of feeding behavior of solidifying Al-7Si-0.3Mg alloy plate castings. Met Trans 21B:715–722, 1990.

30. VK Suri, AJ Paul, N El-Kaddah, JT Berry. Determination of correlation factors for prediction of shrinkage in castings. Part I: Prediction of microporosity in castings: a generalized criterion. AFS Trans 138:861–867, 1994.

5
Microstructure Evolution

Doru M. Stefanescu
The University of Alabama, Tuscaloosa, Alabama

I. GOALS OF MICROSTRUCTURAL EVOLUTION MODELING

Castings are produced with a variety of dimensions, from of a few millimeters up to tens of meters in length. It is natural, therefore, to assume that the important dimensions to use in describing castings are of that magnitude. However, as the microstructure of the casting (the structure which can be seen using an optical microscope) determines the properties of the casting, it, too, is important. In addition, because solidification is the process of moving individual atoms from the liquid to a more stable position in the alloy lattice, the distances which atoms must move during solidification are also important. For these reasons, the effect of solidification on the casting must be analyzed at three different length scales (Fig. 1) [1]:

1. The macroscale (macrostructure): This scale is of the order of centimeters to meters. Elements of the macroscale include shrinkage cavity, macrosegregation, cracks, surface quality, and casting dimensions. Casting properties, and their acceptance by the customer, can sometimes be dramatically influenced by these macrostructure features.

2. The microscale (microstructure): This scale is of the order micrometers to millimeters. In most cases, mechanical properties depend on the solidification structure at the microscale level. To evaluate the influence of solidification on the properties of the castings, it is necessary to know the as-cast grain size and type (columnar or equiaxed), the size of the dendrite arms, the type and the intensity

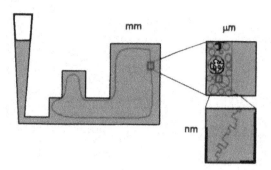

Figure 1   Solidification length scale. (From Ref. 1.)

of chemical microsegregation, and the amount of microshrinkage, porosity, and inclusions.

3. The nanoscale (atomic scale): This scale is of the order of nanometers. At this scale, solidification is discussed in terms of nucleation and growth kinetics, which proceed by addition of individual atoms to the solid. While at the present time there is no database correlating elements of the nanoscale with the properties of castings, an accurate description of liquid/solid interface dynamics requires atomistic calculations. The present knowledge and hardware development does not allow utilization of the atomic scale in applied casting engineering. However, accurate solidification modeling may require at least partial use of this scale during computation.

Modeling implementation into the foundry industry has been led by casting solidification models. The first models did not incorporate fluid flow, and it was simply assumed that once the mold is full, the temperature is uniform across the casting. However, as the technology matured it was soon noticed that this assumption is a source of significant error, in particular for castings with relatively large variations in section size, as exemplified in Fig. 2 [2]. Subsequently, mold filling models were developed and implemented. Today, any casting solidification package that claims state of the art level includes both mold filling and solidification modeling.

Recently, more ambitious goals have been set for casting models. By combining mold filling and solidification models with knowledge based systems for gating and risering, a completely computerized solution for casting design becomes possible [3].

It must be made clear from the beginning that what the foundryman needs and what the models deliver is not the same thing. As far as the foundryman is concerned, the objective of the modeling is to provide infor-

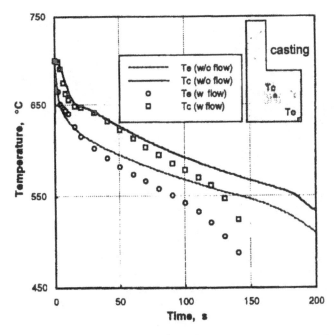

Figure 2 Calculated effect of fluid flow on temperature evolution at the edge and in the center of the casting. Higher cooling rates are predicted when flow is included. (From Ref. 2.)

mation on casting quality. However, after intricate calculations the models can only deliver physical quantities such as temperature, solidification time, composition, pressure, fluid velocity. Matching the foundryman's objectives with models deliverables is still very much an area of research and continuous development.

Some of the quantities calculated by the models can be used directly in the form of maps of temperature, solidification time, fraction of solid, composition, velocity, etc. As is probably clear to the reader, many features of interest to the foundryman, such as microshrinkage, surface quality, mechanical properties, etc. cannot be obtained through direct calculation. Criteria functions are used in many instances to obtain some answers.

More recently, with the development of macrotransport (MT)/transformation-kinetics (TK) codes, it became possible to calculate directly the basic elements of casting microstructure. The goal of this chapter is to briefly review the methodology, performance, and limitations of solidification models designed to predict microstructure evolution during solidification and cooling to room temperature.

II. METHODOLOGY

Microstructures are thermodynamically unstable forms of matter aggregation. They evolve in time because of one or more or the following driving forces:

- Reduction in the bulk-chemical free energy (phase transformations)
- Reduction of the interface energy between phases, grains, or orientation domains (coarsening)
- Relaxation of the elastic-strain energy generated between different phases because of lattice mismatch

The temporal evolution of microstructure may be triggered through the application of external fields, including temperature, stress, electric and magnetic fields.

The approaches that have emerged for the computational modeling of microstructural evolution can be broadly classified into *criteria based, deterministic,* and *probabilistic.* Criteria functions are empirical relationships between the microstructural features, such as microstructure length scale or morphological transitions, and some quantities calculated by the model, such as temperature, temperature gradient, cooling rate. Deterministic modeling of TK is based on the solution of the continuum equations over some volume element. At the end of calculation, microstructure is described by the length scale of the features of interest, such as grain size or phase spacing. Probabilistic modeling introduces elements of randomness in the simulation, and in principle, can produce a graphic display of the microstructure.

While deterministic models have been used successfully to describe the spatio-temporal microstructure evolution of castings of any size, pure probabilistic models are only applicable to relatively small volumes of matter because of restriction on memory and computational time. Consequently, *hybrid* models have also been developed. They combine deterministic and probabilistic modeling.

A. Criteria Functions Modeling

A summary of the criteria functions used to predict casting microstructure is given in Table 1. In this table the subscripts S and L stand for solid and liquid, respectively, G is the thermal gradient, V is the solidification velocity, ΔT is the undercooling, D is diffusivity, $\dot{T}$ is the cooling rate, and t_f is the local solidification time.

The limitations of such an approach can be pointed out through two simple examples. Consider the case where it is required to predict the primary or secondary dendrite arm spacing. The criterion relates the spacing λ_I to the cooling rate obtained from the solidification model and to the type of alloy,

Table 1 Criteria Functions for Microstructure

Microstructural feature	Criterion
SL interface stability	$G/V \geq \Delta T_{SL}/D$
Columnar-to-equiaxed transition	$G_L < G_{min}$
Dendrite arm spacing	Primary arm: $\lambda_I = ct. \cdot \dot{T}^n$
	Secondary arm: $\lambda_{II} = cl. \cdot t_f^{1/3}$
Gray-to-white transition in cast iron	$V < V_{max}$ or $\dot{T} < \dot{T}_{cr}$

Source: Ref. 100.

through experimental data for the constant and the exponent n. For the secondary arm spacing only a constant is required. For example, for Al-4% Cu alloys the constant in the λ_{II} equation is 10^{-6} m/s. The effect of micro- and macrosegregation on the dendrite spacing are completely ignored, even when calculated by the models. Similarly, the paramount role of silicon and manganese segregation in the gray-to-white transition of cast iron is ignored when a simple critical cooling rate criterion is used to predict its occurrence.

B. Deterministic Modeling

Until recently casting properties were evaluated based on empirical relationships between the quantity of interest and some significant output parameter of the macrosolidification model. Most commonly used are temperature and cooling rate $\dot{T}$.

A good description of the cooling rate can be obtained from the energy transport equation coupled with mass and momentum transport. The most important governing equation is

$$\frac{\partial T}{\partial t} + \nabla \cdot (VT) = \nabla \cdot (\alpha \nabla T) + \frac{\Delta H_f}{\rho c} \frac{\partial f_S}{\partial t} \tag{1}$$

where V is transport velocity, α is the thermal diffusivity, ΔH_f is the enthalpy of fusion, ρ is density, and c is the specific heat. This equation can be solved if an appropriate description of the fraction solid, f_S, as a function of time or temperature is available. The heat evolution during solidification included in the source term depends strongly on both the macrotransport (MT) of energy from the casting to the environment, and on the transformation kinetics (TK).

If it is assumed that TK does not influence MT, the two computations can be performed uncoupled. Typically, $\dot{T}$ is evaluated with an MT code, and then, the microstructure length scale that includes phase spacing λ and the volumetric grain number density N are calculated based on empirical equations as a function of $\dot{T}$. This methodology is presently used by classic macrotransport models that solve the mass, energy, momentum, and species macroscopic conservation equations, as discussed in Chapter 2. They are inherently inaccurate in their attempt to predict microstructure because of the weakness of the uncoupled MT-TK assumption. Indeed, since in effect, MT and TK are coupled during solidification, accurate prediction of microstructural evolution revolves around modeling both MT and TK, and then coupling them appropriately. Consequently, the problem is to describe $f_S(x, t)$ in terms of transformation kinetics, and to select appropriate boundary conditions. The governing equation couples MT and TK through f_S, if f_S is calculated from TK.

The first step in building a MT-TK solidification model is to divide the computational space of the casting in macrovolume elements within which the temperature is assumed to be uniform (Fig. 3). In the early approach (e.g., Refs. 4–6) the basic simplifying assumption is that the solid phase has zero velocity (*one-velocity models*), meaning that once nucleated the grains remain in fixed positions. Grain coalescence and dissolution are ignored. The macrovolume element is assumed to be closed to mass and momentum transport but open to energy transport. Then each of these elements is further subdivided in microvolume elements, typically spherical, based on some nucleation law. Within each of these microvolume elements, only one spherical grain is grow-

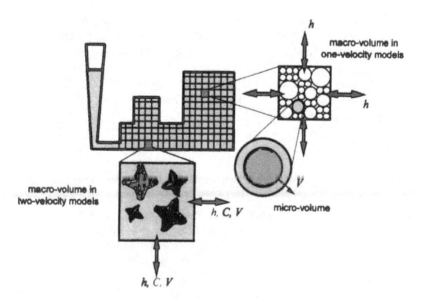

Figure 3 Division of the computational space for deterministic modeling of coupled macrotransport and transformation kinetics.

ing at a velocity V, dictated by kinetic considerations. The isothermal micro-volume elements are considered open to species transport.

Assuming that the macrovolume element is closed to mass and momentum transport, and that the solid is fixed ($V = 0$), the governing equation simplifies to:

$$\frac{\partial T}{\partial t} = \nabla(\alpha \nabla T) + \frac{\Delta H_f}{\rho c}\frac{\partial f_S}{\partial t} \tag{2}$$

It must be noted that the assumption $V = 0$ is always valid once grain coherency is reached, which typically happens at about 0.2 to 0.4 fraction of solid. Thus, for certain cases this is a reasonable approximation.

In more recent models (e.g., Ref. 7), the macrovolume element is open to energy, mass and momentum transport. The solid may move freely with the liquid, thus, these are *two-velocity models*. Within each macrovolume element, solid grains are allowed to nucleate based on some empirical nucleation laws. Since the macrovolume element is open, solid grains can be transported in or out of the volume element. V in Eq. (1) is obtained by coupling with the momentum equations.

Consider now an isothermal macrovolume element. The fraction of solid in this element is made of all the grains that have been nucleated and grown until that time. Thus, the evolution of the fraction of solid in time can be written as the sum of the contribution from new nuclei and that of increased grains volume:

$$\frac{\partial f_S}{\partial t} = \frac{\partial N}{\partial t}v + N\frac{\partial v}{\partial t} \tag{3}$$

where N is the volumetric grain (nuclei) density, and v is the grain volume.

The following conservation equation can be written to describe the nucleation rate [8]:

$$\dot{N} = \frac{dN}{dt} + \nabla \cdot (V_S N) \tag{4}$$

where $\dot{N}$ is the net nucleation rate accounting for the various mechanisms of nucleation, $dN(t)/dt$ is the local nucleation rate (rate of formation of grains per unit volume). The second right-hand term is the flux of grains due to a finite solid velocity. The change in grain volume can be calculated as $\partial v/\partial t = AV$, where V is the growth velocity of the grain and A is its surface area.

In a simplified analysis, it can be further assumed that the grains are of spherical shape. The equation describing the evolution of fraction of solid becomes:

$$df_S = 4\pi\left(\dot{N}\frac{r^3}{3}dt + Nr^2\,dr\right)(1 - f_S) \tag{5}$$

where $r(x, t)$ is the grain radius. The term $(1 - f_S)$ is a correction factor (Avrami or Johnson-Mehl correction) introduced to account for grain impingement occurring once the grains grow to the point where they achieve contact with one another. If instantaneous nucleation is assumed, and if $V_S = 0$, $\dot{N} = 0$ and the evolution of solid fraction can be simply written as:

$$\frac{df_S}{dt} = 4\pi Nr^2\frac{dr}{dt} = 4\pi Nr^2 V(1 - f_S) \tag{6}$$

Integration of this equation results in

$$f_S = 1 - \exp\left(-\tfrac{4}{3}\pi Nr^3\right)$$

The problem is then to formulate $\dot{N}$ and V [or $r(x, t)$]. These calculations are based on specific nucleation kinetics and interface dynamics, to be discussed. If $\dot{N}$ is known, df_S/dt can be calculated and used in Eq. (2) to couple macrotransport with transformation kinetics.

To improve accuracy and/or reduce computational time, several coupling techniques have been developed, including the latent heat method (LHM) [5, 9], the microenthalpy method (MEM) [10], the temperature recovery method, and the micro latent heat method [11]. The LHM fully couples MT and TK and is the most accurate. However, because the time step increment necessary to solve the heat flow equation is limited by the microscopic phenomena, it has to be much smaller than the recalescence period in order to properly describe the microscopic solidification. Thus, much longer computational times are required as compared with codes that do not include TK. The MEM is a partially coupled approach that assumes a constant heat transfer throughout the microsolidification path. While it is not as accurate as LHM, the MEM substantially decreases the CPU time. For a benchmark test CPU was decreased about five times as compared with the LHM, and was only 6% longer than for the classic enthalpy method [11].

C. Probabilistic Modeling

The main techniques presently used for probabilistic modeling of solidification are the *Monte Carlo* (MC) technique, and the *cellular automaton* (CA) technique. They will be described shortly in the following paragraphs.

The MC technique is based on the minimization of the interface energy of a grain assembly. It has been developed to study the kinetics of grain growth [12, 13]. The microstructure was first mapped onto a discrete lattice (Fig. 4). Each lattice site was assigned an integer (grain index) I_j from 1 to some num-

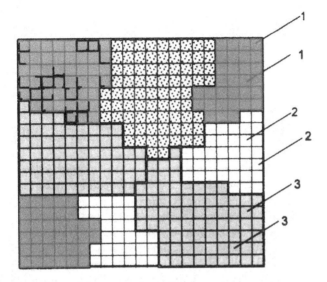

Figure 4 Example of a microstructure mapped on a rectangular lattice; the integers denote orientation and the lines represent grain boundaries.

ber. Lattice sites having the same I_j belong to the same grain. The grain index indicates the local crystallographic orientation. A grain boundary segment lies between two sites of unlike orientation. The initial distribution of orientations is chosen at random and the system evolves to reduce the number of nearest neighbors pairs of unlike crystallographic orientation. This is equivalent to minimizing the interfacial energy. The grain boundary energy is specified by defining an interaction between nearest neighbors lattice sites. When a site changes its index from I_j to that of its neighbor I_k, the variation in energy can be calculated from the Hamiltonian* describing the interaction between nearest neighbors lattice sites:

$$\Delta G = \gamma \sum {}' \left(\delta_{I_j I_k} - 1 \right) \tag{7}$$

where γ is the interface energy and $\delta_{I_j I_k}$ is the Kronecker delta. The sum is taken over all nearest neighbors. The Kronecker delta has the property that $\delta_{I_j I_k} = 0$ and $\delta_{I_j I_j} = 1$. Thus, unlike nearest pairs contribute γ to the system energy, while like pairs contribute zero. The change of the site index is then decided based on the transition probability, which is:

---

* A Hamiltonian cycle is a cycle that contains all the vertices of a graph.

$$
W = \begin{cases} \exp(-\Delta G/k_B \cdot T) & \Delta G > 0 \\ 1 & \Delta G \le 0 \end{cases} \tag{8}
$$

where k_B is the Boltzmann constant. A change of the site index corresponds to grain boundary migration. Therefore, a segment of grain boundary moves with a velocity given by

$$
V = C\left[1 - \exp\left(-\frac{\Delta G_i}{k_B T}\right)\right] \tag{9}
$$

where C is boundary mobility reflecting the symmetry of the mapped lattice and ΔG_i is the local chemical potential difference. Note that this equation is similar to that derived from classical reaction rate theory.

The CA technique, originally developed by Hesselbarth and Göbel [14], was applied first by Rappaz and Gandin [15] to the simulation of grain structure formation in solidification processes. This technique consists in principle in assigning some random nucleation sites of random crystallographic orientation within the computational space, based on probabilistic relationships. Then, deterministic laws, such as $V = f(\Delta T)$, are used to allow these nuclei to grow. The different microvolume elements of the computational space can only be liquid or solid, and change state as the growth vector (computed from a deterministic law) is reaching them. An example of the CA setup for a plate is provided in Fig. 5. The volume is divided into a square lattice of regular cells. The Von Neumann neighborhood configuration was adopted in calculation (i.e., only the nearest neighbors of the cell are considered). The index that defines the state of the cell is zero when the cell is liquid, and an integer

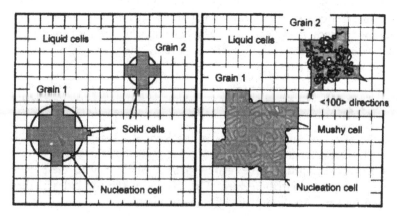

Figure 5 Computational space of a two-dimensional cellular automaton for dendritic grain growth. (From Ref. 16.)

when the cell is solid. The integer is associated with the crystallographic orientation of the grain and it is randomly generated.

For eutectic grains, the velocity is normal to the interface, while for dendritic grains it corresponds to the preferential growth direction of the primary and secondary arms [16]. The extension of each grain is then calculated by integrating the growth rate over time. All the liquid cells captured during this process are given the same index as that of the parent nucleus.

Probabilistic models may attempt to describe *grain growth* or *dendrite growth* (Fig. 6). A grain growth model is reasonable for eutectic solidification or for solid state grain growth. However, dendrites growing in the liquid are not described very well by such models, since within the volume of the dendrite envelope, both liquid and solid coexist. The ratio between the two phases within the dendrite envelope can be characterized through the *internal fraction of solid*.

1. Monte Carlo Grain Growth Models

Chronologically, Brown and Spittle [17] seem to be the first to use the Monte Carlo (MC) approach to model solidification. They were able to qualitatively predict the grain structure of small castings. However, their study was based on a hypothetical material and the correspondence between the MC time step used in the calculations and real time was not clear. Consequently, it was not pos-

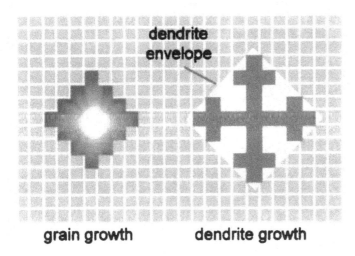

Figure 6 Discretization of computational space for solidification stochastic models.

sible to analyze quantitatively the effects of process variables and material parameters.

A more fundamental approach has been used by Xiao et al. [18]. They simulated microscopic solidification morphologies of binary systems using a probabilistic model (MC) that accounts for bulk diffusion, attachment and detachment kinetics, and surface diffusion. An isothermal two-component system contained in a volume element subdivided by a square grid was considered. Initially, the region was filled by a liquid that consists of particles A and B that occupy each grid point according to a preset concentration ratio. Diffusion in the undercooled liquid was modeled by random walks on the grid. Through variation of interaction energies and undercooling, a broad range of microstructures was obtained, including eutectic systems and layered and ionic compounds. It was shown that, depending on the interaction energies between atoms, the microscopic growth structure could range from complete mixing to complete segregation. For the same interaction energy, as undercooling increases, the phase spacing of lamellar eutectics decreases (Fig. 7). Thus, continuum derived laws, such as $\lambda \cdot \Delta T = ct$, were recovered through a combination between a probabilistic model and physical laws. Further, the composition in the liquid ahead of the interface is very similar to that predicted from the classic Jackson-Hunt model for eutectics, as shown in Fig. 8. However, this model can only be used for qualitative predictions, because the length scale of the lattice a is several orders of magnitudes higher than the atomic size. Thus, this is not a nanoscale model. The model is limited to microscale level calculations, and cannot be coupled to the macroscale because

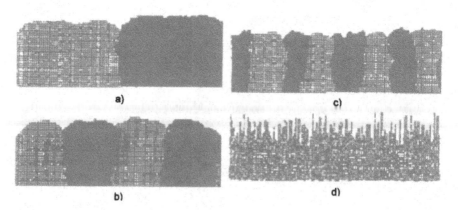

Figure 7 Role of increased undercooling on microscopic growth morphology: (a) $\Delta\mu/kT = 0.1$; (b) $\Delta\mu/kT = 0.5$; (c) $\Delta\mu/kT = 5$; (d) $\Delta\mu/kT = \infty$. $\Delta\mu$ is the chemical potential. (From Ref. 18.)

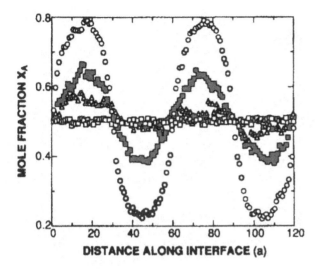

Figure 8 Concentration in mole fraction parallel to the liquid/solid interface at $2a$ (○), $10a$ (■), $20a$ (△), and $50a$ (□) into the liquid; a is the lattice constant. (From Ref. 18.)

of computing time limitations. Nevertheless, the main merit of this work is that it has demonstrated that physical laws can be successfully combined with probabilistic calculation for TK modeling.

To improve the previous models, Zhu and Smith [19] produced a hybrid model by coupling the MC method with heat and solute transport equations. They accounted for heterogeneous nucleation by using Oldfield's [20] model. The probability model of crystal growth was based on a lowest free energy change algorithm. ΔG in Eq. (8) was calculated as the difference between the free energy determined by the undercooling, ΔG_v, and the interface energy existing between liquid/solid and solid/solid grains, ΔG_γ:

$$\Delta G = \Delta G_v - \Delta G_\gamma \equiv \frac{\Delta H_f \, \Delta T}{T_m} v_m + ld(n_{SL}\gamma_{SL} + n_{SS}\gamma_{SS}) \tag{10}$$

where ΔH_f is the latent heat of fusion, ΔT is the local undercooling, T_m is the equilibrium melting temperature, v_m is the volume of the cell associated with each lattice site, l and d are the lattice length and thickness, respectively, n_{SL} is the difference between the number of new L/S interfaces and overlapped L/S interfaces, n_{SS} is the number of S/S interfaces between grains with different crystallographic orientations, and γ_{LS} and γ_{SS} are the interface energies. The time elapsed (MC time step) was calculated using the macroscopic heat transfer method, where the time elapsed corresponds to the amount of liquid that has solidified under given cooling conditions.

Some typical computational results are presented in Fig. 9. A clear transition from columnar to equiaxed solidification is seen when the superheating temperature and therefore the undercooling is increased.

2. Cellular Automaton Grain Growth Models

Rappaz et al. (e.g., Refs. 15, 21) have proposed stochastic models that combine the advantages of probabilistic and deterministic approaches. They coupled a CA model with finite element heat flow (CAFE model). After interpolation of the explicit temperature and of the enthalpy variation at a cell location, the nucleation and growth of grains were simulated using the CA algorithm. This algorithm accounts for the heterogeneous nucleation in the bulk and at the mold/casting interface, for the preferential growth of dendrites in certain directions and for microsegregation. Since in their models a dendritic tip growth law was used and growth velocity was then averaged for a given microvolume element, the final product of the simulation was dendritic grains, not dendritic crystals. In this respect, the method relies heavily on averaged quantities, just as deterministic models do, but can display grain structure on the computer screen.

Unlike Monte Carlo methods, the CA growth algorithms are fully deterministic. They are based on experimental observations of the grain growth in

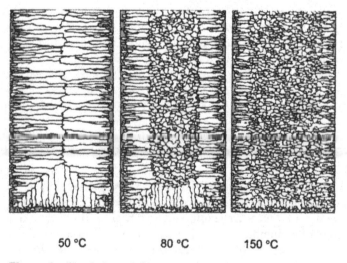

50 °C   80 °C   150 °C

Figure 9 Simulation of the effect of pouring temperature (superheat) on the macrostructure of ingots. (From Ref. 19.)

organic substances and on dendrite tip kinetics models [22]. Locally, the dendritic network propagates along the ⟨100⟩ directions with the same velocity. The computed structure can be either columnar or equiaxed, depending on the local solidification conditions. The use of dendrite tip kinetics models allows a direct correlation of the CA algorithm to time, a classic problem of Monte Carlo models.

One of the most interesting applications of the CAFE models is the prediction of the cellular-to-equiaxed transition (CET) in alloys, with particular emphasis on the role of fluid flow. As shown in Fig. 10, without convection the predicted grain structure is fully equiaxed. It does not reproduce the sedimentation cone and the columnar grain structure observed experimentally. Even

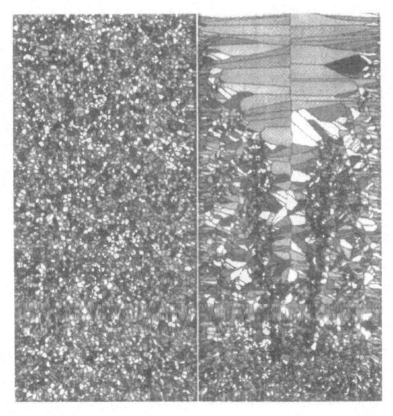

a) without grain movement   b) with grain movement

Figure 10   Simulation of CET in a conventionally cast Al-7%Si alloy. (From Ref. 22.)

such details as the deflection of columnar grains because of fluid flow parallel to the solidification interface can be modeled with the CA technique [23].

The CA models discussed above ignore the growth morphology at the scale of the grain. The graphical output of these models is limited to showing grain boundaries. The complex geometry of the dendrites, i.e., primary and higher order arms, is not modeled. Thus, important features of dendritic solidification, such as internal fraction of solid and dendrite coherency, cannot be described appropriately. However, since the final morphology of the dendritic crystal has a considerable effect on the mechanical properties of the material, the CA models are a significant progress in solidification modeling.

An attempt at modeling the dendrite geometry was made by Stefanescu and Pang [24]. In their microscale model, stochastic modeling at a length scale of 10^{-6} m was coupled with deterministic modeling at a length scale of 10^{-4} m. A deterministic tip-velocity model was used to calculate the advance of the dendrite tip. Arm thickening was also calculated with a deterministic law derived from the dendrite tip velocity law and crystallographic considerations in combination with a deterministic coarsening model. However, the overall growth of dendrite arms was derived from CA probabilistic calculations. Branching of dendrites arm was allowed to occur based on the classic criterion for morphological instability. Thus the dendrite morphology, rather than the grain structure, was simulated. Figure 11a shows a simulated solidification sequence for an Al-4.5% Cu alloy cast with a convective boundary condition ($h = 2.0\,\mathrm{J \cdot K \cdot m^{-2} \cdot s^{-1}}$). The different gray shades represent various crystallographic orientations. It is seen that nuclei initially form on the walls of the mold and grow inward into columnar grains. As the rate of heat extraction is increased ($h = 5\,\mathrm{J \cdot K \cdot m^{-2} \cdot s^{-1}}$), equiaxed grains nucleate and grow in the bulk liquid, away from the mold walls, and a columnar-to-equiaxed transition occurs, as illustrated in Fig. 11b. Finally, when the heat extraction coefficient is further increased to $10\,\mathrm{J \cdot K \cdot m^{-2} \cdot s^{-1}}$, a fully equiaxed structure develops (Fig. 11c).

More recently, Nastac [25] developed a comprehensive stochastic model to simulate the evolution of dendritic crystals (Fig. 12). The model includes time-dependent calculations for temperature distribution, solute redistribution in the liquid and solid phases, curvature, and growth anisotropy. Previously developed stochastic procedures for simulating dendritic grains [26] were used to control the nucleation and growth of dendrites. A numerical algorithm based on an Eulerian-Lagrangian approach was developed to explicitly track the sharp S/L interface on a fixed Cartesian grid. Two-dimensional calculations at the dendrite tip length scale were performed to simulate the evolution of columnar and equiaxed dendritic morphologies including the occurrence of the columnar-to-equiaxed transition.

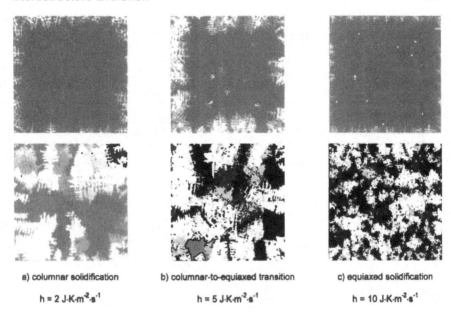

a) columnar solidification   b) columnar-to-equiaxed transition   c) equiaxed solidification

$h = 2 \text{ J·K·m}^{-2}\text{·s}^{-1}$ $h = 5 \text{ J·K·m}^{-2}\text{·s}^{-1}$ $h = 10 \text{ J·K·m}^{-2}\text{·s}^{-1}$

Figure 11 Graphic display of calculated solidification structures of an Al-4.5% Cu alloy, at the beginning (upper row) and end (lower row) of solidification. (From Ref. 24.)

Curvature was used in the formulation of both interface temperature and interface concentration:

$$T^* = T_L + (C_L^* - C_0)m_L - \Gamma \bar{k} f(\varphi, \theta) \tag{11}$$

$$C_L^* = C_0 + \frac{T^* - T_L^{\text{EQ}} + \Gamma \bar{k} f(\varphi, \theta)}{m_L} \tag{12}$$

where $\bar{k}$ is the mean curvature of the S/L interface, $f(\varphi, \theta)$ is a coefficient used to account for growth anisotropy, θ is the growth angle (i.e., the angle between the normal and the x axis), and ψ is the preferential crystallographic orientation angle. The treatment of the interface curvature merits further discussion.

The average interface curvature for a cell with the solid fraction f_S was calculated as:

$$\bar{k} = \frac{1}{a}\left[1 - 2\frac{f_S + \sum_{k=1}^{N} f_S(k)}{N+1}\right] \tag{13}$$

where N is the number of neighboring cells taken as $N = 24$. This includes all the first- and second-order neighboring cells. Equation (13), a modification of

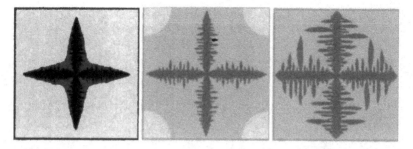

a) Simulated evolution of the growth of an equiaxed dendrite in a Fe-0.6 wt. % C alloy.

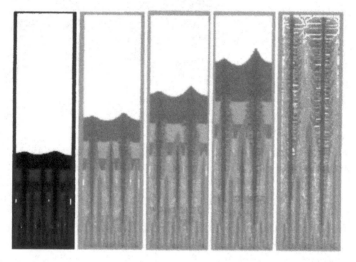

b) Simulated columnar dendritic growth and Nb segregation patterns in unidirectional solidification of

IN718-5 wt. % Nb alloy (assumed multiple nuclei/dendrites).

Figure 12 Examples of simulated dendrite growth for equiaxed and columnar morphologies. (From Ref. 25.)

the method proposed in Ref. 27, is a simple counting-cell technique that approximates the mean geometrical curvature (and not the local geometrical curvature).

The anisotropy of the surface tension was calculated from [28]:

$$f(\varphi, \theta) = 1 + \delta \cos[4(\varphi - \theta)] \quad \text{with } \phi = a\cos\left\{\frac{V_x}{\left[(V_x)^2 + V_y)^2\right]^{1/2}}\right\} \quad (14)$$

where V_x and V_y are the growth velocities in the x and y directions, respectively, and δ accounts for the degree of anisotropy. For four-fold symmetry, $\delta = 0.04$.

Some typical examples of calculation results are given in Fig. 12.

III. PROBLEM FORMULATION IN DETERMINISTIC MODELING

A. Nucleation

To grasp a good understanding of nucleation, a treatment of the problem at the nanoscale level is required. However, for the present discussion, it will be sufficient to consider nuclei as some solid particles of micron size that can serve as substrates for growing grains, without necessarily explaining their nature. The problem is then to establish their population distribution throughout the melt.

It is convenient to classify the types of nuclei available in the melt as resulting from homogeneous nucleation, heterogeneous nucleation, and dynamic nucleation.

Homogeneous nucleation, which implies that growth is initiated on substrates having the same chemistry as the solid, is not very common in casting alloys and will not be considered here.

Heterogeneous nucleation is based on the assumption that the development of the grain structure occurs upon a family of substrates of chemistry different from that of the solid. These substrates have variable potencies and population densities. The formal heterogeneous nucleation theory is based on the assumption that nucleation is only a function of the temperature and potency of an existing nucleant.

However, experiments show that dynamic conditions in the liquid may influence nucleation. Even in the presence of deliberate grain refining additions, there do exist, at all times, other inherent identifiable "nuclei" that originate mostly because of melt convection. This is called *dynamic nucleation*. At least two mechanisms have been proposed to explain it, the big bang mechanism and the crystal fragmentation mechanism.

Because of the complexities of nucleation it is not surprising that estimation of the volumetric density of nucleation sites before and during solidification of casting alloys is not a trivial problem. Evaluation of nucleation laws, to calculate the volumetric density of growing grains, is the weak link in the computer simulation of microstructure evolution during and after solidification. A summary of the models proposed for the quantitative description of nucleation will be given in the following sections.

1. Heterogeneous Nucleation Models

Without thermosolutal convection to transport dendrite fragments from the mushy zone to the bulk liquid, it is reasonable to assume that nucleation of equiaxed grains is based on heterogeneous nucleation mechanisms. While at least two significant methods based on the heterogeneous nucleation theory have been developed (see, for example, Ref. 1), they are empirical in essence, and rely heavily on metal- and process-specific experimental data. The *continuous nucleation model* (Fig. 13a) assumes a continuous dependency of N on temperature. Some mathematical relationship is then provided to correlate nucleation velocity $\partial N/\partial t$ with undercooling ΔT, cooling rate, or temperature. A summation procedure is carried on to determine the final number of nuclei.

The *instantaneous nucleation model* assumes site saturation, that is, that all nuclei are generated at the nucleation temperature T_N (Fig. 13b). Again, an empirical relationship must be provided to correlate the final number of nuclei (grains) in a volume element with ΔT or $\dot{T}$.

A summary of the basic equations and of the parameters that must be assumed or experimentally evaluated is given in Table 2. It is seen that all models require either two or three fitting parameters.

In Oldfield's [20] continuous model a power law function, $N = K_1(\Delta T)^n$, fitting the experimental data on cast iron, was used to evaluate the final number of grains. From this, an equation for nucleation velocity is derived [Eq. (15) in Table 2]. Other continuous nucleation models introduced some statistical functions to help describe the rather large size distribution of grains sometimes encountered in castings. A Gaussian (normal) distribution of number of nuclei

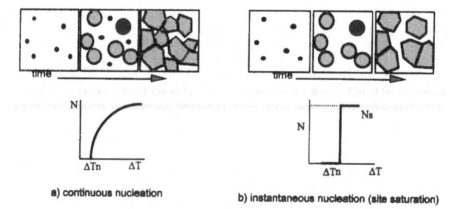

a) continuous nucleation

b) instantaneous nucleation (site saturation)

Figure 13 Schematic comparison between assumptions of instantaneous and continuous nucleation models. (From Ref. 1.)

Table 2 Summary of Nucleation Models

Model	Type	Basic equation	Equation number	Fitting parameters
Oldfield [20]	Continuous	$\dfrac{\partial N}{\partial t} = -nK_1(\Delta T)^{n-1}\dfrac{\partial T}{\partial t}$	(15)	n, K_1
Maxwell and Hellawell [10]	Continuous	$\dfrac{dN}{dt} = (N_s - N_i)\mu_2 \exp\left[-\dfrac{f(\theta)}{\Delta T^2(T_p - \Delta T)}\right]$	(16)	N_S, θ
Thévoz et al. [6]	Continuous (statistical)	$\dfrac{\partial N}{\partial(\Delta T)} = \dfrac{N_S}{\sqrt{2\pi}\,\Delta T_\sigma}\exp\left[\dfrac{(\Delta T - \Delta T_N)^2}{2(\Delta T_\sigma)^2}\right]$	(17)	$N_S, \Delta T_N, \Delta T_\sigma$
Goettsch and Dantzig [30]	Continuous (statistical)	$N(r) = \dfrac{3N_S}{(R_{max} - R_{min})^3}(R_{max} - r)^2$	(18)	N_S, R_{max}, R_{min}
Stefanescu et al. [31]	Instantaneous	$N = a + b\cdot\dot{T}$	(19)	a, b

Source: Ref. 1.

with undercooling was introduced by Thévoz et al. [6], Eq. (17) in Table 2. In this equation, ΔT_N is the average nucleation undercooling and ΔT_σ is the standard deviation. The same distribution was used by Mampaey [29] to model spheroidal graphite iron solidification. However, rather than applying this distribution to the number of nuclei, he applied it to the size of nuclei. To avoid the complication of having to specify θ for heterogeneous nucleation, the width of the substrate was used as a function of undercooling ($K_2/\Delta T$).

Goettsch and Dantzig [30] assumed a quadratic distribution of the number of grains as a function of their size, $N = a_0 + a_1 r + a_2 r^2$. This allows calculation of the number of nuclei of a given radius r, $N(r)$, as a function of the total number of substrates, the maximum grain size R_{max} and the minimum grain size, R_{min} [see Eq. (18) in Table 2].

The main assumption used in the instantaneous nucleation model [31] is that all nuclei are generated at the nucleation temperature. When a unique nucleation temperature is assumed, the nucleation rate can be calculated as [32]:

$$\frac{dN}{dt} = (N_s - N_i)\mu_2 \exp\left(-\frac{\mu_3}{\Delta T^2}\right) \tag{20}$$

where N_s is the number of heterogeneous substrates, N_i is the number of particles that have nucleated at time i, and μ_2, μ_3 are constants. Calculations for eutectic cast iron showed that all substrates became nuclei over a time interval of about 2 s. Therefore, Eq. (20) can be substituted by $\partial N/\partial t = N_s \delta(T - T_N)$, where δ is the Dirac delta function. Integration gives a total number of nuclei of N_s at T_N. While μ_2 and μ_3 affect the nucleation rate, only N_s will determine the final grain density. Thus, the dependency between cooling rate and grain density reflects a direct correlation between cooling rate and the number of substrates. This is illustrated by Eq. (19) in Table 2 where a and b are experimental constants. This is the most common form of the instantaneous nucleation law.

The question is now which nucleation model works best. In principle they all work, since they are based on fitting experimental data. Thus, the issue is which one fits experimental data better. This is debatable. As shown as an example in Fig. 14, the main difference between the Oldfield [20] and the Thévoz et al. [6] models is the use of second- or third-order polynomial, respectively, to fit the data. In other words, they are using two and three adjustable parameters, respectively. Two adjustable parameters seem to be sufficient in most cases.

In general, for alloys that solidify with narrow solidification interval the instantaneous nucleation model is recommended, since it saves computational time. The use of the continuous nucleation model runs into computation complications, related to the definition of the dimensions of the microvolume ele-

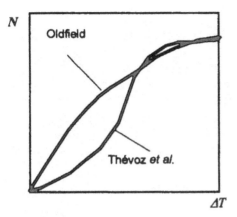

Figure 14 Schematic comparison between dependency of number of nuclei on under-cooling by two continuous nucleation models. (From Ref. 1.)

ment (diffusion distance), when applied to equiaxed dendritic solidification, unless complete solute diffusion is assumed in the liquid.

Experimental evaluation of heterogeneous nucleation laws has been tra-ditionally oversimplified. Typically, the final number of grains at the end of solidification is used to compute a nucleation law. However, as demonstrated through liquid quenching experiments [33], the evaluation a nucleation law from the final grain density may result in inaccurate data, since grain coales-cence plays a significant role. Indeed, the final eutectic grain density in cast iron was found smaller by up to 27% than the maximum number of grains devel-oped during solidification.

2. Dynamic Nucleation Models

For systems where density inversion occurs in the columnar mushy zone, thermo-solutal convection will disperse dendrite fragments into the bulk liquid. Thus, nucleation of the equiaxed grains will depend not only on the potency of heterogeneous substrates, but also on the survival of the dendrite fragments in the melt. Modeling such a nucleation is a difficult proposition not only because of the mathematical complications, but also because of the paucity of informa-tion on the physics of the phenomenon.

Nevertheless, Steube and Hellawell [34] suggest an alternative modeling approach, that considers the crystal fragments resulting from dynamic nuclea-tion. This model should include the following steps: (1) the kinetics of fragment formation or "crystal multiplication"; (2) the transport of fragments from the liquid or mushy zone into the bulk liquid because of natural convection or forced stirring; (3) the survival time of crystal fragments in the bulk liquid above the liquidus temperature; (4) the growth and sedimentation rates of fragments which survive long enough to enter a region below the liquidus temperature. They claim that this alternative approach is physically more correct than that strictly based on heterogeneous nucleation. However, they recognize that it may be difficult to formulate some of the proposed steps with sufficient confidence.

Experimental work on Al-4% Cu alloys [35] showed a strong susceptibility of crystal multiplication to fluid flow. Indeed, the experimentally measured number of grains per unit area could be presented in the following general form:

$$N = f(\text{Re}) + \psi \, \Delta T_{in}$$

where Re is the Reynolds number, ψ is a material constant, and ΔT_{in} is the inlet superheat. This shows increasing refinement of grains with increasing turbulence. It also suggests that the macrothermal state of the fluid rather than the static, local nucleation undercooling is the parameter that mostly influences nucleation.

B. Microsolute Redistribution in Alloys and Microsegregation

During solidification, mass transport occurs through species diffusion and because of momentum transfer (fluid convection). While both influence microstructure evolution, species diffusion is particularly important for a comprehensive theoretical treatment of dendrite growth. Since the dendrite geometry is rather complicated, all models start by assuming some simplified volume element over which calculations are performed (Fig. 15).

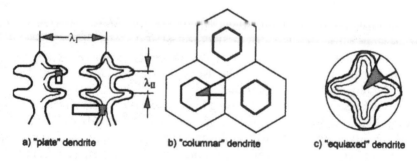

a) "plate" dendrite   b) "columnar" dendrite   c) "equiaxed" dendrite

Figure 15 Schematic representation of models for microsegregation. (From Ref. 41.)

Table 3 Major Assumptions Used in Analytical Microsegregation Models

Model	Geometry	Solid diffusion	Liquid diffusion	Partition coefficient	Growth	Coarsening
Lever rule	No restriction	Complete	Complete	Variable	No restriction	No
Scheil [36]	No restriction	No	Complete	Constant	No restriction	No
Brody and Flemings [37]	No restriction	Incomplete	Complete	Constant	No restriction	No
Clyne and Kurz [39]	No restriction	Spline fit	Complete	Constant	No restriction	No
Ohnaka [38]	Linear, columnar	Quadratic equation	Complete	Constant	Linear parabolic	No
Sarreal and Abbaschian [1C2]	No restriction	Limited	Complete	Constant	No restriction	No
Kobayashi [40]	Columnar	Limited	Complete	Constant	Linear	No
Nastac and Stefanescu [41]	Plate, columnar, equiaxed	Limited	Limited	Variable	No restriction	Yes

Table 4 Equations for Models in Table 3

Model	Equation	Equation number
Lever rule	$C_S = kC_0/[(1 - f_S) + kf_S]$	(21)
Scheil [36]	$C_S = kC_0(1 - f_S)^{k-1}$	(22)
Brody and Flemings [37]	$C_S = kC_0[1 - (1 - 2\alpha k)f_S]^{(k-1)/(1-2\alpha k)}$ $\alpha = 4D_S t_f/\lambda^2$	(23)
Clyne and Kurz [39]	$C_S = kC_0[(1 - 2\Omega k)f_S]^{(k-1)/(1-2\Omega k)}$ $\Omega = \alpha[1 - \exp(-1/\alpha)] - 0.5\exp(-1/2\alpha)$	(24)
Ohnaka [38]	$C_S = kC_0[1 - (1 - \beta k)f_S]^{(k-1)/(1-\beta k)}$ $\beta = 2\gamma/(1 + 2\gamma), \ \gamma = 8D_S t_f/\lambda_I^2$	(25)
Kobayashi [40]	$C_S = kC_0\xi^{(k-1)(1-\beta k)}$ $\{1 + \Gamma[0.5(\xi^{-2} - 1) - 2(\xi^{-1} - 1) - \ln\xi]\}$ $\xi = 1 - (1 - \beta k)f_S,$ $\Gamma = \beta^3 k(k - 1)[(1 + \beta)k - 2](4\gamma)^{-1}(1 - \beta k)^{-3}$	(26)
Nastac and Stefanescu [41]	$C_S^* = kC_0\left[1 - \dfrac{(1 - k)f_S}{1 - (m + 1)(kI_S^{(m+1)} + I_L^{(m+1)})}\right]^{-1}$ $f_S = (R^*/R_f)^{m+1}$ $I_S^{(1)} = \dfrac{2f_S}{\pi^2}\sum_{n=1}^{\infty}\dfrac{1}{(n - 0.5)^2}\exp\left\{-\left[\dfrac{(n - 0.5)\pi}{f_S}\right]^2\dfrac{D_S t}{R_f^2}\right\}$ $I_L^{(1)} = \dfrac{2}{\pi^2}(1 - f_S)\sum_{n=1}^{\infty}\dfrac{1}{(n - 0.5)^2}\exp\left\{-\left[\dfrac{(n - 0.5)\pi}{1 - f_S}\right]^2\dfrac{D_L t}{R_f^2}\right\}$ $(m) = 1$ for plate, 2 for cylinder, 3 for sphere	(27)

A summary of the major assumptions used in some analytical microsegregation models is given in Table 3. The basic equations for these models are presented in Table 4. Many numerical microsegregation models have also been proposed. However, the use of numerical segregation models in MT-TK codes increases dramatically the computational time. Analytical models are by far preferable.

The lever rule, Eq. (21), can be used to describe microsegregation of interstitially dissolved elements such as carbon in iron alloys. The Scheil equation [36], Eq. (22), can be used to calculate microsegregation when solid diffusivity is very small, typically the case for substitutional solutions. However, the diffusion of solute into the solid phase can affect microsegregation significantly, especially toward the end of solidification. For example, little solid state diffusion was found to occur during the solidification and cooling of primary austenite solidified welds of Fe-Ni-Cr ternary alloys, whereas structures that solidify as ferrite were almost completely homogenized as a result of diffusion.

Although Brody and Flemings [37] have proposed a model that assumed complete diffusion in liquid and incomplete back diffusion Eq. (23), they have not solved the "Fickian" diffusion equation. When significant solid state diffusion occurs, mass balance is violated. Consequently, the application of their result is limited to slow diffusion. Ohnaka [38] proposed a model for a "columnar" dendrite [Fig. 15b, Eq. (25)]. Complete mixing in the liquid and parabolic growth were assumed. Based on an assumed profile ($C_S = A + Bx + Cx^2$), an equation for solute redistribution in the solid, that includes Clyne-Kurz's equation [39], was derived. However, the diffusion equation was not directly solved. Prior knowledge of the final solidification time is required. Kobayashi [40] obtained an exact solution (Laguerre polynomial) for the "columnar" dendrite model. Diffusion in solid was calculated but complete liquid diffusion was again assumed. Solidification rate and physical properties, including partition coefficients, were considered constant. Parabolic solidification ($f_S = \sqrt{t/t_f}$, where t_f is the final solidification time) was also assumed.

Nastac and Stefanescu [41] have proposed a complete analytical and a numerical model for "Fickian" diffusion with time-independent diffusion coefficients and zero-flux boundary condition. The model takes into account solute transport in the solid and liquid phases and includes overall solute balance. It allows a comprehensive treatment of dendritic solidification through calculation of the fraction of solid with a MT-TK model. Equations for plate, columnar, and spherical geometry were derived. Note that Eq. (27) is for plate morphology. The spherical geometry can be applied to systems solidifying with equiaxed morphology.

A comparison of microsegregation calculated with different models is given in Fig. 16. It is seen that for the Fe-1% C alloy the lever rule and the

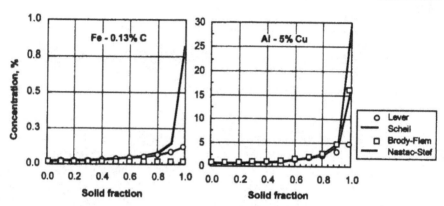

Figure 16   Comparison of the trace of the solid concentration calculated with three models.

Nastac-Stefanescu model predict the same amount of segregation. This is not surprising, since the solid diffusivity of carbon is very high (interstitial diffusion), for practical purposes infinite. Thus, the equilibrium lever rule is satisfactory when modeling Fe-C alloys. The Scheil model predicts much higher segregation, and the Brody-Flemings model much lower. On the contrary, for the Al-5% Cu alloy the prediction of the Nastac-Stefanescu and Brody-Flemings models are closer to the Scheil model because of the very low solid diffusivity of Cu in Al (solid solution diffusion). This infers that the Scheil model is a reasonable approximation for slow diffusing substitutional elements.

C.   Dendritic Growth

Many of the alloys used in practice, such as steel, aluminum-copper alloys, nickel-base and copper-base alloys, are single-phase alloys, which means that the final product of solidification is mostly a solid solution. Depending on the thermal and compositional field, cellular or, in most practical cases, dendritic morphology will occur. For cast alloys typically, the solidification of the primary phase involves two stages. In the first stage unobstructed growth in the liquid occurs. Either columnar or equiaxed dendrites are formed. In the second stage, dendrites come into contact with one another (dendrite impingement or coherency). A dendritic network is formed and liquid flow only occurs between the dendrites (the solid is fixed). The governing mechanism for growth becomes coarsening. Thus, models must describe dendrite growth driven by thermal and solutal supersaturation in the liquid, and dendrite coarsening after dendrite coherency occurs.

1. Dendrite Growth Velocity Models

Both deterministic and probabilistic models have been proposed to describe dendritic growth. Deterministic models differ tremendously in complexity, based on their ultimate goal. They can be classified as follows:

- Tip kinetics models attempt to describe solely dendrite tip kinetics, as determined by the thermal and solutal field, and by capillarity.
- Dendritic array models take into account tip and lateral diffusion.
- Volume averaged models use tip kinetics models and solutions of the thermal and solutal field, in conjunction with volume averaging techniques, to describe growth of equiaxed or columnar dendrites so that the model could be incorporated into solidification computer models;
- Complex geometry models attempt to describe growth kinetics of complex geometry dendrites taking into account the thermal and solutal field, and capillarity. Increasingly popular are the phase-field models that can compute neighboring branches [42–44]. They are, however, limited to qualitative studies because of the large amount of computation required. Consequently, they will not be reviewed here.

Some of the most representative models for dendrite growth will be discussed in the following sections.

2. Tip Velocity Models

Consider a needlelike crystal growing in the liquid. Assume diffusion-controlled growth, which means that the only driving force for growth is the concentration gradient. It can be demonstrated that

$$P_c = \Omega_c \tag{28}$$

where the solutal supersaturation is $\Omega_c = C_L - C_0/C_L(1 - k)$ and the solutal Péclet number is $P_c = Vr_0/2D$.

Thus, the growth velocity of the hemispherical needle can be calculated as:

$$V = \frac{2D_L\Omega}{r} \tag{29}$$

This equation indicates that velocity depends on tip radius r and on supersaturation Ω, which is the driving force. However, velocity is not uniquely defined. In other words, the solution of the diffusion equation does not indicate whether a dendrite will grow fast or slow, but only relates the tip curvature to its rate of propagation. To remove this incertitude it can be assumed that the needle, which can be viewed as a perturbation on the solid/liquid interface will grow with the shortest stable wave length, i.e., $r_s = \lambda_i$. This implies that if the tip radius of the perturbation is smaller than λ_i, the radius will tend to increase, but if it is larger than λ_i, additional instabilities will form and the radius will decrease. It can be demonstrated [41] that the growth velocity of an equiaxed dendrite tip of hemispherical shape in a binary system can be calculated as:

$$V = \mu \, \Delta T^2 \tag{30}$$

with the growth coefficient given by

$$\mu = \frac{1}{2\pi^2 \Gamma} \left[\frac{m_L(k-1)C_L^*}{D_L} + \frac{\Delta H_f}{c\alpha_L} \right]^{-1}$$

where D_L is the liquid diffusivity, Γ is the Gibbs-Thomson coefficient, m_L is the slope of the liquidus line, and C_L^* is the liquid concentration. More complicated equations can be obtained if a paraboloid of revolution shaped tip (self-preserving form) is assumed. However, for cooling rates typical for casting the difference is negligible.

For columnar growth, the simplified equation proposed by Kurz and Fisher [45] can be written in the same form as Eq. (30), with the growth coefficient:

$$\mu = \frac{D_L}{\pi^2 \Gamma m(k-1)C_L^*} \tag{31}$$

Note that when the thermal undercooling component in Eq. (30) is ignored, the growth coefficient for the equiaxed dendrite is two times smaller than that for the columnar dendrite.

All these models require information on the relationship between C_S and C_L. For binary systems the relationship is defined by the liquidus and solidus line on the phase diagram at a given temperature and is described by the partition coefficient k and the slope of the liquidus line m_L. They are both functions of temperature. There is a unique tie line connecting the equilibrium concentrations, and thus only 1 degree of freedom.

However, most alloys of practical significance are multicomponent. For multicomponent alloys, the partition coefficient is a function of the concentration of all the alloying elements. In addition, the liquidus/solidus lines are replaced by liquidus/solidus surfaces. There is more than 1 degree of freedom. Prediction of the solidification path that is of the trace of liquid concentration at the interface from the beginning to the end of solidification is not trivial. The most advanced method to date is the use of the CALPHAD (*calculation of phase diagram*) method [46]. In the CALPHAD method the concentration of solids in equilibrium with a given liquid concentration at a given temperature are obtained by equating the chemical potentials of each component in the existing phases according to standard thermodynamic principles. An example of the output for an Al 356 alloy is given in Table 5 [47].

Assuming weak or no interaction between elements, it is possible to use a simple analytical approach. Equation (30) can be extended to multicomponent alloys as

$$\mu = \frac{1}{2\pi^2} \left\{ \sum_{i=1}^{n} \Gamma_i \left[\frac{m_L^i(k_i - 1)C_L^i}{D_L^i} + \frac{\Delta H_f}{c\alpha_L} \right] \right\}^{-1} \tag{32}$$

Another way of handling multicomponent alloys is the pseudobinary approach [48]. Equivalent liquid concentration $\bar{C}_L$, liquidus slope $\bar{m}_L$, and partition coefficient $\bar{k}$ are defined as follows:

$$\bar{C}_L = \sum_{i=1}^{n} C_L \qquad \bar{m}_L = \frac{\sum_{i=1}^{n}(m_{L_i}C_{L_i})}{\bar{C}_L} \qquad \bar{k} = \frac{\sum_{i=1}^{n}(m_{L_i}C_{L_i}k_i)}{\sum_{i=1}^{n}(m_{L_i}C_{L_i})} \tag{33}$$

These quantities are substituted in Eq. (30). The liquidus and solidus temperatures of the alloy are calculated as:

$$T_L = T_M + \bar{m}_L \bar{C}_0 \quad \text{and} \quad T_S = T_M + \bar{m}_L \bar{C}_f \tag{34}$$

where C_0 and C_f are the initial and final liquid concentrations, respectively.

3. Volume Averaged Dendrite Models

At the present state of the art, it is impossible to produce models that describe the complexities of dendrite growth and are simple enough to have engineering usefulness. Accordingly, a number of simplified models, based on the volume averaging technique, have been proposed.

A summary of the geometry of the equiaxed grains assumed by some models to be discussed is shown in Fig. 17. A dendrite envelope is defined as the interface separating the intradendritic and extradendritic liquid phases. It includes the solid and the intradendritic liquid. The equivalent dendrite envel-

Table 5 Output of CALPHAD (SLOPE subroutine) for an Al 356 Alloy at $T_L = 624$ C

Element	Si	Fe	Cu	Mn	Mg	Zn	Ti
C_L	6	0.2	0.2	0.1	0.35	0.1	0.2
C_S	0.69	0.03	0.023	0.038	0.065	0.047	0.93
m	−6.86	−2.11	−3.19	−1.27	−2.85	−2.06	15.8
k	0.11	0.015	0.12	0.38	0.19	0.47	4.67

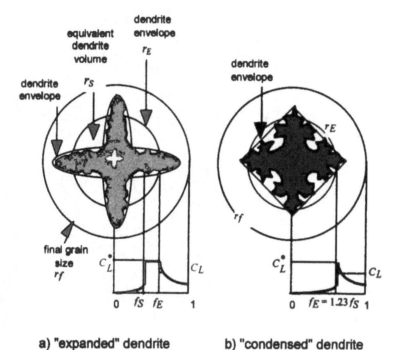

a) "expanded" dendrite   b) "condensed" dendrite

Figure 17 Assumed morphologies and associated concentration profiles of solidifying equiaxed dendrites.

ope is a sphere that has the same volume as the dendrite envelope (r_E in Fig. 17a). The equivalent dendrite volume is the sphere having the same volume as the solid dendrite.

Two main types of dendrites are seen: "condensed" and "extended" dendrites. The condensed dendrites have the equivalent dendrite volume of the same order with the volume of the dendrite envelope ($r_S \cong r_E$). They form typically at high undercoolings. In the case of the extended dendrite the equivalent dendrite envelope is significantly higher than the equivalent dendrite volume ($r_E \gg r_S$). Extended dendrites are common at low undercoolings, when primary arms develop fast with little secondary and higher order arms growth. r_f, the final radius of the grain, is in fact the microvolume element. The main problem to solve is to formulate the radius of the equiaxed grain, or the grain growth velocity.

To calculate the radius of the equiaxed grain, it is necessary to develop a model that describes the growth of a simplified dendrite. From Fig. 17 a number of solid fractions can be defined:

$$f_S = \frac{v_S}{v_f} = \left(\frac{r_S}{r_f}\right)^3 \qquad \text{fraction of solid in the volume element} \qquad (35a)$$

$$f_E = \frac{v_E}{v_f} = \left(\frac{r_E}{r_f}\right)^3 \qquad \text{volume fraction of dendrite envelope} \qquad (35b)$$

$$f_i = \frac{f_S}{f_E} \qquad \text{internal fraction of solid} \qquad (35c)$$

These three relationships fully define the growth and morphology of the dendrite. Spherical envelopes must not necessarily be assumed. Stereological relationships, such as shape factors and interface concentrations, are used for nonspherical envelopes. The time derivatives of these equations define the growth velocity of the dendrite. Present volume average models assume that the movement of the dendrite envelope is governed by the growth model for the dendrite tip. Thus,

$$r_E = \int_0^t \mathbf{V}_E \, dt \qquad \text{until} \qquad r_E = r_f = (4\pi N/3)^{-1/3} \qquad (36)$$

Since r_f is considered known, one more equation is necessary to solve the system of equations in Eq. (35).

In a recent model [49], it was calculated that the final equation for the temporal evolution of the solid fraction is

$$\frac{\partial f_S}{\partial t} = 3f_S\left(\frac{\mathbf{V}_E}{r_E}\frac{1}{\chi_E} + \frac{\mathbf{V}_S}{r_S^i}\frac{1}{\chi_S^i}\right) \qquad \text{with } \chi_E = \frac{4\pi r_E^2}{A_E} \text{ and } \chi_S^i = \frac{4\pi(r_S^i)^2}{A_S^i} \qquad (37a)$$

where $\mathbf{V}_E$ and $\mathbf{V}_S$ are the average normal velocities of the dendrite envelope and the S/L interface, respectively, r_S^i is the radius of the instability, χ_E and χ_S^i are the shape factor of the envelope and of the instability, respectively, and A_E and A_S^i are the interfacial areas of the envelope and of the instability, respectively.

Application of such a model to casting modeling is restricted by the lack of information regarding the shape factors and the interfacial areas. Some simplification can be introduced. For the particular case of the condensed dendrite, it can be demonstrated [49]:

$$\frac{\partial f_S}{\partial t} = \left(\frac{18}{\pi}\right)^{1/3} f_S \frac{\mathbf{V}_E}{r_E} \qquad (37b)$$

In all these equations, the movement of the dendrite envelope is directly related to dendrite tip velocity, that is, $\mathbf{V}_E$ is calculated with a dendrite tip velocity model [e.g., Eq. (30)]. The undercooling is computed as

$$\Delta T = T_M + m \langle C_L \rangle^L - T_{\text{bulk}} \tag{38}$$

where T_M is the melting temperature of the pure metal, T_{bulk} is the bulk temperature defined as the average temperature in the volume element, and $\langle C_L \rangle^L$ is the intrinsic volume average extradendritic liquid concentration.

For details regarding the calculation of V_S and r_S^i the reader is referred to the original paper. The evolution of the fraction solid is related to the inter-dendritic branching and dynamic coarsening (through the evolution of the specific interfacial areas). It is also connected to the topology and movement of the dendritic envelope (through the tip growth velocity and dendrite shape factor).

This model has the capability to calculate dendrite coherency, as shown in Fig. 18 for condensed dendrites. Coherency is reached when $f_E = 1$, that is when the equivalent dendrite envelope reaches the final grain radius.

In a series of papers Wang and Beckermann [50, 51, 52] have proposed a multiphase solute diffusion model for dendritic alloy solidification. Their model accounts for the different length scales in the dendritic structure. A control volume, containing columnar or equiaxed dendrites, is considered to consist of three phases: solid, intradendritic liquid, and extradendritic liquid (Fig. 19). The two liquid phases are associated with different interfacial length scales and have different transport behaviors. Thermal undercooling, melt convection and solid transport were originally ignored, and complete mixing in the intradendritic liquid was assumed. Melt convection and solid transport were however included in the last version of the model. Macroscopic conserva-

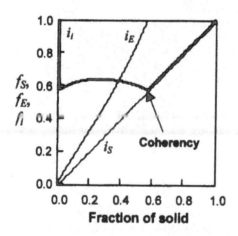

Figure 18 Calculated evolution of the envelope fraction and internal fraction of solid for a cooling rate of 4°C/s during solidification of a Fe-0.6% C alloy. (From Ref. 49.)

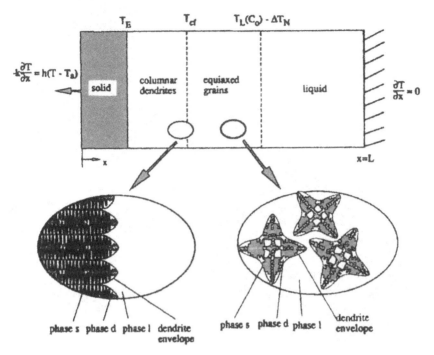

Figure 19 Illustration of the physical problem in the multiphase approach. (From Ref. 51.)

tion equations were derived for each phase, using a volume averaging technique. The governing equations for the microscopic model were summarized as follows:

Dendrite envelope motion: $\dfrac{\partial}{\partial t}(f_S + f_d) = \dfrac{A_E}{v_E}\dfrac{D_L m_L(k-1)C_E}{\pi^2 \Gamma}\left[I^{-1}(\Omega_c)\right]^2$ (39a)

Solute balance of the solid phase: $\dfrac{\partial(f_S C_S)}{\partial t} = C_{Sd}\dfrac{\partial f_S}{\partial t} + \dfrac{A_S}{v_S}\dfrac{D_S}{l_{Sd}}(C_{Sd} - C_S)$ (39b)

Solute balance of the intradendritic liquid: $\dfrac{\partial(f_d C_d)}{\partial t} = (C_E - C_{dS})\dfrac{\partial f_S}{\partial t}$

$+C_E\dfrac{\partial f_d}{\partial t} + \dfrac{A_S}{v_S}\dfrac{D_d}{l_{dS}}(C_{dS} - C_d) + \dfrac{A_e}{v_e}\dfrac{D_d}{l_{dL}}(C_E - C_d)$ (39c)

Solute balance of the liquid phase: $\dfrac{\partial(f_L C_L)}{\partial t} = C_E \dfrac{\partial f_L}{\partial t} + \dfrac{A_L}{v_L} \dfrac{D_L}{l_{Ld}(C_E - C_L)}$

$$(39d)$$

Interfacial solute balance at the Sd interface: $(C_{dS} - C_{Sd})\dfrac{\partial f_S}{\partial t} = \dfrac{A_s}{v_S} \dfrac{D_S}{l_{dS}}$

$$(C_{dS} - C_d) + \dfrac{A_S D_S}{v_S\, l_{Sd}}(C_{Sd} - C_S) \tag{39e}$$

where the subscript d stands for the intradendritic liquid, double subscripts stand for interface quantity (e.g., dS means intradendritic liquid/solid interface), l are the species diffusion lengths. $[I^{-1}(\Omega_c)]^2$ is the inverse Ivantsov function. Other notations are as before. The model for the dendrite tip velocity is essentially that developed by Rappaz and Thévoz [10].

When supplying the expressions for the diffusion lengths, thermodynamic conditions, and geometrical relations, these five equations have five unknowns and thus, can be solved. They represent a complete model for solute diffusion, applicable to both equiaxed and columnar solidification. However, calculation of diffusion lengths and geometrical relations cannot be done without further simplifications. For example, assuming a one-dimensional platelike dendrite arm geometry, it can be shown that $l_{Sd} = d_S/6$, where d_S is the mean characteristic length or the diameter of the solid phase. A complete discussion is not possible within the constrains of the present text. The model can incorporate coarsening and was successfully used to predict the columnar-to-equiaxed transition (CET).

D. Eutectic Growth

During the solidification of eutectic commercial casting alloys such as cast iron and Al-Si alloys, the most common form of the eutectic phase is that of eutectic grains. The growth velocity of eutectic grains can be calculated as

$$V = \mu_V \, \Delta T_{\text{bulk}}^2 \tag{40}$$

The growth coefficient μ_V deviates from that calculated from the Jackson-Hunt theory, mostly because equiaxed solidification is not steady state. It must be evaluated from experiments. Some typical values are presented in Table 6.

Table 6 Growth Coefficients μ_V for Selected Alloys

Alloy	Steady state μ_V, $\text{m} \cdot \text{s}^{-1} \cdot \text{K}^{-2}$	Equiaxed solidification μ_V, $\text{m} \cdot \text{s}^{-1} \cdot \text{K}^{-2}$
Gray cast iron	$9.8 \cdot 10^{-8}$	$(3\text{–}9.5) \cdot 10^{-8}$
White cast iron	$3.96 \cdot 10^{-5}$	$2.5 \cdot 10^{-6}$
Aluminum-silicon	$1.07 \cdot 10^{-5}$	$10^{-6}\text{–}10^{-7}$

IV. DETERMINISTIC MACRO-MICRO MODELING OF SOLIDIFICATION OF SOME COMMERCIALLY SIGNIFICANT ALLOYS

Iron-base alloys, including steel and cast iron, are some of the oldest man-made materials. However, they have enjoyed a remarkable longevity because of their wide range of mechanical and physical properties, coupled with their competitive price. Today, iron-base alloys are by far the most widely used casting materials. While their development has followed the classic path of perception followed by rationalization, it was only relatively recently that these materials have been studied as mathematical systems.

Other alloys, such as aluminum- and magnesium-base alloys, have gained increasing markets because of the drive to decrease fuel consumption by decreasing automobile weight. Yet other alloys like superalloys and in situ composites have been very successful in high temperature applications such as turbine engines. The computational modeling for microstructure prediction of some of these alloys will be discussed briefly in this section.

A. Steel

To simulate the solidification or room temperature microstructure, it is necessary to calculate the time evolution of all relevant phases. For cast carbon steel, it is necessary to tackle many different issues, including solidification of primary dendrites and of various inclusions, as well as micro/macrosegregation.

1. Nucleation of Primary Dendrites

The literature data on nucleation of steel is very limited. For austenite grains the following empirical law was proposed [53]:

$$d_\gamma = 0.336 \left(\frac{dT}{dt} \right)^{-0.5} \tag{41}$$

where d_γ is the final austenite dendrite size in mm and dT/dt is the cooling rate.

2. Growth of Primary Dendrites

The solidification of steel includes the occurrence of a peritectic event. This event can occur through a peritectic reaction or a peritectic transformation. Both mechanisms have been used to build mathematical models for the peritectic solidification of steel (e.g., Refs. 54–56). Typically, these models are rather complicated and may result in substantial increase of the computational time. Fortunately, because of the high carbon diffusivity the peritectic reaction may be assumed to follow equilibrium for modeling purposes.

Another problem is that of the complex thermodynamics of multicomponent alloys. Howe [57] approached the problem of a multicomponent alloy similar to that of eutectic solidification of cast iron, by using the concept of *dominant element equivalent*. Experimental data on carbon and low alloy steel were used to generate an equation for carbon equivalent at the peritectic temperature:

$$CE = -0.123Si + 0.04Mn + 0.06S - 0.018Cr - 0.05Mo + 0.08Ni \quad (42)$$

Assuming equilibrium solidification, the fraction of solid of δ ferrite was calculated based on the equilibrium diagram. The growth rate was calculated from a solute balance equation (diffusion controlled mechanism).

A similar approach may be used to calculate solidification of austenite. If a more detailed analysis is desired, volume averaged dendrite models may be used in conjunction with equilibrium diffusion models for carbon and Scheil models for substitutional solid solution elements.

3. Micro/Macrosegregation

Numerous microsegregation models have been published. Assumptions range from complete diffusion in liquid and no diffusion in solid to diffusion in both liquid and solid (see Tables 3 and 4). When applied to the evaluation of microsegregation in ferrous alloys, additional assumptions must be made on the shape of the volume element for which the calculation is performed (plate, columnar, or equiaxed dendrite). To evaluate macrosegregation, it is necessary to include calculations of momentum transport (composition change because of fluid flow) in addition to the mass transport by diffusion calculations performed by the microsegregation models.

Schneider and Beckermann [58] developed a model for the solidification of multicomponent steels that also includes the peritectic transformation. It is based on the volume-averaged two-phase stationary solid phase assumption, and accounts for bulk-liquid motion. Temperature and concentrations are fully coupled through thermodynamic equilibrium relationships. Macrosegregation patterns for a number of elements were simulated but no validation was presented, other than for the liquidus and solidus temperature of steel. Typical simulation outputs are shown in Fig. 20. The global severity of macrosegregation of an element was found to be linearly dependent on its partition coefficient.

4. Inclusions

Howe [57] described MnS precipitation in steel by removal of the elements from the residual liquid in an appropriate ratio so as not to exceed a prescribed solubility product. Wintz et al. [59] developed a model that predicts the composition and amount of inclusions that form in steel during melt processing and solidification. The model is based on the coupling of microsegregation calculations with a liquid metal–inclusions equilibrium model. The number of moles of element i in the solid $N_i^S(l)$ is calculated as a function of successive molar concentrations at the interface $X_i^S(l)$ with:

$$N_i^S(l) = \frac{1}{\lambda/2}\left[2\alpha_i l X_i^S(l) + (1 - 2\alpha_i)\int_0^l X_i^S(x)\,dx\right] \tag{43}$$

where $\alpha_i = f(D_i^S t_S \lambda^{-2})$ is the modified coefficient for back diffusion, l is the coordinate of the solid/liquid interface, λ is the secondary dendrite arm spacing, D_i^S is the diffusion coefficient of i in the solid, and t_S is the local solidi-

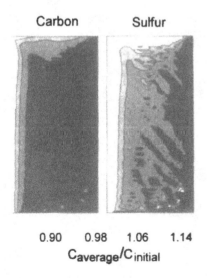

Carbon   Sulfur

0.90   0.98   1.06   1.14
$C_{average}/C_{initial}$

Figure 20 Simulated macrosegregation patterns. (From Ref. 58.)

fication time. Integration of this equation into a multiphase equilibrium program describing the solid metal/liquid metal/inclusions equilibrium makes it possible to calculate the equilibrium concentrations at the interfaces at every fraction of liquid. The model was validated by comparing calculated and experimental amounts of MnS, FeS, and CrS, for steels with Mn/S ratios between 18 and 100.

The model was then used to predict the effect of Fe-rich sulfide inclusions on the solidification behavior of steel. As seen in Fig. 21, when the Mn/S = 50, the FeS content of the sulfides is low ($\sim$ 5%), and remains practically constant throughout solidification. When the Mn/S = 12, the FeS content of the sulfides is higher, increasing from 19 to 37% from the beginning of sulfide precipitation to the end of solidification.

B. Cast Iron

When describing microstructure formation in cast iron during the liquid-solid transformation, we must explain two distinct stages: (1) solidification of the austenite dendrites and/or graphite crystallization from the liquid (before the beginning of eutectic solidification), and (2) solidification of the stable and metastable eutectics, including the various shapes of the carbon-rich phase (graphite or carbide). The main descriptors of the solidification structure of cast iron should include dendrite arm spacing, eutectic grain size, and average graphite lamellar spacing for lamellar graphite (LG) cast iron, nodule count for spheroidal graphite (SG) iron, and micro/macrosegregation.

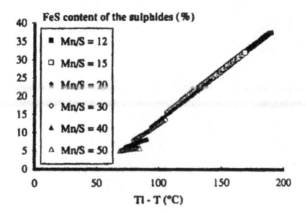

Figure 21 FeS content of sulfides vs. undercooling under the liquidus temperature. (From Ref. 59.)

1. Nucleation and Growth of Austenite Dendrites

Liquid quenching experiments were conducted [60] to evaluate the grain density of primary austenite grains for an iron having 2.98% C and 1.65% Si. The experimental results are shown in Fig. 22. No significant grain coalescence was observed. This is not surprising since the fraction of austenite is small and the grains do not come in contact. At a quenching temperature of 1180°C the correlation between the area grain density (mm^{-2}) and cooling rate (instantaneous nucleation) is given by

$$N_y = 48.12 + 5.33\dot{T} + 0.087\dot{T}^2 \tag{44}$$

The area density of austenite grains was also calculated as a function of undercooling (continuous nucleation) as

$$N_y = 2.45 \, \Delta T^{0.93} \tag{45}$$

Mampaey [61] showed that austenite growth calculations could be based on tip velocity for solid fraction lower than 0.335 and on dendrite arm coarsening models for higher fraction solid.

To evaluate the austenite dendrites tip radius and spacing, Tian and Stefanescu [33] have conducted DS experiments combined with a liquid metal decanting technique. The experimental results plotted on Fig. 23, as well as metallographic observations, indicated that at a growth velocity of 0.65 μm/s a cellular-to-dendritic transition occurred. Two sets of calculated results are also shown on the figure for dendrite tip velocity models (NS is

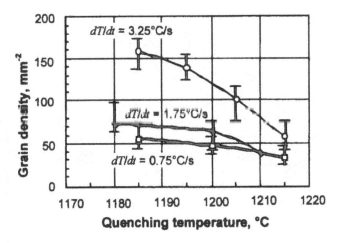

Figure 22 Austenite grain density as a function of quenching temperature and cooling rate. (From Ref. 60.)

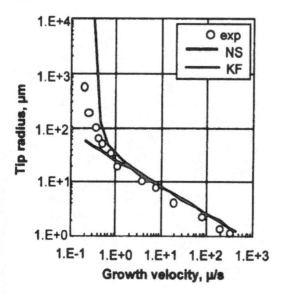

Figure 23 Measured and calculated tip radii of cells and dendrites in a Fe 3.08% C–2.01 % Si alloy solidified under a gradient of 50 K/m. (Experimental data from Ref. 33.)

the Nastac-Stefanescu model and KF is the Kurz-Fisher model). Note that while the simpler NS model fails at growth velocities smaller than 0.8 μm/s, it describes growth well within the range of velocities typical for castings.

2. Crystallization of Graphite from the Liquid

Cast iron is one of the most complex, if not the most complex alloy used in industry, mostly because it can solidify with formation of either a stable (γ-graphite) or a metastable (γ-Fe$_3$C) eutectic. Furthermore, depending on composition and cooling rate several graphite shapes can be obtained at the end of solidification, as exemplified in Fig. 24.

Many theories have been proposed over the years to describe the mechanisms of formation of various graphite shapes, and reviews of these theories have been periodically done (e.g., Minkoff [62], Elliott [63], Stefanescu [64]). It can be generally concluded that the spheroidal shape is the natural growth habit of graphite in liquid iron. Flake graphite is a modified shape, the modifiers being sulfur and oxygen. They affect graphite growth through some surface adsorption mechanism.

A mathematical description of graphite growth from the liquid seems to be out of the question at this stage because of the complexity of the problem. Fortunately, very little graphite growth is generated in the liquid in the case of

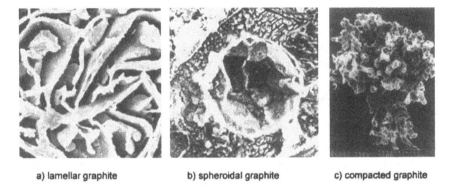

a) lamellar graphite   b) spheroidal graphite   c) compacted graphite

Figure 24 Typical graphite shapes obtained in commercial cast iron.

SG iron (the size of a spheroid before encapsulation in an austenite shell is about 1 μm). Consequently, the effect on fraction solid and latent heat generated is minimal. The mathematics can be circumvented by simply assuming that at the beginning of the eutectic solidification the graphite spheroids have a size in this range.

3. Eutectic Solidification

Based on extensive experimental work [65, 66] a sequence of changes in the eutectic morphology of directionally solidified cast iron was proposed, as shown in Fig. 25. As the ratio between the temperature gradient at the solid/liquid interface and the growth velocity G/V decreases, or the composition C_0 (e.g., Mg or Ce) increases, the solid-liquid interface changes from right to left, i.e., from planar to cellular, and then to equiaxed, while graphite remains basically flake (lamellar). Cooperative growth of austenite and graphite occurs. Further change of G/V or of C_0 brings about formation of an irregular inter-

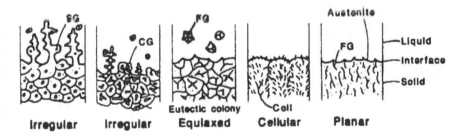

Figure 25 Schematic sequence of structural transitions in directionally solidified cast iron. (From Refs. 66 and 103.)

face, with austenite dendrites protruding in the liquid. Graphite becomes compacted and then spheroidal. Eutectic growth is divorced. From these structures, those that have practical importance are the irregular spheroidal graphite (SG), the irregular compacted graphite (CG), and the eutectic colonies of lamellar graphite.

As shown in Fig. 26, the solidification mechanisms of lamellar and spheroidal graphite cast iron during continuous cooling are quite different. The main models proposed to describe solidification of these irons will be summarized in the following paragraphs.

a. Solidification of the Lamellar Graphite-Austenite Eutectic. Eutectic grain size prediction depends on the quality of the nucleation models. Some typical numbers for the nucleation of lamellar graphite (LG) cast iron of various carbon equivalents (CE) are given in Fig. 27. Most likely, the nucleation potential of an industrial melt will never be predicted from first principles. The solution is to evaluate the nucleation potential of the melt before pouring, and then to bring it to a standard value for which the nucleation constants have been determined. Such evaluation methods include computer-aided cooling curve analysis and the chill test (for cast iron). The major problem with all empirical nucleation models is that they are only valid under the given experimental conditions. In fact, the final grain count measured on the metallographic sample is used as input for the nucleation law. Thus, it is not surprising that calculation of grain size typically gives good results.

Fras et al. [67] pointed out that in some instances fine eutectic grains are found in the thermal center of the casting. It was noted that after the maximum undercooling is reached, nucleation stops during the eutectic arrest, but starts again at the end of solidification, when higher undercooling occurs (secondary

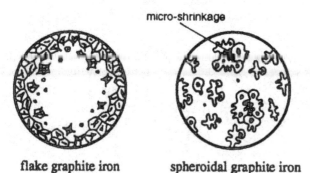

flake graphite iron   spheroidal graphite iron

Figure 26 Schematic illustration of solidification mechanisms of continuously cooled lamellar and spheroidal graphite cast iron.

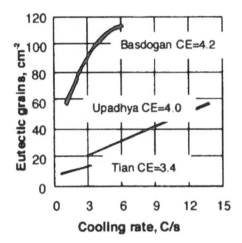

Figure 27 Variation of volumetric eutectic grain density as a function of cooling rate for cast iron. (From Refs. 60, 104 and 105.)

nucleation). The secondary nuclei may have only a short time available for growth. Thus, mixed large grains–fine grains structure will occur. They also demonstrated that such a behavior could be predicted through modeling.

For solidification of regular eutectics a number of relationships have been established between process and material parameters based on the extremum criterion. Eutectic gray iron can solidify with a planar interface. However, the relationship $\lambda^2 \cdot V = $ ct. is not obeyed in the growth of the γ-LG eutectic. Some experimental results are compared with theoretical calculations in Fig. 28. The departure of experimental values from the theoretical line is a clear indication of irregular rather than regular growth in LG cast iron. Lamellar graphite iron is an irregular eutectic and both an extremum and branching spacing can be defined. The average spacing should be higher than the extremum one predicted form Jackson-Hunt theory. Other data for the growth coefficient μ_V are given in Table 6.

Calculation of phase spacing of irregular eutectics such as cast iron is not trivial. Zou and Rappaz [68] attempted a simple approach based on the Jackson-Hunt equation $\lambda^2 V = \mu_\lambda$, where the growth coefficient was chosen to fit the experimental data. For Fe-C-2.5% Si alloys significant discrepancies between predicted and measured average lamellar graphite eutectic spacing was found at cooling rates higher than 0.05–0.066 K/s.

Recently, the general Jackson-Hunt theory for steady state directional solidification was extended to eutectic systems solidifying with equiaxed grains under nonequilibrium conditions (Catalina and Stefanescu [69]). Low Péclet

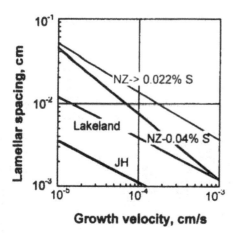

Figure 28 λ-V relationships in LG cast iron. Lakeland [106]—$\lambda = 3.8 \times 10^{-5}$ $V^{-0.5}$ cm; NZ (Nieswaag and Zuithoff [107])—$\lambda = 0.56 \times 10^{-5} V^{-78}$ cm (0.004% S); NZ—$\lambda = 7.1 \times 10^{-5} V^{-0.57}$ cm ($> 0.02\%$ S); JH (Jackson and Hunt— $\lambda = 1.15 \times 10^{-5} V^{-0.5}$ cm (theoretical).

number, isothermal S/L interface, and steady state solutal field around the spherical grain were assumed. A comparison between the values of the average lamellar spacing calculated with this model and those measured on a gray iron eutectic grain is presented in Fig. 29.

Leube et al. [70] used the Magnin-Kurz model for irregular eutectics. According to this model, the graphite lamellar distance in gray iron varies between a minimum and a maximum. The minimum is given by the Jackson-Hunt equation. The maximum distance is obtained from a branching criterion and is calculated as

$$\lambda_{\max} = K \left(\frac{2\Gamma_{Gr} \cos \theta_{Gr}}{f_{Gr} \left(\frac{|m_{Gr}| C_0}{D \tan \theta_{Gr}} \left(\Pi_{Gr} - \frac{P}{f_{Gr}} \right) V + \frac{f_{Gr}}{24} \sigma \right)} \right)^{1/2} \tag{46}$$

where K is a material constant, Γ is the Gibbs-Thomson coefficient, θ is the contact angle, f is the volume fraction, m is the liquidus slope, C_0 is the concentration difference of the eutectic tie-line, D is the liquid diffusivity, Π and P are functions of volume fraction, and G is the temperature gradient. Based on experimental measurements on the metallographic samples, the material constant K was taken to be 4. It is different from the theoretical one since it describes only two-dimensional structures. The distance between two lamellae during grain growth is calculated as

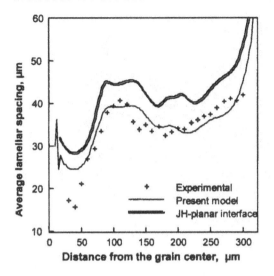

Figure 29 Calculated and measured average interlamellar spacing for an eutectic grain, along the radius from the center of the eutectic grain. (From Ref. 69.)

$$\lambda = R(t, t_{br})2 \sin \gamma \qquad (47)$$

where $R(t, t_{br})$ is an arbitrary radius of the grain set to 0 if the lamellae branch out and γ is the angle between two lamellae. The angle γ is simulated by a random number generator, having values between 7 and 40°. Thus, the lamellae grow as long as the distance between them does not exceed λ_{max}. The model allows calculation of the average length of lamellae for all grains in one volume element, as well as the average of the longest lamellae.

Another interesting contribution of Leube et al. [70] is the simulation of grain morphology. A cross section of the three-dimensional simulation is given in Fig. 30.

b. Solidification of the Spheroidal Graphite-Austenite Eutectic. It is generally accepted that the sequence of solidification of eutectic SG iron is as follows: (1) At the eutectic temperature, austenite dendrites and graphite spheroids nucleate independently in the liquid; (2) very limited growth of spheroidal graphite occurs in contact with the liquid; (3) further graphite nucleation occurs at the austenite-liquid interface; (4) flotation or convection determines the collision of graphite spheroids that have nucleated in the liquid with the austenite dendrites; (5) graphite encapsulation in austenite can occur before or immediately after these collisions; (6) significant graphite growth occurs by carbon diffusion through the austenite shell only after graphite spheroids have attached themselves to austenite dendrites; (7) austenite dendrites grow

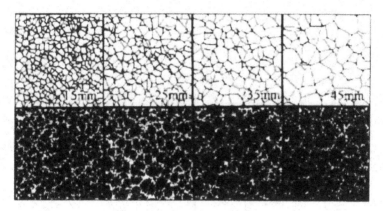

Figure 30 Comparison between experimental and simulated gray iron grain structure at various cooling rates. (From Ref. 70.)

partly due to carbon diffusion and partly because of melt undercooling and supersaturation. This sequence is illustrated schematically in Fig. 31. Note that the eutectic austenite is dendritic and can be hardly discerned from the primary austenite dendrites.

Nucleation of SG iron has been described through empirical relationships by a good number of investigators, e.g., Mampaey [71]. Recently, an attempt at better quantification of the nucleation law (Skaland et al. [72]) resulted in an equation that includes the effect of fading time:

$$N = c \cdot \ln \frac{1.33 + 0.64t_E}{1.33 + 0.64t_S} \qquad \text{mm}^{-3} \tag{48}$$

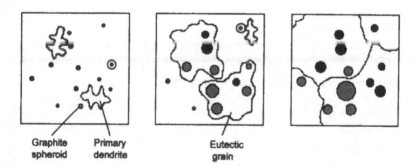

Graphite Primary
spheroid dendrite

Eutectic
grain

Figure 31 Schematic illustration of the sequence of formation of eutectic grains in SG iron for the case of continuous cooling solidification.

where c is a kinetic constant to be evaluated experimentally, and t_S and t_E are the time intervals between inoculation and start and end of solidification, respectively. Graphite nucleation and growth deplete the melt of carbon in the vicinity of graphite. This creates conditions for austenite nucleation and growth around the graphite spheroid. Once the austenite shell is formed, further growth of graphite can only occur by solid diffusion of carbon from the liquid through the austenite. It is rather easy to accept that, once a spheroid is encapsulated in austenite, isotropic, diffusion–driven growth will maintain the spherical shape of graphite. As shown in Fig. 32, the driving force of the reaction is the composition gradient between the austenite/liquid and austenite/graphite interfaces.

The complex austenite–SG eutectic has not been modeled yet in its entire complexity. However, attempts have been made to describe the growth of graphite spheroids together with their austenite shell. Calculations of diffusion-controlled growth of graphite through the austenite shell were originally made based on Zener's growth equation for an isolated spherical particle in a matrix of low supersaturation (Wetterfall et al. [73]).

Su et al. [74], and then many other investigators, successfully incorporated similar diffusion models in numerical solidification codes. To calculate the growth velocity of the austenite shell, the classic law $V = \mu \cdot \Delta T^2$, with the following growth coefficient, is used.

$$\mu_\gamma = D_C^\gamma \left[r_\gamma (r_\gamma / r_G - 1) \left(\frac{m_S - m_L}{m_S m_L} + \frac{C_E - C_{\gamma m}}{\Delta T} \right) \right]^{-1} \frac{m_S - m_\gamma}{m_S m_\gamma} \qquad (49)$$

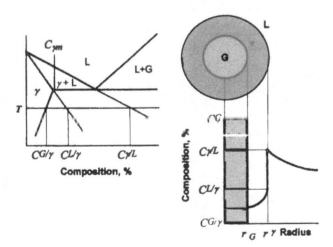

Figure 32 Schematic diagram showing the concentration profile of carbon throughout the liquid and the austenite shell based on the binary phase diagram.

where D_C^γ is the diffusivity of carbon in austenite, and the compositions $C^{i/j}$ and the slopes of the equilibrium lines liquidus (m_L), solidus (m_S), and maximum solubility of carbon in austenite (m_γ) are obtained from Fig. 32. This equation can be further simplified if it is assumed that, as demonstrated experimentally, the ratio between the radius of the austenite shell and that of the graphite spheroid remain constant: $r_\gamma = 2.4 r_G$. The simplified growth rate is [75]

$$\frac{dR_{Gr}}{dt} = 2.87 \cdot 10^{-11} \frac{\Delta T}{R_{Gr}}$$

Once r_γ is known, r_G is calculated from solute balance (Chang et al. [76]):

$$\rho_G \frac{4}{3}\pi \left[\left(r_G^{i+1}\right)^3 - \left(r_G^i\right)^3 \right] C_G = \rho_\gamma \frac{4}{3}\pi \left[\left(r_\gamma^{i+1}\right)^3 - \left(r_\gamma^i\right)^3 \right] C^{L/\gamma} \tag{50}$$

This approach can been extended to the eutectoid transformation for the austenite-ferrite transformation. The austenite-pearlite transformation has also been modeled using the continuous cooling transformation as suggested by Campbell et al. [77]. Thus, the as-cast microstructure of SG iron has been described reasonably well. Some computed and measured results are presented in Fig. 33.

It is somehow surprising that a mathematical description of the eutectic solidification of SG iron that ignores the growth of the austenite dendrites in contact with the liquid can predict reasonably well the microstructural outcome. It is believed that the mass balance equations used in these models are smoothing out the error, by attributing the dendritic austenite to the austenitic

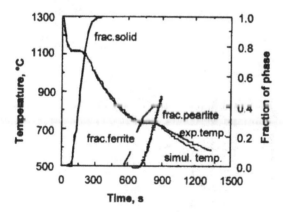

Figure 33 Experimental and simulated cooling curves, and calculated evolution of fraction of phases from liquid to room temperature, for a SG iron bar casting. (From Ref. 76.)

shell. Nevertheless, Lacaze et al. [78] included calculation of the off-eutectic austenite by writing the mass balance as follows:

$$\rho_G \int_0^{R_G} R^2 \, dR + \rho_\gamma \int_{R_G}^{R_\gamma} R^2 \, dR + \left[\rho_L f_L + \rho_\gamma (1 - f_L) \right] \int_{R_\gamma}^{R_0} R^2 \, dR = \rho_L \frac{R_0^3}{3} \tag{51}$$

Here, R_0 is the final radius of the volume element. In previous models, only the first two terms on the left-hand side were included.

To predict graphite size at room temperature, it is necessary to calculate in addition to the graphite size at the end of solidification, its further increase in solid state, during cooling to room temperature and after heat treating. Skaland et al. [72] were successful in explaining the size increase of graphite nodules after heat treatment, using a model based on diffusion controlled growth of graphite, similar to the one described above. For a 30 mm section size, the predicted increase was 8 μm, while the measured one was 13–15 μm.

   c. Solidification of the Compacted Graphite–Austenite Eutectic. Fredriksson and Svensson [79] modeled the growth of CG iron by using two different laws on a cylindrical shape. The growth of radius was assumed to be governed by carbon diffusion through an austenite shell, as for the case of SG iron. The growth in length was described through a linear law: $V = 2.10^{-9} \cdot \Delta T$ (in *m*).

4. Microsegregation

Calculation of microsegregation in SG iron is particularly important since, for certain elements such as Mo or Mn, it is not possible to determine the microsegregation ratio experimentally, because of carbides formation in the intergranular regions. The Nastac-Stefanescu analytical model, that considers diffusion in both liquid and solid, was used to calculate segregation of Mn, Mo, Cu, and Si in SG iron. Some typical results are shown in Fig. 34. Note that, in particular for Cu, the correct distribution is predicted to the end of solidification.

5. The Gray-to-White Structural Transition

The structural gray-to-white transition (GWT) in cast iron is the result of growth and nucleation competition between the stable (gray) and metastable (white) eutectics. Numerical models based on growth competition have been proposed more than 10 years ago (Fredriksson et al. [80], Stefanescu and Kanetkar [5]). The major obstacle was and is the unavailability of data for nucleation of the white eutectic. Simpler models based on the concept of a critical cooling rate above which the GWT occurs were developed and verified by a number of researchers [81, 82].

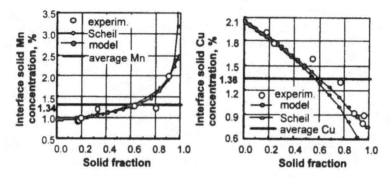

Figure 34 Measured and predicted microsegregation evolution in SG iron. (From Ref. 41.) (Experimental data were taken from Boeri and Weinberg [108].)

A more fundamental numerical model [83] takes into account the nucleation and growth of both the white and gray eutectics, as well as the effects of microsegregation of the third element. The model was incorporated in the commercial software PROCAST. Using a sensitivity analysis of the main process variables, it was found that the amount of silicon in the melt and the nucleation and growth of the gray eutectic have the main positive effects on the GWT. Figure 35 summarizes the combined influence of silicon and cooling rate on the structure of a cast iron having 3.6% C, 0.5% Mn, 0.05% P, 0.025% S.

An interesting point that came out of the analysis of the model is illustrated in Fig. 36. If traditional understanding of the GWT, based on

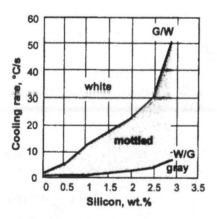

Figure 35 The influence of silicon and initial cooling rate on structural transitions for a 3.6% C cast iron. (From Ref. 83.)

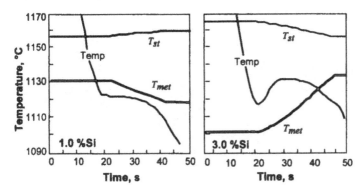

Figure 36 Calculated influence of Si on the equilibrium temperatures and on the cooling curve for a volume element solidifying at an initial cooling rate of 9.3°C/s. (From Ref. 83.)

solidification within the $T_{st} - T_{met}$ interval for gray iron, is used, from the cooling curves it may be implied that at 1% Si the iron solidifies completely white, since apparently, the solidification part of the cooling curve lies completely under the metastable eutectic temperature. However, the calculation of the evolution of the fraction of white and gray eutectic predicts a mottled structure, with about 40% gray eutectic. The reason for this apparent discrepancy is that once nucleated, the gray eutectic will grow under T_{met}. The final structure will result from the nucleation and growth competition between the gray and white eutectic.

Another interesting issue raised by Fig. 36 is that at 1% Si the eutectic interval increases in time, as solidification proceeds. However, at 3% Si the interval decreases. These are effects of silicon redistribution during solidification. When white eutectic solidifies (for 1% Si), silicon is rejected in the liquid and the solidification interval increases. When gray eutectic solidifies (for 3% Si) silicon is incorporated in the solid and the liquid is depleted of silicon. Consequently, the tendency for intergranular carbides formation increases toward the end of solidification.

C. Room Temperature Microstructure

After the first attempt by Stefanescu and Kanetkar [84] to combine a model for the eutectoid transformation with a simple heat transfer calculation, several other investigators proposed heat transport–transformation kinetics models for lamellar [30, 85] and SG iron [76, 86, 87]. The ferrite and pearlite transformation temperatures were calculated with the Thermocalc software:

$$T_{\text{ferr}}(°C) = 739 + 18.4\% \text{ Si} + 2(\% \text{ Si})^2 - 56.4\% \text{ Mn} - 15\% \text{ Cu} \qquad (52)$$

$$T_{\text{pearl}}(°C) = 727 + 21.6\% \text{ Si} + 0.023(\% \text{ Si})^2 - 31\% \text{ Cu} \qquad (53)$$

More recently, a model of ferrite growth that accounts for the role of various alloying elements has been developed [88]. It is a combined diffusion and interfacial mass transfer resistance at the graphite/ferrite interface model. The model has been extended to describe the influence of the major alloying elements used in SG iron, Si, Cu, and Mn.

D. Mechanical Properties

The progress in prediction of room temperature microstructure of cast iron made possible the next development, prediction of some mechanical properties. Space limitation prevents us from elaborating on the subject. However, we will refer the reader to Refs. 85, 89, 90.

E. Aluminum-Silicon Alloys

The Jackson-Hunt parabolic law of irregular eutectic growth, as modified by Kurz and Fisher [91], has been found to be suitable to model the growth of both modified and unmodified equiaxed eutectic grains of irregular eutectics like Al-Si [92]. The modifiers restrain the growth of Si in favorable directions, increasing the twins density, and thus the Si kinetic undercooling. Including the kinetic undercooling of the Si phase, and using weighted averages for the two phases, leads to

$$\overline{\Delta T} = \frac{m_{\text{Si}}\overline{\Delta T^{\text{Al}}} + |m_{\text{Al}}|\overline{\Delta T^{\text{Si}}}}{|m_{\text{Al}}| + m_{\text{Si}}} = \Delta T_c + \Delta T_r + \Delta T_k \qquad (54)$$

with

$$\Delta T_k = \frac{|m_{\text{Al}}|\overline{\Delta T_k^{\text{Si}}}}{|m_{\text{Al}}| + m_{\text{Si}}}$$

where m_{Al} and m_{Si} are the slopes of the liquidus in the Al-Si phase diagram. To obtain an equation uniquely relating ΔT to the growth velocity V, one needs to know ΔT_k^{Si} as a function of V. Magnin and Trivedi [93] derived some values of ΔT_k^{Si} from experimental data. When ΔT_k^{Si} was plotted versus V, it was noticed that a parabolic law fits better the experimental values than the exponential law suggested by Tiller [94]. This can be written as

$$V = \mu_{\text{Si}}\left(\Delta T_k^{\text{Si}}\right)^2 \qquad (55)$$

This means that a law as given by Eq. (30) is still valid in the case of modified irregular eutectics, with a different growth parameter:

$$\mu = \left[\sqrt{K_1 K_2}\left(\phi + \frac{1}{\phi}\right) + \frac{|m_{Al}|}{(|m_{Al}|m_{Si})}\mu_{Si}^{-1/2}\right]^{-2} = \left[\mu_{ex}^{-1/2} + A\mu_{Si}^{-1/2}\right]^{-2} \quad (56)$$

where ϕ is a coefficient relating the lamellar spacing of regular and irregular eutectics: $\lambda_{ir} = \phi\lambda_{ex}$. For Al-Si alloys ϕ has values between 2.2 for rod morphology and 3.2 for lamellar morphology. This law has been verified experimentally by several authors (e.g., Ref. 93) by using directional solidification experiments.

The growth parameter was evaluated experimentally [92]. It was found that it decreases as the degree of modification increases (Fig. 37). During solidification of sodium modified Al-Si alloys, it was noticed that the grain morphology changes from the center to the outside. As shown in Fig. 38, a coarse microstructure surrounds the nucleus while a finer fibrous structure is observed at the periphery of the grain. This observation suggests that a variable growth parameter should be used to simulate the eutectic grain growth. μ is therefore supposed to be initially large and then decreases because of sodium rejection by the solidification interface. Such a growth parameter will impose less undercooling and recalescence on the cooling curve.

This microlevel analysis can be included in a macro model by using an Avrami-type equation for the growth parameter:

$$\mu = \mu_{rod} + (\mu_{lam} - \mu_{rod})\exp\left(-\frac{Vt^{1/2}}{D^{1/2}}\right) \quad (57)$$

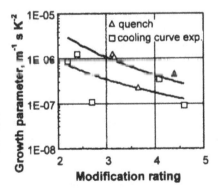

Figure 37 Influence of modification on the average value of the growth parameter (Modification rating: 2 – low; 5 – high.) (From Ref. 92.)

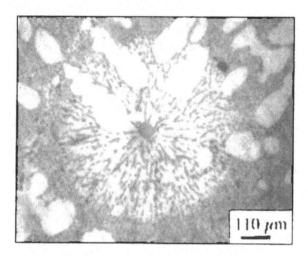

Figure 38 Microstructure of an Al-11% Si eutectic grain core in a sodium modified casting. (From Ref. 92.)

where μ_{lam} is the initial growth parameter (lamellar eutectic $\approx 5 \cdot 10^{-5}\,m\,s^{-1}\,K^{-2}$) and μ_{rod} the growth parameter of the fibrous eutectic ($\approx 5 \cdot 10^{-7}\,m\,s^{-1}\,K^{-2}$). The factor $V/D^{1/2}$ has been used to account for the opposite effects of interface velocity and diffusion in the liquid on the sodium segregation ahead of the liquid-solid interface.

F. Superalloys

The solidification kinetics model described in Sec. III.C.1 was incorporated in PROCAST and used to model the solidification of Inconel 718 [49]. A pseudobinary approach was used to evaluate the liquidus slopes and the partition coefficients as a function of the Nb concentration in the liquid at the interface:

$$k = 0.993 - 0.118(Nb_L^*) + 0.0046(Nb_L^*)^2 \tag{58}$$

$$m = -0.743 \cdot 10^{0.0873 \cdot Nb_L^*} \tag{59}$$

Experimental validation was performed on a casting including a number of plates to allow for variable cooling rate. It was demonstrated that the differences in cooling rate that affect the evolution of fraction solid during solidification (Fig. 39) significantly influences the amount of Laves phases obtained at the end of solidification (Fig. 40). A maximum amount of Laves phase of about 4% was measured and predicted at cooling rates of 6°C/s.

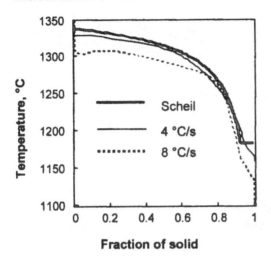

Figure 39 Comparison between calculated evolution of fraction of solid with the Nastac-Stefanescu model and Scheil's model. (From Ref. 49.)

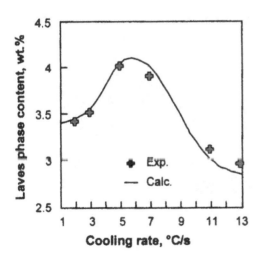

Figure 40 Calculated and experimental Laves phase content as a function of cooling rate. The cooling rate was numerically calculated immediately above the liquidus temperature. (From Ref. 49.)

Boettinger et al. [95] proposed a multicomponent thermodynamic calculation of the phase boundaries between liquid and the various solid phases that can be directly linked to solidification kinetics codes. A subroutine to describe the phase relationship was prepared. The output of this subroutine is

$$C_{L_i} \rightarrow \left(T, C_{S_i}, m_i, k_{ij}\right) \qquad (60)$$

Thus, the solid concentration, the slope, and the partition coefficient can be calculated as a function of the liquid concentration. The authors noted that the product $m_i(1 - k_{ij})$ is not always negative, e.g., Cr and Fe in IN718 and René N4. Such situations occur when the liquidus surface deviates from a linear relationship, e.g., near a minimum in the liquidus.

A second type of subroutine has been used to determine the solidification path using Scheil analysis:

$$(T, C_{0_i}) \rightarrow (C_{L_i}, C_{S_i}, f_S) \qquad (61)$$

where C_{0_i} is the average composition. Some calculation results for an IN718 alloy are shown in Fig. 41. For Scheil solidification, the eutectic reaction $L \rightarrow \gamma + Laves$ is predicted to begin at 1161°C when the fraction solid reaches 0.89. Solidification ends at 1144°C. The predicted final microstructure contains 1.8% Laves. Equilibrium solidification predicts solidification end at 1218°C, with no Laves phase. Since the Laves phase is a nonequilibrium phase, long term annealing should remove it, which is confirmed by practice.

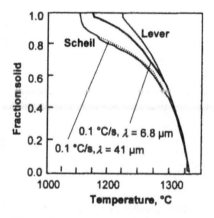

Figure 41   Calculated fraction solid vs. temperature for IN718. The dotted lines are calculation with a diffusion model. (From Ref. 95.)

V. STOCHASTIC MACRO-MICRO MODELING OF SOLIDIFICATION OF SOME COMMERCIALLY SIGNIFICANT ALLOYS

A. Cast Iron

Charbon and Rappaz [96] used a cellular-automaton model to simulate the solidification of SG iron. The computer output, shown in Fig. 42, is quite spectacular, demonstrating the power of the CA approach. However, it must be noted that the simulated grain structure is not correct. Indeed, as discussed previously (see Fig. 31), the eutectic grain of SG iron contains typically more than one graphite spheroid. While the growth velocity is diffusion controlled, and thus the calculation of graphite and austenite growth is correct assuming an austenite shell around the graphite spheroid, it cannot be inferred that each spheroid is associated with an austenite grain.

B. Aluminum Alloys

A stochastic model, based on the coupling of the control volume method for macroscopic heat flow calculation and a two-dimensional cellular automaton model for microstructure evolution, was developed by Cho et al. [97] to predict solidification grain structure in squeeze cast Al-4% Cu and Al-7% Si alloys. Good agreement with experimentally evaluated grain structures is claimed.

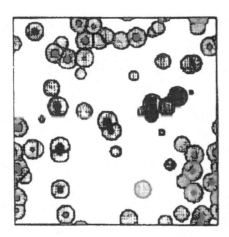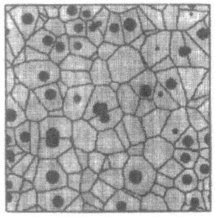

Figure 42   Simulated grain structure for spheroidal graphite iron at $f_S = 0.25$ and 1.

C. Superalloys

Desbiolles et al. [98] have developed a three-dimensional cellular automaton–finite element (CAFE) model that can be used to predict the solidification grain structure in investment cast parts. It is a further refinement of a previously developed 2-D model [21]. Because of the very large number of cells (typically 10^7–10^8) required to describe these parts, special algorithms have been developed to dynamically define the cells during solidification. Typical features predicted by this model include the growth competition between columnar grains (see Fig. 43), grain selection in a pigtail selector, the extension of a grain in an undercooled region of liquid (reentrant corner), as well as polycrystalline growth in directionally solidified turbine blades.

Other researchers [99] have similarly used data obtained from the simulation of heat transfer in conjunction with the cellular automaton technique to model the solidification of investment cast René 77 and IN718.

Figure 43 Calculated grain structure in a DS turbine blade after complete solidification. (From Ref. 108.)

REFERENCES

1. DM Stefanescu. ISIJ International, 35, 6:637, 1995.
2. S Chang, DM Stefanescu. Metall Mater Trans A, 27A:2708, 1996.
3. G Upadhya, AJ Paul. AFS Trans, 102:69, 1994.
4. KC Su, I Ohnaka, I Yanauchi, T Fukusako. In: H Fredriksson, M Hillert (eds.), The Physical Metallurgy of Cast Iron. New York: North Holland, 1984, p. 181.
5. DM Stefanescu, C Kanetkar. In: H Fredriksson (ed.), State of the Art of Computer Simulation of Casting and Solidification Processes. Les Ulis, France: Les Edition de Physique, 1986, p. 255.
6. P Thévoz, JL Desbioles, M Rappaz. Metall Trans A, 20A:311, 1989.
7. CY Wang, C Beckermann. Metall Mater Trans A, 27A:2754, 1996.
8. J Ni, C Beckermann. Metall Trans B, 22B:349, 1991.
9. CS Kanetkar, IG Chen, DM Stefanescu, N El-Kaddah. Trans ISIJ 28:860, 1988.
10. M Rappaz, P Thévoz. Acta Metall 35:1487, 2929, 1987.
11. L Nastac, DM Stefanescu. In: C Beckermann et al. (eds.), Micro/Macro Scale Phenomena in Solidification. New York: ASME, 1992, p. 27.
12. PS Sahni, GS Grest, MP Anderson, DJ Srolovitz. Phys Rev Lett 50:263, 1983.
13. MP Anderson, DJ Srolovitz, GS Grest, PS Sahni. Acta Metall 32, 5:783, 1984.
14. HW Hesselbarth, IR Göbel. Acta Metall 39:2135, 1991.
15. M Rappaz, CA Gandin. Acta Metall Mater 41, no. 2:345, 1993.
16. M Rappaz, CA Gandin, C Charbon. In: Solidification and Properties of Cast Alloys, Proceedings of the Technical Forum, 61st World Foundry Congress. Beijing: Giesserei-Verlag, 1995, p. 49.
17. JA Spittle, SGR Brown. Acta Metall 37:1803, 1989.
18. R Xiao, JID Alexander, F Rosenberg. Phys Rev A, 45,1:571, 1992.
19. P Zhu, RW Smith. Acta Metall Mater 40:683, 3369, 1992.
20. W Oldfield. ASM Trans. 59:945, 1966.
21. M Rappaz, CA Gandin, JL Desbiolles, P Thévoz. Metall Mat Trans 27A:695, 1996.
22. CA Gandin, T Jalanti, M Rappaz. In: BG Thomas, C Beckermann (eds.), Modeling of Casting, Welding and Advanced Solidification Processes VIII. Warrendale, PA: TMS, 1998, p. 363.
23. SY Lee, SM Lee, CP Hong. In: BG Thomas, C Beckermann (eds.), Modeling of Casting, Welding and Advanced Solidification Processes VIII. Warrendale, PA TMS, 1998, p. 383.
24. DM Stefanescu, H Pang. Canadian Metallurgical Quarterly 37(3–4):229–239, 1998.
25. L Nastac. Acta Metall Mater 47:4253, 1999.
26. L Nastac, DM Stefanescu. Modelling and Simulation in Mat. Sci. and Eng., Inst. of Physics Publishing 5(4):391, 1997.
27. R Sasikumar, R Sreenivisan. Acta Metall 42(7):2381, 1994.
28. U. Dilthey, V Pavlik. In: BG Thomas, C Beckermann (eds.), Modeling of Casting, Welding and Advanced Solidification Processes. VIII. Warrendale, PA: TMS, 1998, p. 589.

29. F Mampaey. In: M Rappaz, MR Ozgu (eds.), Modeling of Casting, Welding and Advanced Solidification Processes VI. Warrendale, PA: TMS, 1991, p. 403.
30. DD Goettsch, JA Dantzig. Metall Mat Trans 25A:1063, 1994.
31. DM Stefanescu, G Upadhya, D Bandyopadhyay. Metall Trans 21A:997, 1990.
32. JD Hunt. Mat Sci Eng 65:75, 1984.
33. H Tian, DM Stefanescu. Metall Trans. A, 23A:681, 1992.
34. RS Steube, A Hellawell. In: C Beckermann et al. (eds.), Micro/Macro Scale Phenomena in Solidification. New York: Am. Soc. Mech. Eng., HTD-vol. 218, AMD—vol. 139, 1992, p. 73.
35. JAI Ortega, J Beech. In: M Cross, J Campbell (eds.), Modeling of Casting, Welding and Advanced Solidification Processes VII. Warrendale, PA: TMS, 1995 p. 117.
36. E Scheil. Z Metallk 34:70, 1942.
37. HD Brody, MC Flemings. Trans Met Soc AIME 236:615, 1966.
38. I Ohnaka. Trans Iron Steel Inst Jpn 26:1045, 1986.
39. TW Clyne, W Kurz. Metall Trans A, 12A:965, 1981.
40. S Kobayashi. Trans Iron Steel Inst Jpn 28:728, 1988.
41. L Nastac, DM Stefanescu. Metall. Trans A, 24A:2107–2118, 1993.
42. AA Wheeler, BT Murray, RJ Schaefer. Physica D 66:243, 1993.
43. J Warren, WJ Boettinger. In: BG Thomas, C Beckerman (eds.), Modeling of Casting, Welding and Advanced Solidification Processes VIII. Warrendale PA: TMS, 1998, pp. 613–620.
44. X Tong, C Beckermann, A Karma. In: M Cross, J Campbell (eds.), Modeling of Casting, Welding and Advanced Solidification Processes VII. Warrendale PA: TMS, 1995, p. 601.
45. W Kurz, DJ Fisher. Fundamentals of Solidification. 3rd ed. Trans Tech Publications, 1989.
46. HL Lukas, J Weiss, ET Hening. CALPHAD 6:229, 1982.
47. WJ Boettinger, UR Kottner, DK Banerjee. In: BG Thomas, C Beckerman (eds.), Modeling of Casting, Welding and Advanced Solidification Processes VIII. Warrendale PA: TMS, 1998, pp. 159–170.
48. L Nastac, JS Chou, Y Pang. Symposium on Liquid Metal Processing and Casting. Santa Fe, NM, 1999.
49. L Nastac, DM Stefanescu. Metall Trans A, 27A:4061, 4075, 1996.
50. CY Wang, C Beckermann. Metall Trans A, 24A:2787, 1993
51. CY Wang, C Beckerman. Metall Trans A, 25A:1081, 1994.
52. CY Wang, C Beckermann. Metall Mater Trans A, 27A:2754, 1996.
53. H Jacobi, K Schwerdtfeger. Metall Trans 7A:811, 1976.
54. YK Chuang, D Reininsch, K Schwerdfeger. Metall Trans 6A:235, 1975.
55. H Fredriksson, J Stjerndahl. Metall Trans B, 6B:661, 1975.
56. J Zou, AA Tseng. Metall Trans A, 23A:457, 1992.
57. AA Howe. Applied Scientific Research 44:51, 1987.
58. MC Schneider, C Beckermann. ISIJ International 35(6):665, 1995.
59. M Wintz, M Bobadilla, J Lehmann, H Gaye. ISIJ International 35(6):715, 1995.

60. H Tian, DM Stefanescu. In: TS Piwonka, V Voller, L Katgerman (eds.), Modeling of Casting, Welding and Advanced Solidification Processes VI. Warrendale PA: TMS, 1993, p. 639.
61. F Mampaey. Proc. 62nd World Foundry Congress, Philadelphia, paper 4.
62. I Minkoff. The Physical Metallurgy of Cast Iron. New York: John Wiley & Sons, 1983.
63. R Elliott. Cast Iron Technology. London: Butterworth, 1988.
64. DM Stefanescu. In: ASM Handbook, vol. 15, Casting. Metals Park, Ohio: ASM International, 1988, pp. 168–181.
65. A Rickert, S Engler. In: H Fredriksson, M Hillert (eds.), The Physical Metallurgy of Cast Iron. Proceedings of the Materials Research Society. North Holland, 34, 1985, p. 165.
66. DK Bandyopadhyay, DM Stefanescu, I Minkoff, SK Biswal. In: G Ohira, T Kusakawa, E Niyama (eds.), Physical Metallurgy of Cast Iron IV. Pittsburgh, PA.: Mat. Res. Soc. Proc., 1989, p. 27.
67. E Fras, W Kapturkiewicz, AA Burbielko. In: TS Piwonka, V Voller, L Katgerman (eds.), Modeling of Casting, Welding and Advanced Solidification Processes VI. Warrendale, PA: TMS, 1993, p. 261.
68. J Zou, M Rappaz. In: VR Voller et al. (eds.), Materials Processing in the Computer Age. Warrendale, PA: The Minerals, Metals Materials Soc, 1991, p. 335.
69. AV Catalina, DM Stefanescu. Metall Trans A, 27A:4205, 1996.
70. B Leube, L Arnberg, R Mai. In: BG Thomas, C Beckerman (eds.), Modeling of Casting, Welding and Advanced Solidification Processes VIII. Warrendale, PA: TMS, 1998, p. 463.
71. F Mampaey. 55th International Foundry Congress. Moscow: CIATF, 1988, paper 2.
72. T Skaland, F Grong, T Grong. Metall Trans 24A:2321, 1993.
73. SE Wetterfall, H Fredriksson, M Hillert. J Iron and Steel Inst, p. 323, 1972.
74. KC Su, I Ohnaka, I Yamauchi, T Fukusako. In: H Fredriksson, M Hillert (eds.), The Physical Metallurgy of Cast Iron. Proceedings of the Materials Research Society. North Holland, 34, 1985, p. 181.
75. IL Svensson, M Wessen. In: BG Thomas, C Beckerman (eds.), Modeling of Casting, Welding and Advanced Solidification Processes VIII. Warrendale, PA: TMS, 1998, p. 443.
76. S Chang, D Shangguan, DM Stefanescu. Metall Trans 23A:1333, 1992.
77. Campbell PC, FD Hawholt, IK Drimacombe. Metall Trans A, 22A:2791, 1991.
78. J Lacaze, M Castro, C Selig, G Lesoult. In: M Rappaz, MR Ozgu, KW Mahin (eds.), Modeling of Casting, Welding and Advanced Solidification Processes V. Warrendale, PA: TMS, 1991, p. 473.
79. H Fredriksson, I Svensson. In: DM Stefanescu, GJ Abbaschian, RJ Bayuzick (eds.), Solidification Processing of Eutectic Alloys. Warrendale, PA: Metallurgical Soc, 1988, p. 153.
80. H Fredriksson, JT Thorgrimsson, I Svensson. In: H Fredriksson (ed.), State of the Art of Computer Simulation of Casting and Solidification Processes. Les Ulis, France: Les Edition de Physique, 1986, p. 267.

81. G Upadhya, DK Banerjee, DM Stefanescu, JL Hill. AFS Trans, 62:699, 1990.
82. S Hiratsuka, E Niyama, K Anzai, H Horie, T Kowata, M Nakamura. In: BC Liu, T Jing (eds.), Proceedings of the 3rd Pacific Rim Int. Conf. on Modeling of Casting and Solidification Processes. 1996, p. 130.
83. L Nastac, DM Stefanescu. Trans AFS 103:329, 1995.
84. DM Stefanescu, CS Kanetkar. In: DJ Srolovitz (ed.), Computer Simulation of Microstructural Evolution. Warrendale, PA: Metallurgical Soc, 1985, p. 171.
85. A Catalina, X Guo, DM Stefanescu, L Chuzhoy, MA Pershing, GL Biltgen. In: BG Thomas, C Beckerman (eds.), Modeling of Casting, Welding and Advanced Solidification Processes VIII. Warrendale, PA: TMS, 1998, p. 455.
86. D Venugopalan. Met Trans 21A:913, 1990.
87. V Gerval, J Lacaze. In: J Beech, H Jones (eds.), Solidification Processing 1997. Warrendale, PA: TMS, 1997, p. 506.
88. M Wessen, I Svensson. Met Trans 27A:2209, 1996.
89. H Nakae, T Katsuyama, N Hashihara. In: 62nd World Foundry Congress. Philadelphia, PA: AFS, 1996.
90. IL Svensson, M Wessen, A Gonzalez. In: TS Piwonka, V Voller, L Katgerman (eds.), Modeling of Casting, Welding and Advanced Solidification Processes VI. Warrendale, PA: TMS, 1993, p. 29.
91. DJ Fisher, W Kurz. Acta Met 28:777, 1980.
92. C Degand, DM Stefanescu, G Laslaz. In: I Ohnaka, DM Stefanescu (eds.), Solidification Science and Processing. Warrendale, PA: TMS, 1996, p. 55.
93. P Magnin, R Trivedi. Acta Met Mat 39(4):453, 1991.
94. WA Tiller. In: Solidification. American Society for Metals, 1969, p. 84.
95. WJ Boettinger, UR Kattner, SR Coriell, YA Chang, B Mueller. In: M Cross, J Campbell (eds.), Modeling of Casting, Welding and Advanced Solidification Processes VII. Warrendale, PA: TMS, 1995, p. 649.
96. C Charbon, M Rappaz. Advanced Materials Research. 4–5:453, 1997.
97. IS Cho HF Shen, CP Hong. In: BC Liu, T Jing (eds.), Proceedings of the 3rd Pacific Rim Int. Conf. on Modeling of Casting and Solidification Processes, 1996, p. 19.
98. JL Desbiolles, CA Gandin, JF Joyeux, M Rappaz, P Thévoz. In: BG Thomas, C Beckerman (eds.), Modeling of Casting, Welding and Advanced Solidification Processes VIII. Warrendale, PA: TMS, 1998, p. 433.
99. GK Upadhya, KO Yu, MA Layton, AJ Paul. In: M Cross, J Campbell (eds.), Modeling of Casting, Welding and Advanced Solidification Processes VII. Warrendale, PA: TMS, 1995, p. 517.
100. PR Sahm. In: C Kim, CW Kim (eds.), Numerical Simulation of Casting Solidification in Automotive Applications. Warrendale, PA: TMS, 1991, p. 45.
101. I Maxwell, A Hellawell. Acta Metall 23:229, 1975.
102. JA Sarreal, GJ Abbaschian. Metall Trans A, 17A:2863, 1986.
103. DM Stefanescu, PA Curreri, MR Fiske. Metall Trans 17A:1121, 1986.
104. G Upadhya, DK Banerjee, DM Stefanescu, JL Hill. Trans AFS 98:699, 1990.

105. MF Basdogan, V Kondic, GHJ Bennett. Trans AFS 90:263, 1982.
106. KD Lakeland. BCIRA J 12:634, 1964.
107. H Nieswaag, AJ Zuithoff. In: B Lux, I Minkoff, F Mollard (eds.), Metallurgy of Cast Iron. Switzerland: Georgi Publishing, 1975, p. 327.
108. R Boeri, F Weinberg. Trans AFS 89:179, 1989.

6
Thermophysical Properties

Juan J. Valencia
Concurrent Technologies Corporation, Johnstown, Pennsylvania

Kuang-O (Oscar) Yu
RMI Titanium Company, Niles, Ohio

I. INTRODUCTION

The casting industry around the wold is applying advanced computer simulation technology to understand the critical aspects of heat transfer and fluid transport phenomena and their relationships to metallurgical structures and formation of defects in metal casting processes. The computational tools are enabling the design and production of more economical and higher quality castings. However, the required thermophysical properties input data that is necessary for accurate simulations are frequently unreliable or nonexistent for many alloys of commercial interest. Accurate casting models require integrated and self-consistent thermophysical property data sets for reliable simulation of the complex solidification processes.

Sand, ceramic, and metal molds are extensively used to cast most metals. During the solidification process, the predominant resistance to heat flow is within the mold/metal interface and the mold itself; thus, the primary interest is not the mold thermal history, but the rate at which the heat is extracted from the solidifying metal. Therefore, the heat transfer is the governing phenomena in any solidification process. However, the heat transfer is fundamentally described by conduction, convection, heat transfer coefficient, geometry of the system, and thermophysical properties of both metal and mold material. Analytical and numerical solutions to the heat transfer equations that describe the solidification in metal casting processes are described in Chapter 2 of this book.

Table 1 shows the required thermophysical properties that must be available for input before reliable modeling of a casting process can be accomplished, as well as their influence in the prediction of defects [1]. These data must be evaluated from room temperature through the mushy zone, and to a given superheat. Current commercial software requires that the thermal conductivity, specific heat, latent heat, solidus and liquidus temperatures, and density must be known for heat transfer operations. Viscosity, wetting angle, and surface tension of the molten alloy are required for fluid flow operations. In addition to the metal properties, mold materials properties are also needed to conduct an effective modeling. However, some of needed properties are more critical than others, but this depends on the specific casting process being modeled. For example, the modeling of thin wall castings requires the input of the surface tension and mold wetting angles. Thus, an accurate determination of these properties will be required to simulate the filling of the thin walls in precision aerospace and automotive castings.

Accurate, consistent, and reliable thermophysical property measurements are experimentally difficult. Convection effects in molten samples and their interactions and reactivity with their containers often exacerbate the difficulties.

This chapter presents an overview of the various empirical and semiempirical equations obtained from laboratory measurements and estimation relations that are currently utilized to determine thermophysical properties needed in casting processes.

II. AVAILABILITY OF RELIABLE DATA

Many difficulties are usually encountered in the experimental determination and theoretical estimation of reliable thermophysical properties. In the solid state the property recorded in the technical literature are often widely diverging, conflicting, and subject to large uncertainties. This problem is not only apparent for solid materials, but is particularly acute for materials in the semisolid and liquid state.

The use of thermophysical property data found in the literature for engineering and design calculations of casting processes must not be used indiscriminately without knowing their reliability. It is dangerous to assume that the available thermophysical properties are reliable, since this may cause inefficient computing simulation of the casting process. For instance, studies on thermophysical property effects in steel solidification in green sand molds indicated that errors in the predicting solidification time are more sensitive to the density than to the thermal conductivity errors of the molten metal. Also, significant errors in the solidification time are obtained with errors in the sand properties.

Figures 1 and 2 illustrate these effects [2]. These examples show that variations in the thermophysical properties are serious sources of error. Therefore, prior to using a given set of data, it is very important to critically evaluate and analyze available thermophysical property data, to pass judgment on their reliability and accuracy.

The Center for Information and Numerical Data Analysis and Synthesis (CINDAS) identified, compiled, and performed a critical evaluation of available thermophysical data. CINDAS for over 24 years evaluated the data by using correlation analysis and synthesis of numerical data on the physical properties. This work led CINDAS to generate and provide recommended reference values for diverse materials. CINDAS work is compiled in a number of major publications [3–12].

Recently computer models that can calculate thermophysical properties for various materials in the solid and liquid states have been developed [13–15]. These models that are based on first principles of thermodynamics and kinetics of phase transformations take into account temperature and chemical composition. However, their use is still limited due to the lack of thermodynamic data for materials of industrial interest. Also, sensitivity studies are necessary to truly evaluate the reliability of calculated thermophysical property data from these models in actual casting processes.

III. SPECIFIC HEAT, LATENT HEAT OF TRANSFORMATION, SOLIDUS AND LIQUIDUS TEMPERATURES

A. Specific Heat

Heat capacity is the amount of heat needed to raise the temperature of a given sample of material by 1°C. The amount of heat transferred depends on the mass of the sample. To compare different materials, however, the heat capacities of a unit mass (1 g), i.e., specific heat, need to be used. The specific heat is an extensive property that depends on the amount of species of the system. Specific heat can be defined as a constant volume C_v or as a constant pressure C_p. Theoretical calculation of the specific heat of solid elements as function of temperature, was one of the early achievements of the quantum theory, and is well documented in the literature [16]. However, theoretical and experimental information on the specific heat for metallic alloys of commercial interest is very limited. Also, these data is practically non-existent for alloys being cooled from the liquid through the semisolid state.

Specific heats, particularly at a constant pressure C_p are either measured directly or determined from relative enthalpy H measurements from the following equation:

Table 1 Thermophysical Property Data Required for Metal Casting

Casting process component	Transport phenomena for casting	Thermophysical data required	Computer modeling for process and part design and defect prediction
Furnace Metal	Heat transfer • Conduction • Convection • Radiation	Heat transfer coefficient • Metal/Mold • Metal/Core • Metal/Chill • Mold/Chill • Mold/Environment Emissivity — Metal/Mold/furnace wall Temperature Dependent Parameters • Density • Specific Heat • Conductivity Latent Heat of Fusion Liquidus Solidus	Effective design for • Riser • Chill • Insulation Solidification Direction Solidification Shrinkage Porosity Hotspots
Mold Core Chill	Mass transfer (Fluid flow)	Temperature Dependent • Viscosity • Surface Tension • Density	Effective Design For: • Ingate • Runner • Vents Pouring Parameters • Temperature and • Pouring Rate Mold Filling Time Cold Shut Missruns

SOLIDIFICATION

Insulation

S Microstructure
O Evolution
L
I
D
I
F
I
C
A
T
I
O
N

T Stress Analysis
I
O
N

Phase diagram
Phase Chemical
Composition
Capilarity effect
(Gibbs Thompson coefficient)
Nucleation & Growth
Parameters
Solid Fraction vs Temperature
  Diffusivity
• Solubility

Temperature Dependent Parameters
• Coefficient of Thermal Expansion
• Stress/Strain

Microsegregation
Macrosegregation
Grain size
Grain orientation
Phase morphology
Mechanical properties

Casting Design for:
• Dimension and Distortion
Internal Stresses
Hot Tears and Hot Cracks

Source: Ref. 1.

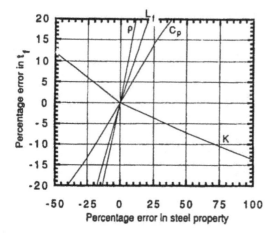

Figure 1 Effect of the error percentage of molten metal thermophysical properties on the steel solidification time. (From Ref. 2.)

$$C_p = \left[\partial \left(\frac{H_T - H_{T0}}{\partial T} \right) \right]_p \qquad (1)$$

In any case, one actually measures the change in enthalpy resulting from a given temperature increase. In metallurgical processes, specific heat can be measured from room temperature to several thousand degrees. An excellent analysis of the various measurement techniques to measure the specific heat can be found in the literature [17]. However, it should be indicated that the

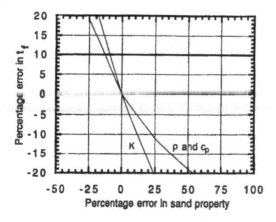

Figure 2 Effect of the error percentage of the sand thermophysical properties on the solidification time of steel. (From Ref. 2.)

difference between relative enthalpy measurements by drop calorimetry and direct C_p measurements by adiabatic or differential scanning calorimetry (DSC) is that in the DSC, only the change in temperature is sufficiently small that the differential of Eq. (1) can be approximated by a finite difference. Because of this reason, the DSC method has rapidly become a very popular method to measure specific heats of metallic materials in a wide range of temperatures in the solid and liquid states.

The DSC is a commercial technique that uses a single heating chamber for both the unknown sample and the standard reference material. The apparatus is careful calibrated with pure metals or sapphire so that the differential thermocouple voltage from the two are directly related to enthalpy differences. DSC instrumentation control, data acquisition, and data evaluation are accomplished by standard computer software. This allows for determination and computation of onset, peak and end of phase transformation temperatures, partial area integration, apparent specific heat, enthalpies of phase transformations, etc. that may occur during heating and/or cooling.

Apparent specific heat is determined by first running a base line over the temperature range of interest with the sample and standard crucibles both empty to establish a zero line for the instrument. The standard reference material and the sample are then run and the apparent specific heat is calculated by the ratio method; that is,

$$C_{p,s} = C_{Pstd} \frac{M_{std}}{M_s} \frac{\Delta V_s}{\Delta V_{std}} \tag{2}$$

where $C_{p,s}$ = sample apparent specific heat, $C_{p,std}$ = apparent specific heat of the standard, M_{std} = mass of the standard, M_s = mass of the sample, ΔV_s = differential potential (in microvolts) of the signal between the base line and the sample and, ΔV_{std} = differential potential (in microvolts) signal between the base line and the standard.

Although, the DSC was designed for a rapid determination of the specific heat, there are some limitations of this technique. Accurate specific heats to 1% are usually achieved to about 725°C. However, as the temperature increases, for instance, to approximately 1500°C, the relative error of the specific heat is of the order of 5% or larger. Other important limitations of the DSC technique are the size of the sample, which is only a few tens of milligrams, the surface tension, and the reactivity of the sample with the container. This is particularly important when dealing with liquid metals. The adequate selection of sample containers and multiplicity of DSC runs to assure repeatability of the data obtained can overcome some of these limitations.

Specific heat data needed for casting processes includes data for the solidifying alloy, mold, material and core material, which are heated by both the release of metal superheat and latent heat of solidification. Relatively small variations in the measured specific heat with heating and cooling rates are observed for both solid and liquid states. However, the kinetics of phase transformations is affected by the rate of heating and cooling. This in turn will produce a shift on the specific heat with temperature. Also, major changes on the specific heat with the heating or cooling rates are observed in the solid-liquid range (mushy region). An example for these effects are given by the DSC curves in Figs. 3 and 4 taken at various heating and cooling rates, respectively, for an aluminum bronze alloy [18].

Note from Figs. 3 and 4 that specific heats of the solid and mushy region do not follow a simple behavior, as that given by an empirical relationship shown by Eq. (3). Instead, the behavior is more complex. The complexity is due to the phase transformations that occurred during heating and/or cooling.

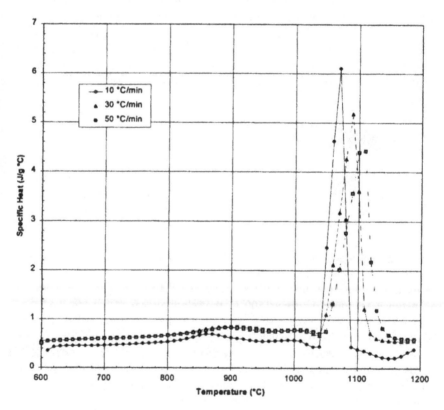

Figure 3 Apparent specific heat for an aluminum bronze alloy determined by DSC at various heating rates. (From Ref. 18.)

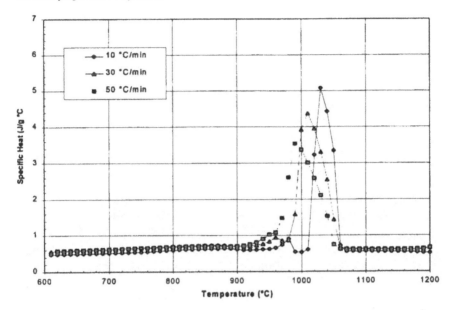

Figure 4 Apparent specific heat for an aluminum bronze alloy determined by DSC at various cooling rates. (From Ref. 18.)

These transformations can also be very complex and are dependent on the heating and cooling rates and on the chemistry of alloy system.

$$C_p = a + bT + cT^2 + dt^{0.5} + \cdots \tag{3}$$

where C_p is the molar specific heat; a, b, c, and d are constants; and T is the temperature.

The specific heat of a liquid metal can also be described by Eq. (3). However, the liquid behavior is simpler, since, in the liquid state, the materials structure consists of temporary, loose aggregations of atoms or molecules that are rapidly reconfigured by thermal energy. The structure dependent aspects of thermophysical properties, such as those associated with phase changes, are therefore largely absent. Changes in specific heat and most thermophysical properties with changing temperature in liquid metals may be gradual and continuous, rather than showing the abrupt effects of phase transitions that take place in the semisolid and solid states. Thus, a reasonable estimate of the specific heat for a liquid alloy $C_{p,L}$ can be calculated from the atomic specific heats of the components of the alloy by using the commonly known Kopp-Neuman rule of mixtures.

$$C_{p,L} = \sum_{i=1}^{n} Y_i C_{p,Li} \tag{4}$$

where $C_{p,L}$ is given in cal/mol K and Y_i is the atomic fraction of element i in the alloy. Equation (4) can also be expressed in weight percent or it can be converted from cal/mol K to weight percent and SI units by the relationship.

$$C_{p,L} = 41.8585 \sum_{i=1}^{n} \left[\frac{C_{p,Li}}{W_i} \right] X_i \tag{5}$$

where $C_{p,L}$ is the specific heat in J/(kg K), $C_{p,Li}$ specific heat cal/mol K, W_i is the atomic mass, and X_i is the weight percentage of component i in the liquid alloy.

As a note of reference the Kopp-Neuman rule has been applied successfully for determining the specific heat of intermetallic compounds in a relatively wide range of temperatures and for several alloy systems. Specific heat data for these systems can be found in reference [19]. In casting processes, the formation of intermetallics is important to consider, since these compounds may appear as eutectics and can coexist with other solid and liquid phases during cooling, thus affecting the overall heat balance of the solidification process.

Table 2 shows the specific heat for selected liquid metals at the melting point T_M and at a given temperature [19, 20]. It can be seen that there are very little changes in the specific heat of the liquid with temperature. Therefore, a reasonable estimate ($\pm 3\%$) can be accomplished using Eqs. (4) and (5). However, the dynamic characteristics of a given casting process require the input of all liquid to solid changes to understand the behavior of the solidifying metal. Thus, in casting processes, determination of the apparent specific heat must be conducted on cooling for the solidifying metal, and on heating for the mold and core materials.

1. Selected Specific Heat Values for Commercial Alloys

The treatment of the specific heat data indicated above represents a "zeroth-order approximation." This is a limited approach due to the complexity of the commercial alloy systems. However, it can be used as a first estimation of the liquid specific heat. Pehlke and coworkers compiled specific heat data for some commercially cast alloys [21]. Some of the data is shown in Table 3. The latent heat of fusion and solidus and liquidus temperatures are also shown in the table. These parameters will be discussed later.

Table 2 Selected Specific Heat Values for Liquid Metals at the Melting Point and a Given Temperature

Element	T_m (K)	C_{pL} (J/g-K)	T (K)	$C_{p,L,T}$ (J/kg-K)	$\Delta H_f^{(20)}$ (Btu/lb)
Al	933	1.08	1173	1.08	167.5
Co	1766	0.59	N/A	N/A	115.2
Cr	2176	0.78	N/A	N/A	—
Cu	1356	0.495	1473	0.495	75.6
Fe	1809	0.795	1900	0.790	117
Mg	924	1.36	1073	1.36	160
Mn	1514	0.838	N/A	N/A	115
Mo	2880	0.57	N/A	N/A	126
Nb	2741	0.334	N/A	N/A	—
Ni	1727	0.620	1900	0.656	133
Si	1683	1.04	1873	1.04	607
Sn	505	0.250	773	0.240	25
Ti	1958	0.700			
Zn	692.5	0.481	873	0.481	46.8

Source: Refs. 19, 20.

2. Selected Specific Heat Values for Mold Materials

The ability of the mold material to absorb heat at a certain rate during casting and solidification of a metal is called heat diffusivity α_m. This parameter is the combination of the thermal conductivity k_m, density ρ_m, and specific heat $C_{p,m}$ of the mold material ($\alpha_m = k_m/\rho_m C_{p,m}$). As the metal releases both superheat and latent heat of solidification, the mold absorbs most of the heat; thus its temperature increases. Therefore, the determination and application of the mold thermophysical properties must be conducted as a function of the temperature. Unfortunately, there is limited thermophysical property data available in the literature for mold materials of interest in the casting industry.

Pehlke and coworkers have compiled thermal diffusivity, density, and specific heat for some commonly used in sand molds [21]. The compiled data is shown in Table 4.

B. Latent Heat of Fusion/Solidification and Solidus and Liquidus Temperatures

At the temperature of fusion of a pure metal the heat supplied to the system will not increase its temperature but will be used in supplying the latent heat of fusion ΔH_f that is required to convert solid into liquid. The opposite could be

Table 3 Specific Heat Data, Latent Heat of Fusion, Solidus and Liquidus Temperaures for Selected Casting Alloys of Commercial Interest

Alloy	Specific heat (kJ/kg K)	Temperature range (K)	ΔH_f (kJ/kg)	T_S (K)	T_L (K)
1 Cr Steel					
0.31C–2Si– 0.69 Mn– 0.073Ni– 0.012Mo–1.1Cr	$C_p = 0.436 + 1.22 \times 10^{-3}T$ $C_p = 0.856$	$1173 \leq I \geq 1683$ $T > 1743$	251	1693	1743
Copper Alloys					
90Cu–10Al	$C_p = 2.746 + 1.667 \times 10^{-3}T$ $C_p = 0.554$	$1303 \leq T \geq 1315$ $T \geq 1315$	228.62	1303	1315
70Cu–30Zn	$C_p = 1.32 + 6.75 \times 10^{-4}T$ $C_p = 0.49$	$1188 \leq T \geq 1228$ $T \geq 1228$	164.8	1188	1228
60Cu–40Zn	$C_p = -0.689 + 1.0 \times 10^{-3}T$ $C_p = 0.489$	$1173 \leq T \geq 1178$ $T \geq 1178$	160.15	1173	1178
70Cu–30Ni	$C_p = 0.348 + 1.286 \times 10^{-4}T$ $C_p = 0.543$	$1443 \leq T \geq 1153$ $T \geq 1513$	N/A	1443	1513
Aluminum Alloys					
A356	0.92 @ 293 K 1.2 @ 823 K 1.1 @ 901 K		440	823	901
Al–4.5Cu	$C_p = 1.287 - 2.5 \times 10^{-4}T$ $C_p = 1.059$	$775 < T > 991$ $T > 911$	N/A	775	991
Al–8Mg	$C_p = 1.583 - 5.29 \times 10^{-4}T$ $C_p = 1.108$	$813 < T > 898$ $T > 898$	N/A	813	898
Magnesium Alloys					
AZ31B Mg–3Al–1Zn– 0.5Mn	$C_p = 1.88 - 5.22 \times 10^{-4}T$ $C_p = 0.979 + 4.73 \times 10^{-4}T$	$839 \leq T \leq 905$ $T \geq 905$		839	905
AZ91B Mg–9A1–0.6Zn– 0.2Mn	$C_p = 0.251 + 1.354 \times 10^{-3}T$ $C_p = 1.43$	$742 \leq T \leq 869$ $T \geq 869$		742	869
KIA Mg–0.7Zr	$C_p = -116.72 + 0.128T$ $C_p = 1.43$	$922 \leq T \leq 923$ $T \geq 923$		922	923
ZK51A Mg–4.6Zn–0.7Zr	$C_p = 1.945 - 6.28 \times 10^{-4}T$ $C_p = 0.90 + 0.516 \times 10^{-3}T$	$822 < T < 914$ $T \geq 914$		822	914
Nickel Alloys					
Hastealloy C	$C_p = 0.281 + 0.283 \times 10^{-3}T$			1534	1578
Inconel 718	$C_p = 0.34 + 0.273 \times 10^{-3}T$ $C_p = 0.725$	$294 \leq T \leq 1423$ $T \geq 1610$	210	1425	1610

Source: Ref. 21.

Table 4 Specific Heat and Density of Mold Materials

Mold material	Dry density (g/cm³)	Specific heat (kJ/kg K)	Temperature range (K)
Silica molding sands			
20–30 Mesh	1.730	$C_p = 0.5472 - 1.147 \times 10^{-3}T - 5.401 \times 10^{-7}T$	$T < 1033$
50–70 Mesh	1.634	$C_p = 1.066 + 8.676 \times 10^{-5}T$	$T > 1033$
80–120 Mesh	1.458		
Silica sand A + 7% Bentonite	1.520	$C_p = 0.4071(T - 273)^{0.154}$	
Olivine sand (5.9% Bentonite)	1.830	$C_p = 0.3891(T - 273)^{0.162}$	
Zircon sand (3.8% Bentonite)	2.780	$C_p = 0.2519(T - 273)^{0.170}$	
Chromite sand (3.9% Bentonite)	2.75	$C_p = 0.3180(T - 273)^{0.158}$	
Graphite (chill foundry grade)	1.922	$C_p = -0.11511 + 2.8168 \times 10^{-3}T$	$T < 505$
		$C_p = 0.6484 + 1.305 \times 10^{-3}T$	$505 < T < 811$
		$C_p = 1.3596 + 4.2797 \times 10^{-4}T$	$T > 811$

said if the system is solidified from the liquid state. Under equilibrium conditions the latent heat of fusion and the latent heat of solidification of pure elements is essentially the same.

The latent heat of fusion and solidification in a binary or multicomponent alloy system occur in a temperature range. On heating, the temperature at which the alloy starts to melt is called *solidus temperature*, and the temperature at which the melting is completed is called *liquidus temperature*. On cooling, the temperature at which the solid starts to form is the liquidus, and the temperature at which the solidification ends is the solidus. Under equilibrium conditions the latent heat of fusion and solidification, as well as the solidus and liquidus temperatures, are essentially the same.

In actual melting and casting processes equilibrium conditions are not followed. This is because during melting the endothermic phase transition is also accompanied by dissolution of the solid phase into the liquid. These processes are ruled by the heat and mass transfer phenomena, which are functions of time. Thus, high heating rates may displace the solidus and liquidus temperatures to high values, as observed in Fig. 3 for an Al-bronze alloy. But on cooling prior to the nucleation of the solid phase, the molten alloy is usually undercooled. The degree of undercooling has a direct effect on the kinetics of the liquid-solid phase transformation and the type of second phases that evolve during solidification. For instance, high undercooling decreases the liquidus and solidus temperatures and may lead to metastable phases, which in turn may also affect the latent heat of solidification.

During solidification the type of structure chosen by a particular system will be that which minimizes the interfacial free energy. According to the simple theory developed by Jackson [22] the optimum atomic arrangements depend mainly on the latent heat of fusion relative to the melting temperature. This theory predicts that there is a critical value of $\Delta H_f / T_m \approx 4R$. A flat interface occurs at $\Delta H_f / T_m$ values larger than $4R$, while values below $4R$ predict a diffuse interface. Most metals have $\Delta H_f / T_m \approx R$ and are therefore predicted to have rough interfaces. On the other hand, some intermetallic compounds and elements such as Si, Ge, Sb, and most nonmetals have high values of $\Delta H_f / T_m$ and generally have closed packed interfaces.

The latent heat, solidus, liquidus, and other phase transformation temperatures are determined using the same techniques as for specific heat and are described elsewhere in the literature [17]. In the case of solidification, the latent heat is easily determined by integrating the apparent specific curve from the beginning to the end of solidification and subtracting the baseline contribution due to the actual specific heat. In addition, the latent heat of an unknown can also be evaluated from the sample with a known transformation enthalpy as

$$\Delta H_f = \Delta H_{\text{std}} \frac{C_{p,s} A_{p,s}}{C_{p,\text{std}} A_{p,\text{std}}} \qquad (6)$$

where $A_{p,\text{std}}$ = peak area of standard transformation, ΔH_{std} = standard transformation enthalpy, ΔH_f = sample latent heat, and $A_{p,s}$ = peak area of sample transformation.

The latent heat of fusion and solidus and liquidus temperatures for various alloys of commercial interest are shown in Table 3. Because of the effects of the heating and cooling rate on these properties, care should be taken when using these values. However, Pehlke's summary represents an initial step in the development of databases for supporting the simulation design for casting processes.

IV. COEFFICIENT OF THERMAL EXPANSION AND DENSITY

A. Coefficient of Thermal Expansion

Any change in temperature of a material has a direct effect on the volume that it occupies. A measure of this change is the coefficient of thermal expansion β, usually expressed by the relationship

$$\beta = \frac{1}{V} \left(\frac{\partial V}{\partial T} \right)_P \qquad (7)$$

where V is the volume at a temperature T at a constant pressure P.

The thermal expansion in solid metals is the result of atomic vibration and/or the motion of atoms around their equilibrium positions in the structure. The structure can expand differently in different directions depending on the type of crystal structure. The coefficient of linear thermal expansion in the ith direction is defined as

$$\alpha_1 = \frac{1}{L_1} \left(\frac{\partial L_1}{\partial T} \right)_P \qquad (8)$$

The sum of three orthogonal linear coefficients of a crystal is equal to its volume coefficient $\beta = \alpha_1 + \alpha_2 + \alpha_3$. The orientation effect is particularly important for single crystal materials such as nickel-base superalloys used in the aerospace industry. However, most metallic and ceramic materials of commercial interest are polycrystalline, and therefore anisotropic in nature.

The temperature dependence of β for solids is very complex but can be approximated by the Grüneisen equation:

$$\beta = \frac{\gamma C_p}{V B_a} \tag{9}$$

where γ is the Grüneisen parameter, C_p is the specific heat at a constant pressure, V is the volume of the solid, and B_a is the adiabatic bulk modulus. Since the product $\gamma/V B_a$ is nearly constant over a wide range of temperature, the value of β has about same temperature dependence as C_p, that is, at low temperatures $\beta \to 0$ as $T \to 0$ and at high temperatures β is slowly increasing. It has been found [19] that most metals expand about 7% on heating from 0 K to their melting temperature. Therefore, a metal with a low melting point T_M has a large coefficient of thermal expansion, and vice versa; some examples are shown in Table 5.

The thermal expansion of liquid metals just above their liquidus temperature or in the mushy region (between the solidus and liquidus) has a very complex behavior. This is due to the short-range structural ordering, which has been very difficult to represent by a theoretical formulation. However, the temperature dependence of β usually takes the empirical form:

$$\beta = a + bT + cT^2 \tag{10}$$

In general, many metals exhibit $\beta = 10^{-4}\,\mathrm{K}^{-1}$ in the liquid phase just above their melting point or liquidus temperature.

A variety of high-speed and noncontact measurement techniques have been employed to investigate the thermal expansion of molten materials at high

Table 5 Relationship Between Melting Temperature and Thermal Expansion for Some Selected Metals

Metal	T_m (K)	β 293–393 K ($10^{-6}\,\mathrm{K}^{-1}$)	ρ at 298 K g cm^{-3}
Zn	692.5	31.0	7.14
Mg	922	26.0	9.84
Al	933.3	23.5	2.70
Ag	1234.8	19.1	10.5
Cu	1356	17.0	8.96
Ni	1728	13.3	8.9
Co	1768	12.5	8.9
Fe	1809	12.1	7.87
Cr	2133	6.5	7.1
W	3693	4.5	19.3

Source: Ref. 19.

temperatures. Rapid resistive self-heating of specimens and a variety of diagnostics, such as high-speed optical pyrometry, laser interferometry, and laser photography, have been used to determine the expansion of both solid and liquid phases as the material is heated through melting [17]. However, thermal expansion in many materials is usually measured using push-rod dilatometers. Although these instruments are commercially available, they must be modified to study solidification phenomena. This is because molten metal interacts with the containers at high temperatures. For instance, with materials such as aluminum alloys that do not form carbides, a graphite piston arrangement has been used successfully. For reactive metals, high alumina or alumina coated with ytria or zirconia containers are being used.

Unfortunately, there is no thermal expansion data available in the open literature for materials of commercial interest in the semisolid or liquid state. However, solid state data is available for some alloys of commercial interest and can be found in the literature [19].

B. Density

Density ρ is defined as the weight per unit volume of a material. The reciprocal of the density $1/\rho$ is the specific volume. The density in the solid state is reasonably determined by the usual or Archimedes method at room temperature, and is available in the literature for most metals [19]. Liquid density data for pure metals can also be found in the literature [19, 23–24]. However, a narrow limit of accuracy exists only for low melting point metals and some of their alloys. The problem in acquiring accurate density data for high melting range alloys of commercial interest is their high temperature handling and reactivity with their containers and environment. Therefore, the availability of data is very limited or nonexistent.

Compared with other fundamental physical properties of liquid metals, such as surface tension and viscosity, density of nonreactive metals can be measured more accurately. To measure the density of liquid metals and their alloys, several techniques have been developed; their description and their limitations on accuracy can be found in the literature [24–28].

Accurate acquisition of density data is highly desirable. This is because density is a variable in the evaluation of the surface tension and viscosity. Further, the evaluation of fluid flow phenomena in a solidifying metal is dominated by the changes of density. Density of liquid metals is also useful to calculate volume changes during melting and solidification and those associated with alloying. However, as a general rule the average density change of non-closed-packed metals on fusion is approximately 3%, while the average volume change of a closed-packed metal does not exceed 5% [29]. Bismuth and gallium are the only exceptions to the general rule, since these elements contract on melting.

Theoretical density for pure metals in the solid state can be calculated from their atomic arrangement in the crystal structure, atomic weight, and atomic radius. However, large changes in density are usually observed in alloy systems. This is because the mass of the solute atom differs from that of the solvent and the lattice parameter usually changes with the alloying. Density of solid alloys as a function of temperature can be calculated from thermal expansion data using the following relationship.

$$\Delta L_{exp} = \frac{L_T - L_0}{L_0} = \alpha T$$

$$\rho_T = \frac{\rho_{RT}}{(1 + \Delta L_{exp})^3} \tag{11}$$

where ΔL_{exp} is the linear expansion and α is the coefficient of linear expansion at temperature T.

Alternatively, the density of a heterogeneous phase mixture ρ_m containing a number n of phases can be estimated using the empirical equation:

$$\rho_m = \frac{1}{\sum_{i=1}^{n}(X_p/\rho_p)} \tag{12}$$

where X_p and ρ_p are the fraction and density of the phase, respectively, at a given temperature.

The density of pure liquid metals as a function of temperature can be reasonably estimated from the following empirical equation [24]:

$$\rho_L = a + b(T - T_m) \tag{13}$$

where ρ_L is given in g/cm^3, a and b are constants, T is the temperature above the melting point, and T_m is the melting point. Both temperatures are in degrees Kelvin.

Selected values for the a and b temperature parameters for various elements were taken from the literature [24] and are given in Table 6. From Eq. (13) and Table 6, it is clear that $\rho_L = a$ at T_m.

The liquid density as a function of temperature of multicomponent systems such as commercial alloys could also be estimated using Eq. (14) and the rule of mixtures.

$$\rho_L = \sum_{i=1}^{n} Y_i \rho_{L,i} - \left[\left(\sum_{i=1}^{n} Y_i \frac{b}{a} \times 10^{-4} \right) \left(\sum_{i=1}^{n} Y_i \rho_{L,i} \right) \right] (T - T_L) \tag{14}$$

Table 6 Density Values at the Melting Point and Dimensionless Values for the Constants a and b for Selected Elements

Element	Melting point (K)	Measured density at T_m, g/cm^3 [19]	a	$-b \times 10^{-4}$
Ag	1233.7	9.346	9.329	10.51
Al	931	2.385	2.390	3.954
Au	1336	17.36	17.346	17.020
B	2448	2.08 at 2346 K	N/A	N/A
Be	1550	1.690	1.690	1.165
Bi	544	10.068	10.031	12.367
Cd	593	8.020	7.997	12.205
Ce	1060	6.685	6.689	2.270
Cr	2148	6.28	6.280	7.230
Co	1766	7.760	7.740	9.500
Cu	1356	8.000	8.033	7.953
Fe	1809	7.015	7.030	8.580
Hf	2216	11.10	N/A	N/A
La	1203	5.955	5.950	2.370
Li	453.5	0.525	0.5150	1.201
Mg	924	1.590	1.589	2.658
Mn	1525	5.730	5.750	9.300
Mo	2880	9.35	N/A	N/A
Nb	2741	7.830	N/A	N/A
Ni	1727	7.905	7.890	9.910
Pb	660.6	10.678	10.587	12.220
Pd	1825	10.490	10.495	12.416
Pt	2042	19.00	18.909	28.826
Re	3431	18.80	N/A	N/A
Sb	904	6.483	6.077	6.486
Si	1683	5.510	2.524	3.487
Sn	505	7.000	6.973	7.125
Ta	3250	15.00	N/A	N/A
Ti	1958	4.110	4.140	2.260
V	2185	5.700	5.36	3.20
W	3650	17.60	N/A	N/A
Zn	692	6.575	6.552	9.502
Zr	2123	5.800	N/A	N/A

Source: Ref. 24.

where ρ_L is the density of the alloy in g/cm^3, Y_i is the atomic fraction of element i in the alloy, a and b are constants, $T > T_L$, and T_L is the liquidus temperature in degrees Kelvin.

V. SURFACE TENSION

The plane of separation of two phases is known as a surface or interface. For a given system at a constant temperature and pressure, the increase in free energy per unit increase in surface area is the *surface tension*, or *interfacial tension*, σ. Liquid metals are notable for their high surface tensions.

Surface tension of liquid metals can be measured using various techniques, which can be found in the literature [30–33]. Surface tension can be related to the heat of vaporization per unit volume, as well as can be related to atomic volumes because metals with large atomic volumes usually have low energies of vaporization.

The fundamental theory of surface tension can be described as the free energy of a system containing an interface of area A and free energy σ per unit area, which can be represented by

$$G = G_0 + A\sigma \tag{15}$$

where G_0 is the free energy of the system assuming that all material in the system has the same properties of the bulk. Therefore the excess free energy is given by σ.

A thermodynamic model approach for multicomponent systems has been developed to calculate the surface tension of liquid alloys [34, 35]. This approach has been based on earlier work on surface tension prediction of binary [36] and ternary [37] solutions. Both calculation approaches were based on the Buttler's equation [38]. In order to calculate the surface tension of a liquid metal by using the Buttler's equation, the surface tensions, the surface areas of the pure constituent elements, and the excess Gibbs energy of the liquid metal must be known. The excess Gibbs energy is the same as that used for calculating the phase diagram and thermodynamic properties. In this thermodynamic approach the description of the alloy system must be established before any property of the multicomponent alloy can be calculated.

It should be noted that this thermodynamic approach to calculate the surface tension has been verified with the experimental data from two well-known Ag-Au-Cu and Cr-Fe-Ni systems. The results obtained were in excellent agreement, and the methodology was used to calculate the surface tension of titanium alloys [39].

We assume that in a multicomponent solution, both the surface and bulk phases are in thermodynamical equilibrium. Thus, the chemical potential of each component in the surface phase is equal to the chemical potential of the corresponding component in the bulk solution plus its surface energy. Based on this thermodynamic requirement, Buttler's equation was derived as

$$\sigma = \sigma_i^\circ + \frac{RT}{S_i} \ln \frac{a_i^m}{a_i} = \sigma_i^\circ + \frac{RT}{S_i} \ln \frac{x_i^m}{x_i} + \ln \frac{\gamma_i^m}{\gamma_i} \tag{16}$$

where σ and σ_i represent the surface tensions of the solution and the pure component i, respectively; S_i is the surface area of component i and it is related to the molar volume of component i, V_i by the expression

$$S_i = bN^{1/3}V^{2/3} \tag{17}$$

where b is a geometric factor (accounts for the atomic arrangement in the surface monolayer, $b = 1.09$ for hcp packing and $b = 1.12$ for bcc packing), N is Avogadro's number; and a_i, γ_i, x_i, and a_i^m, γ_i^m, x_i^m are the activity, activity coefficient, and mole fraction of component i in the bulk and surface phases, respectively.

The activity coefficient of component i in the bulk solution phase γ_i can be calculated from the partial excess Gibbs free energy of the corresponding component by the equation:

$$^{xs}\bar{G}_i = RT \ln \gamma_i \tag{18}$$

The generalized equation for the excess Gibbs free energy of a multicomponent system is given by the expression [34]:

$$^{xs}G(x_1, x_2, \ldots, x_n) = \sum_{i=1}^{n-1} \sum_{j=i+1}^{n} x_i x_j \left[A_{ij} + B_{ij}\left(x_i^{ij} - x_j^{ij}\right) + C_{ij}\left(X_i^{ij} - X_j^{ij}\right)^2 \right] \tag{19}$$

The model parameters for various binary systems are given in Table 7 [40–50]. Also, the activity coefficients γ_i and γ_i^m in the solution are related by

$$\ln \gamma_i^m = \frac{Z^m}{Z} \ln \gamma_i \quad \text{at} \quad x_i^m = x_i, \ i = 1, 2, \ldots, n \tag{20}$$

where Z^m and Z are the coordination numbers of the atoms in the surface and bulk phases, respectively.

Buttler's equation for a ternary system can be expressed in terms of three equations:

Table 7 Model Parameter for Liquid Binary Solutions
$^{xs}G_{ij} = X_{ii}^{ij}X_{ji}^{ij}[A_{ij} + B_{ij}(X_i^{ij} - X_j^{ij}) + C_{ij}(X_i^{ij} - X_j^{ij})^2]$ in Joules

System i-j	A_{ij} (T)	B_{ij} (T)	C_{ij} (T)	Reference
Ti–Al	$-112885.0 + 34.977T$	$-9746.9 - 71.87T$	0.0	[40]
Ti–V	13169.2	0.0	0.0	[40]
Al–V	$-45607.2 + 27.033T$	-1025.1	0.0	[40]
Ti–Cr	1111.0	0.0	0.0	[41]
Ti–Mo	10136.0	0.0	0.0	[42]
Ti–Sn	-97252	-45236.0	0.0	[41]
Ti–Zr	$-27520.0 + 13.4T$	0.0	0.0	[42]
Ti–Fe	$-80411.0 + 17.515T$	16120.0	0.0	[41]
Al–Cr	$-9552.0 + 30.0T$	$2946.0 - 14.0T$	-12000.0	[43]
Al-Mo	-46024.0	0.0	0.0	[44]
Al–Sn	$23810.8 - 15.70T$	$102.3 + 5.455T$	$-6728.6 + 8.49T$	[40]
Al–Zr	$-82055.0 - 25.0T$	$-3311.0 - 2.5T$	10000.0	[43]
Al–Fe	$-87309.1 + 22.831T$	0.0	0.0	[45]
Cr–Mo	$15810.0 - 6.714T$	-6220.0	0.0	[46]
Cr–Sn	$150056.0 - 96.875T$	$-59846.0 + 54.508T$	$-77912.0 + 22.54T$	[40]
Cr–Zr	$-12971.3 + 1.201T$	$8026.0 - 0.743T$	$-9984.9 + 0.924T$	[47]
Cr–Fe	$-14550.0 + 6.65T$	0.0	0.0	[48]
Mo–Sn	80000.0	0.0	0.0	[40]
Mo–Zr	$-55303.5 + 21.132T$	$-26116.5 + 15883T$	0.0	[40]
Mo–Fe	$-6973.0 + 0.37T$	$-9424.0 + 4.502T$	0.0	[48]
Sn–Zr	$21380.0 - 28.942T$	0.0	0.0	[40]
Sn–Fe	$84321.5 - 49.715T$	$46557.0 - 29.698T$	$60681.5 - 52772T$	[40]
Zr–Fe	-81473.0	14525.0	0.0	[49]
V–Fe	$-34679.0 + 1.895T$	-10209.0	0.0	[50]

T is given in degrees Kelvin.

$$\sigma = \sigma_1^\circ + \frac{RT}{S_1}\ln\frac{x_1^m}{x_1} + \ln\frac{\gamma_1^m}{\gamma_1} \tag{21}$$

$$\sigma = \sigma_2^\circ + \frac{RT}{S_2}\ln\frac{x_2^m}{x_2} + \ln\frac{\gamma_2^m}{\gamma_2} \tag{22}$$

$$\sigma = \sigma_3^\circ + \frac{RT}{S_3}\ln\frac{x_3^m}{x_3} + \ln\frac{\gamma_3^m}{\gamma_3} \tag{23}$$

In a ternary solution, $x_1^m + x_2^m + x_3^m = 1$.

The four unknowns (σ, x_1^m, x_2^m, and x_3^m) in these equations can be solved using a simultaneous equations approach. Similar concepts can be applied to calculate the surface tension for a higher order alloy.

It should be noted that in order to calculate the surface tension of the alloy system using Buttler's equation, the input of surface tension of the pure constituents elements in the alloy is needed. Variations on reported surface tension of pure elements are expected due to the experimental techniques used and the levels of impurities in the materials used. However, the surface tension for pure metals as a function of temperature $\sigma(T)$ can be calculated using Eq. (24) and Table 8 [19].

$$\sigma(T) = \sigma_i^\circ + \frac{d\sigma}{dT}(T - T_m) \tag{24}$$

where σ° is the surface tension at the melting point T_m, and $T > T_m$, T, and T_m are in degrees Kelvin.

VI. VISCOSITY

The viscosity is a measure of resistance of the fluid to flow when subjected to an external force. The shear stress τ, or the force per unit area, causing a relative motion of two adjacent layers in a liquid is proportional to the velocity gradient du/dy normal to the direction of the applied force is given by

$$\tau = -n\frac{du}{dy} \tag{25}$$

where the proportionality factor η is termed the viscosity. This concept is known as *Newton's law of viscosity*. Although not tested rigorously, most liquid metals are believed to obey a Newtonian behavior given by Eq. (25). The unit of viscosity is called Poise (P), (1 P = 1 dyne s cm^{-2} = 1 g cm^{-1} s^{-1}). The viscosity of liquid metals is usually expressed in centipoise (1 cP = 0.01 P).

Andrade [51] derived the following semiempirical relation to determine the viscosity of liquid metals at their melting temperatures:

$$\eta = B(MT_m)^{1/2}V^{-2/3} \tag{26}$$

where B is a constant, M is the atomic mass, T_m is the melting point in degrees Kelvin, and V is the atomic volume (cm^3 g-atom^{-1}). The variation of viscosity for most liquid metals, including undercooled liquid, follows Andrade's relationship, expressed as follows:

$$nV^{-1/3} = A\exp\frac{C}{VT} \qquad \text{or} \qquad nV^{-1/3} = A\exp\frac{Ev}{RT} \tag{27}$$

Table 8 Surface Tension for Pure Liquid Metals

Element	σ_i^0 at T_m (mN/m)	$-d\sigma/dT$ (mN/m K)	S_1 [30] (m^2/g atom)
Ag	903	0.16	
Al	914	0.35	46388
Au	1140	0.52	
B	1070	—	
Be	1390	0.29	
Bi	378	0.07	
Cd	570	0.26	
Ce	740	0.33	
Cr	1700	0.32	37672
Co	1873	0.49	
Cu	1285	0.13	
Fe	1872	0.49	36664
Hf	1630	0.21	
La	720	0.32	
Li	395	0.15	
Mg	559	0.35	
Mn	1090	0.2	
Mo	2250	0.30	43609
Nb	1900	0.24	
Ni	1778	0.38	
Pb	468	0.13	
Pd	1500	0.22	
Pt	1800	0.17	
Re	2700	0.34	
Sb	367	0.05	
Si	865	0.13	
Sn	544	0.07	60905
Ta	2150	0.25	
Ti	1650	0.26	47205
V	1950	0.31	41321
W	2300	0.29	
Zn	782	0.17	
Zr	1480	0.20	57021

Source: Ref. 19.

where A and C are material constants, $E_v = R(C/V)$ is the activation energy for viscous flow, and the gas constant R is $8.3144\,\mathrm{J\,K^{-1}}$. Equation (27) can be further simplified to

$$n = n_0 \exp \frac{Ev}{RT} \tag{28}$$

Also, with the exception of Ti, Sb, Zr, Pu, and U, the experimental data for viscosity shows that the B constant in Eq. (26) may be represented in the form [52]

$$\eta = 5.4 \times 10^{-4}(MT_m)^{1/2}V^{-2/3} \tag{29}$$

The variation of the activation energy for liquid metals at the melting point may also be represented by [52]

$$\mathrm{Log}\,E_v = 1.36\log T_m - 3.418 \tag{30}$$

where E_v is in kilocalories.

In another approach Chapman applied the principle of corresponding states to calculate the viscosity of pure metals [53]. In this model there are no assumptions of the structure other than the atoms are spherical and a potential parameter that exits between the atoms separated by the interatomic distance. Chapman deduced a relationship between the viscosity, the energy parameter ε, and the interatomic distance δ, and derived a reduced viscosity η^* relationship, which is a function of the reduced volume V^* and reduced temperature T^*.

$$\eta^*(V^*)^2 = f(T^*) \tag{31}$$

$$\eta^* = \frac{\eta\delta^2 N_0}{\sqrt{MRT}} \tag{32}$$

$$T^* = \frac{\kappa_B T}{\varepsilon} \tag{33}$$

$$V^* = \frac{1}{n\delta^3} \tag{34}$$

where N_0 is Avogadro's number, δ, is the interatomic distance in the closed-packed crystal at $0\,\mathrm{K}$ in angstroms, M is the atomic weight, R is the gas constant, T is the absolute temperature, κ_B is Boltzmann's constant, and n is the number of atoms per unit volume.

The energy parameter ε represents the major difficulty in the calculation of viscosity. Based on the Lenard-Jones energy parameter [54], the parameters for sodium and potassium were determined [55]. Using these parameters for these two metals, Chapman plotted their reduced viscosity as a function of reduced temperature, and observed that the data points produced a smooth curve as predicted by Eq. (31). A polynomial relation as shown by Eq. (35) can also represent these curves.

$$n^*(V^*)^2 = 0.0226\left(\frac{1}{T^*}\right)^3 - 0.0679\left(\frac{1}{T^*}\right)^2 + 0.5129\left(\frac{1}{T^*}\right) \tag{35}$$

Therefore, it was assumed that many liquid metals obey the same interatomic potential energy function. Chapman also estimated the values ε/κ_B for several metals from the viscosity data and found an excellent correlation with the melting temperature. This is expressed by the linear function:

$$\frac{\varepsilon}{K_B} = 5.20T_m \qquad (K) \tag{36}$$

Using Eq. (36) from the melting temperature, Eq. (35) of a pure metal or an alloy, and δ from the molar volume, the viscosity at any temperature may be calculated, including metals with high melting temperatures. Unfortunately, viscosity data for molten alloys is limited or not available, and equations to predict their viscosity have not been developed.

Table 9 shows the viscosity of liquid metals at their melting temperature and their activation energy for viscous flow [19, 56, 57].

VII. ELECTRICAL AND THERMAL CONDUCTIVITY

Thermal and electrical conductivity are intrinsic properties of materials, and they reflect the relative ease or difficulty of energy transfer through the material. The conductivity depends on the atomic/molecular bonding and structure arrangements of the material. Because most engineering materials are comprised of and/or involve more than one phase during their processing, the conductivity depends on the interaction between the intrinsic conductivity of the phases present and the mode of energy transfer between them.

Electrical and thermal conductivity are among the most important properties of metals. The reciprocal of the electrical conductivity is known as the electrical resistivity. Metals are well known for their high electrical conductivity, which arises from the easy migration of electrons through the lattice. The scattering of electron waves in the lattice is the cause of the resistivity in metals. Alloying, temperature, deformation, and nuclear radiation are responsible for disturbances in the lattice arrangement, which causes the decrease in electrical conductivity. Above 100 K over a limited temperature interval, the resistivity may be conveniently expressed by a linear relation of the form:

Table 9 Viscosity of Pure Liquid Metals at their
Melting Temperature and their Activation of Energy for
Viscous Flow

Element	η at T_m (mN s/m^2)	η_0 (mN s/m^2)	E (kJ/mol)
Ag	3.88	0.4532	22.2
Al	1.39	0.1492	16.5
Au	5.38	1.132	15.9
Bi	1.85	0.4458	6.45
Cd	2.28	0.3001	10.9
Ce	2.88	N/A	N/A
Co	4.49	0.2550	44.4
Cu	4.10	0.3009	30.5
Fe	4.95	0.3699	41.4
La	2.45		
Li	0.55	0.1456	5.56
Mg	1.32	0.0245	30.5
Ni	4.60	0.1663	50.2
Pb	2.61	0.4636	8.61
Sb	1.48	0.0812	22
Si	0.94		
Sn	2.00	0.5382	5.44
Ti	5.20		
Zn	3.85	0.4131	12.7
Zr	8.0		

Source: Refs. 19, 56, 57.

$$\rho_{s,T} = \rho_0(1 + \alpha T) \tag{37}$$

where T is the interval between two temperatures T_s and T_0 and α is the temperature coefficient of resistivity.

On melting the conductivity decreases markedly because of the exceptional disorder of the liquid state. However, with some metals such as bismuth, antimony, and manganese, and the semiconductors silicon and germanium, the conductivity markedly increases on melting. Generally, the electrical resistivity of most liquid metals just above their melting points is approximately 1.5 to 2.3 times greater than that of the solids just below their melting temperature [30].

Mott [58] derived an empirical equation to estimate the ratio of liquid/solid conductivity $(\sigma_{e,l}/\sigma_{e,s})$ at the melting point of the pure metal. This equation is based on the simple assumption that the atoms in the liquid metal vibrate slowly, varying mean positions with a given frequency. Mott's equation is expressed as follows:

$$\frac{\sigma_{e,l}}{\sigma_{e,s}} = \frac{v_l}{v_s} = \exp\left(-\frac{80\,\Delta H_m}{T_m}\right) \tag{38}$$

Where v_l and v_s are atomic vibration frequency in the liquid and solid, respectively, ΔH_m is the enthalpy or latent heat of fusion in kJ/mol, and T_m is the melting temperature in degrees Kelvin. Note that $\Delta H_m/T_m$ is the entropy of fusion ΔS_m. Thus, Eq. (38) can be represented in terms of the ratio of electrical resistivity of the liquid and solid (ρ_l/ρ_s) and the entropy of fusion (ΔS_m in cal/mol K).

$$\frac{\rho_l}{\rho_s} = \exp\,(0.33\,\Delta S_m) \tag{39}$$

With the exception of a few metals (e.g., Sb, Bi, Ga, Hg, Sn), the simple relationship proposed by Mott is in very good agreement with experimental measurements.

An extensive review of the various theories for electrical and thermal conductivity can be found in the literature [59–65]. Of particular interest to high-temperature technology such as casting processes is the theoretical relation between the electronic thermal (K_{el}) and electrical σ conductivities, known as the Wiedman-Franz law (WFL), and the constant of proportionality—the Lorentz number L [66]:

$$L = \frac{k_{el}}{\sigma T} = \frac{\pi}{3}\left(\frac{K_B}{e}\right)^2 = 2.45 \times 10^{-8}\ \text{W-ohm/deg}^2 \tag{40}$$

where K_B is the Boltzmann constant and e is the charge on an electron. It is assumed that electronic contribution to the thermal conductivity is predominant when the value of L is close to, or equal to, the theoretical value. Experimental determinations of L in pure metals at room temperature have shown that it varies from 2.23×10^{-8} W-ohm/deg^2 to 3.04×10^{-8} W-ohm/deg^2 for copper and tungsten, respectively. Interestingly, Eq. (40) appears to hold reasonably well for pure metals up to the melting point.

Excellent compilations of transport properties of elements can be found in the literature [19, 67]. However, a comprehensive survey has not been made of the electrical resistivity at high temperatures of alloys of commercial interest. Table 10 shows some the electrical resistivity for some solid and liquid metals at the melting point [19, 30]. The resistivity data for the liquid $\rho_{e,l}$ from the melting point to a given temperature in Table 10 can be calculated by the expression

Table 10 Electrical Resistivities of Solid and Liquid Metals at their Melting Point

Element	T_m (K)	$\rho_{e,s}$ ($\mu\Omega$ cm)	$\rho_{e,l}$ ($\mu\Omega$ cm)	$\rho_{e,l}/\rho_{e,s}$	α ($\mu\Omega$ cm/K)	β ($\mu\Omega$ cm)	T range T_m to: (K)
Ag	1233.7	8.2	17.2	2.09	0.0090	6.2	1473
Al	931	10.9	24.2	2.20	0.0145	10.7	1473
Au	1336	13.68	31.2	2.28	0.0140	12.5	1473
B*	2448	—	210.0				
Be*	1550	—	45.0				
Bi*	544	—	129.0				
Cd	593	17.1	33.7	1.97	Not linear		
Ce*	1060	—	126.8				
Cr*	2148		31.6				
Co	1766	97	102	1.05	0.0612	−6.0	1973
Cu	1356	9.4	20.0	2.1	0.0102	6.2	1873
Fe	1809	122	110	0.9	0.033	50	1973
Hf*	2216		218.0				
La*	1203		138				
Li*	453.5		240.0				
Mg	924	15.4	27.4	1.78	0.005	22.9	1173
Mn*	1525	66	40	0.61	No data		
Mo*	2880		60.5				
Nb*	2741		105.0				
Ni	1727	65.4	85.0	1.3	0.0127	63	1973
Pb	660.6	49.0	95.0	1.94	0.0479	66.6	1273
Pd	1825						
Pt*	2042		73.0				
Re*	3431		145				
Sb	904	183	113.5	0.61	0.270	87.9	1273
Si*	1683		75				
Sn	505	22.8	48.0	2.10	0.0249	35.4	1473
Ta*	3250		118.0				
Ti*	1958		172				
V¹	2185		71.0				
W*	3650		127				
Zn	692	16.7	37.4	2.24	Not linear		
Zr*	2123		153				

Source: Refs. 19*, 30.

$$\rho_{e,l} = \alpha T + \beta \tag{41}$$

The values for constants α and β for the various metals are given in the table. Note also that the electrical resistivity data given in Table 10 is for bulk metals and may not be applicable to thin films.

An accurate measurement of the thermal conductivity is exceedingly difficult in molten metals, and usually high conductivity values are obtained. This is because the convection is caused by the fluid flow of the liquid, which is produced by small temperature gradients in the melt. Because the heat transport in liquid metals and alloys occurs predominantly via electron transfer, the measurement of electrical conductivity appears to be the most accurate method. To this effect, it is appropriate to use the WFL law to estimate the thermal conductivity for the solid and liquid state of a given metal in the region of the melting point. Mills and coworkers [68] combined Mott and WFL equations (38) and (40) to establish a relationship between the thermal conductivity of the solid and liquid.

$$\ln\left(\frac{k_{sm}}{k_{lm}}\right) = K \, \Delta S_m \tag{42}$$

where K is a constant and K_{sm} and k_{lm} are the thermal conductivities for the solid and liquid at the melting point, respectively.

This equation would be useful to estimate the thermal conductivity of a liquid alloy by determining the thermal conductivity (or electrical conductivity) at the melting point and the entropy of fusion. The only limitation is the value of the constant K, which would not have a uniform value for all metals and alloys [68].

Table 11 shows the thermal conductivity for pure metals in the solid and liquid state at the melting point, as well as the estimation of the constant K based on the WFL [68].

Thermal diffusivity is the ability of a material to diffuse, or spread, thermal energy through itself [69–75]. This is determined by combining the material's ability to conduct heat and the amount of heat required to raise the temperature of the material, its specific heat. The density is involved because specific heat is given in units of heat per unit mass, while conductivity relates to the volume of material.

Thermal conductivity k and thermal diffusivity α are measures of heat flow within materials and are related by their specific heat and density. The relationship between k and α is given by

$$k = \alpha C_p \rho \tag{43}$$

where C_p is the specific heat and ρ is the density.

Table 11 Thermal Conductivities at the Melting Point and Entropies of Fusion of Pure Metals [68]

Element	T_m (K)	K_{sm} (W/m K)	K_{lm}	b (W/m K^2)	L_{sm}/L_{lm} Solid	Liquid	ΔS_m J/mol K	K
Ag	1233.7	362	175	200	1.0	1.0	9.15	0.0794
Al	931	211	91	34	1.0	1.0	10.71	0.0785
Au	1336	247	105	300	1.0	1.0	9.39	0.0911
B*	2448							
Be*	1550							
Bi*	544	7.6	12	100	—	—	20.75	0.022
Cd	593	37 ± 3	90	100	1.03	—	10.42	—
Ce*	1060	21	22	125	1.05	1.0	2.99	—
Cr*	2148	45	35	—	—	—	9.63	0.0261
Co	1766	45	36	—	—	—	9.16	0.0243
Cu	1356	330	163	200	1.0	1.0	9.77	0.0722
Fe	1809	34	33	—	0.95	1.06	7.62	0.0039
Hf*	2216	39	—	—	—	—	10.9	—
La*	1203	—	17	—	—	1.0	5.19	—
Li*	453.5	71	43	200	—	—	6.6	0.076
Mg	924	145	79	—	—	—	9.18	0.066
Mn*	1525	24	22	—	1.0	—	8.3	0.0102
Mo*	2880	87	72	—	1.0	1.03	12.95	—
Nb*	2741	78	66	—	1.0	—	10.90	—
Ni	1727	70	60	—	1.03	—	10.11	0.0152
Pb	660.6	30	15	75	1.0	1.0	7.95	0.0872
Pd	1825	99	87	—	—	1.05	9.15	—
Pt*	2042	80	53	—	1.01	—	10.86	0.0392
Re*	3431	65 ± 5	55	—	—	—	17.5	—
Sb	904	17	25	100	—	—	22	—
Si*	1683	25	56	—	—	1.04	29.8	
Sn	505	59.5	27	—	0.90	1.05	13.9	0.0567
Ta*	3250	70	58	—	—	1.0	11.1	0.0169
Ti*	1958	31	31	—	—	1.1	7.28	0.0
V*	2185	51	43.5	—	—	1.0	9.85	0.0161
W*	3650	95	63	—	—	0.98	14.2	0.029
Zn	692	90	50	600	0.90	1.04	10.6	0.055
Zr*	2123	38	36.5	—	—	—	9.87	0.0041

Source: Ref. 68.

Table 11 shows that the L_{sm}/L_{lm} ratio is very close to unity for all solid and liquid metals at their melting points. This is expected due to the experimental errors in the measurement of the thermal conductivity of the liquid. These can be larger than 5% for the thermal conductivity and ±3% for electrical conductivity [68]. Measuring the thermal diffusivity from which the thermal conductivity can be calculated may reduce some of the experimental errors.

Very limited information on the thermal conductivity of cast alloys of commercial interest exists in the literature—particularly in the range closer to the solidus and liquidus or liquid state. Tables 12 and 13 show some of the data available in the literature for ferrous and nonferrous alloys, respectively [21]. Tables 12 and 13 also include the solidus and liquidus temperatures.

In casting processes, the need for thermal conductivity data for mold materials becomes more crucial. Mold materials are usually bulk porous complex materials, and their thermal conductivity is certainly different from the intrinsic thermal conductivity of the individual materials, which form part of the mold. Methods to estimate the thermal conductivity of these materials can be found in the literature [20]. Table 14 shows the thermal conductivity for some mold materials [21].

VIII. EMISSIVITY

Heat transfer by thermal radiation is an important component in casting processes, since heat losses during pouring of the molten metal and heat radiation from the mold contribute to the overall heat balance during solidification of the cast product. Radiation heat transfer is particularly important to investment casting process. Thermal radiation is the energy transferred by electromagnetic waves from the surface of the melt or casting mold. The thermal radiation is comprised by a continuous spectrum of wavelengths forming an energy distribution, which is defined by Plank's law. The integrated form of Plank's equation gives the total emissive power of a body. This is known as the Stefan-Boltzmann equation, and is represented by

$$e = \varepsilon \sigma T^4 \tag{44}$$

Where e is the emissivity and σ is the radiation constant. This is given by

$$\sigma = \frac{2\pi^5 \kappa^4}{15 c^2 h} = 1.37 \times 10^{-12} \, \text{cal/cm}^2 \, \text{s} \, \text{K}^4 \tag{45}$$

Table 12 Thermal Conductivity, Solidus and Liquidus Temperatures for Some Ferrous Alloys of Commercial Interest

Alloy	Thermal conductivity $(\mathrm{W\,m^{-1}\,K^{-1}})$	Temperature range (K)	T_S (K)	T_L (K)
Carbon Steels				
AISI 1008	$K = 13.575 + 0.0113T$	$1122 \leq T \leq 1768$	1768	1808
	$K = 280.72 - 0.1398T$	$1768 \leq T$		
AISI 1026	$K = 15.192 + 0.0097T$	$1082 \leq T \leq 1768$	1768	1798
	$K = 280.72 - 0.1398T$	$1768 \leq T \leq 1798$		
AISI 1086	$K = 8.562 + 0.0145T$	$1073 \leq T \leq 1660$	1660	1754
	$K = 130.36 - 0.0588T$	$1660 \leq T \leq 1754$		
1.2 wt% C	$K = 11.061 + 0.0117T$	$1090 \leq T \leq 1598$	1598	1725
	$K = 92.713 - 0.0393T$	$1598 \leq T \leq 1725$		
Alloyed Steels				
1.0 wt% Cr	$K = 14.534 + 0.0105T$	$1073 \leq T \leq 1693$	1693	1743
	$K = 91.736 - 0.0351T$	$1693 \leq T \leq 1793$		
	$K = 7.847 + 0.0117T$	$1793 \leq T$		
1.5 wt% Mn	$K = 12.556 + 0.01156T$	$1185 \leq T \leq 1777$	1777	1801
	$K = 441.45 - 0.2298T$	$1777 \leq T \leq 1801$		
2.0 wt% Si	$K = 7.089 + 0.0151T$	$1273 \leq T \leq 1761$	1761	1770
	$K = 457.22 - 0.2405T$	$1761 \leq T \leq 1777$		
Stainless Steels				
AISI 304	$K = 12.076 + 0.0132T$	$780 \leq T \leq 1672$	1672	1727
	$K = 217.12 - 0.1094T$	$1672 \leq T \leq 1727$		
	$K = 8.278 + 0.0115T$	$1727 \leq T$		
AISI 316	$K = 11.107 + 0.013T$	$807 \leq T \leq 1644$	1644	1672
	$K = 355.93 - 0.1968T$	$1644 \leq T \leq 1672$		
	$K = 6.597 + 0.01214T$	$1672 \leq T$		
AISI 347	$K = 11.968 + 0.01276T$	$820 \leq T \leq 1672$	1672	1727
	$K = 194.63 - 0.0965T$	$1672 \leq T \leq 1727$		
	$K = 7.081 + 0.01205T$	$1727 \leq T$		
AISI 410	$K = 13.731 + 0.01111T$	$1060 \leq T \leq 1755$	1755	1805
	$K = 160.382 - 0.0724T$	$1755 \leq T \leq 1805$		
	$K = 8.61 + 0.01168T$	$1805 \leq T$		
AISI 430	$K = 12.651 + 0.01147T$	$1026 \leq T \leq 1750$	1750	1775
	$K = 309.43 - 0.158T$	$1750 \leq T \leq 1775$		
	$K = 7.088 + 0.01233T$	$1775 \leq T$		

Source: Ref. 21.

Table 13 Thermal Conductivity, Solidus and Liquidus Temperatures for Some Nonferrous Alloys of Commercial Interest

Alloy	Thermal conductivity $(W\,m^{-1}\,K^{-1})$	Temperature range (K)	T_S (K)	T_L (K)
Copper Alloys				
Cu–10wt% Al	$K = 27.271 + 0.08053T$	$T \leq 1303$	1303	1315
	$K = 72.749 - 5.4817T$	$1303 \leq T \leq 1315$		
	$K = 32.265 + 0.026T$	$1315 \leq T$		
Cu–30wt% Zn	$K = 140.62 + 0.01121T$	$460 \leq T \leq 1188$	1188	1228
	$K = 2430.3 - 1.9161T$	$1188 \leq T \leq 1228$		
	$K = 45.43 + 0.026T$	$1228 \leq T$		
Cu–40wt% Zn	$K = 182.95 - 0.03661T$	$620 \leq T \leq 1173$	1173	1178
	$K = 16479.5 - 13.93T$	$1173 \leq T \leq 1178$		
	$K = 39.724 + 0.026T$	$1178 \leq T$		
Cu–30wt% Ni	$K = 16.041 + 0.0439T$	$T \leq 1443$	1443	1513
	$K = 796.018 - 0.50273T$	$1443 \leq T \leq 1513$		
	$K = -3.799 + 0.026T$	$1513 \leq T$		
Cu–40wt% Ni	$K = 11.3059 + 0.04118T$	$T \leq 1500$	1500	1553
	$K = 11.0189 - 0.6959T$	$1500 \leq T \leq 1553$		
	$K = -2.6191 + 0.026T$	$1553 \leq T$		
Silicon brass	$K = 18.7814 + 0.04612T$	$533 \leq T \leq 1094$	1094	1189
	$K = 376.221 - 0.2806T$	$1094 \leq T \leq 1189$		
	$K = 36.82 + 4.847 \times 10^{-3}T$	$1189 \leq T$		
Aluminum Alloys				
Al–4.5wt% Cu	$K = 192.5$	$573 \leq T \leq 775$	775	911
	$K = 818.67 - 0.808T$	$775 \leq T \leq 911$		
	$K = 52.555 + 0.033T$	$911 \leq T$		
Al–8.0wt% Mg	$K = 24.114 + 0.202T$	$T \leq 813$	813	898
	$K = 1216.46 - 1.2646T$	$813 \leq T \leq 898$		
	$K = 51.19 - 0.033T$	$898 \leq T$		
Magnesium Alloys				
AZ 31B	$K = 67.118 + 0.0656T$	$499 \leq T \leq 839$	893	905
Mg–3Al–1Zn–	$K = 830.9 - 0.845T$	$839 \leq T \leq 905$		
0.5Mn	$K = 3.05 + 0.070T$	$905 \leq T$		
AZ 91B	$K = 18.268 + 0.1121T$	$T \leq 741$	741	869
Mg–9Al–0.6Zn–	$K = 372 - 0.3646T$	$741 \leq T \leq 869$		
0.2Mn	$K = -5.63 + 0.070T$	$869 \leq T$		
K1A die cast	$K = 127.16 + 0.0143T$	$T \leq 922$	922	923
Mg–9Al–0.7Zr	$K = 59214 - 64.071T$	$922 \leq T \leq 923$		
	$K = 11.66 + 0.070T$	$923 \leq T$		

(continued)

Table 13 (*continued*)

Alloy	Thermal conductivity (W m^{-1} K^{-1})	Temperature range (K)	T_S (K)	T_L (K)
ZK 51A Mg–4.6Zn–0.7Zr	$K = 71.96 + 154T - 0.938 \times 10^{-4}T^2$	$T \leq 822$	822	914
	$K = 688 - 0.6722T$	$822 \leq T \leq 914$		
	$K = 9.62 + 0.070T$	$914 \leq T$		
Nickel Alloys				
Hastealloy C	$K = -19.209 + 0.044T$	$T \leq 1070$	1543	1578
	$K = 19.923 + 0.744 \times 10^{-2}T$	$1070 \leq T$		
Inconel 718	$K = 7.9 + 0.0132T + 1.8 \times 10^{-6}T^2$	$294 \leq T < 1423$	1443	1513
	$K = 1.94 \times 10^{-2}T$	$1609 \leq T \leq 1909$		

Source: Ref. 21.

Where κ and h are Boltzmann's and Plank's constants, respectively, and c is the speed of light.

A "black body radiator" is defined as a surface that absorbs all incident radiation and emits radiation of all wavelengths. The black body provides a standard for comparison to define the radiation ability of other surface, by determining the ratio of the total emissive power of the surface to that of the black body when they are at the same temperature. This ratio is known as total emissivity, which varies with temperature, direction of radiation, wavelength, and the surface roughness. The emissivity increases with increasing surface roughness and often increases with increasing temperature.

Because emissivity data throughout the wavelength spectrum is not available for most metallic materials, the following empirical equation has been employed to represent the total emissivity as a function of temperature [76].

$$e_t = K_1 \sqrt{(\rho T) - K_2 \rho T} \tag{46}$$

Where K_1 and K_2 are constants, $K_1 = 5.736$, $K_2 = 1.769$, and ρ is the electrical resistivity in ohm meters.

Equation (46) is in reasonable agreement with experimental data and it shows an increase of e_t with T. However, deviations for some materials at high

Table 14 Thermal Conductivity of Mold Materials

Mold material	Dry density (g/cm³)	Thermal conductivity (W m⁻¹ K⁻¹)	Temperature range (K)
Silica Molding Sands			
20–30 Mesh	1.730	$k = 0.604 - 0.767 \times 10^{-3}T + 0.795 \times 10^{-6}T^2$	$T < 1033$
50–70 Mesh	1.634	$k = 0.676 - 0.793 \times 10^{-3}T + 0.556 \times 10^{-6}T^2$	$T > 1033$
Silica Sand (−22 mesh)			
+7% Bentonite	1.520	$k = 0.946 - 0.903 \times 10^{-3}T + 0.564 \times 10^{-6}T^2$	$T < 1500$
+4% Bentonite	1.60	$k = 1.26 - 0.169 \times 10^{-2}T + 0.105 \times 10^{-5}T^{-2}$	
Olivine Sand	3.18	$k = 0.713 + 0.349 \times 10^{-4}T$	$T < 1300$
+4% Bentonite	2.125	$k = 1.82 - 1.88 \times 10^{-2}T + 0.10 \times 10^{-5}T^{-2}$	
Zircon Sand	2.780	$k = 1.19 - 0.948 \times 10^{-3}T + 0.608 \times 10^{-6}T^2$	
+4% Bentonite	2.96	$k = 1.82 - 0.176 \times 10^{-2}T + 0.984 \times 10^{-6}T^2$	$T < 1500$
Chromite Sand		$k = 941 - 0.753 \times 10^{-3}T + 0.561 \times 10^{-6}T^2$	
Graphite (chill foundry grade)	1.922	$k = 135.99 - 8.378 \times 10^{-2}T$	$T < 873$
		$k = 103.415 - 4.647 \times 10^{-2}T$	$T > 873$
Investment Casting			
Zircon-30% Alumina-20% Silica	2.48–2.54	$k = 3.03 - 3.98 \times 10^{-4}T + 508T^{-1}$	$375 < T < 1825$

Source: Ref. 21.

temperatures can be expected due to the spectral emissivity that changes with the wavelength and direction of emission. In most practical situations, such as in casting processes, an average emissivity for all directions and wavelengths is used. It is also important to note that the emissivity values of metals or other nonmetallic materials depend upon the degree of oxidation and the grain size. Table 15 gives some total normal emissivity for some pristine and oxidized metals. While Table 16 gives the total emissivity for some alloys and refractory materials [77], additional emissivity data for various materials has been compiled and can be found in the literature [78–80].

The data shown in Tables 15 and 16 provide a guideline and must be used with discretion, since in practical applications the values of emissivity may change considerably with oxidation and roughening of the surfaces. Therefore, it is important that the total emissivity should be determined for the actual surface conditions of the materials in question.

Table 15 Total Normal Emissivity of Pristine and Oxidized Metals

Metal	Pristine	Temperature (K)	Oxidized	Temperature (K)
Ag	0.02–0.03	773	—	—
Al	0.064	773	0.19	873
Be	0.87	1473	—	—
Cr	0.11–0.14	773	0.14–0.34	873
Co	0.34–0.46	773	—	—
Cu: Solid	0.02	773	0.24	1073
Cu: Liquid	0.12	1473		
Hf	0.32	1873		
Fe	0.24	1273	0.57	873
Mo	0.27	1873	0.84	673
Ni	0.14–0.22	1273	0.49–0.71	1073
Nb	0.18	1873	0.74	1073
Pd	0.15	1473	0.124	1273
Pt	0.16	1473	—	—
Rh	0.09	1673	—	—
Ta	0.18	1873	0.42	873
Ti*	0.47	1673	—	—
W	0.17	1673	—	—
	0.18	1873		
	0.23	2273		
Alloys				
Cast iron	0.29	1873	0.78	873

* Spectral normal emissivity at 65 μm wavelength.
Source: Ref. 76.

Table 16 Total Normal Emissivity of Various Alloys and Refractory Materials

Alloys	ε	Temperature (K)
Commercial aluminum	0.09	373
Oxidized	0.11–0.19	472–872
Cast iron:		
Solid	0.60–0.70	1155–1261
Liquid	0.29	1873
Commercial copper:		
Heated @ 872 K	0.57	472–872
Liquid	0.16–0.13	1349–1550
Steel:		
Plate rough	0.94–0.97	273–644
Steel liquid	0.42–0.53	1772–1922
Refractory materials		
Alumina:		
85–99.5Al$_2$ O$_3$—0–12SiO$_2$—0–1Fe$_2$O$_3$	0.5–0.18	1283–1839
Alumina-silica:		
58–80Al$_2$ O$_3$—16–18SiO$_2$—0.4Fe$_2$O$_3$	0.61–0.43	1283–1839
Fireclay brick	0.75	1273
Carbon rough plate	0.77–0.72	373–773
Magnesite brick	0.38	1273
Quartz, fused	0.93	294
Zirconium silicate	0.92–0.80	510–772
	0.80–0.52	772–1105

Note: A linear interpolation of the emissivity values with temperature can be done when the emissivity values are separated by a dash.
Source: Ref. 77.

IX. DIFFUSION

Diffusion is defined as the transfer of mass from one region to another on an atomic scale. Diffusion in solids is much smaller than diffusion in the liquid state. In metal systems, the variation of diffusivity between the solid and liquid state differs by 2 to 3 orders of magnitude. For instance, atomic diffusivity of most metals just above their melting temperatures are of the order of 10^{-9} m^2 s^{-1} and increase up to 10^{-4} cm^2 s^{-1} with temperature

The knowledge of diffusion in casting processes is of paramount importance, since the distribution of solute elements between the solid and liquid phases during solidification strongly depends on their diffusion motion. Therefore, in order to understand and mathematically simulate the solidification phenomena in a given casting process, it is important to have knowledge of the diffusion phenomena. This is particularly important at the solidus, solid-liquid range, and just above the liquidus temperature of a given metal system. Unfortunately, there are many problems associated with the experimental determination of diffusion in the semisolid and liquid states, and diffusion data is not readily available.

Currently, extensive diffusivity data can be found for a large variety of metals in the solid state [81]. However, there is practically no data for the semisolid state, and very limited data can be found for the liquid state. Therefore, more reliable and accurate data, particularly for self-diffusivity of liquid metals, is needed to prove existing theories, and to calculate not only solute distribution in a solidifying metal but also in high temperature metallurgical processes.

The flux of matter J that passes per unit time throughout a unit area of a plane perpendicular to the direction of diffusion, is proportional to a concentration gradient of the diffusing atoms is given by the equation

$$J = -D \frac{\partial c}{\partial x} \tag{47}$$

Where c is the concentration, x is the diffusion distance, and D is the diffusion coefficient. This equation is known as Fick's first law for steady state diffusion. In binary systems, diffusion always occurs in the direction of decreasing concentration gradient.

In nonsteady diffusion, the flux changes with the diffusion distance and time, and if diffusivity is independent of concentration. Thus, the diffusion equation is represented by

$$\frac{\partial c}{\partial t} = \frac{\partial}{\partial x} \left(D \frac{\partial^2 C}{\partial x^2} \right) \tag{48}$$

The solutions of this equation depend on the boundary and geometry conditions, and results provide information on concentration as a function of time and system geometry. Detailed methods of solutions of Eq. (48) can be found in the literature [82–84].

An important consideration in diffusion problems is the condition that the atomic movement is completely random in the three dimensions. This is because the thermal vibrations will produce a mean square displacement $\bar{\Delta r}^2$ of the atoms after a time t from their original position. This concept is known as a random walk process and leads to the definition of self-diffusivity by the fundamental relation:

$$D = \frac{\bar{\Delta r^2}}{6t} \tag{49}$$

Equation (49) is usually expressed as

$$r = 2.4\sqrt{(Dt)} \tag{50}$$

The distance $\sqrt{(Dt)}$ is a very important quantity in diffusion problems.

In the solid state, the atoms can diffuse through the lattice by two basic mechanisms. One mechanism is known as substitutional diffusion, in which the substitutional atoms diffuse by a vacancy mechanism. The other is interstitial diffusion, which is defined by the migration of small atoms by forcing their way through the interstitial sites between the larger atoms. Excellent reviews and discussions about these mechanisms can be found in the literature [84, 85] and will not be discussed here. However, it is important to indicate that these mechanisms of diffusion can be represented by the empirical Arrhenius type equation, that is,

$$D = D_0 \exp \frac{-Q}{RT} \tag{51}$$

where the factor D_0 is known as the vibration factor and Q is usually known as the activation energy.

Theoretical calculations have been developed to determining the D_0 factor for both substitutional and interstitial mechanisms of diffusion. These are not discussed here, since they can be found in the literature [83–89]. However, It is important to mention that there is a relative direct linear correlation between the activation energy Q_S with the enthalpy of fusion ΔH_m and the melting temperature T_m for vacancy diffusion [54]. This can be expressed by the equations

$$\frac{Q_S}{(\Delta H_m)} \approx 16 \quad \text{and} \quad \frac{Q_S}{(T_m)} \approx 36 \text{ cal/K (150.624 J/K)} \tag{52}$$

Si, Ge, Ta, Hf, La, and Zr are the exception to these correlations.

Some experimental data reported in the literature for substitutional self-diffusion is given in Table 17 [88].

Note that the normalized activation energy for substitutional self-diffusion Q_S/RT_m for a given crystal structure is approximately constant, and correlates well with Eq. (52) when divided by R (8.317 kJ/mol K).

Unfortunately, diffusivity data for cast materials of commercial interest is not available, and the diffusivity calculations for multicomponent systems are rather complex. For instance, in a binary alloy the rate at which the atoms of the solvent and solute move into a vacant site is not equal and each atomic

Table 17 Substitutional Self-Diffusion Data for Selected Pure Metals

Crystal structure/ class	Metal	T_m (K)	D_0 $(10^{-6}\,m^2/s)$	Q_s (kJ/mol)	Q_s/RT_m	D at T_m $(10^{-12}\,m^2/s)$
bcc/Transition metals	δ-Fe	1811	190	238.5	15.8	26
	β-Ti	1933	109	251.2	15.6	18
	β-Zr	2125	134	273.5	15.5	25
	Cr	2130	20	308.6	17.4	0.54
	V	2163	28.8	309.2	17.2	0.97
	Nb	2741	1240	439.6	19.3	5.2
	Mo	2890	180	460.6	19.2	0.84
	Ta	3269	124	413.3	15.2	31
	W	3683	4280	641	20.9	3.4
bcc/Alkali metals	Li	371	24.2	43.8	14.2	16
bcc/Rare earths	δ-Ce	1071	1.2	90	10.1	49
	γ-La	1193	1.3	102.6	10.4	42
	Al	933	170	142	18.3	1.9
	Ag	1234	40	184.6	18	0.61
	Au	1336	10.7	176.9	15.9	1.3
	Cu	1356	31	200.3	17.8	0.59
fcc	β-Co	1768	83	283.4	19.3	0.35
	γ-Fe*	1805	49	284.1	18.9	0.29
	Ni	1726	190	279.7	19.5	0.65
	Pb	601	137	109.1	21.8	0.045
	Pd	1825	20.5	266.3	17.6	0.49
	Pt	2046	22	278.4	16.4	0.17
hcp	Cd	594	$\perp$ c 10	79.9	16.2	0.94
			$\parallel$ c 5	76.2	15.4	0.99
	Mg	922	$\perp$ c 150	136	17.8	2.9
			$\parallel$ c 100	134.7	17.6	2.3
	Zn	692	$\perp$ c 18	96.2	16.7	0.98
			$\parallel$ c 13	91.6	15.9	1.6
Tetragonal	β-Sn	505	$\perp$ c 1070	105	25	0.015
			$\parallel$ c 770	107.1	25.5	0.0064
Diamond cubic	Ge	1211	440	324.5	32.3	4.4×10^{-3}
	Si	1683	0.9	496	35.5	3.6×10^{-4}

* T_m for γ-Fe is the temperature at which γ-Fe would melt without the intervention of δ-Fe.
Source: Ref. 88.

species must be given its own intrinsic diffusion coefficient. However, Fick's laws for diffusion can be applicable, but an interdiffusion coefficient should be defined:

$$D_i = X_B D_A + X_A D_B \tag{53}$$

where X_A and X_B are the mol fractions for components A and B in the metal solution with the diffusion coefficients D_A and D_B, respectively. Then, D can be substituted by D_i in Eq. (48). This concept was first introduced by Darken [89] and it is known as the Darken equation. In substitutional alloys the interdiffusion coefficient D_i depends on D_A and D_B, because of the atomic size, whereas in interstitial diffusion the diffusion coefficient of the solute (interstitial) is only needed in Eq. (48). Similarly, D_i can be used in the Arrhenius equation (51). However, D_0 and Q are very uncertain and there is no any atomic model for concentrated alloys to make a reasonable estimation.

Availability of diffusion data in liquid metals is very limited due to the experimental difficulty in the acquisition of meaningful data. Several theories to explain and calculate the self-diffusivity of metal in the liquid state have been developed [90–98]. However, those theories based on the rigid sphere model developed equations that appear to have a better correlation with the experimental data. For instance the Longuet and Pole equation [90] describes the self-diffusivity of pure liquid metals:

$$D = \frac{0.25\delta(\pi k T/M)^{1/2}}{(pV/RT) - 1} \tag{54}$$

where δ is the atomic diameter, k is the Boltzman constant, M is the atomic mass, and pV/RT is the compressibility factor.

The atomic diameter δ for the rigid sphere model is given by

$$\delta = (1.146 \pm 0.012) \times 10^{-8} \, V^{1/3} \tag{55}$$

Alder and Wainwright [91] obtained an equation for self-diffusivity at the melting point. They used a backscatter correction factor 0.72 to multiply D and a compressibility factor of 11.2 to obtain:

$$D = 3.27 \times 10^{-6} \left(\frac{T_m}{M}\right)^{1/2} V^{1/3} \tag{56}$$

Using similar equations based on the rigid sphere model, Nachtrieb [92] reviewed the self-diffusivity data for pure liquid metals and found a variation of $\pm 10\%$ when compared with the experimental data. Table 18 shows some of

Table 18 Available Self-Diffusivity Data for Pure Liquid Metals

Element	$D_0 \times 10^{-8}$ (m^2/s)	Q (kJ/mol)	Temperature (K)
Li	14.1	11.8	
	9.4	9.62	
	14.4	12	469–713
	$D = [5.76 + 0.036(T - T_m)] \times 10^{-9}$		
	$D = [6.8 + 0.036(T - T_m)] \times 10^{-9}$		
Cu	14.6	40.61	413–1533
Zn	8.2	21.3	723–873
	12	23.4	693–873
Ag	7.1	34.1	1275–1378
Sn	3.02	10.8	540–956
Pb	9.15	18.6	606–930
	Predicted self-diffusion coefficients $(10^{-9}\,m^2/s)$		
Al	4.87		933
Ni	3.90		1728
Mg	5.63		923
Sb	2.66		903
Fe	4.16		1808

Source: Ref. 100.

the available experimental and calculated data for self-diffusivity of pure liquid metals [99, 100].

Compilation of apparent activation energy for self-diffusivity Q_A of liquid metals indicates that it follows a linear relationship with the melting temperature. This can be expressed by [100]

$$Q_A = 27.6 T_m \text{ (J)} \quad \text{or} \quad 6.6 T_m \text{ (cal)} \tag{57}$$

The modified Stokes-Einstein equation is a relationship between viscosity and diffusion that has shown an excellent agreement for self-diffusivity of liquid metals. This equation can be represented by

$$D = \frac{kT}{\xi(V/N_A)^{1/3}\eta} \tag{58}$$

where ξ is a constant with values between 5 and 6 in cgs units and η is the viscosity.

By combining the modified Stokes-Einstein equation with Andrade's relation [Eq. (25)] for viscosity at the melting point, the coefficient diffusion D_m at the melting point for pure liquid metals can be expressed as

$$D_m = \frac{KN_A^{1/3}}{5.7\xi} \times 10^{-4} \left(\frac{T_m}{M}\right)^{1/2} V_m^{1/3} \qquad (59)$$

In addition, this equation can be rewritten in terms of bulk viscosity η using the hard sphere model with the following equation:

$$D_m = \frac{21.3 \times 10^{-10} T_m V_m^{-1/3}}{\eta} \qquad (60)$$

Surface tension σ, viscosity η, and self-diffusivity D_m at the melting point have been combined to obtain dimensionless parameters for pure metals. These parameters may be very useful in the modeling of solidification processes of dilute alloys. These are expressed as follows [100]:

$$\frac{\sigma\delta}{D_m\eta} = 34.21 \qquad (61)$$

or

$$\frac{\sigma V_m^{1/3}}{D_m\eta} = 29.85 \qquad (62)$$

Very limited work has been conducted on solute diffusion of liquid alloys. This is because there is lack of knowledge of the atomic interactions and distribution functions of liquid alloys. Some calculations using the hard sphere model have been developed for dilute solutions [101, 102]. Unfortunately, there are too many difficulties to have a detailed comparison between the predicted and measured values of solute diffusivities in liquid alloys. This is also aggravated by very large experimental difficulties in the measurement of diffusivity. However, roughly in liquid metal solutions, the coefficients of diffusion of solutes (atoms that form substitutional solutions) range from 10^{-7} to 10^{-8} m^2/s with activation energies of 33.5–50.25 kJ, while the diffusivities of hydrogen in liquid metals range from 6×10^{-7} to 2×10^{-6} m^2/s with activation energies of 4.2 to 21 kJ.

REFERENCES

1. JJ Valencia. Symposium on Thermophysical Properties: Metalworking Industry Needs and Resources, Concurrent Technologies Corporation, Oct. 22–23, 1996. Unpublished work.
2. A Overfelt. Thermophysical property sensitivity effects in steel solidification. Proceedings of a Workshop on the Thermophysical Properties of Molten Materials, NASA Lewis Research Center, Cleveland, OH, Oct. 20–23, 1992, pp. 35–50.
3. YS Touloukian, CY Ho, et al. Thermophysical Properties Research Center Data Book, 3 vols., Purdue University, 1960–1966, 3322 pp.
4. YS Touloukian, ed. Thermophysical Properties of High Temperature Solid Materials, 6 vols. (9 books). New York: MacMillan, 1967, 8549 pp.
5. YS Touloukian, JK Gerritsen, NY Moore, eds. Thermophysical Properties Literature Retrieval Guide, Basic Edition, 3 vols., New York: Plenum Press, 1967, 2936 pp.
6. YS Touloukian, CY Ho, eds. Thermophysical Properties of Matter—The TPRC Data Series, 14 vols., (15 books), New York: IFA/Plenum Data Co., 1970–1979, 16,810 pp.
7. YS Touloukian, JK Gerritsen, WH Shafer, eds. Thermophysical Properties Literature Retrieval Guide, Supplement I (1964–1970), 6 vols., New York: IFI/Plenum Data Co., 1973, 2225 pp.
8. YS Touloukian, CY Ho, eds. Thermophysical Properties of Selected Aerospace Materials. Part I Thermal Radiative Properties; Part II Thermophysical Properties of Seven Materials, Purdue University, TEPIAC/CINDAS, Pt I, 1976, 1058 pp: Pt II, 1977, 242 pp.
9. JK Gerritsen, V Ramdas, TM Putnam, eds. Thermophysical Properties Literature Retrieval Guide, Supplement II (1971–1977), 6 vols., New York: IFI/Plenum Data Co., 1979, 1493 pp.
10. JF Chaney, TM Putnam, CR Rodriguez, MH Wu, eds. Thermophysical Properties Literature Retrieval Guide (1900–1980), 7 vols., New York: IFI/Plenum Data Co., 1981, 4801 pp.
11. YS Touloukian, CY Ho, eds. McGraw-Hill/CINDAS Data Series on Materials Properties, 4 vols., New York: McGraw-Hill, 1981, 1525 pp.
12. Thermophysical Properties of Materials: Computer-Readable Bibliographic Files, Computer Magnetic Tapes, TEPIAC/CINDAS, 1981.
13. J Miettien. Metall Trans 23A:1155–1170, 1992.
14. J Miettien, S Louhenkilpi. Calculation of thermophysical properties of carbon and low alloy steels for modeling of solidification processes. Met Trans 25B:909–916, 1994.
15. YA Chang. A thermodynamic approach to obtain materials properties for engineering applications. Proceedings of a Workshop on the Thermophysical Properties of Molten Materials, NASA Lewis Research Center, Cleveland, OH, Oct. 20–23, 1992, pp. 177–201.
16. C Kittel. Introduction to Solid State Physics. New York: Wiley, 1971.

17. Compendium of Thermophysical Property Measurement Methods, Vol. 1: Survey of Measurement Techniques (KD Maglic, A Cezairliyan, VE Peletsky, eds.), New York: Plenum Press, 1984.

18. C Papesch, JJ Valencia. Effect of heating and cooling rate on an aluminum bronze alloy. NCEMT Report WO 0029 and 0022, Concurrent Technologies Corporation, Johnstown PA, March 1997.

19. CJ Smithels. Metals Reference Book, 7th ed., General Physical Properties, Table 14.1 The Physical Properties of Pure Metals (EA Brandon, GB Brook, eds.) Butterworth-Heinemann, 1992, p. 14-1.

20. GH Geiger, DR Poirier. Transport Phenomena in Metallurgy. Reading, MA: Addison-Wesley Publishing Co., 1973. Data from Appendix II and III, pp. 575–587.

21. RD Pehlke, A Jeyarajan, H Wada. Summary of thermal properties for casting alloys and mold materials. National Science Foundation, Applied Research Division, December 1982.

22. KA Jackson. Liquid metals and solidification. Cleveland, OH: ASM, 1958, p. 174.

23. G Lang. Density of liquid elements. CRC Handbook of Chemistry and Physics, 75th ed. (DR Lide, HPR Fredererikse, eds.). CRC Press, 1994–1995, pp. 4-126 – 4-134.

24. AF Crawley. Density of liquid metals and alloys. International Metallurgical Reviews, Review 180, The Metals Society, 19:32–48, 1974.

25. EE Shpil'rain, KA Yakimovich, AG Mozgovoi. Apparatus for continuous measurement of temperature dependence of density of molten metals by the method of a suspended pycnometer at high temperatures and pressures. Compendium of Thermophysical Property Measurement Methods 2:601–624, 1984.

26. SD Mark, SD Emanuelson. A thermal expansion apparatus with a silicon carbide dilatometer for temperatures to 1500°C. Ceramic Bulletin 37/4:193–196, 1958.

27. LD Lucas. Physicochemical Measurements in Metals Research, Part 2 (RA Rapp, ed.). New York: Wiley Interscience, 1970, p. 219.

28. P Parlouer. Calorimetry and dilatometry at very high temperatures. Recent Developments in Instrumentation and Applications, Rev Int Hautes Temper et Refract 28:101–117, 1992–1993.

29. L Darken, R Gurry. Physical Chemistry of Metals. New York: McGraw-Hill, 1953, pp. 125–126.

30. I Takamichi, RIL Guthrie. The Physical Properties of Liquid Metals. Oxford: Clarendon Press, 1988, p. 188.

31. WK Rhim, AJ Rulison. Measuring surface tension and viscosity of a levitated drop. Report of National Aeronautics and Space Administration Contract No. NAS 7-918, 1996, pp. 1–6.

32. I Egry, G Lohofer, P Neuhaus, S Sauerland. Surface tension measurements of liquid metals using levitation, microgravity, and image processing. Int J Thermophysics 13/1:65–74, 1992.

33. Y Bayazitoglu, GF Mitchell. Experiments in acoustic levitation: surface tension measurements of deformed droplets. J Thermophysics and Heat Transfer 9/4:694–701, 1995.

34. S-L Chen, W Oldfield, YA Chang, MK Thomas. Metall and Mater Trans 25A:1525–1533, 1994.

35. F Zhang, YA Chang, JS Chou. A thermodynamic approach to estimate titanium thermophysical properties. Proceedings of 1997 International Symposium on Liquid Metal Processing and Casting, (A Mitchel, P Auburtin, eds.). American Vacuum Society, 1997, pp. 35–59.

36. KS Yeum, R Speiser, DR Poirier. Metall Trans 20B:693–703, 1989.

37. H-K Lee, JP Hajra, Z Frohberg. Metalkde 83:638–643, 1992.

38. JAV Buttler. Proc Roy Soc A135:348, 1932.

39. J-S Chou, L Nastac, CA Papesch, JJ Valencia, Y Pang. Thermophysical and solidification properties of titanium alloys. National Center for Excellence in Metalworking Technology, Report TR No. 98-87, June 30, 1999.

40. F Zhang, S-L Chen, YA Chang. Modeling and simulation in metallurgical engineering and materials science. Proceedings of the International Conference, MSMM'96 (Z-S Yu, ed.). Beijing, China, 1996, pp. 191–196.

41. JL Murray. Phase Diagrams of Binary Titanium Alloys. Metals Park, OH: ASM International, 1987.

42. JL Murray. Bull Alloy Phase Diagr 2:185–192, 1981.

43. N Saunders, VG Rivlin. Mater Sci Techn 2:521, 1986.

44. L Kaufman, H Nesor. Metall Trans 5:1623–1629, 1974.

45. UR Kattner, Burton. Phase Diagrams of Binary Titanium Alloys. Metals Park, OH: ASM International, 1987.

46. K Frisk, P Gustafson. CALPHAD 12:247–254, 1988.

47. K-J Zeng, H Marko, L Kaj. CALPHAD 17:101–107, 1993.

48. J-O Anderson, N Lange. Metall Trans 19A:1385–1394, 1988.

49. AD Pelton. J Nuclear Materials 201:218–224, 1993.

50. W Huang. Z Metallkde 82:391–401, 1991.

51. EN da C Andrade. Philos Mag 17:497, 698, 1934.

52. ET Turkdogan. Physical Chemistry of High Temperature Technology. New York: Academic Press, 1980, p. 109.

53. T Chapman. AIChE J 12:395, 1966.

54. J Lenard-Jones, AF Devonshire. Proc Roy Soc A165:1, 1938.

55. RC Ling. J Chem Phys 25:609, 1956.

56. JR Wilson. Metall Rev 10:381, 1965.

57. LJ Wittenberg. Viscosity of liquid metals. In. Physicochemical Measurements in Metal Research, vol. IV, part 2 (RA Rapp, ed.). New York: Wiley Interscience, 1970, p. 193.

58. NF Mott. Proc Roy Soc A146:465, 1934.

59. TE Faber. Introduction to the Theory of Liquid Metals. New York: Cambridge University Press, 1972.

60. JM Ziman. Advanced Physics 13:89, 1964.

61. R Evans, DA Greenwood, P Lloyd. Phys Lett 35A:57, 1971.

62. R Evans, BL Gyorffy, N Szabo, JM Ziman. Proceedings of the 2nd International Conference on Liquid Metals, Tokyo (S Takeuchi, ed.). London: Taylor and Francis, 1973.

63. Y Waseda. The Structure of Non-crystalline Materials: Liquids and Amorphous Solids. New York: McGraw-Hill, 1980.
64. S Takeuchi, H Endo. J Japan Inst Metals 26:498, 1962.
65. JL Tomlinson, BD Lichter. Trans Met Soc AIME 245:2261, 1969.
66. L Lorentz. Ann Phys Chem 147:429, 1982.
67. GT Meaden. Electrical Resistance of Metals. New York: Plenum, 1965.
68. KC Mills, BJ Monaghan, BJ Keene. Thermal conductivity of liquid metals. Proceedings of a Workshop on the Thermophysical Properties of Molten Materials, NASA Lewis Research Center, Cleveland, OH, Oct. 20–23, 1992, pp. 519–529.
69. WJ Parker, RJ Jenkins, CP Butler, GL Abbott. Flash method of determining thermal diffusivity, heat capacity, and thermal conductivity. J of Applied Physics 32/9:1679–1684, 1961.
70. RD Cowan. Pulse method of measuring thermal diffusivity at high temperatures. J Applied Physics 34/4:926–927, 1963.
71. JT Schriempf. A laser flash technique for determining thermal diffusivity of liquid metals at elevated temperatures. High Temperatures High Pressures 4:411–416, 1972.
72. KD Maglic, RE Taylor. The apparatus for thermal diffusivity measurement by the laser pulse method. Compendium of Thermophysical Property Measurement Methods 2:281–314, 1984.
73. RE Taylor, KD Maglic. Pulse method for thermal diffusivity measurement. In: Compendium of Thermophysical Property Measurement Methods 1: Survey of Measurement Techniques, 1984, pp. 305–336.
74. Y Qingzhao, W Likun. Laser pulse method of determining thermal diffusivity, heat capacity, and thermal conductivity. FATPC, pp. 325–330, 1986.
75. Y Maeda, H Sagara, RP Tye, M Masuda, H Ohta, Y Waseda. A high-temperature system based on the laser flash method to measure the thermal diffusivity of melts. Int J Thermophysics 17/1:253–261, 1996.
76. CJ Smithels. Radiating properties of metals. In: Metals Reference Book, 7th ed. (EA Brandes, GB Brook, eds.), Butterworth-Heinemann, 1992, pp. 17-1–17-12.
77. J Szekely, NJ Themelis. Rate Phenomena in Process Metallurgy. New York: Wiley Interscience, 1971, chap 9, pp. 251–300.
78. EM Sparrow, RC Cess. Radiation Heat Transfer. Belmont, CA: Brooks/Cole, 1966.
79. HC Hottel, AF Sarofim. Radiative Transfer. New York: McGraw-Hill, 1967.
80. TJ Love. Radiative Heat Transfer. Columbus, OH: Merrill, 1968.
81. CJ Smithels. Metals Reference Book, 7th ed., Diffusion in Metals (EA Brandes, GB Brook, eds.), Butterworth-Heinemann, 1992, pp. 13-1–13-118.
82. J Crank. The Mathematics of Diffusion, 2nd ed. London: Oxford University Press, 1975.
83. W Jost. Diffusion in Solids, Liquids, Gases. New York: Academic Press, 1960.
84. PJ Shewmon. Diffusion in Solids. New York: McGraw-Hill, 1963.
85. DA Porter, KE Easterling. Phase Transformations in Metals and Alloys. New York: Van Nostrand Reinhold Company, 1981, chap. 2, pp. 60–106.

86. BS Chandrasekhar. Rev Mod Phys 15:1, 1943.
87. C Zener. In: Imperfections in Nearly Perfect Crystals (W Shockley, ed.). New York: Wiley, 1952, p. 289.
88. AM Brown, MF Ashby. Correlation for diffusion constants. Acta Metallurgica 28:1085, 1980.
89. LS Darken. Trans Met Soc AIME 175:184, 1948.
90. HC Longuet, JA Pole. J Chem Phys 25:884, 1956.
91. BJ Alder, TE Weinwright. Phys Rev Lett 18:988, 1967.
92. NH Nachtrieb. Physicochemical Methods Measurements in Metals Research (PD Adams, HA Davies, SG Epstein, eds.). London: Taylor and Francis, 1967, p. 309.
93. HA Walls. Physicochemical Methods Measurements in Metals Research (RA Rapp, ed.). New York: Wiley Interscience, 1970, p. 459.
94. CT Vadovic, CP Colver. Phil Mag 21:971, 1970.
95. TE Faber. Introduction to Theory of Liquid Metals. New York: Cambridge University Press, 1972.
96. P Protopapas, HC Andersen, NAD Parlee. J Chem Phys 59:15, 1973.
97. RA Swalin. Acta Met 7:736, 1959.
98. SM Breitling, H Eyring. In: Liquid Metals: Chemistry and Physics (SZ Beer, ed.). New York: Marcel Dekker, 1972, chap. 5.
99. I Takamichi, RIL Guthrie. The Physical Properties of Liquid Metals. Oxford: Clarendon Press, 1988, pp. 201, 217.
100. ET Turkdogan. Physical Chemistry of High Temperature Technology. New York: Academic Press, 1980, pp. 115–118.
101. P Protopapas, NAD Parlee. High Temp Sci 8:141, 1976.
102. M Shimoji. Liquid metals: an introduction to the physics and chemistry of metals in the liquid state. New York: Academic Press, 1976, p. 222.

7
Quick Analysis

Chunqing Cheng
NetScreen Technologies, Inc., Sunnyvale, California

I. INTRODUCTION

Comprehensive models for casting process simulation usually solve the energy and momentum equations and hence, can take into account mold-filling phenomena. These models generate the most accurate and comprehensive information about the casting processes. In theory, these simulations can easily encompass macroscopic models, such as turbulent flow and mass transport, as well as microscopic models, such as solidification kinetics. Details for using comprehensive models have been described in Chapter 2. In general, comprehensive casting simulations involve the following steps:

Solid modeling: Accurately input solid models or convert geometry data through electronic data exchange to the format that a particular simulation package can accept.

Model setup: Generally requires proper casting, geometry discretization, and thermophysical property input.

Simulation: Select and run proper models, for example, heat transfer and solidification models, or mold-filling plus solidification models.

Data mining: Postprocessing on a huge volume of data produced by simulation models. Interactively visualize and analyze three-dimensional results.

These procedures are usually time consuming even if one can afford computer hardware and software costs. The fact that comprehensive simulation codes take considerable effort to implement and may require significant computer resources to execute sometimes makes them difficult to use in a production environment. To solve casting problems effectively and quickly,

many quick-yet-effective approaches can be used. For example, casting solidification times can be used to determine casting shakeout times. In addition,
geometric modulus, in many cases, can be used to determine correct riser sizes.

Therefore, to provide effective solutions to casting problems quickly, a
series of efforts have been made beside comprehensive simulations. This chapter will describe some quick analysis techniques applied to solve casting problems:

- Theoretical analysis
- Geometrical analysis
- Thermal analysis

II. THEORETICAL ANALYSIS

To improve productivity in foundries, one important goal is to predict casting
solidification times so that proper shakeout times can be specified. Too short a
shakeout time cannot ensure the casting quality, while too long a shakeout
time will have an adverse impact on the productivity and, hence, the casting
cost. To estimate solidification times, both casting geometry and heat transfer
problems can be simplified without the introduction of significant errors. In
this section, approaches to quickly estimating casting solidification times by
simplifying both casting geometry and heat transfer problems will be presented.

A. Solidification Times in Sand Casting

Sand casting is the most commonly used casting process to produce large size
castings. Naturally, it is desirable to have a quick yet accurate way to determine shakeout times for sand castings. Figure 1 shows a schematic temperature
profile during solidification of a metal with zero superheat in a semi-infinite
sand mold.

In a typical sand casting process, the thermal conductivity of the sand
mold is much less than that of the metal being cast. Hence, it is reasonable to
assume that the thermal gradient in the metal is negligible compared to that in
the mold. In most cases, heat-affected zones in molds are confined to a layer of
sand only about one-quarter of casting sizes. Thus, sand mold walls can be
considered to be sufficiently thick (i.e., a semi-infinite mold starting from the
casting metal). With these assumptions, the solidified metal front location M
can be derived as:

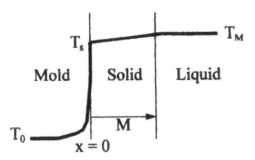

Figure 1 Temperature distribution profile during solidification of a metal with zero superheat in a semi-infinite sand mold.

$$M = \frac{2}{\sqrt{\pi}} \left(\frac{T_M - T_0}{\rho' \, \Delta H_f} \right) \sqrt{k \rho C_p} \sqrt{t} \tag{1}$$

where t is the time; T_0 is the initial uniform mold temperature; T_M is the metal melting temperature; ρ' and ΔH_f are density and latent heat of fusion of liquid metal, respectively; and k, ρ, and C_p, are the thermal conductivity, density, and specific heat of the mold, respectively.

Equation (1) can be used as a starting point to determine shakeout times of castings. For a thin plate, which has "infinite" length and width due to its relatively thin thickness, only half the thickness is needed due to symmetry. Substitution of the half thickness as M (solidification depth) in Eq. (1) leads to the solidification time for the thin plate. For other shapes, solidification times can be estimated based on the thickest section of the casting. The solidification times obtained this way are the minimum shakeout times required for casting productions.

1. Contour Effects on Solidification Times

To account for contour effects or casting solidification times, three basic shapes (infinite plates, cylinders, and sphere) have been discussed in the literature [1, 2]. The definitions of two dimensionless parameters, namely β and γ, are:

$$\beta \equiv \frac{V/A}{\sqrt{\alpha t}} \tag{2}$$

and

$$\gamma \equiv \left(\frac{T_M - T_0}{\rho' \, \Delta H_f} \right) \rho C_p \tag{3}$$

where α is the mold thermal diffusivity and V and A are the volume and surface area of the casting, respectively. The contour effects on casting solidification times can be seen in Fig. 2. For infinite plates,

$$\beta \equiv \gamma\left(\frac{2}{\sqrt{\pi}}\right) \tag{4}$$

For cylinders,

$$\beta \equiv \gamma\left(\frac{2}{\sqrt{\pi}} + \frac{1}{4\beta}\right) \tag{5}$$

And for spheres,

$$\beta \equiv \gamma\left(\frac{2}{\sqrt{\pi}} + \frac{1}{3\beta)} \tag{6}$$

Thus, solidification times can be obtained with the contour effects for three basic shapes. In general, Eq. (6) can be used for other chunky shapes, and Eq. (5) can be used to approximate solidification times of bars with square cross sections. Solidification times obtained through Eqs. (4) to (6) can be used as guides for the estimation of casting minimum shakeout times.

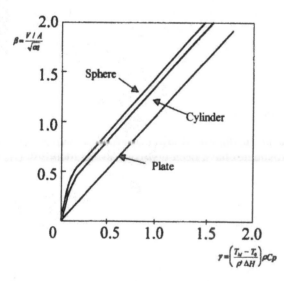

Figure 2 Comparison of freezing times for the three basic shapes in sand molds.

2. Superheat Effects on Solidification Times

To account for the effect of superheat ΔT_s, without invalidating the previous discussions, the latent heat ΔH_f in Eqs. (1) and (3) should be replaced with the effective latent heat $\Delta H'$,

$$\Delta H' = \Delta H_f + C'_{p,l} \Delta T_s \tag{7}$$

where $C'_{p,l}$ is the specific heat of the liquid metal.

B. Solidification Times in Spray Casting

Spray casting can be depicted by a situation in which the mold surface temperature T_s remains constant as shown in Fig. 3. This is a typical Newtonian cooling, and the solidification front position can be expressed by

$$M = \frac{h(T_M - T_0)}{\rho' \Delta H_f a} t - \frac{h}{2k'} M^2 \tag{8}$$

where k' is the thermal conductivity of the metal, h is heat transfer coefficient, and

$$a \equiv \frac{1}{2} + \sqrt{\frac{1}{4} + \frac{C'_p(T_M - T_0)}{3 \Delta H_f}} \tag{9}$$

where C'_p is the specific heat of the metal [1, 3].

Equation (8) can also be applied to quenching, or to the permanent mold surface where a water-cooled channel is used. It is important to note that

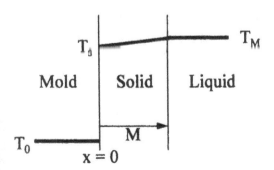

Figure 3 Temperature distribution during solidification with a constant mold surface temperature.

for spray castings, the contour effects are negligible if the temperature gradient in the metal is negligible.

C. Solidification Times in Plastic Injection Molding

In plastic injection molding, temperature gradient in the mold can be ignored without introducing significant error, as shown in Fig. 4. The solidification front is

$$M = 2\beta\sqrt{\alpha't} \tag{10}$$

where α' is the part thermal diffusivity and β is characteristic constant. β can be solved through

$$\beta e^{\beta^2} \operatorname{erf}(\beta) = (T_M - T_0)\frac{C_p'}{\Delta H_f \sqrt{\pi}} \tag{11}$$

Equation (10) is used to estimate solidification times [1, 3].

D. Solidification Times with Competing Heat Transfer Modes

It is not unusual to have multiple dominant heat transfer mechanisms. In these cases, many heat transfer modes compete in casting process. Figure 5 shows the case where both gradients in the mold and in the casting are important, while Fig. 6 shows the case where the mold/metal interface heat transfer is important as well.

For cases approximated by Fig. 5, the following equations apply:

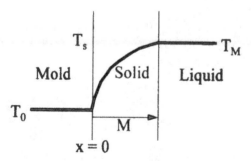

Figure 4 Temperature distribution in a plastic injection molding process.

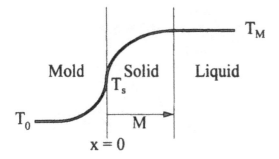

Figure 5 Temperature distribution during solidification without mold/metal interface resistance.

$$\beta e^{\beta^2}\left(\sqrt{\frac{k'\rho'C_p'}{k\rho C_p}} + \text{erf}(\beta)\right) = (T_M - T_0)\frac{C_p'}{\Delta H_f\sqrt{\pi}}$$

to get β (12)

$$\beta e^{\beta^2}\text{erf}(\beta) = (T_M - T_s)\frac{C_p'}{\Delta H_f\sqrt{\pi}}$$

to get T_s (13)

$$\frac{T - T_s}{T_0 - T_s} = \text{erf}\left(\frac{-x}{2\sqrt{\alpha t}}\right)$$

to get the mold temperature distribution (14)

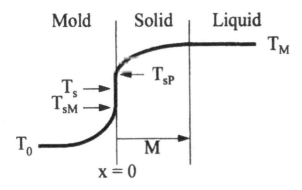

Figure 6 Temperature distribution during solidification with mold/metal interface resistance.

$$\beta e^{\beta^2} = (T_\infty - T_s)\frac{C_p'}{\Delta H_f \sqrt{\pi}} \qquad \text{to get } T_\infty \text{ (imaginary temperature)}$$
$$(15a)$$

or

$$\frac{T_s - T_0}{T_\infty - T_s} = \sqrt{\frac{k'\rho'C_p'}{k\rho C_p}} \qquad \text{to get } T_\infty \qquad (15b)$$

$$\frac{T - T_s}{T_\infty - T_s} = \mathrm{erf}\left(\frac{x}{2\sqrt{\alpha't}}\right) \qquad \text{to get the metal temperature distribution}$$

$$(16)$$

$$M = 2\beta\sqrt{\alpha't} \qquad \text{to get solidification front position or}$$

$$\text{solid thickness} \qquad (17)$$

For cases approximated by Fig. 6, an imaginary reference plane with T_s is first assumed between the two mold/metal interface temperatures, as shown in Fig. 6. Then two fictitious heat transfer coefficients are defined as follows:

$$h_m \equiv \left(1 + \sqrt{\frac{k\rho C_p}{k'\rho'C_p'}}\right)h \qquad \text{for mold side of the plane} \qquad (18)$$

$$h_p \equiv \left(1 + \sqrt{\frac{k'\rho'C_p'}{k\rho C_p}}\right)h \qquad \text{for metal side of the plane} \qquad (19)$$

This problem can be solved in three separate steps: (1) first solve for T_s assuming no resistance interface heat transfer as shown in Fig. 5; (2) use T_s and h_p and Eq. (8) to get T_{sP}, M, and the metal temperature distribution; and (3) use T_s and h_m, and T_{sM} and the following equation:

$$\frac{T - T_s}{T_0 - T_s} = \mathrm{erfc}\left(\frac{x}{2\sqrt{\alpha t}}\right) - e^\gamma \mathrm{erfc}\left[\frac{x}{2\sqrt{\alpha t}} + \frac{h_m}{k}\sqrt{\alpha t}\right]$$
$$(20)$$
$$\text{for short time solution}$$

where

$$\gamma = \frac{h_m}{k}\sqrt{\alpha t}\left[\frac{x}{\sqrt{\alpha t}} + \frac{h_m}{k}\sqrt{\alpha t}\right] \qquad (21)$$

to get mold temperature profiles.

In both cases, tracking the solidification front can lead to the solution of solidification times. Interested readers are referred to the literature [3] for details.

III. GEOMETRIC MODULUS

To help design a casting process, geometrical analysis is usually sufficient. In this section, several popular techniques are presented.

A. Chvorino's Rule

Assuming that surface contours do not have any effect on the heat flux at mold/metal interfaces, for a given mold/metal interface area A, the mold absorbs an amount of heat Q in time t. For a casting of volume V (with no superheat), its solidification time can be expressed in terms of the volume-to-surface area ratio:

$$t = C\left(\frac{V}{A}\right)^2 = C(M_s)^2 \tag{22}$$

where

$$c \equiv \frac{\pi}{4}\left(\frac{\rho' \Delta H_f}{T_M - T_0}\right)^2 \left(\frac{1}{k\rho C_p}\right) \tag{23}$$

and

$$M_s \equiv \frac{V}{A} \tag{24}$$

Equation (22) is often referred to as Chvorinov's rule [4], and C, as expressed in Eq. (23), as Chvorinov's constant. The volume-to-surface area ratio in Eq. (24) is also often referred to as section modulus M_s of the casting.

B. Application of Section Modulus in Casting Designs

When designing a casting process, risers are usually chosen in such a way that the solidification times of the risers are longer than those of the casting parts for proper feeding. From Eq. (22), solidification times are directly proportional to the section modulus of castings. When comparing solidification times, section moduli can be directly compared. Since section moduli are only geometric quantities, comparison of the section moduli makes design problems much simpler. Foundry engineers can design a riser with a larger section modulus to ensure proper feeding of the part without considering the type of casting process to be used.

The simplicity of the section modulus gains wide spread use among casting design engineers. Section modulus works best for casting configurations where none of the mold material becomes saturated with heat. The success of this approach is based on the assumption that the mold material absorbs the same amount of heat per unit mold/metal interfacial area.

C. Point Modulus

It is a well-known fact that the course of casting solidification determines the size and the locations of the casting hot spots. Many researchers [5–7] have tried to extend the simple relationship of the section modulus to predict the course of casting solidification.

Using two-dimensional cases, Neises et al. [5] used the following equation to get the point modulus M_p in casting. By replacing the section modulus in Eq. (22) with the point modulus, solidification times within castings can be obtained. This approach, in general, works well for symmetric parts. For non-symmetric parts, a correction factor has to be introduced:

$$M_p = \frac{1}{\sum_{i=1}^{N} f_i \frac{1}{d_i}} \tag{25}$$

Upadhya et al. [7] used Eq. (25) with $f_i = 0.5$ in three-dimensional cases. Once the point modulus is known, a three-dimensional isochronal map can be produced through Eq. (22).

D. Other Geometrical Methods

Conceptually, geometrical methods are used to find the heaviest section in the part. Hence, Yagel et al. [8] use the minimum distance to the casting surface as the criteria in their work to predict the diecastability. This approach can also be considered as one of the special cases of Eq. (25), as shown below:

$$M_p = \frac{1}{\max(1/d_1, 1/d_2, \ldots, 1/d_i, \ldots, 1/d_N)} \tag{26}$$

Hill et al. [9] used weighted thickness to determine the heaviest section in their work to aid them in their casting designs. Their equations cannot be directly related to Eq. (25). Interested readers are recommended to their original publication [9]. It should also be noted that weighted thickness couldn't be directly related back to solidification times.

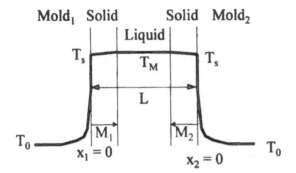

Figure 7 Temperature distribution during solidification of a metal in a mold with two different materials.

IV. THERMAL MODULUS

The geometric modulus approach relies strictly upon the geometry of the casting. Real castings employ chills and insulators to control the progression of solidification. These features are ignored in the geometric modulus approach, which leads to some serious deficiencies in their predictive capabilities. On the other hand, theoretical analysis presented in Sec. II is only good for simply casting environments. Figure 7 schematically shows a thin plate, which is a symmetrical part, in a two-material mold (for example, the drag and cope use different sands, or have different moisture contents). Such a simple case makes the mentioned approaches useless.

To account for thermal effects associated with chills, insulation, and other mold variations, an innovative approach termed *thermal modulus* (TM) [10, 11] is presented in this section. Its basic concept is demonstrated in Fig. 8. The total heat loss is composed of several components of heat loss, which are determined by combining local geometric features and local heat transfer mechanisms. While the total heat loss is not easily obtainable, each heat loss component can be easily obtained, provided that the surface of the element at the mold/casting interface is properly subdivided according to local heat transfer mechanisms.

A. Differences between Section Modulus and Thermal Modulus

In general, geometry based algorithms rely on the distances from the location of interest to various surface segments. The thermal modulus is based on the heat removal rates at various surface segments, Fig. 8. In the thermal modulus

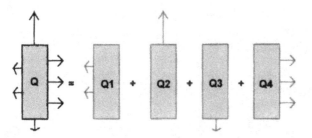

Figure 8 Schematic decomposition of total heat loss rate. (Four sides of the rectangular represent four segments of the casting/mold interfaces.)

approach, the casting is considered as a process in which a certain amount of excess thermal energy Q is removed from a given location. For molten metal, this excess thermal energy is the sum of the thermal energy represented by the metal superheat and the heat of fusion. The rate at which the excess thermal energy is removed from the given location is defined as the thermal modulus (TM) of that location.

B. Advantages of Thermal Modulus

Examination of heat loss components at the mold/casting interface using the TM approach provides quick solutions to the casting heat transfer problems. To determine transient effects, the finite difference method (FDM) and finite element method (FEM) approaches require the calculation of casting and mold temperatures for multiple time increments during the process. For explicit FDM and FEM formulations, information (i.e., process effects) can only be transmitted a distance of one cell for each time increment. Therefore, the boundary conditions at the mold/casting interface can only directly affect the elements at the boundary. Boundary conditions are delayed in the interior as this information is passed along one cell layer at a time for each time increment. This can be visualized through a "domino effect." The impact of the boundary conditions at the interface on the internal elements is through a series of "domino" neighboring elements both in space and in time. For implicit FDM and FEM formulations, however, the impact of the boundary conditions at the interface on the internal elements is implicitly considered through a process of "recursion." Since all the nodes and elements are interconnected and one's behavior is highly dependent on another, the time step cannot be arbitrarily chosen for accurate results. In any case, the lack of time step constraints in a numerical scheme enables implicit formulations to use relatively large time steps to get a faster solution at the cost of accuracy.

The TM model uses a divide-and-conquer approach. First, a much complex interdependent problem is divided into many independent "ideal" problems that can be solved easily. Then, the solutions to these ideal problems are combined to arrive at the answers to the original complex problem. Hence, the speed of the algorithm is not limited by the time step size.

Examination of the heat loss components at the mold/casting interface also overcomes limitations inherited from previous modulus schemes based on geometric considerations alone. For example, consider a chill placed at the bottom of a block to promote directional solidification in a sand mold, Fig. 9. Any approach based on geometric modulus will predict that the hot spot is located at the center of the block, which is obviously incorrect. However, using heat loss components at the mold/casting interface allows one to incorporate more accurate heat transfer mechanisms at all interfaces. The TM approach predicts a hot spot near the top of the casting, which makes sense since the chill extracts heat much faster than the sand. This flexibility allows one to solve problems presented in Fig. 7 easily.

C. Approaches of Thermal Modulus

In the TM approach, the mold/casting interface is divided into several subregions. Since the heat losses can be added together, the individual components of heat loss to each subregion are calculated and added together. The way in which the mold/casting interface is divided will affect the accuracy of the casting solidification time computation. The finer the division, the more accurate the solution. In general, variations of geometric features and local heat transfer boundary conditions result in local heat losses that are different from place to place. Consequently, it is desirable to divide the mold/casting interface according to variations of local heat transfer mechanisms and local geometric

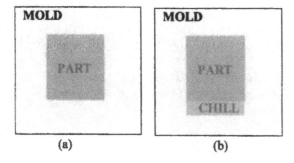

(a)   (b)

Figure 9 A simple casting with two different casting practices: (a) sand casting and (b) sand casting with a chill.

features. Figure 10 shows such a division. Each face could have a different heat transfer mechanism, hence a different rate of heat loss. To demonstrate the flexibility of the current approach, the left side of the block in Fig. 10 is represented by two separate heat loss rates (q_1 and q_2) so that a total of five heat loss components are present. Such a system can be viewed as consisting of five independent subsystems. The heat transfer rate from each subsystem is computed independently of the rest. The total heat loss added from all five subsystems is equivalent to that of the original system. This relation can be expressed as

$$q = \sum_i q_i \qquad (27)$$

where q is the overall rate of heat loss and q_i is the rate of heat loss component corresponding to surface segment i at the mold/casting interface. Assuming $\bar{q}$ is the time average of q, the total excess energy Q is

$$Q = \bar{q}t \qquad (28)$$

where t is the solidification time for the location of interest and $\bar{q}$ is the thermal modulus by definition. Equation (28) shows that the larger the TM, the smaller the solidification time.

Before obtaining the TM, the rate of heat loss component of each sub-region must be calculated. In one implementation [11], a series of disks (spheres in the three-dimensional case) are considered so that the centers of the disks coincide with the location of interest. The size of each disk (sphere) is determined by the geometry of the cast part, as shown in Fig. 11. In other words, the radius of each disk (sphere) equals the distance from the location of interest to the mold/casting interface under consideration. With such divisions, combining all corresponding portions of these disks (spheres) can approximately duplicate the entire casting geometry. Each component of the rate of heat loss can be

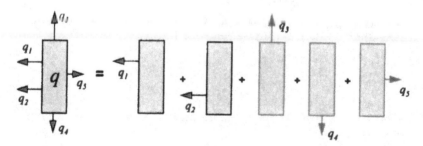

Figure 10 Schematic decomposition of a complex heat transfer system into several subsystems; one surface can be subdivided into several segments.

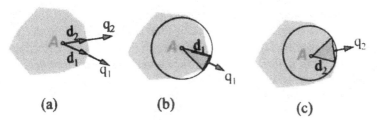

Figure 11 Schematic representing individual heat loss component: (a) two individual heat loss components of interest with subregional surfaces of distances d_1 and d_2, respectively; (b) simplified case to get heat loss component q_1; and (c) simplified case to get heat loss component q_2.

easily obtained through the solution of a one-dimensional axisymmetrical heat transfer problem.

To obtain the heat loss component for a particular subsurface segment, the disk (sphere) corresponding to that subsurface segment is used. In such a system, the disk (sphere) is assumed to be the same material as that of the cast part, while the surroundings are considered to be the same as that of the mold material in contact with the casting in this subsurface region. The interface condition between the casting and mold is identical to that defined in the original casting system. These assumptions lead to an identical heat transfer boundary condition everywhere for the entire disk (sphere). Therefore, the heat loss component can be easily obtained as a one-dimensional axisymmetrical problem.

In any local region, the heat transfer mechanisms can be classified as (1) mold-diffusion dominant (typical of sand casting), (2) part-diffusion dominant (typical of plastic injection molding), (3) interfacial-heat-transfer dominant (typical of diecasting), and (4) combinations of these mechanisms. Proper application of any of the first three mechanisms will allow further simplification of the heat transfer problem, as shown in Sec. II.

The parameters of this one-dimensional heat transfer system consist of those corresponding to the local set of thermal parameters in the original casting system, including initial temperature, latent heat, superheat, and the dominant heat transfer mechanism. Therefore, the TM methodology is sensitive to the original casting material properties and solidification parameters.

Once all individual heat loss rate components are obtained through the series of one-dimensional axisymmetrical problems as mentioned above, the TM for the location of interest can be easily obtained. Thus, for any location of interest, the TM algorithm does not directly deal with the complex boundary conditions of the casting as a whole; instead, the algorithm handles a number of uniform and simple boundary conditions to find each local heat loss rate component. The time average of these values is then added to arrive at the TM (an overall heat loss rate) for the location of interest. This approach accounts for many different mold materials and interface conditions, thereby overcoming the limitations of the geometric approach.

By relating the TM to the energy balance, i.e.,

$$\tilde{q} = \frac{dH}{dt} = \frac{\partial H}{\partial T}\frac{\partial T}{\partial t} + \frac{\partial H}{\partial f_s}\frac{\partial f_s}{\partial t} \tag{29}$$

the temperature history at any location of interest can be obtained. In Eq. (29), H is the enthalpy and f_s is the fraction of solidified metal.

D. Comparisons of Thermal Modulus Results and Heat Transfer Results

The thermal modulus approach has been implemented based on finite difference grids in a generic, three-dimensional code [11]. In this implementation, the actual casting phenomena are approximated by well-established one-dimensional and quasi-one-dimensional heat transfer solutions. To further simplify the problem, the mold is considered to be of infinite thickness.

Thermophysical and solidification properties of materials shown in Tables 1 and 2 are used in both the thermal modulus program and a FDM based code [12].

Figure 12 shows a low loss launch valve plug casting for aircraft carriers with the complete gating configuration. The configuration of this steel casting was optimized from a series of design changes based on FDM simulation results [13]. This system of four chills and four insulator sleeves was success-

Table 1 Thermophysical Properties of Materials Used in Simulations

Material	Heat capacity $(cal\,g^{-1}\,C^{-1})$		Density (g/cm^3)		Thermal conductivity $(Cal\,s^{-1}\,cm^{-1}{}^\circ C^{-1})$	
	Solid	Liquid	Solid	Liquid	Solid	Liquid
Steel	0.25	0.25	7.8	7.8	0.1	0.1
Sand	0.18		2.7		0.005	
Copper chill	0.09		8.971		0.9514	
Insulator	0.25		2.49		0.00304	
Air		0.01		0.0011		56.24

Source: Ref. 12.

Table 2 Solidification Properties for Steel

Pouring temperature	1620°C
Liquidus	1520°C
Solidus	1400°C
Latent heat	65 cal/g

Source: Ref. 12.

fully used to cast several parts. The dimensions of the configuration and process details can be found in Ref. 13.

Figure 13 shows the temperature maps from both the three-dimensional FDM heat transfer code and the TM code. The TM code predicts a temperature distribution similar to that of the FDM heat transfer code. Both codes predict hot spots at the same locations.

Figure 14 shows cooling curves from both the FDM heat transfer code and the TM code. It can be seen that for locations A and B, both models predict similar thermal histories. However, to generate these results with 238,134 (78 × 43 × 71) nodes, the FDM heat transfer code required 6 h 40min on a Silicon Graphics, Inc. (SGI) Indigo2 workstation, whereas the TM code required 3 min 30 s on the same workstation. The TM code in this case performed 110 times faster than the FDM code and generated similar results.

E. Implication of the Thermal Modulus

The major advantage of using the TM model is the significant reduction in CPU time. This allows modeling engineers to handle significantly larger models

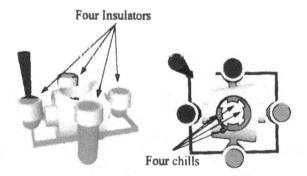

Four Insulators

Four chills

Figure 12 Final casting design with the rigging system indicating locations of chills and insulators for a low loss valve plug casting.

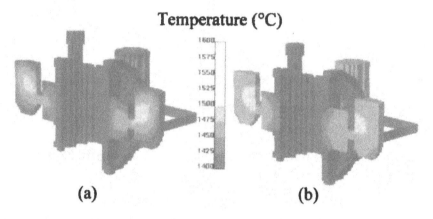

Figure 13 Simulated temperature distributions in the low loss launch valve plug casting: (a) FMD heat transfer code and (b) thermal modulus code (1200 s after pour).

than traditional FDM and FEM methods can handle. On the other hand, since the TM model can rapidly predict hot spot locations, it can be used as a powerful filter for comprehensive FDM and FEM software packages to reduce the number of tedious heat transfer and computational fluid dynamics calculations.

Traditional design algorithms based on the geometric modulus approach cannot account for common casting variables, such as chills and insulators.

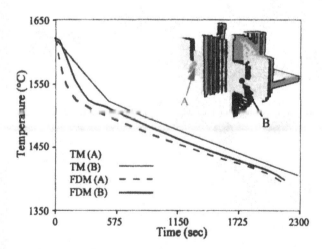

Figure 14 Simulated thermal histories (cooling curves) at two different locations of the low launch valve plug casting.

The TM model considers various common casting variables. Therefore, the thermal modulus should be used as a kernel for casting expert systems to help casting design engineers optimize their rigging systems.

Figure 15 shows how a typical casting process can be optimized using the TM model. In Fig. 15a, a low loss launch valve is cast in a sand mold. There are two hot spots in the casting. Once two chills are added to the bottom portion of the casting, the bottom hot spot disappears, Fig. 15b. Finally, after an insulator sleeve is added around the riser, the top hot spot moves into the riser, Fig. 15c, resulting in a sound casting. Using the TM model, various combinations can be conducted very quickly, thus leading to an optimized casting process.

Since the thermal modulus approach produces casting thermal history, other casting thermal parameters, such as cooling rates, local solidification times, and thermal gradients, can be calculated [14, 15]. Furthermore, the propensity for casting defects [16] and the casting part microstructure [17] can also be predicted based on the casting thermal parameters and temperature history. This means that the TM model can generate casting thermal data equivalent to that from the comprehensive heat transfer approach with a CPU time that is 2 orders of magnitude shorter.

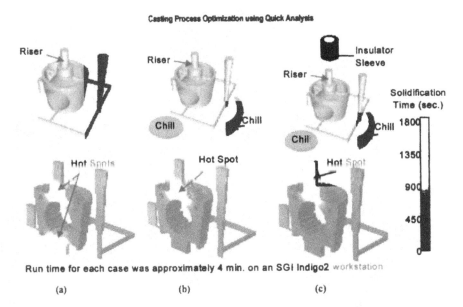

Figure 15 Thermal modulus algorithm is used to optimize casting process: (a) two hot spots in casting, (b) one hot spot in casting, and (c) no hotspot in casting.

V. MOLD FILLING

Quick fill models are intended to simplify the casting flow models and approximate the temperature (energy) loss during the mold filling stage.

A. Bernoulli Approach

When casting can be approximated by a series of networks of simple geometry identities (blocks, cylinders, and spheres), the Bernoulli approach can be used to simulate mold filling. Wang [18] has used this approach in his work.

Wang [18] first decomposed an entire diecasting system, including the shot sleeve, runner, and in-gate, die cavity and vents, into individual sections. These sections are connected to form a flow network, Fig. 16. A relation is developed to calculate the pressures at the two ends P_A and P_B, the flow rate $q_{n_1 n_2}$ in each section, and the solidification in each section, Fig. 17.

At any intersection point, say A, net mass gain is zero:

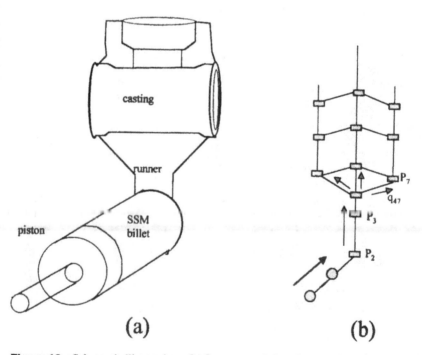

Figure 16 Schematic illustration of a flow network for SSM (semisolid metalworking) die filling: (a) actual die setup; (b) schematic illustration of flow network.

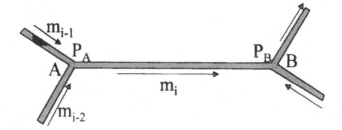

Figure 17 Schematic illustration for approach of flow network die filling analysis.

$$m_{i-2} + m_{i-1} - m_i = 0 \qquad\qquad\qquad\qquad\qquad (30)$$

For any individual section i, say AB, mass flow rate is driven by pressure drop:

$$m_i = \frac{(P_A - P_B)D_i^2 A_i}{L_i f} \qquad\qquad\qquad\qquad (31)$$

where P is the liquid metal pressure, m is the liquid metal flow rate in the section i, f is the viscosity, L is the section length, D is the section diameter or thickness.

By combining this pressure-flow relation, Eq. (31), with the mass balance requirements at the intersections, Eq. (30), a set of simultaneous linear equations can be developed. Once the pressure and flow rate are obtained, the die cavity flow front at various times can be easily estimated. Figure 18 shows a plot of the flow front mapped onto the die cavity geometry.

B. Mixed Approach

For a shape casting, it is difficult to predetermine flow paths. Hence, it is not easy to apply the Bernoulli flow model throughout if flow paths cannot be identified. To get a quick answer to some of fluid flow problems, a mixed approach is proposed. The following assumptions are employed in the mixed approach:

1. *Mechanical energy balance*: In this stage, potential energy can be converted into kinetic energy, or vice versa. This is mainly applied to vertical flow paths so that freely falling liquid can be modeled, along with any liquid that is shot upward. During this part of the simulation, the flow path can be determined. The Bernoulli model is then applied.
2. *Projectile travel*: If the liquid metal has a horizontal velocity component, then the flow path will be determined by the combination of this horizontal velocity component and the gravitational force until

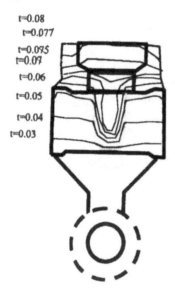

t=0.08
t=0.077
t=0.075
t=0.07
t=0.06
t=0.05
t=0.04
t=0.03

Figure 18 Flow front locations at various filling times (seconds) estimated by flow network schemes for the casting shown in Fig. 16.

the liquid metal hits a "wall," i.e., the casting mold inner surface. This ensures that the correct trajectory is predicted.

3. Once the known velocity components are exhausted, all possible free surface directions are checked. If bottom adjacent cells are empty, these cells are filled first to maintain expected fluid behaviors.

4. If only adjacent empty cells are in the same level or above the current flow level, the Monte Carlo method is applied to the empty cells in the same plane.

5. If only adjacent empty cells are above the current flow level, the flow moves upward.

These principles are implemented in a generic three-dimensional code based on FDM-grids [11]. Some of results from this implementation are shown in Fig. 19.

VI. SUMMARY

In this chapter, several quick analysis techniques have been presented to provide solutions to the foundry industry. Although limited to simple shape cast-

Quick Fill Code   FDM Code

Figure 19 Simulated liquid metal front locations in a metal matrix component casting by QuickFill code and FMD code.

ings, analytical solutions are helpful to determine casting shakeout times. Geometrical analysis would ensure sound casting designs in many cases, even though heat transfer mechanisms are not in the picture. On the other hand, thermal modulus can account for the effects of geometric variations as well as for various heat transfer mechanisms by treating complex heat transfer modes via many simple heat transfer modes. As a result, the thermal modulus works well with common foundry techniques that make use of risers, chills, and insulators.

The major advantage of using the TM model is the significant reduction in CPU time compared with the comprehensive FEM and FDM approaches presented in Chapter 2. The TM model can generate data equivalent to that generated from comprehensive heat transfer approaches in a time that is 2 orders of magnitude shorter.

REFERENCES

1. CM Adams. Thermal considerations in freezing. Liquid Metals and Solidification, ASM, Cleveland, Ohio, 1958.
2. CM Adams, HF Taylor. Trans AFS 65:170–176, 1957.
3. GH Geiger, DR Poirier. Transport Phenomena in Metallurgy. Reading, PA: Addison-Wesley Publishing Company, 1982, pp. 329–360.
4. N Chvorinov. Theory of solidification of castings. Die Giesserei 27:17–24, 1940.

5. SJ Neises, JJ Uicker, RW Heine. Geometric modeling of directional solidification based on section modulus. AFS Transactions 95:25–30, 1987.

6. R Kotschi, L Plutshak. An easy and inexpensive technique to study solidification of castings in three dimensions. AFS Transactions 89:601–610, 1981.

7. G Upadhya, CM Wang, AJ Paul. Solidification modeling: a geometry based approach for defect prediction in castings. Light Metals 1992, Proceedings of Light Metals Division at 121st TMS Annual Meeting in San Diego, CA (ER Cutshall, ed.), pp. 995–998, 1992.

8. R Yagel, SC Lu, AB Rebello, RA Miller. Volume-based reasoning and visualization of diecastability. Proceedings of Visualization '95 in Atlanta, GA (GM Neilson and D Silver, eds.), 1995, pp. 359–362.

9. JH Hill, JT Berry, S Guleyupoglu. Knowledge-based design of rigging systems for light alloy castings. AFS Transactions 99:91–96, 1991.

10. C Cheng, KO Yu. Innovative approach for modeling the heat transfer during casting solidification. Proceedings of the International Symposium on Computational Fluid Dynamics And Heat/Mass Transfer Modeling In: The Metallurgical Industry in Montreal, Quebec, Canada (SA Argyropoulos and F Mucciardi, eds.), 1996, pp. 56–67.

11. C Cheng. Quick and approximate methods and quick analysis software. NCEMT Report, July 16, 1997.

12. RAPID/CAST, Concurrent Technologies Corporation, Johnstown, PA.

13. G Upadhya. Producibility validation of low loss launch valve plug casting. NCEMT Report, August 30, 1995.

14. T Uchida, M Morikawa, S Saito. A method of shrinkage prediction and its application to steel casting practice. E Niyama, AFS Int Cast Metals Inst J 7:52–63, 1983.

15. YW Lee, E Chang, CF Chieu. Modeling of feeding behavior of solidifying Al-7Si-0.3Mg alloy plate casting. Met Trans 21B:357–362, 1990.

16. VK Suri, C Cheng, AJ Paul. Casting porosity prediction: a comparison of some criteria functions with experimental observations. TMS Annual Meeting, San Francisco, CA, February 1994.

17. R Choo, G Upadhya, C Cheng, VK Suri. MK-82 cast ductile iron test cylinder and bomb body casting analyses. NCEMT (National Center for Excellence in Metalworking Technology) Report, July 19, 1995.

18. C Wang. PIM software. NCEMT (National Center for Excellence in Metalworking Technology) Report, July 16, 1997.

8
Electronic Data Interchange

Gerald M. Radack
Concurrent Technologies Corporation, Johnstown, Pennsylvania

I. INTRODUCTION

To create the computational grid for casting process simulation, geometry information on the regions being simulated is needed. Geometry information may be available either in the form of a drawing (blueprint) or in the form of an electronic geometry model. If geometry is obtained in the form of a drawing, then a geometry model must be recreated before the mesh can be generated. Building a geometry model from a drawing is a complex, labor-intensive, and high cost operation. Almost all OEMs use computer-aided design/computer-aided manufacturing (CAD/CAM) software, and thus have electronic geometry models of the parts that they order from foundries. Obtaining these geometry models electronically can result in significant cost and time savings. However, electronic data interchange among various CAD/CAM systems and between the CAD/CAM systems and casting modeling systems is a technological challenge.

The objectives of this chapter are to:

- Introduce the basic technical background regarding electronic data interchange
- Introduce currently available software technologies for performing electronic data interchange, and introduce emerging technologies
- Point out technical problems
- Provide some tips for practical application of electronic data interchange

The emphasis will be on transfer of three-dimensional solid geometry.

II. DEFINITIONS

This section contains some definitions that are used throughout the chapter.

A *product* is any object made by a process, either natural or man-made.

Product data is information about a product, stored in electronic form.

An *application* is a set of processes that exchange product data among themselves. Design and manufacture of castings is an example of an application.

An *entity* is the computer representation of some geometric concept such as surface, curve, or point.

An *algorithm* is a procedure for solving a computation problem. An algorithm is broken into a series of steps, with rules specifying the order of the steps and for recognizing when the procedure is complete. Algorithms are somewhere in between pure mathematics and computer programs. Programmers must implement algorithms in programming languages in order to create working programs.

A *data structure* is an arrangement of data elements used to represent some concept or item. For example, the data structure to represent a point in 3-space could consists of three floating point numbers (the x, y, and z coordinates). A data structure can also incorporate other, simpler data structures, and can include constraints on the components. For example, the data structure to represent a line in 3-space could consist of two instances of the data structure for point, representing the two endpoints of the line, with the constraint that the endpoints must not be the same.

III. THE NEED FOR DATA EXCHANGE

CAD systems are becoming essential tools for foundries. Starting with the desired geometry of the final product, foundry engineers can use the CAD system to design gate and riser systems, patterns and molds. The geometry created in a CAD system can also be input to analysis tools. In order for a company to get the maximum benefit from its investment in CAD, it is important that data be created only once, and reused, not reentered, wherever needed. For example, if a company is using CAD software to design the product and tooling, and is using a finite element analysis package to simulate the solidification process, then it should be able to transfer the geometry from the CAD software to the simulation package. It should not have to reenter the geometry from scratch into the simulation package.

Many customers of foundries have embraced CAD, including key industries such as aerospace and automotive. These customers now make bid packages, including the CAD geometry, available electronically to suppliers. Foundries that can read the geometry into their own CAD systems, rather than reading drawings and reentering the design, have a clear cost and time advantage.

Thus, there is a need to exchange data electronically both between similar systems (e.g., between a customer's and a foundry's CAD systems) and between dissimilar systems (e.g., between a foundry's CAD system and a casting analysis system). While this book is on casting analysis, a company's strategy for data management and data exchange is rarely driven by analysis alone, and the maximum benefit is achieved when the same information is used widely. So we will examine data exchange in the broader context of the casting enterprise.

As is the case with humans, data exchange can only occur between two systems if they "speak the same language" or if a translator is available to convert between the two languages that the systems understand. The simplest form of "language" for communicating geometry, one that is supported by all CAD/CAM systems, is a file format with a given set of rules for syntax and semantics. A more sophisticated form of language, supported by many CAD/CAM systems, is a subroutine library or message-passing protocol that allows two programs to communicate directly rather than passing files. In this chapter, we will focus on the file transfer form of communications.

Unfortunately, each CAD/CAM system has its own way to represent geometric information both internally and in an external file—its own "native language." Conceptually, the simplest solution to the data exchange problem would be for all parties who need to exchange data have the same CAD/CAM system. However, due to the number of CAD/CAM systems on the market and the lack of a clear market leader, this option is rarely practical. Some OEMs in the automotive industry have provided incentives for their suppliers to use the same CAD/CAM system that they do. However, since each OEM uses a different CAD/CAM system, unless a foundry is content to supply to only one OEM, it will still need to support a separate CAD/CAM system for each customer or use translators.

IV. ELECTRONIC REPRESENTATION OF GEOMETRY

In order to understand data exchange issues, it is necessary to know something about how geometry is represented electronically by CAD/CAM and related systems. A discussion of the mathematical foundations of computer geometry is beyond the scope of this book, so we will only introduce a few key concepts here and rely on intuition for the rest.

From the perspective of CAD/CAM systems, geometric objects are composed of points, either in the Euclidean plane $\mathcal{R}^2$ or 3-space $\mathcal{R}^3$. Finding efficient ways to keep track of the points that make up an object and developing algorithms to perform desired operations efficiently on objects represented by the chosen data structure are two key challenges facing CAD/CAM system researchers and designers. But since all electronic representations of geometry depend on numbers, we first discuss briefly how numbers are represented in the computer.

A. Computer Representation of Numbers

Modern computers store data in the form binary digits, or bits. Each bit can have a value of 0 or 1. A group of n bits can have 2^n different values. Computer memory is generally organized into groups of 32 bits, called words.

Computer hardware can process two numeric data formats: fixed and floating point. The fixed point format represents a number as a sign bit followed by a series of data bits. Fixed point numbers can be unsigned (only positive values) or signed (positive and negative values). A single word can represent an integer in the range of 0 to $2^{32} - 1$ (unsigned), or 2^{-31} to $2^{31} - 1$ (signed).

The floating point format (sometimes called real numbers) represents a number using the equivalent of scientific notation, as: $\pm m \times 2^e$ or $\pm m \times 10^e$, where m is called the mantissa and e is called the exponent. The number of bits allocated for the mantissa and exponent, which varies between different computer families, will determine the range of magnitudes and the precision that can be achieved with floating point numbers. It is important to note that floating point numbers in a computer differ from the real numbers of mathematics in some important ways:

- Since there are an infinite number of real numbers and only a finite number of values that can be represented by any computer's floating point data type, many real numbers cannot be represented exactly and must be approximated. Irrational numbers such as π can never be represented exactly using floating point.
- There is a largest and smallest number that can be represented by a computer's floating point data type, whereas there is no largest or smallest real number in mathematics.

In addition to the imprecision of the representation itself, additional errors will creep in during floating point arithmetic calculations when a program is executed.

The floating point data type generally comes in two sizes: single precision, which uses a total of 32 bits, and double precision, which uses a total of 64 bits. Some CAD/CAM systems use even more bits to get higher precision. However, many algorithms used in CAD/CAM systems are highly sensitive to numerical inaccuracies, and even small inaccuracies can cause problems. Adding more bits to the floating point representation is not a panacea, and incurs the cost of greater memory usage, slower processing, and larger file sizes.

For more information on this subject, see Ref. 1.

B. Classes of Representations

Computer representations of geometry can be classified as evaluated and unevaluated. An evaluated representation consists of data directly corresponding to the shape. An unevaluated representation, on the other hand, keeps track of information that can be used to generate the shape. As an example, suppose we want to represent the shape of a polygon in $\mathcal{R}^2$. We can give a series of (x, y) coordinates for the vertices of the polygon (an evaluated representation), or we can give a set of linear inequalities of the form $ax + by + c > 0$, which must be solved to find the polygon (an unevaluated representation). In the following subsections, we discuss boundary representation, the major evaluated representation in use for CAD today, and constructive solid geometry, a major unevaluated representation.

The introductions to representation types in the following subsections are by necessity sketchy. For further details, see Ref. 2 or 3.

C. Boundary Representation

A boundary representation (B-rep) system explicitly keeps track of the boundary of the part. Since the boundary of a solid, its surface, separates the interior from the exterior, if you know the boundary, you know what points are in the interior of the solid. In principle, a B-rep could consist of a single mathematical equation for the surface of the object. However, for most real-world shapes, such an equation cannot be defined conveniently or at all. Therefore, we resort to building the boundary of a solid from a set of surface patches. The simplest surface patch is the triangle. Any surface shape can be approximated by a sufficiently large number of small triangles. More generally, any polygon shape can be used. A boundary representation made up of flat polygons is called faceted and the individual polygons that make up the boundary are called facets. By allowing the surface patches to be curved and allowing them to have "holes," the number of surface patches needed to represent a typical object is greatly reduced. This can improve performance of the system considerably. The shapes for curved surfaces can be either based on "primitives" (shapes built-in to the system such as spheres and cylinders) or "free-form" (arbitrary shapes that are created, for example, by fitting a surface to a set of data points). A variety of schemes are used to specify free-form surfaces, but they can be reduced to sets of equations for which the surface points constitute the solution. CAD/CAM systems are not general equation solvers—all surfaces within a system must be represented in a mathematical form that the system can understand. Thus, in many cases, an approximation must be used that approximates but does not exactly give the desired shape.

A boundary representation data structure generally divides the shape definition into two aspects: the geometry and the topology.

The geometry aspect of the data structure consists of pure shape information, ignoring how the shapes fit together to form a solid model. The primary elements of geometry are:

- Points, represented by x, y, z coordinates
- Curves, represented by equations in $x, y,$ and z with 1 degree of freedom
- Surfaces, represented by equations in $x, y,$ and z with 2 degrees of freedom

The topology aspect of the data structure tells how the points, curves and surfaces fit together to form the boundary of a solid. The primary elements of the topology of a solid model are closed surfaces, faces, edges, and vertices.

The boundary of a solid is represented by one or more closed surfaces. One closed surface represents the boundary of the outside of the solid; the other surfaces, if they exist, represent the boundaries of any voids in the solid. A closed surface is split up into as many faces as are needed for ease of representation. Associated with each face is a surface patch that is conceptually a piece that has been cut out of a larger (possibly infinite) surface.

The interface between two adjacent faces is an edge. Associated with each edge is a curve segment that is conceptually a piece that has been cut out of a larger (possibly infinite) curve. A curve segment is characterized by its curve geometry and its endpoints.

A B-rep CAD system keeps track of all solid models simply by their boundaries. Advantages include:

- Complex shapes can be created easily and treated in a consistent way.
- Geometry can be easily imported from a variety of sources.

Disadvantages include:

- Relationships between shapes used in the design process are not maintained. Thus, if the operator asked the CAD system to combine two shapes A and B together using a modeling operation (e.g., intersection) to form a new shape C, the system would apply mathematical algorithms to compute the B-rep of shape C. Once the B-rep of C has been computed, the link between C and A and B is lost. Any later changes to the shape of A or B would not be reflected in C.

- B-reps of complex shapes are seldom mathematically exact. This can lead to errors, especially when data are transferred between different systems that make different assumptions about the representation.

D. Constructive Solid Geometry

Constructive solid geometry (CSG) is an alternative to B-rep for representing complex shapes. In a CSG system, as in a B-rep system, the operator builds up the desired shape from simpler shapes using operators, starting with primitives. The difference is that if the CSG system is asked to perform a modeling operation on A and B to form C, it does not explicitly calculate the B-rep of C. Rather, it stores the fact that C is formed from A and B. For example, if C is the intersection of A and B, then C is represented in the computer as the following statement:

$C = $ intersection (A, B)

The CSG representation of an object is a set of statements of the above form, collectively called the tree. Whenever any operations must be performed on a solid (e.g., computing the volume, creating a finite element mesh, or displaying on a computer screen) they are done directly from the tree. See Fig. 1 for an example of a CSG representation.

E. Parametric Representations

Parametric representations are similar to CSG in that they are maintained by the CAD/CAM system in an unevaluated form. The difference is that parameters controlling size and location of objects and the various modeling operations can be based on variables, expressions and constraints entered by the designer instead of just numeric values. This makes it easy to change the model as the design process progresses, or to represent entire families of products with a single model just changing a few variables. Most mid to high end CAD systems today have at least some parametric capabilities.

F. Hybrid Representations

Many CAD systems today have elements of both CSG and B-rep to try to reap the benefits of both—they maintain a tree representation, but a B-rep is automatically computed for each node in the tree. They can also use an arbitrary B-

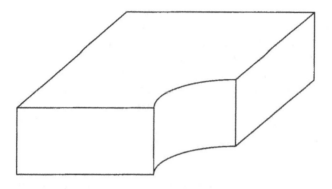

```
# create a box B with width 4, depth 5, height 2
# centered at the origin
B = box(4.0, 5.0, 2.0)
# create a cylinder C with diameter 3, height 2
# centered at the origin
C = cylinder(3.0, 2.0)
# transform cylinder C so that it is centered at (2,-2.5,0)
# call the result Ct
Ct = translate(C1, 2.0, -2.5, 0.0)
# subtract the volume of cylinder Ct from box B
# call the result Obj
Obj = difference(B, Ct)
```

Figure 1 A solid with its CSG representation. Lines starting with "#" are comments and not part of the representation.

rep solid as a primitive for CSG operations. Such hybrid systems avoid some of the disadvantages of pure CSG and B-rep systems, at the cost of more complex data structures.

V. DATA EXCHANGE

In order to produce a casting, geometry information must be exchanged between many parties. Each of these parties may be using a variety of CAD/ CAM systems, analysis systems, or other computer programs.

A. Native File Formats

Each CAD/CAM system has a "native" file format for storing geometry information. The structure of the data within a native file format generally corresponds closely to the data structures that the vendor uses within the CAD/CAM system. Because of the close correspondence between the internal data structures and the file structure, files can be read and written quickly, with no loss of accuracy.

No two vendors' native formats are the same, for a variety of reasons:

- Each CAD/CAM or analysis system has a different set of features, requiring a different set of data structures to support it efficiently.
- Even ignoring such differences, each vendor devises its own set of data structures in order to try to optimize speed and reliability of its system, make development of the system and add-ons easier, etc.
- Industry-standardized three-dimensional geometry file formats were considered inadequate for representing the data within a commercial CAD/CAM system.

Vendors also sometimes change the native file format for their software when they release a new version.

B. Translators

A translator is simply a program that converts from one geometry data format to another. For simplicity, we will assume in this discussion that each translator reads in a file in one CAD/CAM system's native format and writes out a file in another CAD/CAM system's native format. It is also possible for a translator to interface to a CAD/CAM through a mechanism other than reading and writing files, e.g., making direct access to the system's database while the system is running. This simplification does not alter the general results or conclusions of this chapter.

Translators may perform other services such as checking the input file for errors, fixing modeling problems, and printing statistics about the solid model being translated.

Translators can be obtained from a variety of sources. The primary sources for translators are the CAD/CAM vendors themselves, and third party software companies who specialize in developing translators. If one needs to translate from format A to format B and no translator is available between the two systems, then there are two options:

- Find an intermediate format C such that there is a translator from A to C and from C to B. This could be another proprietary format or a neutral format (see Sec. VI).
- Develop a custom translator, either in-house or by contract with a software development organization. This option is affordable only to very large organizations.

C. Selecting a Translator

In the fortunate case where more than one translator is available to translate between systems A and B, a company is faced with deciding which one to use. Selecting a translator is much like selecting any piece of software. An evaluation plan with specific criteria for choosing a winner should be decided upon before beginning the selection process. The following general criteria should be considered:

- *Quality of translation*: Are geometric models valid and usable after reading into system B? Methods for determining this will be discussed in Sec. IV.C.
- *Performance*: Does the software translate quickly enough, without making undue demands on system resources (e.g., memory)?
- *Software usability*: How hard is it to understand and operate the software? Does the software have adequate on-line help and documentation?
- *Training*: Is training required? If so, are adequate training courses available, at convenient times and locations? Will the software vendor teach the course on-site? Is videotape or on-line training available?
- *Software support*: What support is available (800 number, e-mail, etc.)? Will answers be provided in a timely manner? Is there a charge per incident or a flat fee?
- *Cost.*

Instead of purchasing a translator, an organization can outsource file translation. Several companies specialize in CAD translation and provide services such as on-line translation over the World Wide Web, batch services for translating large sets of files, and consulting to fix problems that cannot be handled by translation software automatically. There is at least one company providing free translation between several formats over the Web. Before using an on-line translation service, a potential user should ensure that the service provider adequately protects the privacy of proprietary data.

D. Point-to-Point Translation vs. Neutral Formats

The most obvious way to exchange data between a variety of systems is to use a set of point-to-point translators. A point-to-point translator is a special purpose program that directly translates the file format of program A to the file format of program B. (See Fig. 2.) This can lead to a proliferation of translators. For each pair of programs A and B, one would need a translator from A

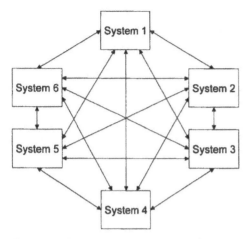

Figure 2 Point-to-point data transfer between n programs requires $n(n-1)$ translators. Each arrowhead represents a translator.

to B and from B to A. This means that if there are n programs, there could be up to $n(n-1)$ translators. Disadvantages of this approach include:

- *Cost*: Each translator must be purchased (or written in-house) and maintained, and users must be trained in how to operate it.
- *Configuration management problems*: Every time the vendor changes the file format of program X, the translators to and from program X must also be updated. Such updates typically lag the release of the program, resulting in a temporary inability to transfer data (or to use the latest release of a program) until the translator is available. Also, coordinating software updates in a large installation can be a time-consuming and painful process.

Instead of translating directly between every pair of systems as needed, we can add an extra step of translating to a common "neutral" format (see Fig. 3). Instead of employing translators to and from every program's format, translators are employed to and from the neutral format only. This reduces the number of translators to $2n$.

The neutral format should be any format powerful enough to hold all the data input to or output from each program involved in the data exchange. Although there could be separate neutral formats for different stages of the product life cycle (e.g., design and analysis), the trend toward concurrent engineering implies a need for a common, integrated neutral format with all data needed throughout the life cycle of the part. The STEP standard (see Sec. VI.D) is intended to meet this need.

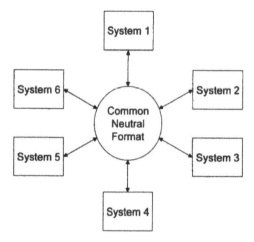

Figure 3 Transfer using a common neutral data format reduces the number of translators required to $2n$. Each arrowhead represents a translator.

A translator that translates from some proprietary format to a neutral format is called a preprocessor. A translator that translates from a neutral format to a proprietary format is called a postprocessor.

E. Translation Testing

We can consider the translation process to consist of two systems, system A and system B, the translator software, and the procedures used to perform the translation. System B can be either a CAD/CAM system, a mesh generator, or any other software that needs a solid model as input. The procedure used to perform the translation includes any settings in system A, system B, or the translator that may affect the result of the translation. Before making a purchasing decision on a translator, the entire translation process should be tested to ensure that it produces satisfactory results. It is also recommended that the testing be repeated before a new version of any of the involved software products is put into production.

Assume that a solid model has been created in system A, saved to system A's native file format, translated from system A's native file format to system B's native file format, and read into system B. The following criteria can be used to assess the quality of the translation process.*

---

* The remainder of this subsection is based on Ref. 4.

1. Does system B load geometry from the file without error?
2. Does system B recognize the geometry loaded as a solid model?
3. Does the solid model loaded into system B represent the same shape as the solid model saved from system A?
4. Can further modeling operations be applied to the model in system B (if system B is a CAD system)?
5. Is the quality of the model adequate for use by downstream applications (e.g., casting analysis)?

Item 1 is trivial to test.

Item 2 is easy to test if system B is a CAD system. Most CAD systems have built-in commands for checking the validity of a solid model.

To check item 3, a number of tests can be performed, such as:

- Take screen shots from various angles in both systems and see if the object looks the same.
- Take various cross sections on both systems, and see if the cross sections look the same.
- Compute metrics such as the surface area, volume, center of gravity, and moments of inertia on both systems and see if they are "close enough." The amount of acceptable deviation will depend on the intended use for the model. Organizations will have to decide how much deviation they can accept. In most cases, the values should differ by less than 1%.
- Compute metrics for various cross sections on both systems and compare them as in the previous item.

To test item 4, one can perform various CAD operations on the geometry in system B, for example, cutting holes in the model.

To test item 5, one could, for example, run the casting analysis software and see if accepts the geometry and produces reasonable results.

There is model checking software available for some CAD formats from third party software vendors. Such software is particularly useful if system B cannot perform the CAD operations described for items 3 and 4 above, e.g., if system B is a casting analysis program and no CAD system is available that understands system B's input format. Even if system B is a CAD system, such software can be very useful in identifying and isolating the sources of problems.

F. Causes of Translation Problems

A problem translating directly from system A to system B may be due to one of the following causes:

- The model was created using poor modeling practice in system A.

- There is an error in system A's code for saving to a file.
- There is an error in the translator.
- There is an error in system B's code for reading from a file.
- The version of system A or system B is newer than that supported by the current version of the translator.
- The user must set certain parameters in system A, the translator, or system B, in order to get translation to work correctly.

1. Numerical Accuracy Problem

One frequent source of translation problems is incompatibilities in the way different systems handle the numerical accuracy problem.

Geometric models produced by CAD/CAM systems are almost always inexact because:

- Most real numbers (in the mathematical sense) cannot be represented exactly using the floating point format (see Sec. IV.A).
- The data representation (e.g., the set of mathematical forms supported by the system for representing surface or curve geometry) is inherently unable to represent the needed shape exactly.
- The algorithms used by the system to evaluate the results of various modeling operations are designed to produce approximations, not mathematically exact results. This could be because (1) there is no known closed-form solution for the result of a modeling operation; or (2) there is a known closed-form solution, but it requires the use of a mathematical form that is not supported by the CAD system or is too costly to compute.

The above presents a dilemma for programmers, since their code is based on mathematical formulas that assume all numbers are represented exactly. These formulas do not take into account the inexact nature of both the input values (taken from the solid model, which is an approximate representation of an ideal shape) and the computer arithmetic used to perform the calculations.

A typical manifestation of this problem is that when performing a computation, two values that should be equal will actually differ by a small amount. Programmers compensate by using a small floating point value **E**, the error tolerance, whenever comparing values. For example, suppose that a programmer is implementing an algorithm that needs to compare points $\mathbf{p} = (p_x, p_y, p_z)$ and $\mathbf{q} = (q_x, q_y, q_z)$ to see if they are the same. Rather than testing whether the coordinates of the points are identical, i.e., testing that

$$p_x = q_x \qquad p_y = q_y \qquad p_z = q_z$$

he would instead test that the computed distance between the points is less than **E**, i.e., that

$$\text{dist}(\mathbf{p},\ \mathbf{q}) = \left((p_x - q_x)^2 + (p_x - q_x)^2 + (p_x - q_x)^2\right)^{1/2} < \mathbf{E}$$

Because of this practice, some anomalies can creep into a solid model, for instance:

- Small gaps between surface patches that are supposed to be adjacent on the boundary of the solid.
- Points that are supposed to be on a surface (e.g., a plane) falling slightly off it.

These types of anomalies generally do not cause problems in the system in which the solid model was created, since each CAD system is more or less self-consistent; that is, it knows how to deal with the anomalies that it itself generates. However, different systems have different values for **E** and operate under different sets of assumptions concerning the model, and therefore allow a different set of anomalies and different code to compensate for those anomalies. So when a model is read into a system different from the one that created the model, the receiving system could fail to load the geometry, could load it but declare the model not to be a valid solid, etc. In general, there is little problem transferring a solid if the receiving system has a larger value of **E** than the sending system. It is when the receiving system has a smaller value of **E** that problems are most likely to arise. Often it is possible to "clean up" the model in the receiving system, but this can be a labor-intensive process. Sometimes the user can set a value for the error tolerance before reading in a model. A translator can also attempt to clean the model as part of the translation process. This activity needs to be monitored carefully, however, since attempts to fix such problems automatically may lead to incorrect results.

For discussion of the problems of numeric accuracy for geometric computing from a theoretical standpoint, see Refs. 5 and 6.

2. Data Representation Incompatibilities

Another common source of translation problems is incompatibilities in the way different CAD/CAM systems represent geometry. For example, if an object is modeled in system A using CSG and system B only supports B-rep, then the translator will have to convert the model to B-rep, losing the tree structure in the process.

Going from a more powerful representation to a less powerful one generally requires that surfaces be split up. To understand why, consider as an analogy the representation of curves using polynomials. It is a well-known theorem of mathematics that a curve of the form $y = f(x)$, where f is a polynomial, can have as many inflection points as the degree of the polynomial minus 1. Suppose that curve C has 9 inflection points and is represented in system A with a single polynomial of degree 10. Suppose further that system B only supports polynomials of degree 3 (a "less powerful" representation). Then the translator from A to B will have to split the curve into several segments so that each segment can be represented by a polynomial of degree 3.

This decomposition of geometry leads to larger data sets, which take longer to process, require larger files for storage, etc. It also increases the chance of numerical accuracy problems creeping into the model (see Sec. V.F.1).

G. Testing

If a neutral format is used to transfer data from system A to system B, there will be two translators involved (see Fig. 4a):

- A translator from system A to the neutral format (the "preprocessor for A")
- A translator from the neutral format to system B (the "postprocessor for B")

Once an error is discovered during testing, isolating its source becomes an exercise in problem solving. Figure 4b shows a complete test set-up. By testing various paths through the diagram, sources of errors can be isolated. An

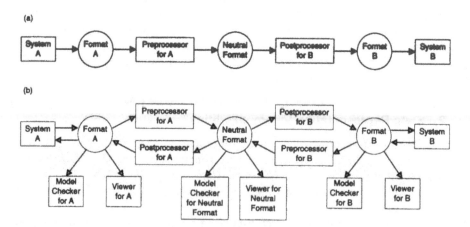

Figure 4 (a) Data transfer using a neutral format. (b) Complete data transfer test setup.

important test is the roundtrip test: saving a model from system A and translating to the neutral format, then translating back and reading the file back into system A. This can be extended to go all the way from system A to system B then back. One can also start a roundtrip test with the neutral format, by using preexisting files that are known (or at least believed) to be good. Such files are often provided with translators, or on the Web sites of companies in the translation business. For greatest benefit, such files should have been generated by software other than the software under test.

A "model viewer" is a program that is designed to display geometry from a given file format, but not to perform any CAD/CAM operations. A "model checker" is a program specifically designed to check a geometric model for errors and questionable items. For maximum benefit, such testing software should come from a separate "code base" than the software under test. In general, this means that it should be from a company other than the ones providing the CAD/CAM systems and the translators. However, one must be careful because there are software components on the market for geometry manipulation and translation that are commonly used by CAD/CAM and translator software vendors. The same software components may be incorporated into two products from different vendors, in which case the code base for the products would overlap.

Any errors in translation from system A to system B that are caused by the translators themselves, and not by poor modeling practice or errors in systems A and B themselves, can be classified as follows:

1. Differing but valid interpretations of the neutral format specification
2. All others (misunderstandings of the neutral format specification down to outright bugs in the code)

Using a neutral format that is specified unambiguously can eliminate errors of type 1.

Using translators that are certified by an independent testing laboratory as conforming to the neutral format can reduce the likelihood of errors of type 2.

H. Binary and Character Formats

Data can be exchanged in either character or binary format. Using a binary format, data structures are exchanged more or less as they are represented in the computer. So a word in the data (representing, say, a real number) would be written out as 4 bytes (32 bits) without first being converted into a sequence of characters recognizable by a human.

If a character format is used, the data are first converted to a series of readable characters before being written to the file. A binary format file is generally more compact than the corresponding character format file, and can be input and output faster by a computer. However, binary format file cannot be read directly by a human; a special program must be run to display the contents in human-readable form. Different computers represent data differently in binary form (for example, one computer might use 7 bits for the exponent of a floating point number, while another might use 9). As a result, trying to exchange data in binary form between different computer platforms can result in errors unless the software uses a standard, platform-independent format—such as the ANSI/IEEE 754 standard for floating point numbers. Doing the necessary conversion to this platform-independent format can cancel some of the time savings of using a binary format.

A character format file can be read directly by humans and even corrected if necessary using widely available software (a text editor or word processor). Instead of using a binary format when compactness is required, one can use a compressed character format, i.e., run a character format file through a compression algorithm [7]. General-purpose file compression programs or archive utilities can be used for this purpose.

It is recommended that a character format be used if possible.

VI. NEUTRAL FORMATS

In this section, we look at three neutral formats that are in widespread use for data interchange: stereolithography (STL) format, Initial Graphics Exchange Specification (IGES) and Standard for the Exchange of Product Model Data (STEP).

A. Standards

A neutral format should ideally be a standard. Standards are specifications published by organizations that are chartered by governments as standards making bodies. These groups require that a specification undergo thorough review and achieve international consensus approval before becoming a standard. Standards generally change less frequently than other formats, and are more rigorously defined. Standards represent the broadest set of interests— views of users, technology providers, and other interested parties are solicited and considered during the standards development process. These factors allow companies to protect their investment in data represented according to the standard. Also, there is an increasing emphasis on the use of international standards in international trade.

B. Stereolithography (STL) Format

Stereolithography is the most commonly used rapid prototype technology and uses STL file format to represent article geometry. The STL file format is probably the simplest format for three-dimensional geometry exchange. STL is not a formal standard but has become a part of the industry "folklore" and is simple enough that no maintenance of the specification is necessary. It is widely understood by CAD/CAM, rapid prototyping, and analysis systems.

STL is a faceted boundary representation in which the facets are all triangular. An STL file consists of a sequence of facet definitions. Each facet definition has the coordinates of the facet's normal and its three vertices. Figure 5 shows a tetrahedron and the STL character format file that represents it. A binary STL format also exists. Note the duplication of coordinates in the STL file. Each vertex appears in three times in the STL file, since it is contained on three facets.

Because of its extreme simplicity, it is very easy to construct software to read and write the STL format.

A major disadvantage of the STL file format is that it does not include any information on connectivity between the facets. Because the vertices of each facet are represented separately in the STL file, there is no guarantee that vertices which should coincide will do so exactly (see Fig. 6). This can lead to errors in subsequent processing of the STL data. For example, consider an

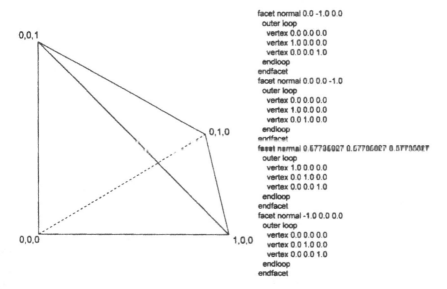

Figure 5   A tetrahedron and its STL file representation.

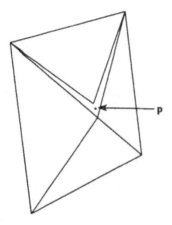

Figure 6 Gaps in the surface caused by displacement of vertices due to numerical or computational errors.

algorithm that needs to compute the intersection of the surface in Fig. 6 with a line perpendicular to the plane of the drawing. If a program is unlucky enough to choose a line that goes through point p, it will not find any intersection, resulting in incorrect results or outright failure of the program. In addition, connectivity between facets is not maintained in the data structure, making evaluation of many modeling operations more difficult and error prone. This is a special case of a more general problem that occurs with almost any CAD system and geometry representation (see Sec. V.F.1), but STL's lack of connectivity between facets makes the problem worse.

STL shares another problem with all faceted representations: the large number of facets that are needed to represent curved surfaces.

Because of the above problems, STL is not generally used for CAD to CAD exchange.

C. Initial Graphics Exchange Specification (IGES)

IGES is an American National Standard (ANS USPRO/IPO-100) for exchange of product definition data. IGES started out as a standard for exchanging two-dimensional drawings electronically between computer-aided drafting systems. IGES has been extended to include three-dimensional surfaces, finite element data, tolerances, piping definition, and layered electrical product definition.

The initial version of IGES was published in 1980. Version 5.3 was released in May 1997.

IGES uses a character file format. However, it is not very readable by humans, and is also rather inefficient for processing by computers.

Each entity in an IGES file is assigned to a layer. Layers correspond to the transparent sheets that engineers traditionally used for CAD drawings. By overlaying different sheets on top of a base, one can see different aspects of the drawing. For example, mechanical, electrical, and plumbing systems could be on different layers of a drawing.

IGES contains the following classes of entities:

- Curve and surface geometry (two- and three-dimensional)
- Annotation
- CSG
- Entity relationships
- Drawings and views
- Finite element

Some of the problems that have been experienced with IGES include [8]:

- Complexity and slowness of code to read and write IGES physical files, due to the complex file structure
- Ambiguities in the standard, leading to errors when transferring data between different systems (because the software developers for the vendors interpreted the standard differently)
- Lack of support for solid models
- Lack of an effective means to extend the standard to satisfy needs of specific applications and to cover product data from throughout the product life cycle
- Errors in communication, due to the receiving system supporting a different subset of IGES entities, or a different number of layers, from the sending system
- Lack of any means for testing translators for conformance to the standard

D. Standard for the Exchange of Product Model Data (STEP)

STEP is an International Organization for Standardization (ISO) standard (ISO 10303, "Product Data Representation and Exchange") that is designed to be the "next generation" product data standard. It expands the scope of IGES while avoiding the problems that occurred with IGES. In particular:

- STEP is designed from the ground up to represent all aspects of a product, not just geometry and drafting information. The objective of the STEP project is to create a single standard for representation, storage, and exchange of all product data, across all discipline views and across the entire life cycle of a product.

- Data structures in STEP are specified abstractly, independent of the implementation technology (e.g., the file syntax). This allows new technologies to be "plugged in" without rewriting large parts of the standard.

- The standard is structured so that it can be extended to meet new requirements without disturbing existing users. There is an emphasis on upward compatibility, so that existing files can be read in the future.

To STEP, a product can be any manufactured or natural object. STEP is being used by industries as diverse as shipbuilding, aerospace, road building, furniture, and building construction.

The desire to create a single standard encompassing all product data was motivated by the following observations:

- Concurrent engineering requires that all aspects of a product and its entire life cycle be considered simultaneously during product development. To accomplish this, the data concerning all aspects of the product and all life cycle stages must be integrated together holistically.

- There is no clean separation along discipline or product lines. For example, many industrial products now have sophisticated computers and software embedded in them; designers of printed circuit boards must consider both electrical and mechanical characteristics.

- Software vendors would like to sell families of products to as wide a market as possible. Using the same exchange standard allows commonality of code, thus reducing development costs.

Although STEP supports the representation of a wide variety of product and process data, STEP does not prescribe what data must be exchanged. It is up to the parties doing the exchange to agree on what will be included in an exchange file.

The initial release of the standard was in the fall of 1994. Work to extend and enhance the standard is continuing at a rapid pace.

For more background on STEP, see Refs. 8–10. Up-to-date information about the status of the STEP standard can be found on the World Wide Web site of the ISO subcommittee developing the standard (http://www.nist.gov/sc4).

1. Structure of STEP

The STEP standard is rather large and complex and is published as a series of documents called "parts." We will briefly describe the aspects of STEP that are of most interest to someone trying to use the standard.

The STEP standard is based on a set of set of generic data structures called the Integrated Resources (IRs), which are independent of application, industry, etc. For example, there is an IR for "line," but line can have many meanings depending on the context. Application Protocols (APs) describe how subsets of the IRs are used to meet the needs of specific applications (disciplines, industries, etc.). Patterns of IR usage that occur across multiple APs are codified in documents called Application Interpreted Constructs (AICs). For example, there are AICs for faceted B-rep and CSG. Since the AICs are incorporated by reference in the AP documents, faceted B-rep will be represented in essentially the same way in each AP, making it possible to exchange common data between translators supporting different APs.

Translators are expected to satisfy the requirements of specific APs, not "STEP" as a whole. Accompanying each AP is an Abstract Test Suite (ATS) which provides a series of tests that an independent laboratory can use to judge whether a translator conforms to the AP. The most widely used AP today is probably STEP part 203, "Configuration Controlled Design," commonly called "AP 203." APs often define subsets called "conformance classes." For example, see Table 1 for the AP 203 conformance classes. Note that the correspondence to AICs is conceptual since AP 203 does not currently use the AICs (they were invented after AP 203 was published).

2. Status of STEP

Since its initial release, there have been many demonstrations of STEP-based data exchange. Several major OEMs are using STEP in production for at least some of their data interchange.[*]

AP 203 is the maturest AP that supports three-dimensional geometry, and is best supported by commercial translators at this writing. AP 203 should meet most needs for transfer of geometry into casting analysis systems. One major deficiency is that it does not support CSG. This will probably be added to a second edition, currently in the planning stages. Another problem is that given AP 203's focus, conforming translators must include configuration control information; however, some CAD/CAM systems do not maintain such information. This will be fixed in AP 203 through the addition of new con-

---

[*] For a collection of press releases concerning the use of STEP by several OEMs in the automotive and aerospace industries, see http://pdesinc.aticorp.org/news.html.

Table 1 Geometry Support Within STEP

AIC part number	AIC title	Corresponding AP 203 conformance class
501	Edge-based wireframe	3
502	Shell-based wireframe	3
503	Geometrically bounded 2D wireframe	
507	Geometrically bounded surface	2
508	Non-manifold surface	
509	Manifold surface	4
510	Geometrically bounded wire frame	2
511	Typologically bounded surface	
512	Faceted boundary representation	5
513	Elementary boundary representation	
514	Advanced boundary representation	6
515	Constructive solid geometry	

formance classes requiring minimal configuration control information. Also, AP 205, "Mechanical Design Using Surface Representation," will allow pure geometry exchange in the future, but it is currently still in development and not yet supported by translators.

However, it is important to note that moving geometry data between different APs is not difficult provided that they support the same geometry AICs.

The IGES/PDES Organization (IPO), the body responsible for developing the IGES standard, and for leading U.S. efforts for STEP standards development, has determined as a matter of policy that version 6.0 will be the final release of IGES with functional enhancements. Upon release of IGES 6.0, IPO efforts will be directed toward migrating users from IGES to STEP. The migration from IGES to STEP will likely be gradual; vendors will likely continue to sell and support IGES translators for some time to come; and IPO plans to support IGES with minor changes to correct errors in the standard.

E. Choosing a Neutral Formats

The choice of neutral format will depend on cost, availability and quality of translators, capabilities of customers and suppliers, etc.

Table 2 contains a comparison of STL, IGES, and STEP. Some important considerations are:

- STL and IGES support exchange of surfaces. STEP supports exchange of solids as well as surfaces.

Table 2 Comparison of STL, IGES, and STEP

Representation	STL	IGES	STEP
Most advanced 3D geometry type supported	Faceted	Surfaces	B-rep solids, CSG
Suitability for further modeling operations	Low	Medium	High
Accuracy of geometry representation	Less accurate	More accurate	More accurate
Types of information supported	Geometry	Geometry, drafting presentation	Geometry, tolerances, presentation, process plans, materials, analysis, etc.
Status as standard	Not a formal standard	U.S. standard, widely accepted around the world	International standard
Maturity	Highly mature	Mature	Portions still undergoing development; boundary representation is well tested.

- STL supports only faceted surfaces, whereas IGES and STEP support curved surfaces.
- STL supports only geometry. IGES excels at exchange of drawing information. STEP is the only neutral format that can handle a wide variety of "nongeometry" data from across the product life cycle.
- If one is using IGES as a neutral format, it is important to select translators that support the same subset of IGES entities.
- If one is using STEP as a neutral format, it is important that all translators selected support at least one common conformance class of a common AP.

In many cases, an organization may find it necessary to support more than one neutral format. However, if a single neutral format were to be chosen today, STEP AP 203 is a good choice for many. Of the three neutral formats described, it provides the best support for three-dimensional solid models, and the best growth path allowing a manufacturing enterprise to capture key data about product and process in an integrated repository supporting concurrent engineering.

VII. STEP APPLICATION PROTOCOL FOR CASTING (AP 223)

The objective of STEP AP for casting (AP 223) is to provide a neutral format for data relating to cast parts, from customer requirements through delivery [11]. The focus of AP 223 is on data needed throughout the life cycle of a cast part, including the specification, analysis, process engineering, manufacturing, and delivery phases. In addition to information maintained by a foundry, AP 223 supports communication between a customer and a foundry, and between a foundry and its suppliers.

A. Units of Functionality

AP 223 is organized into the Units of Functionality (UoFs) described below:

The *product specification* UoF allows specification of details about a cast part, or a part used to make a casting. The latter includes mold and die components and assemblies, patterns, pattern masters, cores, inserts, flasks, and secondary tooling. The cast part can be represented at different life cycle stages, e.g., customer part design, raw casting, and finished casting. Information that can be included in a product specification includes geometry, material composition, tolerances, applicable specifications and standards, process requirements (including heat treatments), property requirements (e.g., tensile strength), and inspection, test, and reporting requirements.

The *product management* UoF allows representation of versions of product and process specifications, approvals, and references to external items (computer files, pictures, specifications and standards, etc.). It also allows representation of configuration management concepts such as requests for change, request for clarification, and design update.

The *substance composition* UoF allows representation of materials data in two ways: composition in terms of chemical elements and composition in terms of physical structure (macro- and microstructure). The UoF can be used to represent this information at different life cycle stages:

as specified by the customer, as predicted by analysis, as built (the result of laboratory tests), etc.

The *process plan* UoF permits the representation of the process used to make a casting.

The *tolerance* UoF is used to represent tolerances on a part. For cast parts, certain special tolerances can be applied, as specified in ISO standards 8073 and 10135.

The *geometry* UoF is used to represent geometry of parts. Two-dimensional and three-dimensional boundary representations are supported.

The *property* UoF is used to represent property values, and relationships between property values.

The *quality assurance* UoF is used to store information about actual manufactured parts, including inspection and test results. Data can be stored on individual parts. Information on lots and heats can be maintained as well.

The *simulation* UoF allows representation of records of simulation runs, including simulation inputs and outputs. AP 223's data model for simulation is designed to be a general model which can handle any kind of simulation data—finite element, finite difference, geometric, etc.

B. Status of AP 223

As of July 1999, AP 223 existed as a working draft [12]. It must go through several ISO ballot cycles, advancing to Committee Draft and then to Draft International Standard before becoming an International Standard. Software vendors are typically reluctant to implement a STEP AP until it is close to approval as an International Standard. Publication of AP 223 as an International Standard is not likely to happen until 2002 at the earliest. In the mean time, AP 223 can be used in an experimental environment by researchers and industrial groups with the information technology resources to implement the necessary software themselves

VIII. SUMMARY

There are two basic approaches to exchange of solid model geometry between different systems: direct translation and neutral formats. Neutral formats have the advantage of reducing the number of translators needed to go between n systems from $n(n - 1)$ to $2n$, fostering reuse of the data across many applications. The STEP standard is a neutral format that not only has robust three-dimensional solids support, but also supports a wide variety of product and process data, enabling concurrent engineering applications and allowing organizations to reap the maximum benefit from their data.

REFERENCES

1. I Flores. The Logic of Computer Arithmetic. Englewood Cliffs, NJ: Prentice-Hall, 1963.
2. ME Mortenson. Computer Graphics: An Introduction to the Mathematics and Geometry. New York: Industrial Press, 1989.
3. ME Mortenson. Geometric Modeling, 2nd ed. New York: Wiley, 1997.
4. PR Wilson. STEP Interoperability Methodology for Solids Data Version 1. NISTIR 5794, National Institute of Standards and Technology, Gaithersburg, MD, 1996.
5. JE Hopcroft, PJ Kahn. A paradigm for robust geometric algorithms. Algorithmica 7:339–380, 1992.
6. CM Hoffman. The problems of accuracy and robustness in geometric computation. Computer 22(3):31–41, March 1989.
7. G Held. Data Compression: Techniques and Applications, Hardware and Software Considerations, 2nd ed. New York: Wiley, 1987.
8. S Bloor and J Owen. Product Data Exchange. London: UCL Press, 1995.
9. J Fowler. STEP for Data Management, Exchange and Sharing. Twickenham, UK: Technology Appraisals, 1995.
10. J Owen. STEP: An Introduction. Winchester, UK: Information Geometers, 1993.
11. GM Radack, CY Wang, AJ Paul, GK Sigworth. New STEP Standard Streamlines Exchange of Casting Data, Modern Casting 87(4):40–42, April 1997.
12. Product data representation and exchange: Application protocol: Exchange of design and manufacturing product information for cast parts. Working draft of ISO 10303-223, ISO TC184/SC4/WG3 N728, 1998.

9
Sand Casting

Michael L. Tims
Concurrent Technologies Corporation, Johnstown, Pennsylvania

Qizhong Diao
ASAT Inc., Fremont, California

I. INTRODUCTION

Sand casting is the most economical method of manufacturing multiton castings. Castings of a few pounds can also be economically made from the process; however, other casting methods, including permanent mold and die casting, are less expensive for large production runs. Sand castings are typically chunky in shape, have a rough surface texture, and have a relatively low-dimensional accuracy relative to other casting methods. The process is amenable to a wide variety of nonreactive alloys. Postprocessing typically includes heat treatment to acquire desirable mechanical properties, finish machining to ensure accuracy of critical dimensions, and inspection to ensure desired properties and shape.

The principles of simulation are addressed in other chapters of this text. Here we concentrate on specific applications of computer simulation to sand casting. The process derives its name from the primary material comprising the mold—sand (typically silica-based), which is lightly fused together with clay and a binder. Though green sand mold casting is the primary focus of this chapter, most of the discussion will also apply to dry sand and other sand casting processes.

II. BASIC FEATURES OF A SAND CASTING

Figure 1 outlines the steps used to make a typical sand casting. Specific process design details depend upon customer requirements for shape and finished properties. Limitations or standard practices within the foundry also dictate process design.

A. Pattern

Obviously, the process must yield a casting of the desired shape. To obtain this shape, a pattern is prepared. Patterns are tools used to make cavities in the molding material into which molten metal can be poured to form the cast part. A general rule of thumb is that the pattern should be made as close to the desired casting shape as possible although extra features are applied for pattern removal and to improve part castability. Patterns are typically made from wood, although plastic, styrofoam, and metal may also be used. To accommodate removal from the sand, a draft of a few degrees (2 or 3 degrees is common in sand castings) is applied to the pattern. This draft should be included in numerical models when it adds significant thickness locally to part features. Patterns are typically split into two halves along a plane known as the parting plane. These planes are typically horizontal, but they may also be vertical and they may even be offset from one end of the part to the other to more easily accommodate castings having complex shapes. Each half of the pattern is then fastened to opposite sides of a match plate, which may be a simple piece of plywood. A flask is then placed over one side of the match plate. Sand is placed into the flask and firmly packed around the pattern to form one half of the finished mold. The pattern is then removed. The sand packing procedure is then repeated using the other side of the match plate assembly. After curing the sand, the two mold halves are clamped together and the mold is ready for casting.

Patterns need to be built to compensate for (1) shrinkage of metal during solidification and (2) thermal contraction of the metal during cooling to room temperature. In other words, patterns should be made larger than the anticipated casting dimensions by an amount equal to this shrinkage. Rules of thumb used for estimating the amount of shrinkage (called *shrink rules*) depend on the type of metal and the molding process used. For example, the shrink rule typically applied to gray iron or aluminum with hand packed sand is 1/8 in. to 1 ft. That is, each dimension of the pattern is scaled larger than the finished part by 1/8 in. for each linear foot of any dimension in the part. The shrink rule for gray iron in Disamatic molds, on the other hand, is only 1/16 in. to 1 ft, due to the stronger mold resulting from this process. Keep in mind that shrink rules represent an average amount that any casting dimension typically

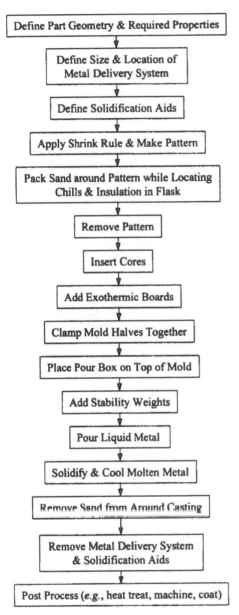

Define Part Geometry & Required Properties

Define Size & Location of
Metal Delivery System

Define Solidification Aids

Apply Shrink Rule & Make Pattern

Pack Sand around Pattern while Locating
Chills & Insulation in Flask

Remove Pattern

Insert Cores

Add Exothermic Boards

Clamp Mold Halves Together

Place Pour Box on Top of Mold

Add Stability Weights

Pour Liquid Metal

Solidify & Cool Molten Metal

Remove Sand from Around Casting

Remove Metal Delivery System
& Solidification Aids

Post Process (*e.g.*, heat treat, machine, coat)

Figure 1   Process steps in making a typical sand casting.

changes during the solidification and cooling stages. Actual shrinkage is difficult to predict since it depends upon a number of factors including casting geometry, localized mold strength, and solidification sequence within the part. As a result, during manufacture of production castings, different amounts of linear shrinkage occur throughout the casting. For parts with critical dimensional requirements, it is not unusual to produce a few trial parts to determine the actual shrink. Based upon the observed shrinkage from these trial castings, the pattern shape is modified for production runs. Computer simulation, which couples calculations of stress, deformation, and solidification, is a promising tool to predict actual shape changes in complex castings. This is an important area of current research in the casting community. After further characterization of these phenomena, improved simulation accuracy and reliability will be available to the casting process designer.

B. Gating System

The gating system (i.e., liquid metal delivery system) must also be considered in the design of the process. Figure 2 shows a typical gating system, which includes a pouring basin, sprue (often referred to as a downsprue), an optional filter, runner, and ingates. The gating system must also be an integral part of the pattern design discussed above. Since gating of a given casting is often modified to improve casting quality, the associated pattern segments are typically made from materials that are easy to form. Aluminum, plastics, and wood are common, although it is not unusual to use steel or even copper gating patterns for well-proofed designs.

Metal is poured from a ladle into a pouring basin. The pouring basin is a cutout on top of the mold surface and holds a pool of liquid metal that cushions the metal stream prior to entry into the mold cavity. Often the pouring basin is lined with a ceramic insert to eliminate potential sand erosion—a source of defects in sand castings—as the metal stream continues to pass through this region. After passing over a dam, the metal enters the sprue. The liquid metal accelerates due to gravity as it falls through the sprue. If left to freely fall (e.g., through a constant-diameter sprue), the cross-sectional area of the metal stream would be reduced according to the following formula, which is based upon acceleration due to gravity g and continuity of material.

$$\frac{A_x}{A_T} = \frac{1}{\sqrt{1 + 2gx/V_T^2}} \tag{1}$$

where A_x is the cross-sectional area of the freely falling liquid stream at a distance x from the top of the liquid stream, A_T is the cross-sectional area of the liquid at the top of the liquid stream, and V_T is the downward velocity of

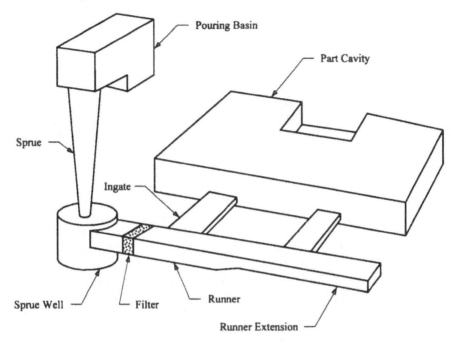

Figure 2 Typical metal delivery system.

the liquid at the top of the metal stream. If a constant diameter sprue were used, a large exposed free surface area of the metal stream would result. That coupled with the movement of the metal stream would lead to aspiration of gas from the permeable sand mold, which would cause oxidation and therefore contamination of the incoming metal. To avoid this problem, sprues are tapered by several degrees to ensure that the actual cross section is less than that predicted by a freely falling liquid stream as given in Eq. (1). For extremely tall castings, several sprues and collection stations may be present along the height of the mold. This technique allows the metal to fall only a short distance in each stage of its decent to the bottom of the metal delivery system. In this way, the metal stream is not permitted to gain too much decent speed and the sprue segments can be more conveniently sized.

Sprues are usually enlarged at their base. This enlargement is called a sprue well or sprue base. It is typically three times larger in cross section than the bottom of the sprue. Like the pouring basin, the job of the sprue well is to provide a liquid metal cushion to minimize mold erosion from the incoming liquid metal stream as the flow is redirected from the vertical to the horizontal direction. After passing through the sprue well, the metal enters the runner system. Runners distribute metal to the various entry points into the cavity containing the shape of the finished part. Ceramic filters are sometimes placed in runners. In addition to sifting out loose sand particles, filters also remove inclusions (i.e., small metallic oxide, nitride, or carbide particles) and dross (metal oxide scum) from the incoming metal. Filters also restrict flow and are a source of pressure drop in the metal stream. Another method of eliminating incoming defects is to provide extensions to the runners (see Fig. 2). Typically, the metal that first passes through the mold is filled with loose sand and dross. As the metal first fills the runner, its momentum thrusts the material at its leading edge into the runner extensions where it is trapped and hopefully contains a large portion of the associated defects.

The metal enters the part cavity from the runners through ingates. Ingates should be designed to avoid jetting, should be located so that the incoming metal does not impact and wear an opposing mold surface, and should be located to minimize the height that the liquid metal must fall to fill the cavity upon its entry. Frequently the top of the gate is below the top of the runner to minimize inflow of dross, which usually floats since it is less dense than liquid metal. If a pressurized gating is desired, then the sum of the cross-sectional area of all ingates should be less than that of the sum of the runner cross-sectional areas, which in turn is less than the cross-sectional area of the bottom of the sprue. Pressurized gating systems are useful in reducing interaction of the liquid metal with the gas (usually air) initially in the mold cavity and surrounding sand.

C. Solidification Aids

After the mold cavity is filled, the liquid metal must cool to the freezing point, solidify, and then cool to room temperature. If heat were removed in a uniform manner from all casting surfaces, the thinner sections would freeze and cool the most rapidly. Thicker sections would take longer to solidify and all such castings would solidify from the outside surface towards the inner regions. Such a progression of solidification would result in defects as a result solidification shrinkage common to the vast majority of metals and their alloys. Solidification shrinkage defects appear as holes in castings (like swiss cheese, except that solidification shrinkage cavities are very irregular in shape) and greatly reduce mechanical properties (especially fatigue, fracture, and toughness performance) of finished parts. Typical solidification shrinkage for metal alloys is shown in Table 1. To eliminate solidification shrinkage (also known as macroshrinkage), foundry process engineers rely upon a variety of solidification aids.

Table 1 Typical Solidification Shrinkage for
Metal Alloys

Metal	Total volumetric shrinkage upon solidification (%)
Carbon steel	2.5–3.0
1% Carbon steel	4.0
White iron	4.0–5.5
Gray iron	−1.6–2.5
Ductile iron	−2.7–4.5
Copper	4.9
Cu · 30Zn	4.5
Cu · 10Al	4.0
Aluminum	6.6
Al · 4.5Cu	6.3
Al · 12Si	3.8
Magnesium	4.2
Zinc	6.5

1. Risers

Strategically located exterior reservoirs of liquid metal can be used to combat solidification shrinkage. As metal within the part cavity shrinks, these reservoirs can supply makeup liquid metal to ensure that a solid casting is produced. Such reservoirs, called risers or feeders, must meet four basic requirements.

1. The riser must remain molten until after the section of the casting to which it feeds is completely solidified.
2. The riser must operate at a higher fluid pressure than that of the casting section to which it is attached. This is typically accomplished by placing the riser at a higher elevation than the casting and letting gravitational forces provide the desirable pressure gradient from the riser to the casting.
3. The riser must contain enough metal to compensate for the volume change needed by the casting during solidification. In other words, the riser cannot be drained out before the casting has completely solidified.
4. The riser contact must not freeze before the riser.

Risers are usually connected directly to the top of a casting (a *top* riser) or to the side of a casting (a *side* riser). They generally are attached to the thickest segment of a casting. Risers may extend to the top of the mold (an *open* riser) or they may be completely enclosed by the mold (a *blind* riser). Sometimes, hot metal is poured directly into open risers, especially after the rest of the mold has been filled. This practice ensures that the hottest metal is located in the risers, which can help keep the risers molten for a longer period of time. Occasionally, molds are completely filled by pouring directly into open side risers. However, this practice often leads to increased levels of certain defects and is therefore discouraged for any casting requiring a nominal quality level.

Most risers are so-called conventional risers, which are an integral part of the mold cavity surrounded by sand. Foundries prefer to minimize the cost of final grinding and cleaning of the castings. Therefore the riser contact with the casting, called a riser neck, is usually made with a smaller cross section than that of the riser. This design limits the locations in the mold where a riser can be placed. In order to avoid the sand surrounding the neck from being pulled out upon removal of the pattern, the riser must be located on the parting line. This is true for both horizontal and vertical parting planes. This often creates problems for horizontally parted castings having top risers. Several methods are commonly used to alleviate this problem.

1. Eliminate the riser neck and use a riser having a blunt cone shape. This requires a significant amount of work to remove the riser from the finished part, however.

2. Use a ceramic insert in the sand mold to form the riser neck. A small landing is added to the pattern where the riser neck is to be located. After removal of the pattern, the washer-shaped insert is placed on the landing, as illustrated in Fig. 3. Inserts are commercially available in a series of standard sizes. This is a low cost and easy way to put a riser almost anywhere on the casting.

3. Use insulation around the risers. High-temperature, low-conductivity materials are available for this purpose. With the use of insulation, the volume of risers can be reduced by about one-third.

4. Use exothermic risers. These are sleeves or wafer-board sections that line the surface of risers. Upon reaching a certain temperature, an exothermic chemical reaction occurs within the material, which keeps the metal in the risers hotter for a longer period of time. One kind of exothermic riser, for example, uses aluminum particles that react when its temperature is above 540°C. The use of exothermic risers can reduce the required riser volume by about half. Since the sleeves must be purchased from a supplier, the direct savings from reduced riser size are often minimal during use of insulated or exothermic riser sleeves. Their real benefit, however, is the increased versatility that they offer for locating risers on castings. The rate at which heat is generated from these exothermic materials is not uni-

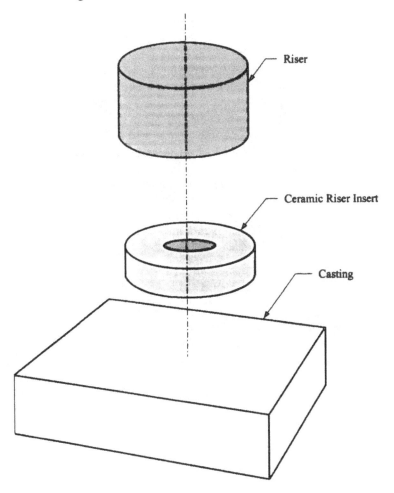

Exploded view shown

Figure 3 Illustration of riser inserts

form over time and is generally not well characterized by the manufacturers. Consequently, modeling of exothermic riser materials becomes a bit of a challenge.

5. Place exothermic powder on top of open risers. These materials are generally shoved on by hand after mold filling is completed. Exothermic powders behave in a manner like that of exothermic risers.

As mentioned above, a riser should have enough metal to feed the volume change in the casting during solidification. It should also be hot enough to keep the metal in the riser molten until after the part has solidified. For most metals, which shrink 3–7% in volume during solidification, a riser that is big enough to keep itself hot will have adequate volume to feed the casting without becoming drained before the part is solid.

Some castings are designed with single side risers that are used to feed only one casting. In other designs, one side riser is used to feed two castings. Periodically, a single side riser is used to feed three or four castings. It is a common practice in foundries to use a much larger side riser to feed two or more castings than one used to feed a single casting of the same size and shape. This wasteful practice is unnecessary in most instances for correctly sized risers. In fact, single side risers used to feed multiple castings can sometimes be made smaller since such risers are surrounded by hot metal instead of cold sand. This results in a lower heat loss rate, which enables feeding for a longer period of time.

2. Chills

As risers help to promote progressive solidification by keeping segments of a casting molten for longer periods of time, chills are designed to locally remove heat more quickly then sand does. Chills, therefore, force solidification to occur more rapidly in selected regions of the casting. This characteristic can be used to promote a solidification sequence that progresses from the bottom of the casting upwards and outwards toward the risers. Such a solidification pattern enhances the likelihood of obtaining a sound casting (i.e., one free of solidification shrinkage).

Chills are usually made from metal. Steel, iron, and copper are commonly used. Other materials may also be used including ceramic blocks to provide a lower, more sustained rate of heat removal then that offered by metallic chills. Chills may be buried under a thin layer of sand or they may be in direct contact with the casting. Typically, direct contact chills are coated with a ceramic-based mold wash to avoid welding of the casting to the chill. Chills are usually placed at or near the bottom of the mold to ensure upward progression of solidification. In addition, regions that are difficult to riser (such as internal features) may be chilled instead. Another application of chills is at tee sections where small "kiss" chills can eliminate hot spots and the associated solidification shrinkage that may otherwise form there. Chills are not usually placed on downward facing surfaces since this orientation can result in chills falling out of place and damaging the mold or casting.

Chills are typically solid brick-shaped segments that are individually placed into the sand. In this way the chills can be used many times for a wide variety of casting shapes. For high-production parts and/or precision sand castings, specially contoured chills may be produced. In either case, chills of varying thickness promote a proportional amount of localized cooling.

To obtain a smoother surface on the finished cast part, chills may be buried under a thin layer (typically 0.5–2.0 cm) of molding sand. Obviously, the sand layer delays the impact of the chills, which is acceptable for most thick casting sections. These buried chills usually, however, have little effect on the required cooling time prior to removing the casting from the sand (i.e., shakeout). Direct contact chills that have been coated with mold wash also cause a minor delay in the chilling effect. Since mold wash is put on in thin layers, its thermal effects are generally not explicitly included in numerical models. The interface heat transfer coefficient between the casting and mold h_i can easily be modified to account for the effect of the mold wash as follows:

$$h_{MW} = \frac{1}{1/h_i + L_{MW}/k_{MW}} = \frac{h_i k_{MW}}{k_{MW} + h_i L_{MW}} \qquad (2)$$

where h_{MW} is the modified interface heat transfer coefficient between the casting and mold wash covered chills, L_{MW} is the mold wash thickness, and k_{MW} is the mold wash thermal conductivity.

3. Insulation

Insulating bricks are sometimes used to locally reduce the rate of heat loss from a casting. These bricks are usually placed in the top portions of the mold to help keep these regions of the casting hotter longer. In this way, the top portions of the casting can continue to feed the shrinkage occurring in the lower segments of the casting during solidification. Of course, risers should then be used to feed these upper casting sections to ensure their soundness. Like chills, insulating bricks are generally contoured to the shape of the casting and are often buried under a thin layer of sand. Unlike chills, insulating brick are rarely made into large segments that contour to the shape of the casting. Insulating bricks are typically used only once and then discarded.

D. Vents

Prior to filling the mold, gas—typically air—is located in the mold cavity. To avoid trapping gas in the casting and to eliminate certain types of mold fill defects, the gas must be easily pushed out of the mold cavity as the metal enters. Green sand has relatively high gas permeability. Thus, the gas can easily escape through the sand. When additional vents are needed, they can be connected directly to the mold cavity or they can be located within a small distance

(~ 0.5 cm) from the mold cavity in the sand. The former is more effective but will leave a mark on the casting that must later be ground off.

Mold venting has not been widely simulated for sand castings. However, if necessary this phenomenon can be modeled as compressible gas flow through a porous media. Note that the incoming metal increases the gas pressure by compressing the gas both through a reduction in volume as well as through an increase in temperature. Additionally, transient pressure effects in the sand and the associated increase in the local mass of the gas may be accounted for by a source term. Values for tortuosity factors should be determined experimentally and will likely vary widely based upon the size distribution of the sand particles, the type and proportion of binder and clay used, the amount of water remaining in the sand, and the level of compaction used to form the sand mold.

E. Cores

Internal features often cannot be made using standard patterns as described above. To accommodate internal features, cores are typically used. Cores are shaped inserts and are made from sand having special binders. They are prepared separately from the rest of the sand mold and are placed into the mold cavity after the exterior impression is made by the patterns. In addition to the shape of the internal feature, cores have extensions, called core prints, that fit into ledges created by the pattern. Core prints allow accurate positioning of the core as the mold segments are joined together.

Cores may also be employed to (1) create external features, which are back-drafted from the parting line, (2) provide tight corners that cannot be easily or accurately shaped by sand, or (3) impart better finish to selected surfaces. There are a wide variety of core making processes, including hot core, warm core, cold core, and iso-cured core. A different type of sand is sometimes used in the core to take advantage of different venting and/or thermal characteristics. Hot or warm cores are cured by applying heat to the surface. This allows creation of hollow cores by draining loose sand out of the center after curing (i.e., hardening the binder) the outer skin of the core. Finally, cores, especially solid ones, are denser than green sand resulting in higher heat conductivity. This characteristic can be effectively used to create cores that act as mild chills to promote a desirable progression of solidification.

Small metallic inserts, called chaplets, are sometimes used to help support the central portions of long cores. Chaplets also help keep the cores from shifting, especially during mold filling. They are usually made from the same alloy as the casting and become integral to the final part. They typically do not have a significant impact on the solidification response of the casting. Incomplete fusion between the chaplet and the liquid metal may occur if the metal stream is partially solidified prior to contacting the chaplets or if the surface of the chaplet is contaminated with grease, rust, or other materials.

F. Sand

A variety of sands is available to the sand founder. Silica sand is the most commonly used. Faceted grains work best, since they hold their position in the mold better than rounded grains. The sand particles are held together with blends of clay, water, and other chemical binders. The water must be driven off to allow the sand to harden prior to pouring of the casting. More common today are chemically bonded sands, which provide stronger molds that improve dimensional consistency and reduce the likelihood of cracking due to rough handling.

Other sand types commonly used in the foundry are chromite, olivine, and zircon. Chromite sand is much denser than silica sand and is often used to provide localized chilling since it conducts heat better than silica sand. In other instances, steel shot may be mixed with the sand to obtain the same type of effect. Magnetic removal can then be used to separate the shot from the sand.

In all cases, the thermophysical and thermomechanical properties of sand vary significantly depending upon grain size and distribution, binders, degree of packing, humidity, and other factors. Phelke et al. [1] have catalogued thermophysical properties of sand. These properties provide a good starting point for thermal analysis. To obtain improved properties for a specific set of process conditions and sand type, measurements must be made.

The old method of packing sand around the pattern by hand has generally been replaced by automated methods. Hand packing is still used to make test castings or low volume parts. The disadvantage of hand molding, aside from low productivity, is the potential nonuniformity in how the sand is packed. Modern molding lines use pressurized air to blast sand around the pattern. The blasting effect delivers the sand and also squeezes it to form the desired amount of compaction. The process is computer controlled and is used to make molds at high productivity. The high pressure also makes the sand mold denser and stronger.

G. Dead Weight

When full, the mold is acted upon by the static pressure (called metallostatic pressure) caused by the freestanding liquid metal. The local pressure p acting on the mold caused by the static liquid metal can be expressed as

$$p = \rho g h + p_f \tag{3}$$

where ρ is the liquid density, g is the acceleration due to gravity, h is the depth from the free surface of the liquid, and p_f is the pressure at the free surface of the liquid. The metallostatic pressure distorts the mold. Equation (3) can be used to define the pressure boundary condition acting on the interior of the mold after mold filling is complete. Unless additional external pressure is applied or a vacuum exists, the effect of atmospheric pressure can be ignored which results in a p_f value of zero.

Because sand is less dense than the metal used in the casting, buoyancy forces act to push the top portion of the mold upward. To keep the mold segments from separating, they are clamped or bolted together. In addition, dead weight is sometimes placed on top of horizontally parted molds to reduce mold flotation and mold distortion.

III. COMMON SAND CASTING DEFECTS

The primary objective for completing a casting simulation is to accurately predict the behavior of the process and to use this information to enhance the likelihood of producing a casting having the desired shape, mechanical properties, and any other performance characteristics. Central to defining the process design using simulation methods is prediction of commonly encountered defects. Numerical simulation is a very powerful tool for predicting a wide range of common casting defects. Many defects, however, are not well understood nor are they easily modeled with today's techniques. This section focuses on several common sand casting defects, defines their common causes and cures, and discusses the state of the art in accurately predicting each phenomenon.

A. Solidification Shrinkage

Of the many sand casting defects, solidification shrinkage (i.e., shrinkage porosity) is the one most easily and most commonly characterized in computer simulations. This defect is a result of the significant reduction in volume that accompanies the freezing of nearly every metal alloy. This is quite the opposite of what happens when water freezes! Typical volumetric reductions for metal alloys range from 3 to 7%. A few alloys expand during some phases of the solidification process. For example, some iron alloys will initially shrink during solidification and then expand as carbon flakes or nodules form in the latter stages of solidification. Discussions of appropriate methods to handle these alloys are available in the literature and will not be discussed here.

Solidification shrinkage defects occur when isolated regions of liquid metal develop within the part cavity. These isolated regions do not have any feed paths to risers or other exterior metal feeders. Consequently, as these isolated pools solidify, their volume shrinks resulting in voids within the part. Such isolated pools are called *hot spots* and the resulting voids tend to develop near the top of the pools as solidification progresses. The type of alloy influences the shape and distribution of solidification shrinkage voids. Short freezing range alloys tend to have a few large voids (macroshrinkage) that are easily seen by the naked eye. Long freezing range alloys, on the other hand, will often have many hundred small voids (microshrinkage) scattered over a wide region of the casting.

In computer simulations, one can observe hot spots by tracking the temperature contours within the casting. As a good first approximation, such an analysis needs only to include thermal effects. More costly mold filling analyses are rarely needed to simulate solidification shrinkage development. The analyst typically focuses on appropriate solidification aids to ensure that as solidification progresses, it advances toward the risers or other external features. Appropriately sized and correctly located chills, risers, insulation, and/or exothermic materials can usually solve solidification shrinkage problems. In other instances, padding may be needed on the part, while in extreme cases, part redesign may be required for economical production.

B. Gas Porosity

The solubility of gas in liquid metal is significantly greater than that in solid metal. As the melt is prepared, it readily absorbs surrounding gases—most notably oxygen, nitrogen, and hydrogen. Foundrymen have developed several different treatments to the liquid metal to reduce the amount of dissolved gas prior to pouring. When these methods are ineffective in removing a sufficient amount of gas or when the metal absorbs gas during pouring and solidification, another type of void, called gas porosity, may develop in castings. Gases absorbed during pouring or solidification are thought to come from two primary sources. They may be initially present in the sand or mold cavity. Secondly, they may be generated as a result of the chemical breakdown of materials within the mold, including mold wash and/or binders. Turbulent flow during mold filling often results in increased pickup of gas by the liquid metal within the mold cavity.

During solidification, much of the absorbed gas is rejected by the solidifying material. The exact amount represents the difference between the solubility limits in the liquid and solid states. In so doing, the concentration of the gas in the remaining liquid continues to increase until the solubility limit of the liquid is reached. Additional solidification beyond this point results in rejection of the excess gas and bubbles begin to form in the liquid. These bubbles are spherical in shape and appear over a broad area corresponding to the liquid/

mushy pool shape at the point when the liquid solubility limit has been reached. Typically, these bubbles are assumed to form at the base of secondary dendrite arms. The bubble diameter is assumed to be equal to that at the base of neighboring secondary dendrite arms. Actual bubble development must also overcome surface tension effects. The pressure inside a gas bubble is greater than that on the outside. This pressure difference Δp can be computed as

$$\Delta p = \frac{2\gamma}{r} \tag{4}$$

where γ is the bubble surface tension and r is the bubble radius. Equation (4) can be used to establish the number of gas bubbles in the solidifying solid through use of the perfect gas law. The liquid pressure on the exterior of the bubble can be computed from Eq. (3). Instantaneous diffusion of the gas into the liquid is often assumed. Actual diffusion effects may also be included to capture the spatially varying concentration gradient in the liquid. However, this calculation adds another field variable to the solution and the actual diffusion of gas is complicated by convection and local microstructural features.

Gas porosity is sometimes difficult to distinguish from microshrinkage. In fact, the voids left behind by shrinkage porosity are sometimes filled with gas. In general, shrinkage porosity will develop when the riser rules above are violated or when long freezing range alloys are cast. When the mold design leads to a good progression of solidification, any resulting microvoids are generally related to high levels of dissolved gas in the molten metal. Overall, the technical community needs to develop more comprehensive methods of predicting gas porosity in sand castings. Afterward, more robust predictions will be available in the various commercial casting analysis codes.

C. Lack of Fill

A general class of casting defects is known as lack of fill defects. Typically, these problems are caused by a significant loss of heat from the metal stream during filling. If a sufficient amount of heat is lost, the metal can partially freeze prior to complete mold filling. As the metal continues to freeze, individual dendrites can fuse together and resist any further metal movement. The point at which the material can no longer flow is called the coherency point. Not much is clearly understood about how to predict this condition. It is clear, however, that the coherency point depends upon the alloy, the local solid fraction, the structure of the solidified particles, and the local shear strain rate of the fluid. The coherency point typically ranges from a solid fraction of 0.4 to 0.8, although higher values are also possible under conditions of high shear strain rate.

Lack of fill defects are especially prevalent during filling of long, thin sections. Here surface tension effects, combined with the fluid wetting angle, dictate the fluid pressure that must be overcome to push the material into the thin section. Equation (4) above can be used to provide a worst case estimate of the necessary pressure with the bubble radius equaling one-half of the channel thickness.

Freezing of the incoming material for a uniformly thick section can be estimated using simple one-dimensional approximations from basic heat transfer theory. For constant temperature surroundings,* uniform temperatures can be assumed for parts having small volume, large surface areas, high thermal conductivity, or low rates of heat loss from their surface. These conditions are summarized by the Biot number (Bi), which is defined as follows for a casting.

$$\text{Bi} = (\text{volume-to-surface area ratio}) \left(\frac{h_i}{k_c}\right) \tag{5}$$

where h_i is the heat transfer coefficient between the mold and the casting, k_c is the thermal conductivity of the cast material, and the volume-to-surface area ratio for a long, thin, constant thickness channel is the channel thickness L divided by 2. When $\text{Bi} < 0.1$, the part has little temperature variation throughout and a method known as the lumped capacitance method (LCM, also known as negligible internal resistance) can be applied to determine the time dependent temperature of the part [2]. For typical mold filling situations, this condition is satisfied for casting thickness of about 0.5 in. or less. The time-dependent temperature T of the leading edge of the molten metal front under these conditions is

$$T = T_{\text{mold}} + (T_0 - T_{\text{mold}}) \exp\left(-\frac{2h_i t}{\rho L c_{p,\text{app}}}\right) \tag{6}$$

where T_{mold} is the temperature of the mold, T_0 is a modified initial molten metal temperature, t is time, ρ is the density of the cast material at the liquidus temperature T_L, and $c_{p,\text{app}}$ is the apparent uniform specific heat of the cast

---

* Carefully note this necessary condition of constant temperature surroundings for the present discussion. It is an altogether appropriate assumption for mold filling conditions, especially when evaluating how quickly a metal flow front freezes in a long, thin channel. After mold filling is completed, however, the mold surface temperature quickly heats up and the constant temperature surroundings condition is violated. Therefore, the analysis methodology that follows would not be appropriate to determine solidification time. For these instances, solutions using semi-infinite solids [2] are appropriate.

material in the mushy region. Note that for this simple hand calculation, the latent heat of fusion is assumed to be uniformly distributed over the freezing range. The modified initial molten metal temperature is defined as

$$T_0 = T_L + \frac{c_{p,L}}{c_{p,\text{app}}}\left(T_{\text{pour}} - T_L\right) \tag{7}$$

and

$$c_{p,\text{app}} = \frac{c_{p,L} + c_{p,S}}{2} + \frac{\Delta H_{fg}}{T_L - T_S} \tag{8}$$

where $c_{p,L}$ is the specific heat of the liquid metal at T_L, T_{pour} is the pouring temperature, $c_{p,S}$ is the specific heat of the solid metal at the solidus temperature T_S, and ΔH_{fg} is the latent heat of fusion (i.e., not including sensible heat). Use of Eq. (6) requires a single value of specific heat. To accommodate the significant difference in the liquid state specific heat and the apparent specific heat in the mushy region, the modified initial molten metal temperature is introduced in Eq. (7). This artificial temperature represents an increase in the actual liquidus temperature to account for the heat available during cooling of the liquid.

D. Laps

When two or more cold metal streams converge, a potential for lap defect formation exists. A lap is a region where such flow fronts have not completely fused together resulting in a weak seam in the casting. Free surfaces having dross or other contaminates are especially prone to laps. No widely accepted formulation for predicting lap formation has been proposed. Modeling of this defect requires a mold fill simulation. Observation of how the various flow fronts develop and converge can lead to a qualitative understanding of when laps form. Alternate gating designs can be evaluated to determine which ones lead to (1) a minimal number of converging free surfaces, (2) convergence in innocuous regions of the part, or (3) an increase in free surface temperature or mixing turbulence at the time of flow front convergence. Laps formed by flow doubling back upon the free surface of the metal (such as when the flow front runs directly into a vertical wall) are not as easily predicted.

E. Hot Tears, Hot Cracks, Residual Stresses, and Distortion

These defects have their origin in the thermal contraction/expansion of the casting and mold materials. Hot tears occur just as a region is completing solidification. The remaining thin film of liquid ruptures as a result of the stresses encountered during cooling, resulting in a discontinuity in the finished casting. Hot cracks occur when the casting is hot, yet solid. Residual stresses and distortion describe the state of the casting after cooling to room temperature. These phenomena are related to strains associated with thermal expansion, sudden volume changes as a result of freezing, and solid state transformations. Mechanical interaction between the casting and mold, as well as temperature gradients within the casting contribute to these strains. Such phenomena are a topic of current research interest. Frankly, today's casting analysis codes do not yet do a reliable job of predicting these behaviors for a wide variety of conditions. Although general trends can usually be inferred from such analyses, today's codes, especially those using rectangular or brick elements, should not be relied upon to provide anything more than general guidance. The interested reader is referred to Chapter 3 of this text on stress analysis.

F. Macrosegregation

As alloys solidify, the composition of the solid metal is typically different than that of the liquid. The compositional difference results in rejection of certain alloying elements to the liquid. Consequently, the composition of the liquid changes during the freezing process. For a binary alloy, this compositional change continues until saturation of the alloying element occurs within the liquid. After this, the solid material freezes with a composition consistent with that of the liquid—frequently in a solid phase consisting of two different chemical compositions. For multicomponent alloy systems, alloy saturation occurs at different times for each of the alloying elements. Detailed analyses of multicomponent alloy systems are therefore significantly more complex to simulate.

A casting with significant macrosegregation will have mechanical properties that vary throughout. Little can be done to correct macrosegregation problems once they are present in a casting. The analysis community has, however, gained a good handle on predicting this phenomenon. By careful design of the solidification aids, numerical models can be useful in reducing the severity of macrosegregation or at least controlling the chemistry in the critical areas of the casting. In severe cases, redesign of the casting may be necessary to ensure uniform chemistry and thereby obtain uniform properties in the finished casting.

Diffusion of the rejected alloying elements into the liquid phase tends to be a relatively slow process. As a result, accurate simulation of macrosegregation requires calculation of alloy diffusion effects. Movement of the alloying elements in the liquid by convection should be accounted for in the local compositional calculations.

G. Microsegregation

Microsegregation is closely related to macrosegregation described above. This phenomenon occurs at a much finer scale, however. As dendrites form, small pockets of liquid metal are often trapped in the spaces between these dendrites. Local chemical compositional differences in the solidified part result. Microsegregation is often treated through a homogenization heat treatment following casting. Today's casting analysis codes are equipped to model this phenomenon.

H. Mold Erosion

This problem has not been widely evaluated by numerical methods. However, many of the basic data needed are present in a mold fill simulation. The shear stress on the mold from the moving fluid, along with the impact of the metal stream, loosen and carry away sand particles. When severe, mold erosion will alter the shape of the part cavity and therefore change the shape of the casting. More importantly, however, the loose sand particles typically become entrapped in the liquid metal and may be solidified into the casting. This results in a significant reduction of mechanical properties, most notably ductility, fatigue strength, and fracture toughness. Smooth transition of flow from one segment of the casting to another will reduce the tendency for mold erosion. In terms of simulation, the turbulent kinetic energy should be minimized in those areas susceptible to mold erosion. Quite clearly, the degree of compaction of the sand plays a significant role in determining the severity of mold erosion. Therefore, it becomes difficult to provide exact numerical values for the analyst to use when attempting to minimize or eliminate mold erosion problems.

I. Inclusions

Inclusions are a concern any time that molten metal is present. Inclusions are metal oxide, nitride, carbide, calcide, sulfide, or other harmful foreign particles that naturally develop as a result of reactions with the atmosphere or other materials used during the metal refining operations. Inclusions are usually hard and often result from inadequate cleaning of the metal. They reduce mechanical properties of the finished part. Ceramic filters are the best method of removing inclusions from metal. In addition, special gates—called swirl gates—can be used to trap most types of inclusions. These gates are cylindrical cavities that are put in the runner system. The incoming liquid enters at a tangent and the outgoing material also leaves at a tangent. This creates a swirling action, which causes the lighter inclusion particles to centrifuge to the center where they are trapped and do not enter the part cavity. Particle tracking, available in most commercial software packages, allows the analyst to track movement of incoming inclusions. To obtain accurate solutions, simulations must include effects of density differences between the inclusion particles and liquid metal. Furthermore, to establish the probable location of these particles in finished parts, one should also include natural convection effects after mold filling is completed.

IV. MOLD FILLING CONSIDERATIONS

While most of the problems encountered in castings are related to solidification phenomena, several issues arise in the mold filling area as well. Solution of mold fill problems requires use of the Navier-Stokes equations, as discussed in Chapter 2 of the text. The present section is intended to provide some general guidance in mold fill analyses of sand castings. Many of the desirable attributes of good mold filling design are discussed above in Sec. III on common sand casting defects. In general, significant opportunities for further application of numerical methods to mold filling of sand castings still remain both in terms of the types of problems that can be addressed as well as a broader application of the analysis tools to solve industrial problems.

A. Turbulence

Turbulence is a common concern during mold filling especially when it leads to increases in the free surface area of the liquid metal stream. Turbulent flow results in increased metal oxidation, mold erosion, and certain lap defects. To avoid turbulence, smooth transitions between mold segments are desirable. In addition, ingates that flow into broad regions can reduce turbulence. For gating into thin sections, tangential (or at least angled) gates are preferred. Another reason for the popularity of filters is to choke the metal flow and reduce the turbulence of the metal stream in the runners. In all cases, the process design should ensure that the metal falls a minimal distance from the ingate while filling the mold cavity. Campbell [3] recommends a maximum fill velocity of 0.5 m/s to avoid the consequences of turbulent flow regardless of the alloy. Codes that compute turbulence effects are common. Use of the data to enhance gating designs is uncommon, however.

B. General Rules for Filling of Sand Castings

1. Filling should be smooth and uniform.
2. Pressurized gating systems ensure that the gates and runners are full of metal after their initial filling. This can be accomplished by keeping the cross-sectional area of ingates less than that of the runners, which should be less than that at the bottom of the sprue. Gating systems that are kept full of metal result in less free surface turbulence, gas entrapment, oxidation, and erosion of the mold.
3. The pour cup should remain full of metal during the pouring process. This ensures an uninterrupted flow of metal resulting in a constant pressure head in the metal delivery system.
4. The metal delivery system (along with solidification aids) should be designed for easy removal from the solid casting.
5. The gating should be designed to ensure that hot metal is available to fill all regions of the mold. Premature freezing of the metal during filling leads to a wide variety of casting defects. Simulation of such defects requires a coupled thermal-fluids-solidification analysis.
6. Filling should be accomplished in a manner that allows optimum production using the available ladles, melt furnaces, and other foundry equipment. Although many casting analysis codes correctly predict the sequence of mold filling events, many of them lack robustness in predicting the actual fill times for a wide variety of cast shapes.

V. HEAT TRANSFER CONSIDERATIONS

Two main factors determine the thermal effectiveness of a mold design. First, the progression of solidification must ensure that the casting freezes from the bottom upward and towards exterior feeders. Secondly, the thermal effectiveness of the mold design is based upon the ability to provide a desirable microstructure—both grain size and shape and microstructural constituents.

The vast majority of simulations in sand castings consider only heat transfer effects. This is because the vast majority of sand casting problems are related to thermal effects. Thermal-only analyses are also significantly less expensive to complete, require only about 10% of the CPU time, and are available from a wider array of analyses codes.

A. Sand

The thermal properties of the sand significantly impact the thermal response of the casting. Variations in the mixing and molding practices impact the thermophysical properties of the sand. This, in turn, creates variations in the thermal response of the casting and impacts its final mechanical properties. Chief among the variations is the amount of water used to prepare green sand molds and the amount of water remaining in the sand when the casting is poured. Other complicating factors in the simulation of sand castings are the effects associated with (1) evaporation of residual moisture in the sand mold and (2) chemical breakdown of sand binders. The heat associated with these events can be captured with an energy source term in the sand elements. Additional complications arise, however, in accounting for the associated variation in thermophysical and mechanical properties of the sand as a result of these phenomena. At best, only inconsequential errors in analyses have occurred in thermal analyses as a result of using constant thermophysical properties for the sand. However, the change in mechanical properties of the sand probably accounts for a significant amount of the discrepancy between numerical simulations and experimental observations for those problems involving hot tears, hot cracks, residual stresses, and distortion.

B. Heat Transfer During Mold Filling

The chief mode of heat loss from the liquid metal during mold filling is by convection with the colder mold. Convective heat transfer coefficients during mold filling range in the several hundred to a few thousand $W/m^2 \cdot K$. Radiation from the free surface may also be important for large castings that take several minutes to fill.

C. Interface Between the Casting and Mold

Although proper thermal management will not guarantee a sound casting (other factors must also be considered), it remains the focus of most casting analyses. Thermal effects during mold filling aside, the major focus of the problem is establishing a good definition of the interaction between the casting and mold. Most codes treat the interface as a convective boundary condition. This is done for mathematical convenience within the models rather than an actual description of the heat transfer mechanism at work in this interface. The Grashof number is too small [2] to expect convection to be occurring in this thin gap between the casting and mold. In reality, the heat is transferred by a combination of radiation and conduction across the stagnant gas gap between the casting and mold. Best estimates of this interface heat transfer coefficient come from experimental observation (although values from sources such as Poirier and Geiger [4] can provide useful approximate values). Especially powerful for determining these coefficients are the inverse heat transfer methods. In the absence of such data, the following equation can be used to provide an estimate of the interface heat transfer coefficient h_{int}.

$$h_{int} = \left(\frac{\varepsilon_M \varepsilon_C}{\varepsilon_M + \varepsilon_C - \varepsilon_M \varepsilon_C} \right) \sigma (T_C^2 + T_M^2)(T_C + T_M) + \frac{k_G}{L_G} \tag{9}$$

where ε_M is the emissivity of the mold, ε_C is the emissivity of the casting, σ is the Stefan-Boltzmann constant (5.67×10^{-8} W/m$^2 \cdot$ K^4), T_C is the *absolute* temperature of the surface of the casting, T_M is the *absolute* temperature of the mold surface, k_G is the thermal conductivity of the gas contained in the gap, and L_G is the gap thickness. Note that values for T_C and T_M do change as solidification and cooling continue. Although the emissivities of the casting surface and sand are not well defined, values in the 0.5–0.7 range are expected. Surface temperatures for both the casting and mold are computed as part of the normal calculation process. Values for k_G can be found for a variety of gases in standard heat transfer textbooks, while values for L_G must be estimated from experience (or calculated from a coupled stress analysis). After a skin has solidified on the surface of the casting, the minimum value for L_G should not usually be less than the root mean square (RMS) of the mold surface roughness. The gap thickness is both temporally and spatially variant. It has been observed that alloys having wide mushy zones, like aluminum, tend to form large gaps, while alloys with narrow mushy zones, like iron, usually have smaller gaps. For high temperature materials, such as iron and steel, the radiation term [the first term on the right-hand side of Eq. (9)] dominates, which reduces the need for an accurate gap thickness until long after solidification is complete.

D. Exterior Boundary Conditions

Once pouring is completed, most sand castings are left to cool and solidify in the open environment. Occasionally, some form of exterior forced cooling is also applied. However, the exterior boundary conditions rarely have a significant impact on the solidification behavior of a sand casting unless the sand layer between the casting and flask is thin relative to the thickness of the casting. When left to cool naturally, the convective heat transfer coefficient on the exterior surface of sand casting molds is typically between 10 and 20 W/m$^2 \cdot$ K. Figure 4 shows the approximate proportion of convection and radiation from a surface with surroundings of 20°C and a surface emissivity of 0.7. These results also assume that the surroundings

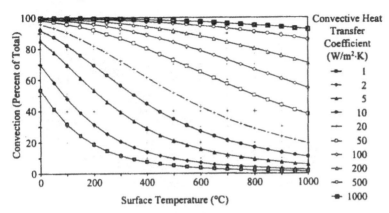

Figure 4 Effect of temperature on proportion of convection and radiation from a heated surface.

are black, and that the surface does not have a direct radiation path from one portion of its surface to another. This figure can be used as a guide to establish the relative importance of each mode of heat transfer from the exterior of a mold. Should radiation become important (up to about 60% of the total heat loss), then the following expression can be used to estimate an equivalent effective total heat transfer coefficient from the surface h_{tot}

$$h_{tot} = h_c + \sigma\varepsilon(T_s^2 + T_\infty^2)(T_s + T_\infty) \qquad (10)$$

where h_c is the actual convective heat transfer coefficient, ε is the mold surface emissivity, T_s is the mold surface *absolute* temperature, and T_∞ is the *absolute* temperature of the surroundings.

E. Considerations in the Thermal Simulation of Sand Castings

1. Mold wash retards the onset of solidification as a result of its low thermal conductivity. Mold wash generally does not have a significant effect on shakeout time, however [5].

2. Stacked chills do not have the same effect as a single chill of the same overall thickness [5]. This difference is a result of the gap between consecutive chill layers, which act as thermal resistors to heat flow.

VI. SUMMARY

To date, significant work has been completed in heat transfer, fluid flow, and solidification simulations of sand casting. The technical community has a good handle on development of hot spots leading to macroshrinkage defects. Design of solidification aids (most commonly risers and chills) can be confidently guided by the simulation results. Macrosegregation, microsegregation, and solidification microstructure can generally be computed with confidence using most commercial codes. Some codes predict gas porosity; however, the algorithms are generally limited in their robustness. Effects associated with stress and strain are not well characterized primarily as a result of limited fundamental knowledge of material behavior around the solidification temperature. As always, better fundamental understanding (along with accurate material properties) leads to improved predictive capabilities. Material properties are well characterized and well catalogued for only a few materials, however. Research opportunities remain strong in the areas of predicting mold erosion, binder breakdown, definition of coherency point, and accurate prediction of fill-related defects. Future versions of commercially available codes will include robust algorithm for simulation of many of these technically challenging problems.

REFERENCES

1. RD Pehlke, A Jeyarajan, H Wada. Summary of thermal properties for casting alloys and mold materials. NTIS Report NSF/MEA-82028, December 1982.
2. FP Incropera, DP DeWitt. Fundamentals of Heat Transfer. New York: John Wiley & Sons, 1981.
3. J Campbell. Castings. Oxford, UK: Butterworth-Heinemann Ltd., 1995.
4. DR Poirier, GH Geiger. Transport Phenomena in Materials Processing. Warrendale, PA: TMS, 1994.
5. ML Tims. A one-dimensional analysis of casting mold parameters. Proceedings of Modelling of Casting, Welding and Advanced Solidification Processes—VI, Palm Coast, FL, March 1993. Warrendale, PA: TMS, pp. 701–708.

10
Lost Foam Casting

Chengming Wang
Concurrent Technologies Corporation, Johnstown, Pennsylvania

I. LOST FOAM CASTING PROCESS

The lost foam casting process is also known as evaporative pattern casting (EPC), full mold casting, polycast, or replicast. In this process (Fig. 1), liquid metal is poured directly onto a refractory-coated foamed polymer pattern, which is buried in loose sand. The polymer pattern undergoes thermal degradation and is progressively liquefied, vaporized, and replaced by the molten metal, which solidifies and produces the casting. The most commonly used foam materials in EPC process are expanded polystyrene (EPS) and polymethyl methacrylate (PMMA). Currently, the EPC process is utilized to manufacture a wide variety of ferrous and nonferrous components catering primarily to the automotive industry. There are about 30 foundries in North America that are committed to the lost foam casting process [1–4].

II. ADVANTAGES AND DIFFICULTIES ASSOCIATED WITH LOST FOAM CASTING

The lost foam casting process provides many operational advantages such as simplicity, great design versatility, high casting yield, high production rate, readily automated production, good surface finish and mechanical properties, and reusable sand. However, the lost foam casting process involves many extremely complicated physical phenomena. Extensive trial-and-error procedures are required to determine the appropriate casting conditions. The quality of the casting is determined by many factors, including the characteristics of the materials, gating design, and selection of control parameters. The materials

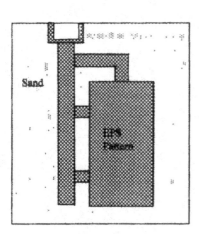 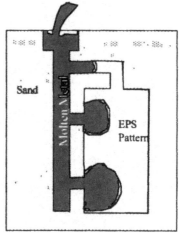

(a) Mold of lost foam casting

A refractory-coated polymer (EPS)

pattern is buried in loose sand

(b) Pouring molten metal into mold

The EPS pattern undergoes thermal

degradation, progressively liquefied,

vaporized and replaced by the molten

metal

Figure 1   A schematic illustration of the lost foam casting process.

involved are the EPS pattern, glue, sand, casting alloy, and coating. The process design includes the ingate sizes and locations, casting orientations, and compaction of sand. The control parameters include pouring temperature, sand temperature, pouring rate, and metallostatic head. To produce a good quality casting, a perfect match of all of these factors is required, and the operational window is rather small. If the process is not properly designed, defects such as laps and folds, inclusions, entrapment of liquid and gaseous pattern material, and lustrous carbon may form. Process simulation can help design engineers in optimizing the process to minimize defect formation and shorten the design-to-manufacture cycle time.

III.   ROLE OF SIMULATION

Since the physical phenomena in EPC process are extremely complicated and the technology is still relatively novel, there are not many reliable simulation programs available. Most past studies on lost foam casting process have been experimental in nature. Some semiempirical formulations regarding the metal flow speed and gas pressure generated from pattern decomposition have been reported [1, 2]. However, the actual functions of simulation are limited because of the following reasons:

- The coupled phenomena, including pattern degradation, fluid flow, and heat transfer, are too complex for any computer program to perform a comprehensive simulation.
- Mold filling and solidification are only one part of the whole process, and many other important factors such as EPS pattern density, property of the glue, coating permeability, and compaction of sand affect casting quality.
- There is still not enough fundamental understanding about the various physical phenomena in the lost foam casting process.

Lost foam casting process is an interdisciplinary area which requires cooperation among people with different expertise. Simulation alone can not solve all of the problem. However, in combination with systematically designed experiments, process simulation can provide general rules for the process design, and guidelines for solving the problems in lost foam casting.

IV. PHYSICAL PHENOMENA INVOLVED

In conventional empty cavity casting processes, the flow fronts of the liquid metal flow are free surfaces, because most of the time they are truly free, though sometimes air pressure in the cavity may slow down the flow. In lost foam casting, as soon as the molten metal is in contact with the EPS/PMMA pattern, the pattern will decompose into gases and liquid of styrene or their mixture, depending upon the temperature of the liquid metal (Fig. 2). The liquid metal flow speed is limited by the pattern decomposition rate. The pattern decomposition is a chemical reaction which is assumed to be a fast mechanism. Since the only "force" for pattern decomposition is the thermal energy transferred from the liquid metal front, the actual pattern decomposition rate is controlled by the liquid metal-pattern interface heat transfer rate. In other words, the liquid metal flow speed, which is the same as the actual pattern decomposition rate, is dependent on the heat transfer rate from the liquid metal front to the pattern. The above argument is based on the assumption that metallostatic head, which drives the liquid metal flow, is high enough to overcome the internal pressure of styrene gases ahead of the liquid metal front. The internal gas pressure depends on the rate of gas generation as well as

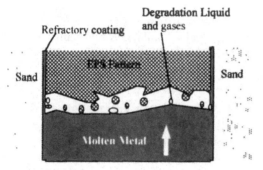

Figure 2   Reactions at the interface between molten metal and EPS pattern.

on the rate of gas escape. The rate of gas generation depends on the pattern density, casting-section thickness, and liquid metal temperature, whereas the rate of gas escape depends on the permeability of the coating layer and sand mold, as well as on the available surface area for gases to escape. The size of gap between the liquid metal front and pattern depends on the dynamic balance of the internal gas pressure and metallostatic head.

V.   MODELING OF LOST FOAM CASTING PROCESS

From the above discussion, it is seen that the lost foam casting process involves many complicated physical phenomena, which include heat transfer, mass transfer, fluid flow, chemical reaction, solidification. These phenomena occur simultaneously in a transient manner within a very short period of time during the mold filling stage. There are so many parameters involved that a comprehensive simulation of the entire problem is an extremely difficult task. Any effort trying to micromodel the process, such as formation of liquid drops of EPS material or molecular movement of the gases, would be both impractical and unnecessary. Instead, simplified approaches which are based on the data from systematically designed experiments and production experiences have been proven to be very effective [2, 5, 6]. In many cases, using simple experiments to determine the key parameters is more reliable and less expensive than using massive computational techniques. The simplified modeling approach can be implemented in the existing casting simulation code very easily. The following sections discuss the fundamentals of the simplified approaches and how the existing simulation codes can be modified to simulate the lost foam casting process.

A. Interface of Liquid Metal and EPS Pattern

The major difference between the lost foam casting process and conventional empty mold casting processes is the phenomenon occurring at the interface between the liquid metal and EPS pattern. The fluid flow and heat transfer in other regions can be modeled in the same way as those in the empty mold casting processes. In fact, the simulation of lost foam casting process can use the same simulation program as conventional empty mold casting processes, except a special boundary condition has to be applied at the interface between the liquid metal front and EPS pattern. The solution method for the simulation of conventional empty cavity casting processes has been discussed in detail in previous chapters. The most critical task in modeling the lost foam casting process is to determine the velocity of the flow front and the heat transfer rate on the liquid metal surfaces. In certain cases, the liquid metal flow velocity is the same as the actual pattern decomposition rate where the liquid metal front is in contact with EPS pattern. However, in other cases, the liquid metal flow velocity may be slower than the actual pattern decomposition rate. If that happens, large gaps may form between the liquid metal front and EPS pattern, and the collapse of the sand wall may occur. Nevertheless, the liquid metal front can not move faster than the pattern decomposition rate.

Factors determining the pattern decomposition rate are:

1. Heat transfer rate from the liquid metal front to EPS pattern
2. Removal rate of liquid and gaseous degradation products
3. Gas pressure of EPS degradation, which depends on factors 1 and 2 above
4. Height of metallostatic head
5. Friction between the mold wall and liquid

In certain cases, heat transfer rate is the controlling factor. In other cases, gas removal is the controlling factor. Heat transfer rate is a function of metal temperature, gap size between the liquid metal front and EPS pattern, and orientation of the interface. Removal rate of liquid and gaseous degradation products depends on the gas pressure and the permeability of coating and sand. Heat transfer mechanisms include radiation, convection, and conduction. The rate of heat transfer can be determined as follows (Fig. 3):

$$q_0 = q_R + q_C + q_W \frac{A_W}{A_0} \tag{1}$$

where q_0 = total heat flux from the liquid metal front
q_R = heat flow by radiation from the liquid metal front to EPS pattern

sand(3)

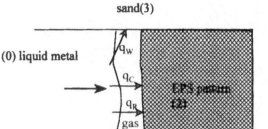

Figure 3 Heat flow at the interface of the liquid metal and EPS pattern.

q_C = heat flow by convection and conduction from the liquid metal front, through gases, to EPS pattern, or through direct contact between the liquid metal front and pattern

q_W = heat flow from the liquid metal front to coating and sand by radiation and convection

A_0 = surface area of the liquid metal front

A_W = surface area of wall contacting with gas

Heat flow rate received by EPS pattern is

$$q_2 = q_R + q_C = q_0 - q_W \frac{A_W}{A_0} \tag{2}$$

Radiation heat flow from the liquid metal front to EPS pattern is

$$q_R = F_{\text{Eff},02}\sigma(T_0^4 - T_2^4) \tag{3}$$

where subscript 0 stands for the liquid metal, and 2 for EPS pattern. T is temperature; σ is the Stefan-Boltzmman constant,

$$\sigma = 1.355 \times 10^{-12}\,\text{cal s}^{-1}\,\text{cm}^{-1}\,\text{K}^{-4} = 0.1712 \times 10^{-8}\,\text{Btu h}^{-1}\,\text{ft}^{-1}\,\text{R}^{-4}$$

$F_{\text{Eff},02}$ is the effective view factor between the liquid metal front and EPS pattern found from the following equation [7]:

$$\frac{1}{F_{\text{Eff},02}} = \frac{1 - \varepsilon_0}{\varepsilon_0} + \frac{1}{\tau_{0g2}F_{02} + \varepsilon_g/2} + \frac{1 - \varepsilon_2}{\varepsilon_2} \tag{4}$$

In above equation, ε_0 and ε_2 are emissivities of the liquid metal front and EPS pattern surfaces; τ_{0g2} is gas tranmissivity; F_{02} is the view factor between the liquid metal and EPS pattern. Note that $F_{\text{Eff},02}$ depends on F_{02}, which is a function of the shape of interface and the gap size between the liquid metal and EPS pattern. When the liquid metal front is very close to EPS pattern, F_{02}

approaches to 1.0. When the liquid metal front is far from EPS pattern, $F_{Eff,02} \ll 1.0$. Typical thermophysical properties of EPS pattern are shown in Table 1.

The combined convective and conductive heat transfer rate through gas from the liquid metal front to EPS is

$$q_C = h_C(T_0 - T_2) \tag{5}$$

where h_C is the heat transfer coefficient. h_C is the result of very complex reactions between the liquid metal, gases and EPS pattern, which cannot be easily calculated by any numerical method. Instead, h_C should be determined from experiments. In fact, h_C is the dominating factor in determining the heat flow from the liquid metal front to EPS pattern. By combining Eqs. (2), (3) and (5), we get

$$q_2 = q_R + q_C = F_{Eff,02}\sigma(T_0^4 - T_2^4) + h_C(T_0 - T_2) \tag{6}$$

In terms of equivalent heat transfer coefficient between the liquid metal front and EPS pattern,

$$q_2 = h_{eff,02}(T_0 - T_2) \tag{7}$$

$$h_{eff,02} = F_{Eff,02}\sigma\frac{T_0^4 - T_2^4}{T_0 - T_2} + h_C \tag{8}$$

The transient pattern decomposition rate (velocity) u_2 is determined from the conservation of energy at the liquid metal front by the equation

$$u_2 = \frac{q_2}{\rho \, \Delta H_{EPS}} = \frac{h_{eff,02}(T_0 - T_2)}{\rho \, \Delta H_{EPS}} \tag{9}$$

where ρ is the density of the pattern, ΔH_{EPS} is the heat of vaporization for pattern decomposition; and u_2 is the pattern decomposition rate (velocity). The rate of heat loss from the molten metal surface is

$$q_0 = q_2 + q_W \frac{A_W}{A_0} \tag{10}$$

Table 1 Typical Values of Thermophysical Property of EPS Pattern

Emissivity	0.6
Gas transmissivity	0.88
Density	$0.02 \, \text{g/cm}^3$
Heat of fusion	240 cal/g

Among the different mechanisms of radiation, convection, and conduction, which one is the dominating mechanism in heat transfer between the molten metal front and EPS pattern? Using the typical values of the emissivity shown in Table 1 [2, 6, 7], we get $F_{Eff,02} = 0.42$. In this case, $q_R \approx 0.6$ cal cm^{-2} s^{-1} °C^{-1}) in aluminum casting (equivalent to $u_2 = 0.2$ cm/s), and $q_R \approx 1.6$ cal cm^{-2} s^{-1} °C^{-1} in gray iron casting (equivalent to $u_2 = 0.5$ cm s^{-1}).

Experiments have shown that the magnitude of the pattern decomposition rate, u is between 5–15 cm s^{-1} for aluminum casting [2, 6]. This implies that $q_2 \approx 24 \sim 72$ cal cm^{-2} s^{-1}. Therefore, $q_C \approx 97\% \sim 99\%$ of q_2 and is the dominating mechanism. Note that the above statement is true only when the molten metal front and EPS pattern are almost in contact with each other. This condition is referred to as the *contact* situation. When the molten metal front and EPS are far apart, radiation and gas convection are the dominating mechanisms. This is referred to as the *noncontact* situation. When the metallostatic head is not large enough to overcome the gas pressure and wall friction, the molten metal will move slower than EPS pattern decomposition rate, and a *noncontact* situation will occur. On the other hand, if the metallostatic head is high and the flow rate of the molten metal is large, liquid metal is always in contact with EPS pattern. In order to avoid sand collapse, the liquid metal should always be kept in contact with EPS pattern. However, numerical simulation should be able to handle both situations.

B. Modeling the Interface of Liquid Metal Front and EPS Pattern

1. Schemes for Contact Situation

As previously mentioned, the lost foam casting process can be simulated by modifying the existing casting mold filling code to take into account the special boundary conditions at the molten metal front and EPS pattern interface. The noncontact situation and the transition from contact to noncontact situations will be discussed later. In contact conditions, the simulation can be approached in one of the following two ways, depending on how the experimental data is processed.

Scheme 1. Determining heat transfer coefficient, $h_{eff,02}$, from experimental data, and using Eqs. (7) and (9) to determine q_2 and u_2. For given coating, sand, and pattern properties, $h_{eff,02}$ can be considered as

$$h_{eff,02} = f\left(T, P, \frac{d}{M}\right) \quad \text{(contact situation)} \quad (11)$$

where T and P are temperature and pressure at the liquid metal front, d is the gap size between the liquid metal front and EPS pattern, and M

is the modulus of the casting section. M is defined as the ratio between the volume and surface area in the local region. M equals half thickness for a plate section, or half of the radius for a cylindrical shape. The decomposition rate u_2 is found from Eq. (9).

Scheme 2. Determining u_2 from experimental data, and using Eqs. (7) and (9) to determine q_2. For given coating, sand and pattern properties, u_2 can be considered as

$$u_2 = f\left(T, P, \frac{d}{M}\right) \quad \text{(contact situation)} \tag{12}$$

With either scheme mentioned above, the liquid metal velocity u_0 can be determined by $u_0 = u_2$, because a contact situation is assumed. The heat loss from the molten metal front is

$$q_0 = q_2 + q_w \frac{A_W}{A_0} \approx q_2 \tag{13}$$

2. Transition from Contact to Noncontact Situation

When the metallostatic head is not large enough to overcome the gas pressure and wall friction, the molten metal front will move slower than the EPS pattern decomposition rate, and a noncontact condition will develop. On the other hand, if metallostatic head is high and the flow rate of the molten metal is large, the liquid metal can always be in contact with the EPS pattern. The transition from contact to noncontact situation occurs when the liquid metal flow can not keep up with the pattern decomposition. Mathematically, this happens when the pressure of liquid metal adjacent to EPS pattern drops to ambient pressure, i.e., the boundary condition $u_0 = u_2$ is applied. To maintain continuity, a smaller u_0 has to be applied, i.e., $u_0 < u_2$, and in this way the separation of liquid metal front and pattern occurs. This is referred to as *ambient pressure transition*.

A more sophisticated approach would be to calculate the gas pressure from the rate of decomposition and the rate of removal of the decomposed pattern material. However, this is not an easy job, because the decomposed material is a mixture of gas and liquid. The exact volume of the mixture is a complicated function of many factors. Besides, the permeability of the coating is a complex function of the amount of residuals which has been already absorbed. As a matter of fact, as discussed below, this kind of sophisticated approach may not be necessary.

In real production, the coating is always permeable to gases and has the ability to absorb certain amount of liquid residuals. If a separation (noncontact) situation occurs (i.e., $u_0 < u_2$) and q_2 drops to the magnitude of

$2\,\text{cal}\,\text{cm}^{-2}\,\text{s}^{-1}$, which is significantly smaller than the value of $q_2 = 25\,\text{cal}\,\text{cm}^{-2}\,\text{s}^{-1}$ in the contact situation, the pattern decomposition rate will be reduced by 90%. Since the gas generation rate is slowed down, the pressure will quickly drop to the ambient value because of the permeability and absorptivity of the coating. If the noncontact situation is caused by gas pressure buildup, the reduction of the gas generation rate will increase the liquid metal velocity. Consequently, the contact situation will be resumed. The process is repeated until the mold is full. In that case, the problem can be handled as contact, and the transition from contact to noncontact as mentioned above would be sufficiently accurate. On the other hand, if the non-contact situation is caused by geometric nature or premature freezing, the ambient pressure transition approach as mentioned above should be even more valid, because the problem is essentially the same as conventional empty mold casting.

The above arguments may not apply to the situation where the permeability of the coating is not sufficient. If that happens, sand collapse may no longer be a problem because of the strength of the coating, but the formation of a void due to gas pressure buildup may occur. Simple experiments can be conducted to determine whether the coating is permeable enough to avoid defect formation. An indication of this problem would be that the liquid metal moves extremely slowly, because of the high gas pressure in the void. In this situation, a simulation is not necessary, and the coating properties should be improved.

The transition from noncontact to contact situation occurs when the liquid metal flow catch up with pattern decomposition. Mathematically, this happens when u_0 increases to u_2 and the pressure of liquid metal adjacent to EPS pattern increases to above the ambient pressure.

3. Simulation of Noncontact Situation

The noncontact situation is the case where a gap forms between the liquid metal front and EPS pattern during mold filling. As mentioned above, an ambient pressure condition can be applied at the free surfaces of the liquid metal. The velocity and heat transfer at the surface of the liquid metal front is determined in the same way as conventional empty mold casting. The heat transfer rate received by the pattern is

$$q_2 = q_R = F_{\text{Eff},02}\sigma(T_0^4 - T_2^4) \tag{14}$$

The decomposition rate of EPS pattern is determined by Eq. (9).

VI. SAMPLE ANALYSES RESULTS

Reference 2 presented the algorithm and the results of simulation of the fluid flow and heat transfer in a very simple geometry lost foam casting. A scheme similar to scheme 1 in the contact situation was used.

In Ref. 5, scheme 2 was used to simulate the fluid flow and heat transfer during the mold filling of a simple geometry lost foam casting. The simulation was made by modifying an existing finite difference method (FDM) casting simulation code by imposing proper boundary conditions on the metal flow front described as above. The momentum, heat, and mass transport are solved in the same way as in the original FDM code. An important prerequisite for simulation is the experimental data. A significant number of experiments have been conducted for aluminum alloys cast by the EPC process. Extensive research has been conducted by Shivkumar [1, 6], who provided a wealth of information on the flow characteristics of aluminum alloys during the EPC process. On the other hand, experimental data for gray iron and ductile iron are sparse. Therefore, example runs have been made for aluminum alloys; the results are presented in the following paragraphs. Nevertheless, the techniques presented here are generic in nature and can be applied to a wide variety of alloys.

The modified FDM code [5] has been used to simulate the filling and cooling of a number of casting parts (Aluminum 319). Some details about the input parameters are as follows:

	Casting metal	Mold	EPS pattern
Range of melting temperature, C	604–540		
Initial mold temperature, °C		30	
Pouring temperature, °C	750		
Density, $kg\,m^{-3}$	2600	1600	20
Specific heat, $J\,kg^{-1}\,K^{-1}$	960	1045	
Thermal conductivity, $J\,sec^{-1}\,m^{-1}\,K^{-1}$	117	0.5	
Heat of fusion, $J\,kg^{-1}$	388,000		
Heat of degradation of pattern, $J\,kg^{-1}$			1,003,000

Pattern decomposition rate, $m\,sec^{-1}$ (deduced from the data given in Ref. 6, is $u_2 = -0.034 + 0.96p + 0.012T$, where p is the liquid metal pressure head (m) and T is temperature (C) at metal-pattern interface, respectively.

Case 1. The first case is a simple plate casting ($0.28 \times 0.15 \times 0.013\,m$). The plate is equipped with two side ingates and one top ingate. Hollow sprue and ingates are used to reduce the frictional energy loss. This is chosen as an example because the experimental results are available from Refs. 1 and 6. The experimental results about the times of arrival of the metal at different locations in the plate are shown in Fig. 4. Computer simulation results of the flow front locations at different times are shown in Fig. 5. Comparison of the measured and calculated filling shows satisfactory agreement. Note that different starting times were used in the calculated and the experimental results, and thus only the relative values are of interest. It is clear that a junction of two metal fronts is created in the central region of the plate. The regions where surface laps are likely to form are identified by light shades in Fig. 6.

Case 2. The second case is similar to case 1, with the following differences: (a) a hollow pattern for the casting plate as well as hollow sprue and ingates are used; (b) the thickness of the plate is increased to 0.022 m. The computed results for the velocity and temperature distribution during mold filling are presented in Fig. 7. The computed flow front locations at different times are remarkably different from those shown in Fig. 5.

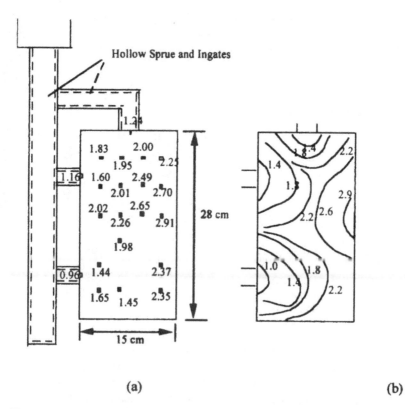

(a)   (b)

Figure 4   Measured times of arrival of the metal front for a plate pattern.

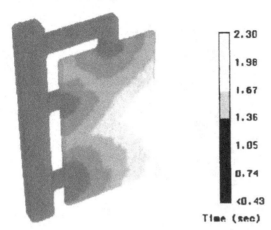

Figure 5 Computer simulation results of the flow front locations at different times for the plate shown in Fig. 4.

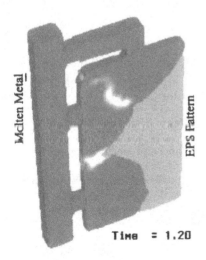

Figure 6 Calculated results showing the regions where surface laps are likely to form during filling.

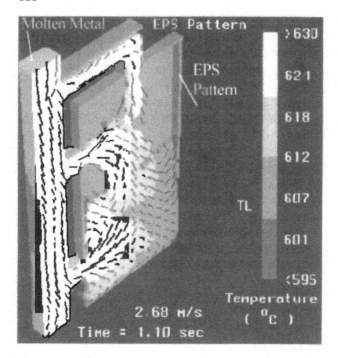

Figure 7   Velocity plot at a cut surface during filling of a plate casting with a hollow pattern

REFERENCES

1. S Shivkumar, B Gallois. Physico-chemical aspects of the full mold casting of aluminum alloys. Part II: Metal flow in simple patterns. AFS Transactions, 1987.
2. HL Tsai, TS Chen. Modeling of evaporative pattern process, part I: metal flow and heat transfer during the filling stage. AFS Transaction, 1988, pp. 881–890.
3. MJ Lessiter. Lots of activity taking place among lost foam job shops. Modern Casting, April 1997, pp. 28–31.
4. MJ Lessiter. Today's lost foam technology differs from yesteryear. Modern Casting, April 1997, pp. 32–35.
5. CM Wang, AJ Paul, WW Fincher, OJ Huey. Computational analysis of fluid flow and heat transfer during the EPC process. AFS Transaction, 1993, pp. 897–904.
6. S Shivkumar. Fundamental characteristics of metal flow in the full-mold casting of aluminum alloys. PhD thesis, Stevens Institute of Technology, 1987.

7. GH Geiger, DR Poirier. Transport Phenomena in Metallurgy. Reading, MA: Addison-Wesley Publishing Company, Inc., 1980, p. 396.

8. SSS Abayarathna, HL Tsai. Modeling of evaporative pattern process. Part II: Determination of possible carbon pickup. AFS Transaction, 1989, pp. 645–652.

9. SSS Abayarathna, HL Tsai. Modeling of evaporative pattern process. Part III: Heat/mass transfer in sand mold and its effect on casting solidification. AFS Transaction, 1989, pp. 653–660.

11
Investment Casting

Dilip K. Banerjee
GE Global Exchange Services, Gaithersburg, Maryland

Kuang-O (Oscar) Yu
RMI Titanium Company, Niles, Ohio

I. INTRODUCTION

The investment casting process allows the casting engineer to make full use of enormous capability and flexibility inherent in the process. Parts produced by this process are truly functional, cost effective, and aesthetically pleasing. A wide range of parts from a few grams to more than 1000 kg can be competitively produced with this process. A detailed discussion of many of the advantages of investment casting process is provided in Ref. 1. Some of these advantages are:

Complexity: A wide range of both internal and external complexities can be achieved. Complex parts are produced at lower costs with improved functionality.

Freedom of Alloy Selection: Any castable alloy can easily be used. The alloys that are difficult to machine or forge can be cast by this process in a cost-effective manner.

Close Dimensional Tolerances: Castings are produced very close to the final dimensions. This eliminates the need for elaborate machining operations.

Availability of Prototype and Temporary Tooling: Direct machining of wax patterns or use of quick, inexpensive tooling methods allows for timely collaboration between the foundry and the designer. This allows for production of components, which are functional and manufacturable.

Reliability: The wide use of investment casting to produce key components of airframe and jet engines demonstrates that it is a very reliable process.

Although the following discussion applies in general to all alloys, primary emphasis will be given to the process with respect to superalloys and titanium alloys.

II. INVESTMENT CASTING PROCESS

The basic investment casting process has been the same for centuries. There have been significant enhancements since the late 1920s. Application of superalloys was developed as part of production of dental components.

Figure 1 is a schematic illustration of the investment casting process [1]. The first step is to make an exact replica or a pattern of the component in wax or plastic. Allowance for solidification shrinkage should be taken into consideration in pattern design. Sometimes the part may contain internal passages. Preformed ceramic cores (Fig. 2) are inserted in die cavities around which pattern material (wax or plastic) is injected. Several patterns may be assembled in a cluster or tree configuration and gates, runners, and risers are

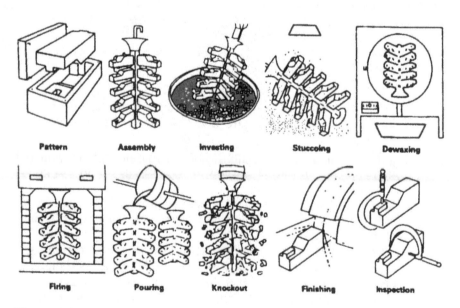

Figure 1 Schematic illustration of the investment casting process. (From Ref. 1.)

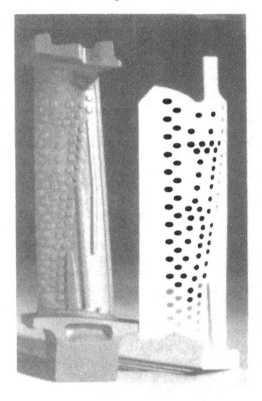

Figure 2 Typical ceramic core configuration and cutaway of air cooled turbine airfoil. (From Ref. 1.)

then added to form a complete wax assembly (Fig. 3). Care should be taken to assemble the cluster or tree in a way which can ensure the proper filling of molten metal in cavities during casting. Molds are usually produced by first immersing the entire assembly into a ceramic slurry. In order to strengthen the mold, a dry, granular ceramic stucco is applied immediately. These steps are continued several times for developing a rigid mold. Next, the assembly is dried. The wax is then melted out. Next, the mold is fired to increase the strength. In order to control solidification during the casting operation, mold wraps or insulating blankets are often used to selectively wrap the ceramic mold outside surface. Figure 4 shows circular molds to cast single crystal turbine airfoils [3].

Depending on the types of alloys to be cast, the melting and casting operations can be performed either in air or vacuum. Aluminum alloys, steels,

Figure 3   Four castings made from an assembled wax tree. (From Ref. 2.)

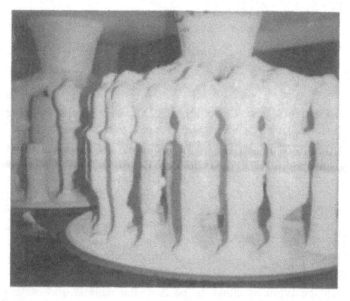

Figure 4   Circular molds for casting single crystal turbine airfoils. (From Ref. 3.)

and cobalt-base alloys are typically cast in air, whereas nickel-base superalloys and titanium alloys are cast in vacuum. Because of their excellent control capabilities on molten metal superheat and chemical composition, induction melting furnaces (either in air or vacuum) are typically used for melting metals. Titanium alloys are melted by VAR (vacuum arc remelting) skull melting and induction skull melting processes [4].

Both static and dynamic casting modes are used for investment casting process. For conventional structural castings, almost all alloys are cast in static mode. One exception is titanium alloys. Due to the strong chemical reaction between the molten titanium and ceramic mold and core, titanium alloys are typically cast with very low molten metal superheats and ceramic mold preheat temperatures. This situation causes difficulty for liquid metal to fill the entire mold cavity during the mold filling process. In order to facilitate mold filling, preheated ceramic molds are sometimes kept under high speed spinning during the liquid metal pouring and subsequent solidification steps. This casting method is called centrifuge casting. In this condition, centrifugal forces generated by the high speed spinning can force the liquid metal to fill the mold cavity and result in sound castings. Another exception is the TCS (thermally controlled solidification) process which uses the withdrawal mode similar to that used for directionally solidified turbine airfoils to produce thin wall equiaxed grain structural castings [5].

Depending on the resultant casting grain structures, turbine airfoils are cast in either static or withdrawal mode. Equiaxed grain turbine airfoils (Fig. 5) are typically cast in static mode. For castings with columnar grains or single crystal structure, dynamic casting mode (withdrawal process) is used (Fig. 6). During casting, an induction-heated graphite susceptor or resistance heater is positioned around the ceramic mold and it heats the mold to a specified temperature. The mold is usually open at the bottom and rests on a water-cooled copper chill. After pouring, the copper chill and ceramic mold are withdrawn at a predetermined rate away from the heat source. The upper portion of the mold will remain hot but the bottom portion of the mold will start to cool by radiation. This allows for maintaining a fixed thermal gradient during the alloy solidification and promotes the formation of columnar grains. The initial columnar grain structure in the starter block (which directly contacts with the water-cooled copper chill) proceeds to fill the entire mold cavity resulting in a columnar grain airfoil (Fig. 5). A single crystal casting (Fig. 5) is obtained by inclusion of a "grain selector" or single crystal seed above the starter block, which permits only a single grain to pass through. For equiaxed grain structural castings that are produced by the TCS process, the mold is not open at the bottom to avoid the direct contact between the molten metal and the water-cooled copper chill. As a result, no initial columnar grains will form. The withdrawal speed

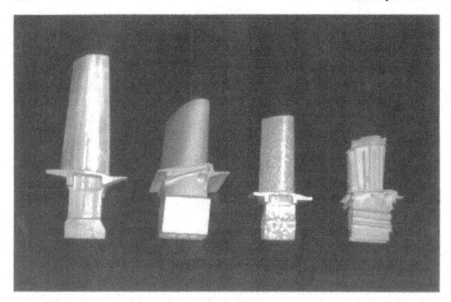

Figure 5 Grain structures of turbine airfoils. From left to right: columnar grains, single crystal, equiaxed grains, internal cooling passage.

of the TCS process is usually significantly faster than the conventional directional solidification withdrawal process. This results in a moderate temperature gradient which is high enough to facilitate the feeding capability for reducing microporosity but is still low enough to maintain an equiaxed grain structure for the resultant casting.

III. APPLICATION OF INVESTMENT CASTINGS

Investment castings are commonly made by nickel-base superalloys, titanium alloys, aluminum alloys, cobalt-base alloys, and steels. The most important application for investment castings is in the aerospace industry. Nickel-base superalloys are used for jet engine structural castings as well as turbine airfoils. The latter includes both jet engine for aircraft propulsion and industrial gas turbine for power generation. Structural castings always have equiaxed grain structures (Fig. 7) whereas turbine airfoils have equiaxed grains (EQ), directionally solidified (DS) columnar grains, and single crystal (SX) types of structures (Fig. 5). All three types (EQ, DS, and SX) of turbine airfoils are commonly used for both jet engine and industrial gas turbine applications. Titanium castings are used as structural

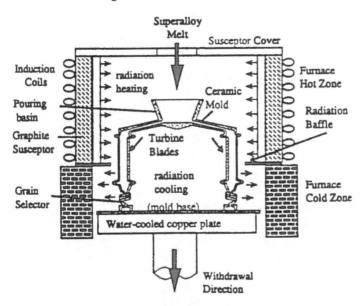

Figure 6 Schematic drawing of directional solidification process for casting columnar-grained and single-crystal turbine blades. (From Ref. 3.)

Figure 7 Superalloy equiaxed grain structural casting made from an assembled wax tree similar to that shown in Fig. 3. Due to symmetry, only half of the casting is shown. (From Ref. 2.)

components for both jet engines and airframes. Recently, titanium alloys are also being used to cast golf club heads. However, most of the golf club heads are still made from stainless steel. Aluminum castings are primarily used as the airframe structural components. Cobalt-base alloys, stainless steel, and titanium alloys are the most used alloys for medical implant components. Cobalt-base alloys are also used for jet engine turbine airfoils. However, their use is quite limited in comparison with nickel-base super-alloys. Table 1 is a summary of applications for different investment castings.

IV. MODELING OF INVESTMENT CASTING PROCESS

Characteristics that are unique to the investment casting process include high temperature ceramic mold, casting in the vacuum, centrifugal force for mold filling, withdrawal process for promoting directional solidification and/or feeding capability, and strict requirements for casting grain structure. All these characteristics will require special treatments for modeling approaches.

A. Mold Geometry Generation and Meshing

The step in geometric model building that is unique to the investment casting process is the mold (shell) generation. The mold thickness is typically 6–13 mm and is relatively uniform around the casting. Thus, the mold has a similar degree of geometric complexity as the casting. The mold elements may be

Table 1 Common Applications of Investment Castings

Applications	Alloys				
	Nickel	Cobalt	Titanium	Aluminum	Steel
Jet engine structure	x		x		
Jet engine turbine airfoil	x (EQ, DS, SX)	x (SX)			
Industrial gas turbine airfoil	x (EQ, DS, SX)				
Airframe structure			x	x	
Medical implant		x	x		x
General structure	x	x	x	x	x
Golf club head			x		x

generated automatically by extruding the casting outside surface by the desired mold thickness and then generating elements in the expanded region automatically. Care must be taken to ensure that intrusions of elements into adjoining elements are avoided. In addition, the thickness of the ceramic mold is not completely uniform. In general, the corner region can be either thicker or thinner than the average mold thickness (Fig. 8). Thus, the extrusion operation needs to have a capability to account for this mold thickness variation. Many of the commercial software packages have the capability to generate a mold of desired dimension automatically [6].

B. Heat Transfer

The heat transfer problem for the metal is handled by solving the energy conservation equations with appropriate interface and boundary conditions. A coupled thermal/flow model will treat the heat loss during mold filling and afterward during the solidification process. In general, the model should handle heat conduction in the core and the mold, convection and radiation across the metal/mold interface, radiation heat transfer and natural/forced convection (if cast in air) at the mold outer interface. Because nickel-base superalloy and titanium alloy castings are typically cast in a vacuum environment, radiation heat transfer is the only method under this condition for heat loss from mold surface to the water-cooled furnace shell. Since there is a simultaneous exchange of heat flux between different parts of the mold's outer surface and the mold outer surface and furnace shell interior, enclosure type radiation with proper view factors is involved. For the DS and SX

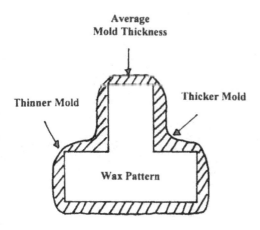

Figure 8   Thickness variation of an investment casting ceramic mold.

casting simulation, the view factors will change continuously during the with-drawal process [7].

The radiation effects may be calculated using the latest techniques of "net radiation of gray body." The view factors should be calculated auto-matically during the withdrawal process. The shadowing effects should also be taken into consideration. The enclosure may move with respect to cast-ing in DS and SX production and therefore, the view factors need to be updated. The faces of the castings in a cluster, which see each other and the enclosure should participate in the radiative heat exchange process. This is especially useful in high-temperature applications where self-radiation from one part of the mold to another part of the mold is important. For a circular cluster with symmetry, only one casting and its correspond-ing section of furnace enclosure need to be modeled (Figs. 9 and 10). Many

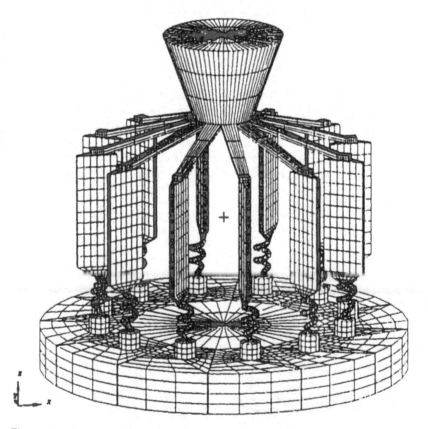

Figure 9   Cluster configuration of a plate casting. (From Ref. 8.)

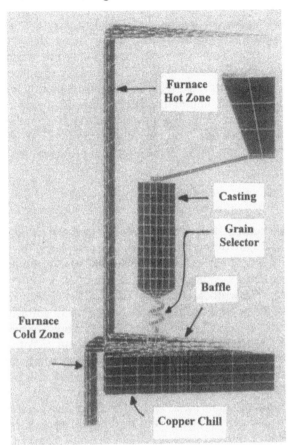

Figure 10 Finite element model for a plate casting in the circular cluster shown in Fig. 9. (From Ref. 8.)

commercial software packages have built-in capability for view factor radiation calculation [6].

In sand casting process, metal chills are commonly used to increase local cooling rate and create a preferred solidification pattern. In investment casting process, however, blanket insulation materials such as kaowool are commonly used to wrap specific areas of the ceramic mold in order to slow down the local cooling rate and establish a preferred solidification pattern. During the simulation, effects of the kaowool wrap may be modeled by suitably adjusting the mold surface emissivity or heat transfer coefficient. The values of these adjusted emissivity or heat transfer coefficient need to be established from experimentally obtained mold surface heating and cooling curves, e.g., thermo-couple data. In addition to slowing down the cooling rate during casting soli-dification, kaowool wrap also slows down the mold local cooling rate during the transfer of high-temperature mold from a mold preheat furnace to a metal casting furnace and creates a temperature gradient in the empty mold before the metal pouring. This will provide an additional effect on creating preferred solidification patterns during casting. Thus, the most accurate way to model the effects of kaowool wrap is to model both the mold transfer from the preheat furnace to casting furnace and the subsequent casting processes inside the casting furnace.

C. Fluid Flow

There are two fluid flow phenomena in a casting process: free surface flow during mold filling and natural convection after the mold is completely filled. For investment castings, the effect of natural convection on the casting quality is minimal. On the other hand, the mold filling process has major effects on the quality of castings produced. One important characteristic of the investment casting process is its capability to produce thin wall castings with intricate geometries. For jet engine turbine airfoils, the casting wall thickness can be as thin as 0.5 mm (0.020 in.) whereas the mold thickness is about 6.25 mm (0.25 in.). Under this condition, the ratio of the thermal mass between the ceramic mold and the molten metal is quite large. As a result, a small increase in the mold temperature during metal-mold initial contact can result in a relatively large temperature loss in liquid metal and no-fill defect may result. No-fill (Fig. 11) is the condition where the liquid metal does not fill a portion of the mold. No-fill may occur due to metal freeze off when the alloy cools to a temperature such that fraction solid and viscosity increase do not permit filling. This condition can be easily modeled by using the temperature distribution that is predicted by the mold filling model. When a local liquid metal temperature is lower than a specified temperature before the entire mold is filled, no-fill defect io aooumod to ooour at that looation. Aloo, tho no fill may ooour whon tho combined effects of pressure, wettability of the mold, and liquid metal surface tension prevent fine features in the mold from filling. In investment casting foundries, many iterations are often made by changing the gating geometry and processing parameters. A design is accepted when the no-fill problem is eliminated. The fluid flow simulation serves also as a powerful tool in designing an optimum gate design that would reduce pour weight significantly in com-parison to that predicted by traditional design.

For accurate predictions, mold filling modeling must be coupled with heat transfer analysis. The output of the mold filling analysis includes mold filling sequence and metal and mold temperature distributions. The combina-

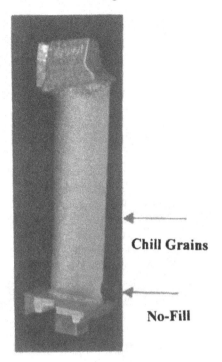

Figure 11 A superalloy equiaxed grain turbine blade with no-fill and chill grains defects.

tion of mold filling sequence and metal temperature distribution can be used to predict entrapped gas porosity (if cast in the air) and weldline type of defects. In addition, the calculated metal and mold temperature distributions at the end of the mold filling are used as the initial metal and mold temperatures for the subsequent solidification analysis.

D. Stress Analysis

The thermally induced stresses are usually calculated in all parts of the model (e.g., casting, mold, core etc.) with elastic, elastoplastic, and elasto-viscoplastic constitutive models that are described in Chapter 3.

One has to input properties of each material, e.g., Young's modulus, Poisson ratio, yield stress, hardening law, coefficient of thermal expansion for performing a stress analysis. The stress analysis comes into picture once the casting is filled and a layer of solidified material forms at the metal/mold interface. The model should have a contact algorithm to handle the interfaces between the casting and mold. One can predict residual stresses, plastic deformation, hot tears, cold cracks, and the final shape of the solidified part by using a proper stress model. For an accurate prediction, the model should include the complex cracking behavior of the ceramic mold.

The heat flow across the casting/mold interface is affected by a gap formation. As the metal solidifies, it contracts away from mold and gaps of varying widths are formed at different locations along the casting/mold interface. The heat transfer coefficient at the casting/mold interface should be updated with the progressive changes in the gap widths.

Nonuniform cooling and/or hindered shrinkage will lead to the formation of thermal stresses in the casting. When the level of thermal stress is higher than the yield stress of the material, plastic deformation occurs. Plastic deformation results in geometric distortion and is detrimental to the shape accuracy of the casting [9]. Presence of imperfections in the casting and mold material may help cracks to develop and propagate with the favorable increase of thermal stresses during solidification at those locations. If the cracking happens during solidification, it is called hot tears. Cold cracks is the term commonly used for the case when cracking happens after the solidification is completed. High amount of plastic deformation and residual stresses could also cause recrystallization during the solution heat treatment following the casting process and allow recrystallized grain defects to form in DS and SX castings [10].

Suitable displacement boundary conditions need to be specified, as a rigid body movement of all calculated materials has to be constrained. The predicted values of stresses and strains depend on the set up of the boundary conditions. For highly asymmetric geometries like the turbine blades, the correct choice of the boundary conditions is quite difficult. Often stress calculations are performed for the casting only. The time dependent mechanical interactions between casting, mold, and core are often neglected for making calculations simple. In some cases, the solution accuracy is enhanced when a steady constraint between casting, mold, and core is assumed.

Stress analysis simulation for a complete metal/mold system with large number of elements converges slowly and takes much longer to complete. Models of the casting alone may be quick and effective in some cases. Figure 12 shows results of calculation of stress distribution for a flap casting alone and the casting with the mold. This type of study may be used to determine the causes for the formation of cracks.

To accurately model the formation of stresses and related defects during casting, the effects of casting metal dimensional change and material strength difference between the casting metal and ceramic mold need to be included. During stress analysis, for those casting processes with metal

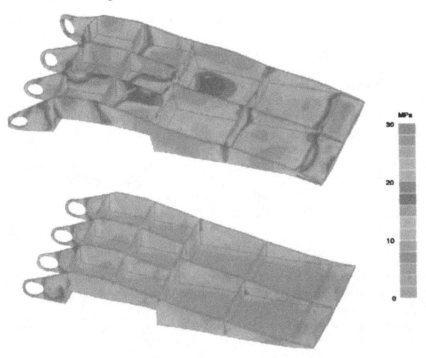

Figure 12 Calculated stress distribution for a flap casting. Top: intrinsic stress model, no external influence was included; bottom: the part was included in the investment casting mold. Note the high stress levels around the hole. (From Ref. 11.)

molds (permanent mold casting, die casting, squeeze casting, and semi-solid metalworking), it is relatively accurate to assume that the mold strength is significantly stronger than the casting metal and, hence, only the effect of casting metal dimensional change need to be considered. For investment casting process, because the strengths of casting metal and ceramic mold at solidification temperatures are fairly close, effects of both casting metal dimensional change and material strength difference between metal and mold have to be accurately taken into consideration. In fact, some investment casting foundries control the mold material chemical composition and/ or micro/macrostructure (e.g., with induced porosity) to produce a mold material with an appropriate strength. In this way, the mold will have adequate strength to withstand the physical operation before the casting is poured yet the mold will be easily crushed during and/or after solidification to avoid the formation of thermal stresses and related defects in the resultant castings.

V. DEFECTS FORMATION AND PREDICTION

The definition of defects depends on the quality requirement of castings. Investment castings used for aerospace applications have more stringent quality requirements than most of the other types of castings. As a result, many microstructure features such as dendrite arm spacing (DAS) and grain structures, which typically are not specified as quality requirements for other castings, are considered as a part of quality requirements for investment cast aerospace components. Investment castings are still subject to the requirements for other common casting defects such as macroshrinkage, microporosity, hot tears, cold cracks, and dimensional distortion, etc. The modeling approach for these common defects have been discussed in Chapters 3 and 4. Here, only the formation of microstructure related defects will be emphasized.

A. Quality Specification

The biggest difference between the quality requirements of investment castings and other types of castings is the specification of casting microstructures [12]. Table 2 is a summary of typical microstructure requirements for investment castings. In general, aluminum castings have a requirement for DAS due to its close relationship with casting mechanical properties [13]. Equiaxed grain structural and airfoil castings have requirements in both grain size and morphology. Grain sizes which are either bigger or smaller than the specified size range will be considered defects. In addition, grains with high aspect ratio, i.e., columnar grains are also considered defects. For DS turbine airfoils, the size as well as the growth direction of columnar grains are strictly specified. In addition, all equiaxed shape grains, e.g., equiaxed grains, freckles, and recrystallized grains, are considered rejectable defects. For SX turbine airfoils, all grains are treated as defects. Furthermore, casting crystallographical orientation is also specified. Figure 13 is a schematical illustration of various grain defects in single crystal turbine airfoils [14].

B. Modeling of Microstructure Related Defects

Microstructure evolution in castings is a critical issue in solidification modeling of castings. Microstructure is associated with the microsegregation pattern and the appearance of phase. The permeability of the mushy zone is determined by the dendrite structure which also has influence on the microporosity formation and the mechanical properties of the cast part. At the macroscale (0.01–1 m), heat exchange, the fluid flow, and the transport of solute species due to convection effects has to be considered. At the mesoscale (0.1–10 mm), the formation of grains is influenced both by the macroscopic aspects and by the

Table 2 Microstructure Requirements for Investment Castings

Alloys	Structural EQ castings	Turbine airfoils		
		EQ	DS	SX
Superalloys	Grain size Columnar grains	Grain size Columnar grains	Grain size Grain direction Equiaxed grains Freckles Recrystallized grains Misoriented grains	Columnar grains Equiaxed grains Freckles Recrystallized grains bigrains Multigrains High angle boundaries Slivers Zebra Crystallographical orientation
Titanium	Grain size Columnar grains	—	—	—
Aluminum	DAS	—	—	—

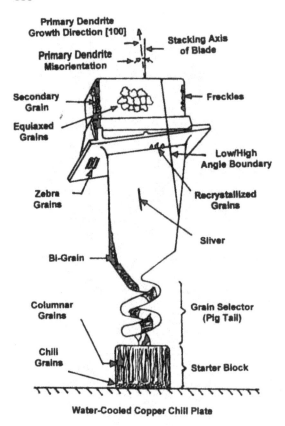

Figure 13 Schematical illustration of various grain defects in single crystal turbine airfoils. (From Ref. 14.)

nucleation and growth of grains (microscopic). At the microscopic level (1–100 μm), the kinetics of dendrite tips or eutectic lamellae/fibers and the coarsening phenomena are considered. This has been mostly done under steady state conditions without considering the presence of convection. At the atomic level (0.1–1 nm), the microstructure is associated with the phenomena occurring at the scale of the diffuse liquid/solid interface of metallic alloys.

Three different modeling techniques are used to predict casting microstructure evolution and defects formation: traditional macroscopic heat transfer modeling, deterministic modeling, and stochastic modeling. Heat transfer modeling [10, 12, 15, 16] is based on casting macrothermal history and can be easily performed by using postprocessing. Often a microstructural defect map based on G (temperature gradient) and R (solidification rate) is constructed to facilitate the prediction of defects. Based on theoretical analysis and experiments, different combinations of these parameters can correlate reasonably well with certain types of defects (Fig. 14). These maps are generally constructed using simple castings (e.g., cylinders, plates) by relating the resultant microstructure with thermal parameters. The advantages of heat transfer modeling are ease of use and a fast turnaround time. However, because it does not provide any information about microstructure evolution and defects formation, traditional heat transfer modeling cannot precisely predict how the microstructures and defects evolve and form. Deterministic modeling includes the macro/micro modeling and nucleation and growth laws; thus, it can provide more information about microstructure evolution and defects formation than traditional macroscopic heat transfer modeling. Stochastic modeling is based on the Monte Carlo and Cellular Automata techniques. The main idea of stochastic model is to follow the evolution of the individual grains considering nucleation, growth kinetics, and a preferred growth direction. It can compute three-dimensional grain structures and one can section along any plane of interest to depict a direct picture of microstructure. The major disadvantages of stochastic modeling are that it requires a large hard disk memory and a long simulation time.

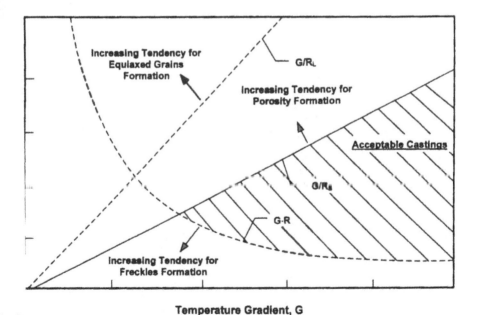

Figure 14   Defect map for single crystal castings. (From Refs. 10, 12. 16.)

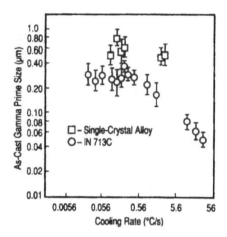

Figure 15 Measured as-cast gamma prime size as a function of calculated cooling rate. (From Refs. 10, 12.)

All three modeling techniques are useful tools to predict the casting microstructure evolution and defects formation, if used appropriately. The details of deterministic modeling and stochastic modeling techniques are described in Chapter 5. Here, only the application of these modeling techniques to predict various microstructures and defects that are unique to investment castings will be described.

1. As-Cast Gamma Prime Size

The as-cast gamma prime size is generally a function of casting cooling rate [10]. Figure 15 shows that for IN 713C (an equiaxed grain alloy), for high values of cooling rate, the as-cast gamma prime size increases with decreasing cooling rate. At low cooling rates, however, the gamma prime size appears to be independent of the cooling rate. Single crystal alloys, e.g., René N5, usually have higher Al and Ti contents which result in higher gamma prime solvus temperatures and larger as-cast gamma prime sizes than equiaxed grain alloys.

2. Dendrite Arm Spacing

For dendritic solidification, DAS is the most convenient parameter for representing the degree of microsegregation of castings. The fineness of DAS is directly related to LST (local solidification time) or average cooling rate during solidification. As heat is extracted faster from castings (i.e., shorter LST and higher average cooling rate), the size of the resultant DAS becomes smaller and the microsegregation effects are diminished.

For aluminum alloy, it has been shown that a linear correlation can be established between the ultimate tensile strength (UTS) and the secondary dendrite arm spacing of D357-T6 castings. As a result, DAS is specified as a quality requirement for aluminum castings [13]. However, the DAS-UTS relationship varies with the heat treatment process and chemical composition of the alloy. Therefore, a preliminary evaluation using attached coupons on the casting is necessary before actual DAS measurements on the casting are meaningful.

During casting, primary dendrite grows directly opposite to the heat flow direction, and primary dendrite arm spacing (PDAS) can be measured in the transverse plane [17] that is perpendicular to the primary dendrite growth direction (Fig. 16). The growth direction of secondary dendrites is perpendicular to the primary dendrite growth direction and the secondary dendrite arm spacing (SDAS) can be measured (Fig. 17) along the primary dendrite growth direction [17]. In DS and SX castings, primary dendrites can grow up to 50 cm long. Thus, the structures of both primary and secondary dendrites are very well developed and the PDAS and SDAS can be easily measured. Because of this, DAS data of many alloys (even for some of those equiaxed grain alloys) were measured from directionally solidified castings [18]. LST and average cooling rate during solidification can be obtained by macro heat transfer analysis. Yu et al. [15] demonstrated that DAS vs. LST relationship established from simple shaped cylinders can be used to predict DAS distribution in complex shaped superalloy SX turbine airfoils accurately. For aluminum alloy EQ

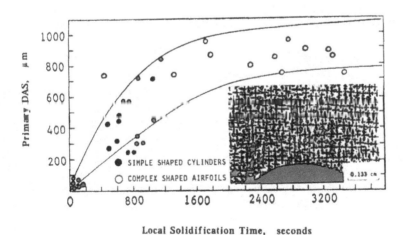

Local Solidification Time, seconds

Figure 16 Measured primary dendrite arm spacing of the alloy René N5 as a function of calculated local solidification time. (From Ref. 15.)

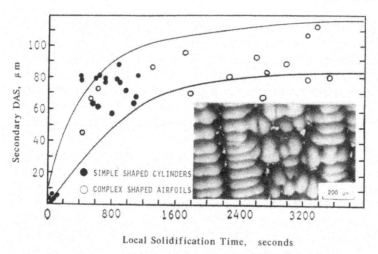

Local Solidification Time,   seconds

Figure 17 Measured secondary dendrite arm spacing of the alloy René N5 as a function of calculated local solidification time. (From Ref. 15.)

castings, the dendrite structure is generally not clearly defined like those DS and SX castings. Thus, the procedure for the measurement of DAS is more complicated [13].

From the micro modeling point of view, the microstructure resulting from dendrite growth is governed by solute diffusion and curvature-kinetics contribution. One has to solve the Ficks' laws of diffusion along with a solute flux balance equation and assume equilibrium solute partitioning at the solid/liquid interface. The stabilizing curvature contribution can be accounted for by using any of the two methods as discussed below. Saito et al. [19] tracked the position of the solid/liquid interface with boundary elements using a Greens function approach. In the recently popular phase field approach [20], a fixed FDM (finite difference method) grid is used. The diffuse solid/liquid interface spreads over several mesh points. An additional set of equations is solved to control the evolution of the curvature at the solid/liquid interface. Both of these two approaches may be used to describe the formation of dendrite pattern in a single grain. However, the computational time required for solving these equations over the entire casting is so large that it is practically impossible to use these methods on a three-dimensional commercial casting.

Dendrites grow after nucleation along ⟨100⟩ directions in cubic metals. The growth velocity is a function of the undercooling at the tip. This can be calculated by assuming marginal stability criterion and a stationary solute profile around a parabolic tip. The nucleation is heterogeneous in most cases. At small undercooling, it has been observed that grains nucleate rapidly once a critical temperature is reached (similar to instantaneous nucleation). This relationship follows a Gaussian distribution and can be obtained from carefully controlled experiments. Within the envelope of a grain, solute mixing is complete in the liquid region and microsegregation models can be used to describe the internal fraction of solid. For eutectics, the approach is the same except that the grains are almost spherical in a uniform temperature field and are fully solid.

3. Grain Structures in EQ Castings

Because the macro grain structure of EQ castings is equiaxed grains, the presence of any columnar grains will be considered as rejectable defects. The macro heat transfer condition of the casting needs to be controlled so that equiaxed-columnar transition does not occur. It is well known that casting grain morphology is controlled by the G/R ratio, where G is temperature gradient and R is solidification rate (Fig. 18). When the G/R value is higher than a critical value, columnar grains form. G, R, and G/R values can be easily calculated and plotted using a heat transfer modeling approach. Another important feature of grain structure is the grain size distribution. Grain size is related to LST and average cooling rate [12]. Very short LST and high average cooling rate will result in very fine grains, e.g., chill grain defect (Figs. 11 and 19). The formation of chill grains can be modeled by the combination of mold filling and heat transfer analysis.

By using the CA-FE/FD (Cellular Automata–finite element/finite difference) simulation, three-dimensional grain structures can be produced and viewed at any two-dimensional metallographic section. Grain structures such as columnar grains, grain size distribution, and morphology of eutectic grains can be determined. In CA-FE/FD models, FEM/FDM models are used to calculate macro heat flow whereas CA technique is used to calculate the evolution of the grains. For nickel-base superalloy EQ (both structural and airfoil) castings, cobalt aluminate is commonly used as the face coat material for the ceramic mold. During casting, cobalt aluminate dissolves into the molten metal and serves as inoculant to form nuclei for crystallization and formation of uniform EQ grain structure. Thus, the nucleation law to be used for the CA calculation needs to take into account the heterogeneous nucleation effect from cobalt aluminate. The CA-FE technique is especially useful in investment castings as current models, which usually do not include convection effects, can reliably predict the grain structure because convection effects are not highly pronounced due to relatively moderate inlet velocities used in investment castings.

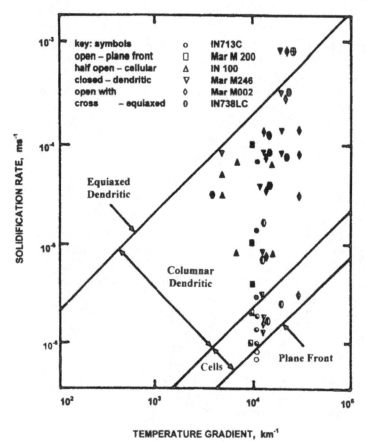

Figure 18 Solidification morphologies of directionally solidified superalloy castings. (From Ref. 18.)

4. Grain Structures in DS and SX Castings

In DS and SX castings, the macro heat transfer condition needs to be controlled so that only columnar-dendritic solidification will occur and, hence, columnar grains (DS castings) and single crystal structure (SX castings) will form. In heat transfer modeling, this situation can be modeled by G/R plot. Locations in castings with G/R values that are lower than the specified value will form equiaxed grain defects. By using CA-FE simulation, the evolution of grain structure can be depicted. Figure 20a shows the prediction of grain structure in a directionally solidified turbine blade. Figure 20b displays the grain structure in a single crystal turbine blade. Nucleation is allowed to

Figure 19 Enlarged view of airfoil trailing edge depicting chill grain formation. (From Ref. 18.)

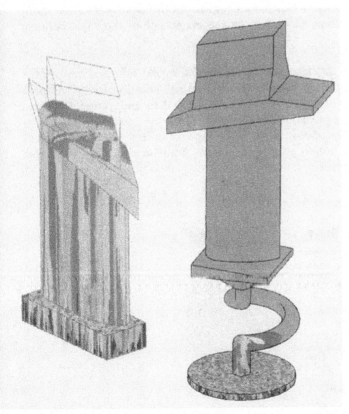

Figure 20 Simulated grain structure for directionally solidified (left) and single crystal (right) turbine airfoils. (From Ref. 21.)

take place at the small plate at the bottom of the casting next to the chill. Due to the presence of the pigtail selector, only one grain eventually survives the competition among many columnar grains nucleated.

Once a columnar-dendritic solidification condition is established, the next step is to control the contour and orientation of the solidification front which can be easily modeled by using isotherm or isochron plots. Grain defects that are functions of the contour and orientation of the solidification front include HAB (high-angle boundaries), LAB (low-angle boundaries), bigrains, multigrains, misoriented grains, and MOD (misoriented dendrite direction).

a. HAB and LAB.   These are grain defects in SX castings. During casting, primary dendrites grow opposite the direction of the heat flow, perpendicular to the mushy zone contour. If the mushy zone contour is curved and not flat, primary dendrites at different locations of the mushy zone will have different growth directions. When these dendrites of different orientation meet, HAB or LAB may form depending on the magnitude of the angle between dendrites (Fig. 21).

b. Bigrains and Multigrains.   Sometimes a second solidification front is also present in the casting. Usually this second solidification front is located near the trailing edge of the airfoil because it is thin there and, hence, it tends to cool faster than the rest of the airfoil. As a result, two grains with different crystallographic orientations form. This type of effect is called bigrains (Fig. 22). If more than two grains are present simultaneously in the casting, it is called multigrains.

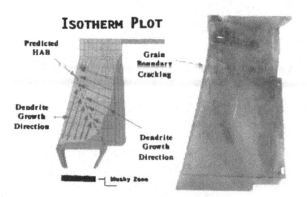

Figure 21 Comparison of model predicted and experimentally inspected high angle boundary in a single crystal casting. (From Ref. 15.)

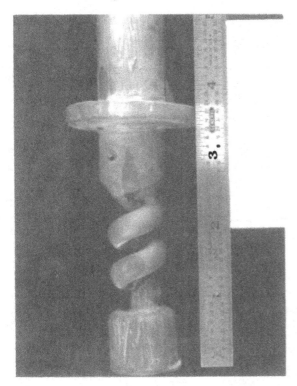

Figure 22 Bi-grain in a single crystal casting. (From Ref. 22.)

    c. Misoriented Grains and MOD. In both DS and SX castings, the casting crystallographical orientation, i.e., ⟨001⟩ direction, is required to be parallel to the casting stacking axis. In DS castings, this requirement is assured by the specification of grain orientation. If the angle between the columnar grain orientation and casting stacking axis (**Fig.** 13) is bigger than a specified value, castings will be rejected. Columnar grains with growth direction that is not parallel to the casting stacking axis are called misoriented grains. One mechanism for misoriented grains to occur is when temperature in a location ahead of the interface drops below the liquidus temperature allowing new grains to nucleate there. Usually these grains have a random orientation but can still grow as columnar grains due to the imposed temperature gradient. The other reason for the formation of misoriented grains is when the solidification front (liquidus isotherm) is not perpendicular to the casting stacking axis so the primary dendrites and hence columnar grains growth direction form an angle with the casting stacking axis. For SX castings, Laué back reflection x-ray technique is used to determine the deviation of the primary dendrite growth direction from the stacking axis of the casting. When deviations are beyond a specified value, castings are rejected. These deviated dendrites are called MOD. Since the primary dendrite and columnar grain growth direction is directly opposite to the heat flow direction and perpendicular to the mushy zone contour, the liquidus isotherm and isochron (time to reach the liquidus temperature) plots can be used to predict the misoriented grains in DS castings and MOD in SX castings.

Even if the macro heat transfer conditions are well controlled so that columnar-dendritic solidification and flat solidification front contour are obtained, many spurious grains may still form during DS and SX casting. This is because it is difficult to maintain G and R within specified ranges when the castings are complex and relatively large. Also, as crystals grow into expanding parts of the mold or spread onto shelves, significant variations occur in the values of these parameters. These spurious grains include zebras, freckles, slivers, and recrystallized grains. Figure 23 shows several types of spurious grains form in a René N5 single crystal casting. This slab has two different ramps and platform angles. A low-angle grain boundary occurred on the shallower ramp angle. Spurious grains seemed to nucleate on the platform that required more vertical growth of the grains. Also, this casting exhibited freckle chains with sliver grains nucleating off some of the freckle chains.

During the growth of superalloy SX castings, stray grains can arise via several different mechanisms. The action of these mechanisms can be predicted on the basis of the thermal (and sometimes also the solutal) fields in the vicinity of the solidification front. In general, the ability to define the exact conditions when these mechanisms occur is not as well developed as the ability to calculate the thermal (and sometimes solutal) fields.

Nucleation of stray grains requires undercooling of the melt ahead of the primary dendritic solidification front. This can easily occur in reentrant corners and on ledges or platforms, where the temperature gradient ahead of the interface can be negative. Undercoolings sufficient to produce nucleation can also occur even with positive temperature gradients if the geometry is one, which requires the dendrites to grow rapidly.

A computationally intensive model such as the CAFE model may be used to predict the occurrence of stray grains in reentrant corners. However, simple analytical models have also been developed to predict this effect by using the models in postprocessing mode in a commercial FEM package. This approach may be advantageous to use for simple casting geometries and thermal field.

d. Zebra Grains. Zebra grains arise when dendrites grow rapidly into an undercooled region, as on a platform, and then undergo partial fragmentation as a result of recalescence. The degree of undercooling which occurs on

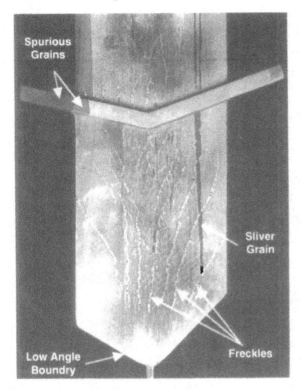

Figure 23   Spurious grains in a René N5 single crystal casting. (Courtesy of Howmet Research Corporation.)

a platform can be predicted on the basis of an analysis of the thermal field and the kinetics of growth along a path, which follows the crystallographic axes. However, there are no quantitative models available that can predict how much undercooling and recalescence are required to produce zebra grains.

e. Freckles.   Freckles are dark etching nearly circular spots that are rich in carbides or carbide-forming elements. A typical freckle chain in a low-pressure turbine blade is displayed in Fig. 24. Current understanding is that freckles or channel segregates originate due to convective instabilities above or in the mushy zone. Instabilities develop in a positive thermal gradient when the segregated interdendritic liquid is less dense than the original liquid. In the unstable mode, low-density liquid develops from the mushy zone in the form of plumes or fingers. These are fed by flow of segregated liquid which may

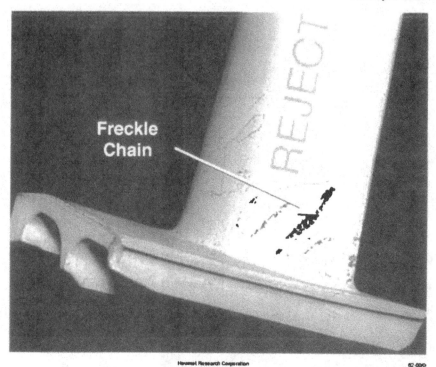

Figure 24 Freckle chain in a single crystal casting. (Courtesy of Howmet Research Corporation.)

cause delayed growth and remelting developing into narrow, open channels in the mushy zone below each finger.

Freckles form when the cooling rate of the casting drops below a critical value [10, 23]. This is because a lower cooling rate results in a coarse DAS which, then, increases the tendency for convective flow instability [24]. The tendency of freckles formation is strongly alloy dependent. For nickel-base superalloys, elements like Ti, Al, and Ta are segregated toward interdendritic liquid. Because the densities of Ti and Al are lower than the densities of bulk liquid of superalloys, an alloy with a higher content of Ti and Al will result in a lower density of interdendritic liquid and higher tendency to form freckles. On the other hand, because the density of the Ta is higher than the superalloy bulk liquid density, the segregation of Ta toward the interdendritic liquid will increase the density of interdendritic liquid and reduce the tendency to form freckles [25, 26]. Another element that tends to increase the freckles formation tendency is W, which usually segregates toward the solid dendrite and decrease the density of the inter-dendritic liquid.

An accurate modeling of freckle formation needs to predict the inter-dendritic fluid flow [27, 28] and the onset of channel formation in the mushy region. Szekely and Jassal [29] pointed out that to have accurate results, the thermosolutal convection in the all-liquid zone, as well as the interaction of the convective transport phenomena between the all-liquid zone and the mushy zone of a solidifying alloy need to be modeled. Recently, a continuum model has been used in simulation of solidification and freckle formation [30–34]. In this continuum model, a set of equations of conservation of mass, energy, and solute concentration is solved in conjunction with the momentum equations. The mushy zone is treated as a porous medium of variable porosity (i.e., volume fraction of liquid). The fraction of liquid varies from zero (all-solid region) to 1 (all-liquid region) in such a way that, when the volume fraction of liquid is zero, no fluid motion is possible and the system reduces to the energy equation. When the volume fraction of the liquid is 1, the equations automatically become the Navier-Stokes and transport equations for an all-liquid region. The equations are solved in the whole domain, with no tracking of internal interfacial (e.g., liquid-mushy) conditions. The formation of liquid plumes in the form of chimney convection emanating from channels (where freckles are formed) within the mushy zone can be predicted (Figs. 25 and 26). These models can be used to evaluate the effects of solidification conditions (e.g., cooling rate, temperature gradient, solidification rate, and mushy zone shape) and alloy composition on the tendency of freckle formation.

f. Slivers. Slivers are grains forming streaks in the microstructure [10]. They are usually aligned close to the primary direction, but misoriented in the transverse direction (Fig. 27). Two possibilities have been proposed for the formation of slivers. One possibility assumes slivers nucleating off some of the stray grains such as freckles (Fig. 23). The other possibility suggests that slivers nucleate from nonmetallic inclusions that contact with the metal.

For the nonmetallic inclusion related slivers, the formation tendency can be modeled by the isochron plot [15]. For a given inclusion content of the input material, the actual inclusion content during the casting depends on the chemical reaction rate between molten metal and ceramics (mold and core). A higher reaction rate results in a decrease of active elements such as Y in the metal and an increase in metal inclusion content. The later effect then increases the tendency for the formation of slivers. Three most important factors controlling the kinetics of the metal-ceramic reaction are casting geometry (surface to volume ratio), reaction temperature, and metal-ceramic

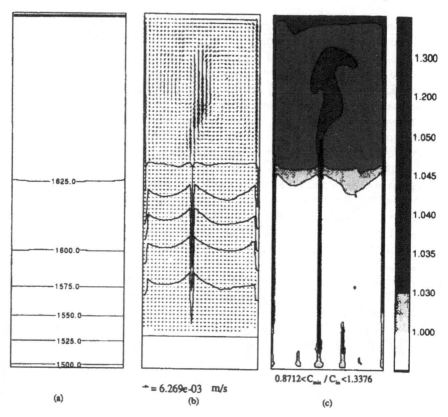

$\leftarrow$ = 6.269e-03 m/s

$0.8712 < C_{min} / C_{in} < 1.3376$

(a)   (b)   (c)

Figure 25 Calculation for the formation of freckles in a single crystal casting (a) isotherms in degrees Kelvin, (b) velocity vectors and solidification contours (increasing in 20% increments from top), and (c) normalized mixture concentration pattern of Ti. (From Ref. 24.)

contact time. For a given casting geometry, the liquidus temperature isochron (time to reach the liquidus temperature) represents the contact time between liquid metal and ceramics. A longer contact time results in a bigger active element (Y) loss and a higher sliver formation tendency.

The liquidus temperature isochron can also be used to predict the thickness of the alpha case in titanium castings. In titanium investment castings, the contact between the molten titanium and the ceramic mold increases the casting surface oxygen content. High oxygen content promotes the formation of alpha phase. As a result, the oxygen-rich layer at the surface of titanium castings is commonly called alpha case. Alpha case is brittle and detrimental to casting mechanical properties; it needs to be removed by the chemical

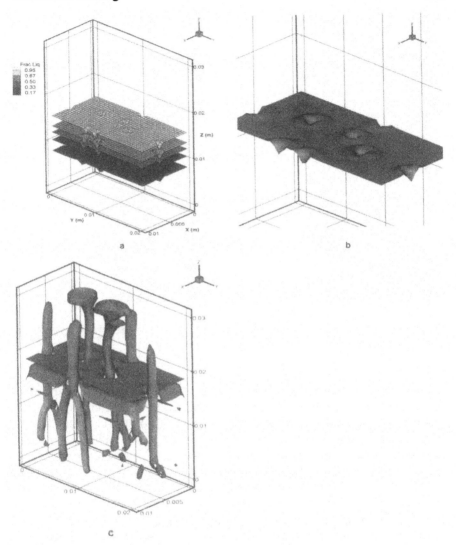

Figure 26 Simulation results of directionally solidified Ni-6Al-6Ta-5W: (a) meshed isosurfaces of various volume fraction of liquid; (b) isosurface fraction of liquid 0.98 showing volcanoes at the channel exists; and (c) trace paths left by the liquid plume channels. (From Ref. 26.)

Figure 27 Transverse (left) and longitudinal (right) views of a sliver in a single-crystal casting. (From Ref. 10.)

milling process. Depending on the actual thickness, the removal of the alpha case sometimes causes problems for foundries in controlling the casting dimension. The thickness of the alpha case depends on the chemical reaction rate between the molten titanium and the ceramic mold, and hence is strongly temperature dependent, increasing rapidly with the increase of reaction temperature. Consequently, titanium investment casting foundries typically keep the mold preheat temperatures low to avoid the formation of an excessive thickness of the alpha case. The liquidus temperature isochron represents the contact time between the molten titanium and the ceramic mold and thus is a good indicator for the final thickness of the alpha case in the resultant casting.

   g. Recrystallized Grains. Recrystallized grains result from stress imparted to the casting either during the casting withdrawal process (Fig. 28) or during post-solidification processing and handling (Fig. 29). These grains nucleate during solution heat treatment and appear on the surface of the casting. The formation tendency of recrystallized grains increases with the increase in casting residual stress and can be simulated by stress modeling.

Figure 28 No fill, hot tearing and recrystallized grains resulted from stress imparted to a single-crystal casting during solidification. (From Ref. 22.)

VI. ROLE OF INVERSE MODELING

The accuracy of the simulation of a complex investment casting will depend immensely on the exactness of the input data used. Two major types of data are needed for a simulation: (1) thermophysical properties (thermal conductivity, specific heat, latent heat, etc.) of casting, mold, and cores and (2) boundary conditions (heat transfer coefficient, heat flux, emissivity, etc.).

Inverse modeling allows for determination of thermophysical quantities or boundary conditions by coupling the numerical conduction solution techniques with experimental temperature measurements at few key locations in casting and mold. The inverse model uses the geometry, thermal history at certain specified locations, initial conditions, known thermophsysical data,

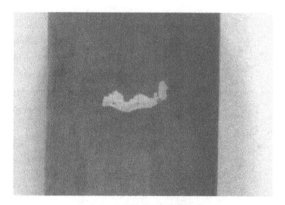

LONGITUDINAL

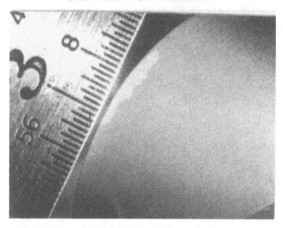

Figure 29 Longitudinal (top) and transverse (bottom) views of a recrystallized grain resulted from stress imparted to the casting during post-solidification processing and handling. (From Ref. 10.)

known boundary conditions to determine an unknown property or a boundary condition. Usually an iterative procedure is used to obtain the optimum value of the unknown parameter, which gives the best agreement between the measured and calculated values at the key locations (where experimental data are input). The details of this procedure can be obtained in Ref. 35.

Although accurate data can be calculated by this technique, this technique is often slow and requires many iterations and may be impractical to use in a complex casting with hundreds of thousands elements. At the present time, viscosity data cannot be practically obtained by this method.

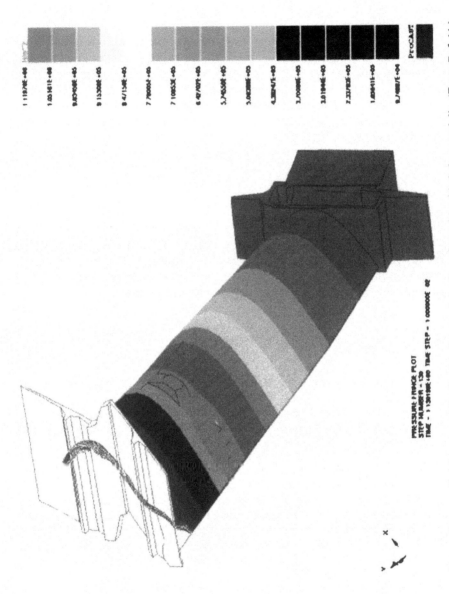

Figure 30 Predicted pressure distribution for wax injection of a turbine blade in a steel die. (From Ref. 11.)

VII. WAX INJECTION

Most investment casting processes use injected wax or plastic pattern. It is important to predict the behavior of the materials during the injection process. This type of capability will reduce the hand finishing operations usually done on patterns. The non-Newtonian viscosity behavior may be approximated by a Carreau-Yasuda formulation [6]. Accuracy of the temperature dependent thermophysical properties used for the wax or plastic material is important for a better prediction. Significant cost savings can be achieved by modeling the wax injection process routinely. A typical pressure prediction for wax injection in a steel die is shown in Fig. 30.

VIII. SUMMARY

Casting process modeling has been used extensively by investment casting foundries to solve routine production problems. The approach for modeling various types of investment castings has been discussed. Future development efforts should emphasize on reducing the simulation cycle time and enhancing the accuracy to predict and eliminate various casting defects.

REFERENCES

1. RA Horton. Metals Handbook, Vol 15, Materials Park, OH: ASM International, pp. 253–269.
2. KO Yu, JJ Nicholas, L Hosamani. Solidification modeling of Alloy 718 structural castings. Superalloys 718, 625, 706 and Various Derivatives (EA Loria, ed.). Warrendale, PA: TMS, 1994, pp. 177–188.
3. JA Oti, KO Yu. Production processing of investment cast complex shaped NiAl single crystal airfoils. Proceedings of the International Symposium on Structural Intermetallics, Seven Springs, 1993, pp. 505–512.
4. DJ Chronister, SW Scott, DR Stickle. Induction melting of titanium, zirconium, and reactive alloys. Proceedings of Vacuum Metallurgy Conference (LW Lherbier, GK Bhat, eds.). Iron and Steel Society, 1986, pp. 7–10.
5. S Shendye. Thermally controlled solidification. Manufacturing Technology for Aerospace Materials: A Technology Demonstration and Information Exchange, Arlington, VA, April 20–21, 1999.
6. ProCAST Reference Manual, UES, Inc., Dayton, OH, 1999.
7. BG Thomas, DD Goettsch, KO Yu, MJ Beffel, M Robinson, D Pinella RG Carlson. Modeling the directional solidification process. Modeling of Casting, Welding, and Advanced Solidification Processes—V, Davos, Switzerland (M

Rappaz, MR Ozgu, KW Mahin, eds.). Warrendale, PA: The Metallurgical Society, vol. 5, 1990, pp. 603–610.

8. KO Yu, JA Oti, WS Walston. Solidification modeling of NiAl single crystal castings. High-Temperature Ordered Intermetallic Alloys V, vol. 288, Proceedings of Materials Research Society 1992 Fall Meeting, Boston, 1992, pp. 915–920.

9. M Fackeldey, M Diemer, M Meyer ter Vehn, PR Sahm. Recent advances in the application of a combined heat, stress and microstructure simulation on the casting process of a single crystal turbine blade. Solidification Processing, Proceedings of the 4th Decennial International Conference on Solidification Processing (J Beech, H Jones, eds.). July 7–10, 1997, University of Sheffield, UK, pp. 41–44.

10. KO Yu, MJ Beffel, M Robinson, DD Goettsch, BG Thomas, RG Carlson. Solidification modeling of single crystal investment casting. AFS Transactions, 1990, pp. 417–428.

11. BA Mueller, RK Foran, A Hines, D Hirvo, T Simon, JS Tu. Investment casting applications of process modeling. Solidification Processing, Proceedings of the 4th Decennial International Conference on Solidification Processing (J Beech, H Jones, eds.). July 7–10, 1997, University of Sheffield, UK, pp. 170–174.

12. KO Yu, JJ Nichols, M Robinson. Finite-element thermal modeling of casting microstructures and defects. JOM 6:21–25, 1992.

13. SAE ARP 1947. Determination and acceptance of dendrite arm spacing of structural aircraft quality D357 aluminum alloy castings. Society of Automotive Engineers, Warrendale, PA, 1996.

14. KO Yu, M Robinson. Monocrystal turbine blade scale-up. AFML F33615-80-C-5008, Final Report, April 1987.

15. KO Yu, JA Oti, M Robinson, RG Carlson. Solidification modeling of complex-shaped single crystal turbine airfoils. Superalloys 1992 (SD Antolovich et al, eds.). Warrendale, PA: TMS, 1992, pp. 135–144.

16. JS Tu, RK Foran. The application of defect maps in the process modeling of single-crystal investment casting. JOM 6:26–30, 1992.

17. HD Bordy, AF Giamei. Effect of hafnium additions on the solidification behavior of directionally solidified superalloys. AFWAL-TR-81-4123, Contract No. F33615-75-C-5204, Final Report, October 1981.

18. M McLean. Directionally solidified materials for high temperature service. The Metals Society, London, 1983.

19. Y Saito, G Goldbeck Wood, H Muller-Krumbhaar. Phys Rev 38A:2148, 1988.

20. JD Warren, WJ Boettinger. Acta Metall Mater 43:689, 1995.

21. CA Gandin, M Rappaz, JL Desbiolles, E Lopez, M Swierkosz, P Thévoz. 3D modeling of dendritic grain structure in a turbine blade investment cast part. Solidification Processing 1997 (J Beech, H Jones, eds.). University of Sheffield, UK, pp. 289–294.

22. KO Yu. Correlation of microstructural morphology and defects occurrence with solidification conditions in investment cast René N5 single-crystal cylinders. Appendix of Phase IX—Advanced Turbine Airfoil Casting Technology, Task 2: Casting Simulation, AFWAL, No. F33615-85-C-5014, Fifth Interim Technical Report, February 1988–July 1988.

23. SM Copley et al. The origin of freckles in unidirectionally solidified castings. Metall Trans 1:2193–2204, 1970.
24. MC Schneider, JP Gu, C Beckermann, WJ Boettinger, UR Kattner. Modeling of micro- and macrosegregation and freckle formation in single-crystal nickel-base superalloy directional solidification. Metall & Mater Trans 28A:1517–1531, 1997.
25. TM Pollock, WH Murphy. The breakdown of single crystal solidification in high refractory nickel-base alloys. Metall & Mater Trans 27A:1081–1094, 1996.
26. SD Felicelli, DR Poirier, JC Heinrich. Modeling freckle formation in three dimensions during solidification of multicomponent alloys. Metall and Matls Trans 29B:847–855, 1998.
27. R Mehrabian, M Keane, MC Flemings. Interdendritic fluid flow and macrosegregation: influence of gravity. Metall Trans 1:1209–1220, 1970.
28. S Kou, DR Poirier, MC Flemings. Macrosegregation in electroslag remelted ingots. Elec Furnace Proc 35:221–228, 1977.
29. J Szekely, AS Jassal. Experimental and analytical study of the solidification of a binary dendritic system. Metall Trans 9B:389–398, 1978.
30. DR Poirier, JC Heinrich, SD Felicelli. Simulation of transport phenomena in directionally solidified castings. Proceedings of the Julian Szekely Memorial Symposium on Materials Processing (HY Sohn, JW Evans, D Apelian, eds.). Warrendale, PA: TMS, 1997, pp. 393–410.
31. SD Felicelli, JC Heinrich, DR Poirier. Numerical model for dendritic solidification of binary alloys. Numer Heat Transfer, Part B, 23:461–481, 1993.
32. SD Felicelli. PhD dissertation, The University of Arizona, Tucson, AZ, 1991.
33. SD Felicelli, DR Poirier, JC Heinrich. Macrosegregation patterns in multicomponent Ni-base alloys. J Crystal Growth 177:145–161, 1997.
34. SD Felicelli, JC Heinrich, DR Poirier. Three-dimensional simulations of freckles in binary alloys. J Crystal Growth 191:879–888, 1998.
35. M Rappaz, JL Desbiolles, JM Drezet, CA Gandin, A Jacot, P Thévoz. Application of inverse methods to the estimation of boundary conditions and properties. Modeling of Casting, Welding, and Advanced Solidification Processes VII (M Cross, J Campbell, eds.). Warrendale, PA: TMS, 1995, pp. 449–457.

12

Permanent Mold Casting

Chung-Whee Kim
EKK Inc., Walled Lake, Michigan

I. INTRODUCTION

Usually, the permanent mold casting process means that the metal mold (die) casting process occurs with gravity pouring. In this chapter, however, permanent mold casting processes are interpreted as all metal mold (die) casting processes regardless of the way the molten metal is being introduced. The reason for this interpretation is that all metal mold casting processes have similar solidification characteristics. Thus, the traditional permanent/semi-permanent mold casting, tilt-pour, and low-pressure casting processes are all included in this discussion. In terms of solidification simulation, high-pressure die casting and squeeze casting are similar to permanent mold casting. However, due to the significant difference in the mechanism to introduce molten metal into the die cavity, the mold filling analysis for high-pressure die casting and squeeze casting is quite different from that of the permanent mold casting processes and is discussed separately in Chapter 13.

In this chapter, flow analysis, which is unique to each metal mold casting process, will be discussed first, followed by a discussion of solidification analysis.

II. FLOW ANALYSIS

Although the solidification characteristics of metal mold casting processes are similar to each other, from a simulation engineer's point of view, the manner in which molten metal is introduced into the mold/die is unique for each process.

Thus, the flow (filling) analysis for each metal mold casting process needs to be discussed separately.

A. Permanent/Semipermanent Mold Casting

In the traditional permanent mold casting process, the molten metal is introduced into the mold cavity by gravity similar to sand mold casting. Thus, the modeling approach and practical application of flow analysis for the permanent mold casting process is similar to those of the sand mold casting process. Mold filling analysis for sand mold casting is discussed in Chapter 9. In this chapter, a few points unique to the metal mold casting process will be discussed.

A ceramic filter or screen is often used in permanent mold casting to remove inclusions in the molten metal. This filtration process alters flow pattern and pressure distribution during mold filling. Thus the effects of filters and screens should be included in mold filling analysis.

Back pressure often can be ignored in flow analysis for sand mold casting due to the relatively high sand mold permeability. Back pressure is important, however, in permanent mold casting, unless the cavity is well ventilated. Mold filling analysis can be used to determine how to establish a proper mold ventilation system. The inclusion of the effect of back pressure in mold filling analysis complicates the calculation, but results in improved accuracy. Back pressure stabilizes the fluid front, and suppresses the numerical instabilities at the liquid metal front.

The liquid metal flow pattern in permanent mold casting is difficult to control and sometimes can cause difficulties in numerical convergence during mold filling analysis. A gravity filling system is difficult to analyze numerically because neither inertia nor viscous force is dominant. These factors dictate the solution of a full Navier-Stokes equation with dynamic free surfaces.

B. Tilt-Pour Casting

The tilt-pour casting process also uses gravitational force to introduce molten metal into the die cavity. However, it has more control on flow rate than the permanent mold casting process. The mold for the tilt-pour casting is rotated by mechanical force to achieve a quiet, quick filling. The rotational timing and direction, as well as the gate size and number, are important for maintaining a smooth filling.

The mold filling analysis for the tilt-pour casting process is, from a numerical analysis point of view, easier than any other casting processes due to the following: (1) The mold filling flow of the tilt-pour casting process is usually quieter than that of gravity permanent mold casting, (2) there are no time-dependent boundary conditions, and (3) all of the molten metal is in the analysis system at the beginning of the process. Thus, no inlet boundary conditions are required. In addition, inertia due to rotation (changing gravity direction) is negligible. This indicates that only the direction of gravity needs to be changed during numerical analysis in accordance with the rotation of the mold.

Figure 1 shows a shaded finite element model of an a minivan aluminum cross-member casting and rigging system. The size of this 10 kg casting is approximately 1 m length by 50 cm width. One unique feature of this relatively large aluminum automobile component is that the parting line is not flat, but rather elevated about 25 cm along the casting.

The model shown, which has 560,000 nodes and elements, includes casting, gating, ladle, and four mold components. Mold filling conditions for two-, three-, and five-gate models were simulated. First, the five-gate

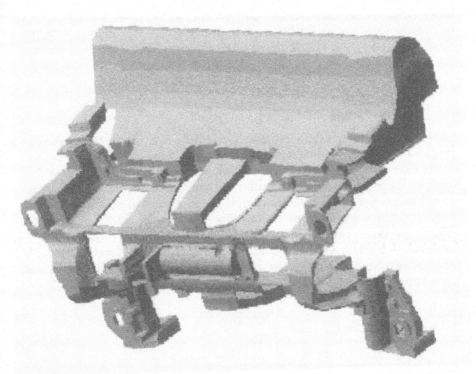

Figure 1 Shaded finite element model of a minivan aluminum cross-member casting and rigging.

Table 1 Tilt Pour Timing

Angle (degrees)	Time (s)
0–10	5
10–45	12
45–90	3

model was constructed and then two outermost gates were disconnected to create the three-gate model. Finally the center gate was disconnected for the two-gate model. Table 1 shows the speed and timing of rotation for the 20 s tilt cycle. The fluid material properties used in the computation are listed in Table 2.

Simulation results for the two-gate model show a rather quiet filling pattern (Fig. 2). The figure also shows that two strong streams meet at the bottom of the casting (under the center gate area). The analysis with the five gates is shown in Fig. 3, indicating that the outermost gates were filled first and the center gate was filled next. This flow pattern results in a situation where the two center gates are back-filled from the outer gates. The analysis with the three-gate model (Fig. 4) indicates that this rigging system is the most promising among the three systems investigated for this casting.

C. Low-Pressure Casting

In the low-pressure casting process, a relatively low pressure (~5 PSI) is applied to the sealed furnace causing the molten metal to be slowly pushed up into the mold cavity. The amount of molten metal and the speed at which it enters the mold cavity are controlled by the time dependent furnace pressure. For example, Fig. 5 shows a pressure vs. time relationship for aluminum automobile engine block casting. The low-pressure casting process can achieve a greater control on the mold filling sequence than other casting processes. As a result, quiet filling is usually achieved making this process desirable to produce high-integrity castings such as automobile suspension components.

Table 2 Fluid Material Properties

	Aluminum	Units
Density	2.48	g/cm^3
Viscosity	0.03	poise

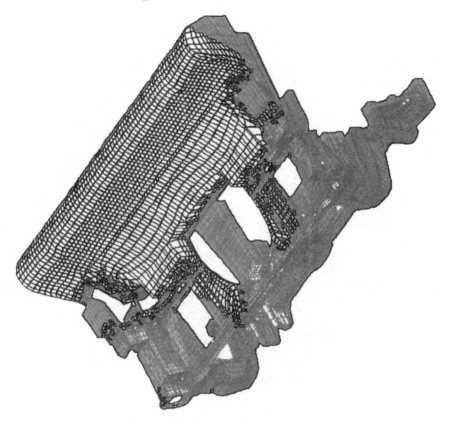

Figure 2 Flow pattern for two-gate model (11 s after pouring).

In the mold filling simulation, the boundary conditions dictate the use of the time dependent furnace pressure, instead of a fixed inflow velocity. This is due to the furnace pressure controlling the flow rate in the low-pressure casting process. These boundary conditions can lead to a large computation time. Pressure (propagated at the speed of sound) boundary conditions are usually more difficult in numerical convergence than velocity boundary conditions. An assumed constant inflow velocity has been assigned in some analyses due to this difficulty. This is not a valid assumption because the in-gate velocity is not considered as input data in the low-pressure casting process but rather as a part of the analysis results (i.e., output). In other words, we do not know the time dependent in-gate velocity a priori .

Figures 6 and 7 show velocity vector plots of a low pressure aluminum cast wheel at 2.3 and 3.7 s after raising the pressure, respectively. A time

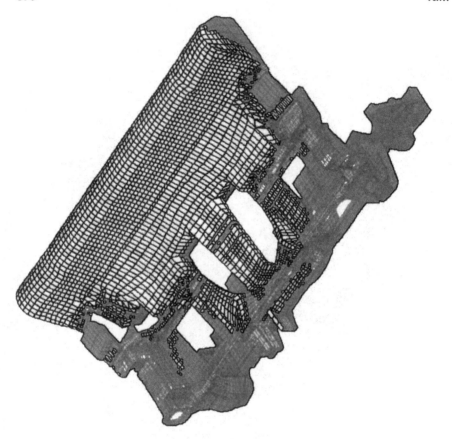

Figure 3 Flow pattern for five-gate model (11 s after pouring).

dependent pressure similar to that shown in Fig. 5 was assigned. Although a fountain was formed over the inlet, an overall smooth filling was achieved. It is apparent that the liquid metal velocity at the entrance of the sprue changed with time and was nonuniformly distributed across the inlet.

III. SOLIDIFICATION ANALYSIS

As mentioned before, the solidification characteristics of various metal mold casting processes are similar to each other but significantly different from those of nonmetallic mold casting processes. These differences come not only from the significant differences in thermal properties of mold materials, but also from the processes themselves.

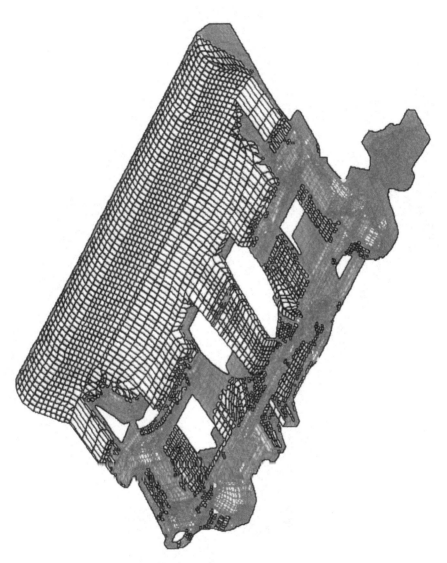

Figure 4 Flow pattern for three-gate model (11 s after pouring).

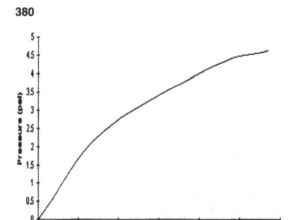

Figure 5 Pressure vs. time for low-pressure casting an aluminum automobile engine block.

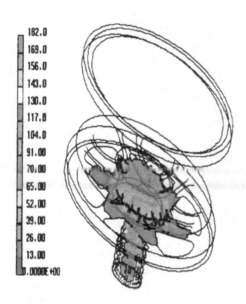

Figure 6 Velocity vector for a low-pressure aluminum cast wheel at 2.3 s after raising the pressure.

ELAPSED TIME = 3.70019

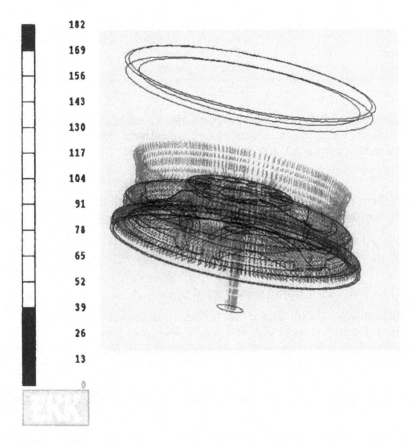

182
169
156
143
130
117
104
91
78
65
52
39
26
13
0

Figure 7 Velocity vector for a low-pressure aluminum cast wheel at 3.7 s after raising the pressure.

In the following sections, we will discuss (1) the effects of the mold thermal properties, (2) the importance of cyclic analysis to reach a periodic quasi-steady condition, and (3) the effects of the mold geometry.

A. Thermophysical Properties of Molds

For reference purposes, the thermal properties of H13 die steel (used for the following examples in this chapter) are listed in Table 3, along with those of a sand mold. The thermal layer or heat penetration depth $\delta(t)$ is defined as the

Table 3 Thermal Properties of Mold Materials

	H13 die steel	Sand mold	Units
Density	8.1	1.8	g/cm^3
Specific heat	0.115	0.217	$cal\,g^{-1}\,°C^{-1}$
Conductivity	0.062	0.0016	$cal\,cm^{-1}\,s^{-1}\,°C^{-1}$

distance beyond which, for practical purposes, there is no heat flow [1]. This is schematically shown in Fig. 8. Assuming one-dimensional temperature distribution is a cubic polynomial, $\delta(t)$ can be represented as [1]

$$\delta(t) = \sqrt{8\alpha t} \tag{1}$$

where α is thermal diffusivity and t is time. Using Eq. (1) with thermal property values that are listed in Table 3, one can obtain

$$\delta(t) = \sqrt{0.532t} \qquad \text{cm for a steel mold} \tag{2}$$

and

$$\delta(t) = \sqrt{0.0328t} \qquad \text{cm for a sand mold} \tag{3}$$

Equations (2) and (3) indicate that the heat penetration depth of a metal mold is approximately 4 times that of a sand mold. After 5 min, the heat will penetrate only 3.14 cm into the sand mold, compared to 12.56 cm for the metal mold. This is one of the reasons that during sand mold casting process model-

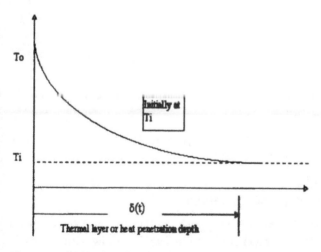

Figure 8 Definition of thermal layer. (From Ref. 1.)

ing, the mold thickness is usually assumed to be infinite. Under this assumption, the heat transfer from the mold surface to atmosphere can be neglected. In the metal mold casting process, however, this assumption is not true and the effects of both mold geometry and surface heat transfer need to be included in the model.

B. Cyclic Analysis

In metal mold casting, the molds are used repeatedly. As a result, the molds develop a unique temperature distribution pattern during the initial cycle of the casting process. Based on the thermal balance, they gradually reach a set of periodic quasi-steady state conditions which have a great influence on final casting quality and productivity. To obtain an accurate solidification analysis and understand the thermal effect of the casting process, we need to reproduce these quasi-steady conditions accurately. In the solidification simulation of a permanent mold casting, a cyclic analysis with casting parameters, such as the liquid metal pouring temperature, dwell time, open time, and spray amount, needs to be completed prior to the final solidification analysis [2].

In metal mold casting, cooling channels are often introduced in the mold to avoid overheat conditions and to improve casting quality and productivity. However, in some instances heat dams (artificial gaps) are introduced to produce locally high temperatures thereby promoting directional solidification. Furthermore, hot oil lines are sometimes introduced to stabilize local temperature. Heat loss from the die surface to the atmosphere is usually small but cannot be neglected if an accurate heat balance in a large time span (quasi-steady state condition) is to be obtained.

The coating on the mold cavity surface, which is selectively applied, greatly affects the final solidification conditions. This coating creates a thermal resistance between the casting and the mold. The mold parting line and the interface between various die components also creates thermal resistance. In addition, the shape and size of mold components, as well as the type of mold materials, influence the quasi-steady state conditions. All these physical phenomena must be accurately represented in the numerical model to result in an accurate thermal analysis and, hence, the final solidification simulation.

Figure 9 shows a die section temperature distribution at the moment liquid metal is being introduced into a die cavity. Casting, gating system, and die components (including holding dies), as well as water lines, are all included in the model [2]. Initially, a uniform temperature of 150°C was assumed for all die components except the holding die which had a lower temperature (50°C). Various interfacial heat transfer coefficients (inverse of thermal resistance) used in this example are listed in Table 4. Interfacial heat

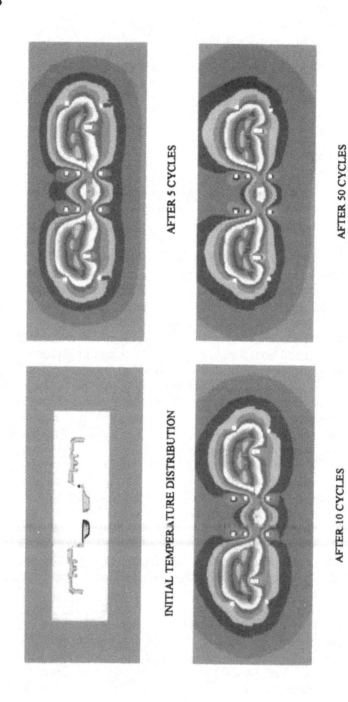

AFTER 5 CYCLES

AFTER 50 CYCLES

INITIAL TEMPERATURE DISTRIBUTION

AFTER 10 CYCLES

Figure 9 Die section temperature distribution before the liquid metal is being injected into the die. (From Ref. 2.)

Table 4 Interfacial Heat Transfer Coefficients

	Value $(\text{cal cm}^{-2}\,\text{s}^{-1}\,\text{C}^{-1})$
Between casting and die	0.4
Spray die	0.02
Between water line and die	0.143
Between oil and die	0.05
Between die and die (bolted)	0.119
Between die and die (machine clamp)	0.0239
Air and die	0.0012

transfer is modeled by using zero thickness interface elements which will be discussed in the next section.

After 5 casting cycles, a unique temperature distribution developed from the initial uniform temperature distribution toward a quasi-steady state condition. The temperature distribution after 10 casting cycles shows that a 5-cycle analysis is not long enough to reach a steady state condition. The 10-cycle analysis results are comparable with the results of the 50-cycle analysis. Thus we may conclude that this casting process would reach a quasi-steady condition after 10 casting cycles. A temperature history near the die surface (Fig. 10) illustrates how the die reaches its quasi-steady thermal condition. Based on high pressure die casting experience, it is proposed that approximately 4 cycles per 10 cm of cavity steel thickness is required for the die to reach its periodic steady state condition.

C. Interface and Coating

The coating on the internal mold surface, which is selectively applied, greatly affects the final solidification condition in metal mold casting contrary to sand mold casting. The poor thermal conductivity of the sand mold is the dominant factor in slowing down the heat transfer rate and making the mold internal surface coating of secondary importance in the sand mold casting process. However, the coating on the metal mold can create a significant thermal resistance between the casting and the metal mold thereby altering the solidification pattern and time. Consequently, the effects of the coating must be included in the numerical analysis. In addition, the mold parting line and the interface between various die components create thermal resistance. This resistance reduces the heat transfer rate and limits the effectiveness of the cooling lines and, hence, must be included in the numerical analysis.

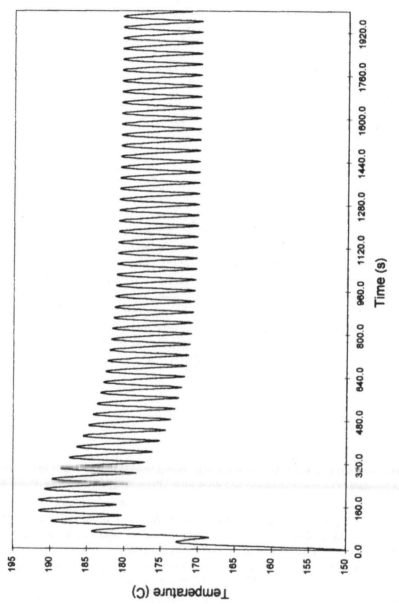

Figure 10 Temperature history at a point near the die surface. (From Ref. 2.)

Although the effect of this thermal resistance can be modeled by using an effective thermal conductivity for the first layer mold elements, the zero thickness element approach is better suited for this purpose. In particular, to calculate an accurate die surface temperature, which is different from the casting surface temperature at the same location (Fig. 11), the zero thickness element is an essential tool.

The temperature distribution of the die, particularly at the die surface, has an important influence on the formation of casting defects [3–5]. As a result, die surface temperature distribution is commonly used as one of the die design criteria. Successful use of an appropriate die surface temperature distribution to avoid mis-run problems on the transmission case of an electric car has been reported [5].

D. Mold Dimensions

Metal mold geometry influences the final solidification results through the heat balance. Figure 12 shows a one-quarter section of a low-pressure aluminum wheel casting. A cursory examination indicates that a one-tenth model of the original casting could be used, based on casting geometrical symmetry. Mold geometry, however, has a quarter symmetry. Thus, a quarter of the casting and mold was modeled in this example. Dark spots in Fig. 12 indicate possible

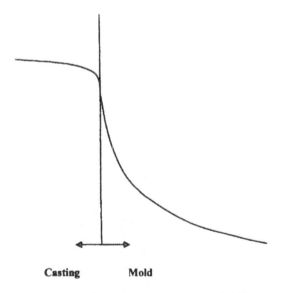

Casting   Mold

Figure 11   Schematics of temperature distribution at metal-mold interface.

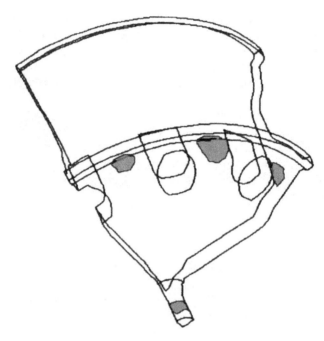

Figure 12   Solidification result of a one-quarter section of an aluminum wheel casting made by low-pressure casting process; dark spots indicate possible shrinkage locations.

shrinkage porosity locations. It can be seen that the predicted shrinkage amount is not the same in each spoke. Figure 13 illustrates that the mold surface temperature distribution at ejection does not have radial symmetry because of the mold shape (Fig. 14). Since the thick section has a larger heat capacity than the thin section, the thin section has a higher temperature than the thick section. The nonuniform mold temperature distribution has an effect on the final solidification pattern of the casting. This example shows that the mold (die) geometry is an important factor in the solidification analysis of the metal mold casting process.

IV. CONCLUSION

Thermal and filling simulations of the permanent mold casting process can be accurately performed using finite element analysis.

To conduct a solidification analysis of the permanent mold casting process, accurate geometries and thermophysical properties of the casting and the

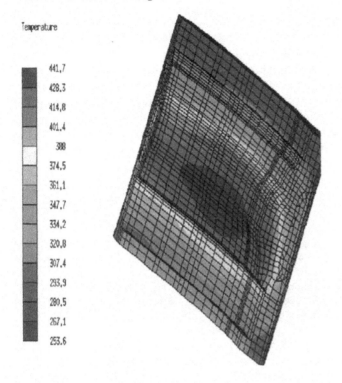

Figure 13 Temperature distribution on the side mold of the casting shown in Fig. 12.

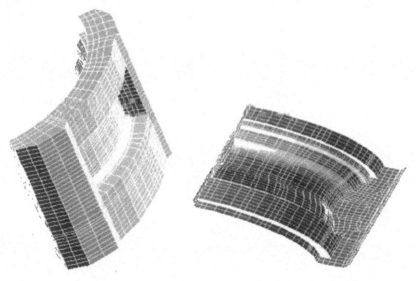

Figure 14 Side mold geometry for the aluminum wheel shown in Fig. 12.

mold must be included in the model. Also, many casting cycles must be evaluated during the simulation so that the temperature distribution in the mold is allowed to reach a quasi-steady state condition.

The mold filling simulation of permanent/semipermanent mold casting processes should include the effects of the filter, if one is used to clean the molten metal, the air back pressure in the cavity, and any vents. For the tilt pour process, the rotational timing and direction should be included in the model. In the low-pressure casting process, proper boundary conditions should be used to account for the effects of the time dependent furnace pressure.

REFERENCES

1. N Ozisik. Heat Conduction. New York: Wiley-Interscience, 1980, pp. 335–339.
2. C-W Kim. A cyclic analysis of permanent mold casting. In: Modeling of Casting and Advanced Solidification Processes VI, Warrendale, PA: TMS, 1993, pp. 749–756.
3. C Mitcham. Process modeling of an aluminum alternator cover for Delphi E & E. Rosemont, IL: North American Die Casting Association, November 21, 1996.
4. C Mitcham. Process modeling of an Aluminum generator frame: EKK. Die Casting Engineer 41/3: May/June 1997.
5. S Mahaney, C-W Kim. Modeling of die cast process: a finite element method approach. Die Casting Engineer 40: November/December 1996.

13
Die Casting

Horacio Ahuett-Garza,* R. Allen Miller, and Carroll E. Mobley
The Ohio State University, Columbus, Ohio

I. INTRODUCTION

In the die casting process, nonferrous parts of complex geometry are mass produced to near net shape. Aluminum, magnesium, zinc, and copper alloys are commonly used in this process. Products range from small valve fittings and housings to large transmission casings. Die castings generally provide structural support of some form and in many cases must meet containment or pressure tight requirements.

Thin walls are a common feature of die castings. To a great extent, the nature of the process is determined by this characteristic. Relatively short solidification times, of the order of 1 min or less, are usually observed. The thinnest regions of the casting solidify faster in as little as a few hundredths of a second.

To prevent the solidification mechanisms from hindering metal flow into the cavity of a die, short fill times are required. Cavity fill times are typically a few hundredths of a second. Under these conditions, flow regimes at the gate are turbulent. Jet and atomized flows are not uncommon and even recommended for certain applications [1].

Large cavity pressures, of the order of 70 MPa (10,000 psi), are typical of this process. Such pressures are needed not only to guarantee that the flowing metal reaches even the smallest features of a cavity, but also to force metal into voids generated during the solidification process (risering techniques common to other casting processes cannot be applied in die casting

**Current affiliation*: ITESM Campus Monterrey, Monterrey, Mexico

because of the geometric characteristics of the product). Because of the functions that die castings perform, porosity caused by either solidification shrinkage or gas entrapped during fill is a determining factor in the quality of a casting.

Die casting dies lie at the heart of the casting operation. They are designed and manufactured to operate in a harsh environment where extreme thermo-mechanical loads are present: large-cavity pressures and high temperatures, as well as large clamping forces that keep dies together during the solidification process (dies must open to allow for part ejection; therefore they must be kept clamped during the casting cycle). For this reason, high-strength materials are used in dies. Due to the high costs and lead times associated with their fabrication, dies are reusable through long life cycles. Throughout their lifetime, a single die may produce as many as 500,000 parts or more.

II. COMPUTER SIMULATION APPLIED TO DIE CASTING

Computer simulation involves the application of numerical models to predict the evolution and performance of the different physical processes that take place during a die casting cycle:

- Fluid flow during cavity fill
- Heat transfer within the molten metal and die
- Solidification, that is, the release of latent heat as well as the volumetric and phase changes that accompany this process
- Thermoelastic response of the die and casting: stresses, strains, and displacements

The purpose of a simulation is to provide an insight into the evolution of these physical mechanisms. Based on a careful analysis of simulation results, decisions can be made as to how the different process conditions and die design parameters can be manipulated to improve the quality of a particular casting operation. Caulk has correlated the types of computer simulations needed for each physical process with the die casting problems that each one addresses [2]. This correlation is presented in Table 1.

From the industry's perspective, the goals of a simulation and subsequent analysis are to reduce lead times and scrap rates, and increase die life and dimensional accuracy of the product. Clearly, achieving these goals is not a straight forward task.

Table 1 Correlation of Die Casting Problems with Types of Computer Simulation Needed for Prediction

Analysis Type	Die Casting Problem															
	Process								Casting							
	Bolt Breaking	Die Cracking	Erosion	Flash	Heat Checking	Soldering	Sticking	Tearing	Dimensions	Drags	Gas Porosity	Galls	Poor Fill	Seams	Shrink Porosity	Thin Wall
Casting Temperature							▨	▨		▨		▨	▨	▨		
Die Temperature		▨			▨	▨	▨			▨						
Solidification											■				■	
Cavity Fill				▨							■			▨	▨	
Casting Distortion									▨							
Die Distortion	▨	▨		▨	▨				▨							▨

■ Traditional   ▨ Unique to Die Casting

III. COMMERCIAL SIMULATION SOFTWARE

In the North American market there are numerous computer systems that can be used to analyze the physical processes that take place in die casting. A list of the major attributes of these simulation programs is presented in Table 2. ABAQUS* and ANSYS[†] have been included in this table as examples of commercial codes that can be used to model the structural response of the die, a capability that falls beyond the scope of the rest of the systems in the table (except for the case of GM's DieCas).

Recently, the North American Die Casting Associations organized a forum in which representatives from the different casting simulation packages introduced their software via the analysis of a casting. The comments that follow are based primarily on the material presented at this event [3].

Two computer codes, MAGMAsoft[‡] and ProCast[§], were developed for the analysis of casting processes in general. They are the major players in the field of solidification analysis in the North American market. Much of their advertised success comes from case histories in gravity and low-pressure casting. Nevertheless, their use in the die casting industry continues to increase.

---

* Hibbit, Karlsson & Sorensen, Inc. Pawtucket, RI.
[†] Ansys, Inc. Houston, PA.
[‡] Magma Foundry Technologies, Inc. Arlington Heights, IL.
§ UES, Inc. Dayton, OH.

Table 2 Major Attributes of Commercial Simulation Codes

	Magmasoft	ProCast	EKK	Flow 3D	DieCast	2D-BEM	MetlFlow
Solver							
FEM		x	x				
FDM	x						
BEM				x			
Closed form					x	x	x
Platform							
PC			x			x	x
Workstation	x	x	x	x	x		
Filling simulation							
Turbulent flow	x	x	x	x			
Jet flow				x			
Coupled with heat	x	x		x			
Backpressure		x		x			
Entrapped gas				x			
Solidification							
Shrink porosity	x		x		x		
Thermoelastic analysis					x		

EKK*, the newest code in the market, was designed with die casting in mind, but it can be used to analyze other casting processes as well.

While MAGMAsoft uses a finite difference formulation to solve the heat transfer and fluid flow problems, ProCast and EKK use a finite element approach to generate the same results.

Flow 3D† is a system designed primarily for the purpose of analyzing complex fluid flow processes: from the flow of Newtonian fluids under isothermal conditions in laminar regimes to slurries and mixtures of gas and liquids under nonisothermal conditions in turbulent and jet flow regimes. The original niche for this code was the analysis of flow in heat exchangers of nuclear reactors. It has not been tailored for the analysis of casting; however, its ability to model complex flow regimes may find an application in the analysis of filling processes in die casting.

General Motors uses DieCas, a proprietary computer code developed in house for the prediction of thermal fields and thermally induced deflections in dies [4, 5]. This system is rather unique in both its approach (its analysis predicts thermal fields under quasi-steady state conditions) and the numerical method that it uses (boundary element formulation). In its latest versions, models to predict the effects of heat released during fill are being incorporated in the code [6].

Research work by Australia's Commonwealth Scientific Research Organization (CSIRO) has resulted in the development and commercialization of two programs that facilitate the design of cooling lines, runner systems and gates, and process parameters for injection.

One of these codes is called 2D-BEM, a two-dimensional boundary element code that assists in the design of cooling lines. With this software, designers can estimate the effects that cooling line placement and sizing has on the thermal fields of a die at different cross sections. Informed decisions can then be made to approach an optimum cooling line design [7].

The other software is Metlflow, a system that computes process conditions (injection times, pressures) based on the location and size of runners and gates [8]. Computations are based on the analysis of fluid flow in tangential runners using classical hydraulics formulations [9]. Neither one of these codes relies on complex formulations or high-performance numerical engines. For this reason, their analysis is somewhat limited in scope and accuracy. Nevertheless, these systems provide first order of magnitude results that can be used to make educated design choices.

---

* Walled Lake, MI.
† Los Alamo, NM.

The latest trend in the development of analysis tools focuses on the needs of designers at early stages. An example is Castview*, a tool that can be used to verify the compatibility of a part design with the process of die casting. Part geometries are voxelized and empirical relations are used to verify thick regions in the die and casting, as well as fill related problems. This tool is capable of generating results in a small fraction of the time it takes to analyze a part with a commercial system. Its results are primarily qualitative in nature, and are intended to provide information that can be used to filter out designs of castings and dies early in the design process.

Generic systems like ABAQUS and ANSYS provide a different option. These systems may be used for the analysis of heat transfer, thermoelastic deflections of dies, and even temperature induced distortions of castings. However, a considerable amount of effort is needed to prepare a model for anyone of these tasks. The analysis of fluid flow or solidification of the kind found in die casting is generally beyond the scope of their capabilities. On the other hand, these systems have an extensive library of elements that can be used to analyze a variety of related physical processes.

A. Capabilities and Limitations of Commercial Systems

All of the computer systems presented in Table 2 require a considerable investment. Most of them need to be run on workstations or high-end PC's to produce results in reasonable times. Typically, the price of the software, as well as training for an analyst, is higher than what is invested in the hardware. There seems to be a consensus among users that, in order to justify the investment, there must be enough demand for simulation results to keep an analyst working full-time [3]. The premise is that if there is not enough demand for simulation results to keep an analyst busy on a full time basis, the resource is severely underutilized.

Systems designed for the analysis of casting or die casting address fluid flow, heat transfer, and solidification mechanisms. These systems have certain capabilities that reduce the amount of work needed to prepare and analyze a model:

- They are designed to handle and create multiple boundaries and materials within a single model, i.e. casting and dies / molds.
- Special tools and routines are provided to facilitate the definition of loads and boundary conditions typical of the casting processes.
- They provide a database of material properties and process conditions that have been developed for casting applications.

---

*Developed at Ohio State. Currently commercialized by the North American Die Casting Association.

- They are prepared to present results in certain ways that are meaningful to the application, such as the last region to fill within the cavity or the last region to solidify in the casting.

In spite of the fact that these codes are prepared to facilitate preprocessing of a model, preparation of a model for simulation is still a time-consuming task. As already mentioned, a typical analysis focuses on the delivery of molten metal to the cavity, the solidification of the part, and the removal of heat through the mold. In general, these systems produce results that can be used to design certain features of the dies, such as cooling lines (location and size) and develop the necessary process parameters (injection and intensification pressures, operating temperatures, etc.) [8, 10, 11]. Designers can play with different scenarios to balance filling, or move shrink porosity to noncritical regions. In addition, gate design or vent location can be evaluated with the results of a simulation.

As indicated in the previous section, commercial computer systems have been shown to produce reliable results in the case of sand casting, permanent mold casting and other low pressure processes. However, their ability to model die casting adequately on a routine basis is still under scrutiny.

From the standpoint of the practical application of these systems to the analysis of die casting, the resolution of results is an important issue. That is, in the context of the process conditions that can be controlled in die casting, what is the minimum degree of accuracy required from simulation results to improve the quality of the casting operation? The answer to this question plays an important role on the demands placed on the model and the quality of the physical data used in such models. Experience has shown that a resolution of about 70 °C in thermal fields and about 0.004 in. in die displacement fields is typical of current modeling capabilities. This is perhaps borderline with the ability of die casters to control part dimensions and thermal conditions in the field. Justifiably, there exists a certain degree of skepticism regarding the benefits of using computer modeling in die casting.

B. Issues in the Application of Commercial Simulation Systems for Die Casting

The development of adequate models for die casting is complicated by the complex nature of the process. The basic mechanisms of heat transfer, fluid flow, and mechanical response of the die occur simultaneously. To a certain degree, these mechanisms are coupled. For example, heat transfer is affected by fluid flow mechanisms and vice versa.

Similarly, there is an intrinsic dependency between deflections of dies and the heat transfer characteristics of the interfaces between components.

Commercial solidification programs are incapable of predicting die deflections as temperature patterns evolve. Consequently, there is an inherent inaccuracy in the manner in which heat transfer is modeled because the changes in the heat transfer characteristics of the domain caused by the developing thermal patterns are not accounted for in the computations of heat flow. More important perhaps, it is not known what the significance of this limitation is.

A different issue is presented by how the solidification process is handled. Typically, the same solidification models are used to analyze conventional casting process and die casting. In conventional casting processes the only external force during solidification is gravity, while in a die casting process solidification occurs in the presence of relatively large pressures. The question remains as to whether it is possible to ignore the effects that pressure has on the evolution of latent heat and the thermal resistance of the casting/mold interface without causing significant inaccuracies in the results.

Development and analysis is complicated by the wide range of temporal and dimensional scales found in die casting. For example, geometric features of a cavity can be as small as a millimeter in dies that may be a meter thick. By the same token, while filling mechanisms develop completely within a few hundredths of a second, it takes about a minute for solidification to be complete and it may take a die several hours to reach quasi-steady state conditions (thermal periodicity).

Perhaps the major weakness of current systems lies in their inability to model the cavity fill mechanisms. This is not a minor issue, given the importance that this process has on the quality of the casting. Modeling cavity fill requires the prediction of the transient, hydrodynamic behavior of a fluid in a regime that surpasses the onset of turbulence while releasing heat [12]. The fluid itself is a combination of solid particles suspended in a liquid that in some microscopic regions may flow through a partially solidified grid (mushy zone). At a macroscopic level, molten metal must displace air as it flows into the cavity. The air pressure varies as a function of both its temperature and the rate at which it is evacuated from the cavity. Among the specific problems that complicate obtaining a full solution, one can find.

- The inherent difficulty in solving the Navier-Stokes equations, particularly the problems associated with the nonlinear convective terms of the momentum and energy equations.
- The presence of multiple phases: liquid metals, solidified particles and entrapped gases.
- The need to provide a time averaging solution of the equations in the turbulent regime. Solution of the Reynolds stresses requires the addition of two partial differential equations to account for kinetic turbulent energy and its rate of dissipation in the domain.

- The need to predict fill fronts simultaneously with other primary variables: velocity, pressures, etc., in a three-dimensional space; jet and atomized flows are clearly more complex.
- The iterative nature of the numerical process needed to solve the transient problem.

All of the solidification simulation systems are capable of predicting flow patterns. However, to one degree or another they rely on gross simplifications to address the aforementioned issues. For this reason, questions are raised about the accuracy of their simulation results. For example, it is clear that the prediction of the evolution of gas porosity during fill is beyond the capability of any of these systems, at least in the time frames that the process of die design and fabrication imposes. The same can be said about the prediction of back pressures as the liquid fills the cavity. These shortcomings necessarily affect the ability of these codes to predict fill fronts, which in turns affects other predictions. Under current practices, some of these issues may not be significant for engineering purposes. However, these issues represent an obstacle for improving the quality and acceptance of simulation results.

The development of models for certain thermal loads also presents a challenge. For example, preheating of dies by means of partial shots, heating with torch or heat radiators cannot be handled at all. Perhaps more important, the effects of the application of lubricant cannot be predicted by these systems. There is simply no knowledge about the rate of heat transfer that takes place in any of these instances.

Finally, it is clear that current computer systems limit themselves to the analysis of thermal and fluid flow phenomena. As mentioned, commercial solidification systems are not designed to handle structural analysis of any kind. While commercial software can predict thermal fields in the part at any moment of the casting cycle, the prediction of postejection casting dimensions or the distortion of the cavity or the structure of the die are beyond the capabilities of their models. In this context, the fact that their architectures are not as open as those of generic packages causes problems, because transporting their thermal predictions to other programs for further analysis may not always be possible. This is particularly troublesome in those cases in which a stress/strain/displacement analysis of a die is necessary.

While the prediction of final casting dimensions is still a topic of research, elastic deflections of dies may be simulated with structural analysis packages within the resolution that was defined in previous sections. Given the fact that die distortions affect cavity dimensions, the prediction of die distortions is an intermediate step toward the prediction of final part dimensions. Other factors that affect casting dimensions are the relief of stresses generated as the part solidifies while being constrained within the cavity, the relief of stresses generated by the ejection process and the thermal distortion of the casting as it cools down to room temperature.

IV. THERMOELASTIC DEFLECTIONS OF DIE CASTING DIES

It is known that dies deflect during the casting operation. Field experiments have reported significant slide and platen movement [13]. Deflections have a direct impact on the quality of the die casting operation. Product dimensions are clearly affected by the deflections of the cavity. Die deflections are also a source of operational concerns. In severe cases, deflections may affect the performance of sliding cores or the ability of a die to eject a casting.

A major concern is the ability of a die to remain closed during the injection/intensification process. The analysis of die deflections can help improve the quality of a design by providing information that can be used to:

- Improve the match between a die and a machine
- Define the proper size of a die
- Improve the design of shutoff areas, that is, the regions of the die that surround the cavities

To illustrate the state of the art in the die design as well as the potential benefits of the analysis of die deflections, the design of shutoff areas is used as an example. In general, casters spend a great deal of effort in the manufacture of shutoff areas to guarantee a perfect fit between mating components. The goal is to keep molten metal within the cavity region throughout the casting cycle. The current practice is to fit shutoff areas by hand, at room temperature. Under current design practices no analysis is done to predict thermal growth of the die and compensate for it during the manufacture of shutoffs. Instead, any misalignment/misfit caused by thermal growth is expected to be overcome by the locking action of the casting machine. Metal that escapes from the cavity because of die misfit usually ends up forming a thin strip of material that follows the contour of the parting surface. This strip is known as flash and requires extra operations for its removal. Flash buildup within the structure of the die can cause moving mechanisms to jam or even break. To prevent flash buildup, casters are forced to make frequent stops to clean die cavities, thus affecting the productivity of the casting operation and possibly affecting casting quality.

Die manufacturers have very limited tools for dealing with die deflections. For example, designers apply a constant scaling factor to compensate for thermal growth of the cavity. This practice neglects the nonuniformity of the thermal patterns within the die, as well as the directionality of the shrinking mechanisms that take place while the part resides within the cavity. Deflections caused by the mechanical loads (clamping force, cavity pressure) are rarely accounted for at the design stage, except by oversizing the support structures of the die. The most common approach for addressing deflection related problems is to leave excess material in the cavity so that adjustments can be made after exploratory runs.

These practices are a clear reflection of the degree of understanding of the behavior of the die under the loads produced by the casting operation. Part of the problem is that dies are complex enough that their response is not easy to predict intuitively. Compounding the problem is the fact that casters deal directly with process conditions such as melt temperature, clamping force, injection and intensification pressures, cycle times, timing of intensification, and so on. Mapping process conditions to loads is not a straightforward task, especially in view of the interactions among the different loading mechanisms.

Simulation tools offer the potential to help casters predict the behavior of a die while it is still at the design stage. In particular, computer models may allow the designer to estimate the behavior of the die under different process conditions (assuming of course that a proper correlation can be made). As explained before, there are no analysis/simulation tools that can address the full range of loads that a die encounters.

A. Simulation of Die Casting Die Deflections

To illustrate the mechanical behavior of a die under typical die casting conditions, a simulation was prepared. Figure 1 presents the steps needed to perform the computer simulation. Contrary to the case of solidification models prepared with commercial systems, the structural analysis of a die must be performed in a system that is not designed for modeling the casting environment. Consequently, loads and boundary conditions need to be defined explicitly. The definition of multiple materials and interfaces needs to be done explicitly too, a time-consuming task.

In general, it is possible to model thermal and mechanical boundary conditions with relative ease. Similarly, mechanical loads such as hydrostatic cavity pressures and clamping forces may be used to capture maximum deflections of a die. On the other hand, the definition of thermal loads presents a challenge because of the existence of a transient behavior prior to the attainment of periodic conditions (i.e., a die starts at room temperature, and as the operation progresses its temperatures builds up until quasi-steady state conditions are reached). Furthermore, there is not a whole lot of information about the nature of all thermal loads (heat from solidification, lubricant spray, etc.).

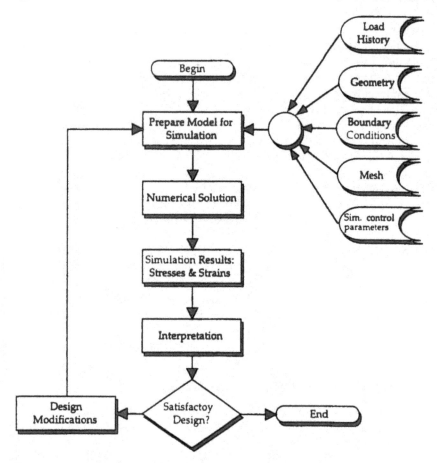

Figure 1 Flow chart for a typical simulation session.

As an approximation, heat from solidification can be modeled based on the assumption that the total heat released by a casting is given by:

$$H = C_l(T_m - T_l) + \text{heat of fusion} + C_s(T_s - T_e) \tag{1}$$

where: H = total heat released by the part
C_l = specific heat for liquid casting material
C_s = specific heat for solid casting material
T_m = melt temperature
T_l = liquidus temperature
T_s = solidus temperature
T_e = temperature at end of cycle (280°C)

The rate of heat release can then be approximated by an exponentially decaying function, where the initial rate is several times larger than the rate at the end of the thermal cycle. The shape of the heat release is given by

$$Q = K^* t b^{mt} \qquad (2)$$

where Q = heat rate
K = arbitrary constant
b = base (in our example, 10 was used)
m = parameter that defines the rate of release
t = time

This form approximates empirical data reported by several studies, including those of Papai and Mobley [14]. The parameters K and m need to be adjusted until thermal fields similar to those reported by field measurements or simulations performed in one of the specialized heat transfer modeling systems are obtained.

The application of lubricant spray may also be modeled as a heat flux from the die to the atmosphere. In this example, the amount of heat removed by spraying of the cavity was assumed to be 15% of the heat released by the part, which is in the range of values reported in the literature [14]. Clearly, the accuracy of results is limited by the quality of the available data, be it experimental or simulated.

Figure 2 presents the geometry of the open/close die whose deflections were simulated. Table 3 summarizes the assumed process parameters. The properties of aluminum Al380 was used to approximate the total heat released by the part. Load histories were defined based on the sequence shown in Table 4. The pressure in the cavity was assumed to be hydrostatic. Other environmental parameters and constants were collected from the literature [6, 14–18].

IDEAS* was used for preparation of the models and for displaying simulation results. ABAQUS was used to carry out numerical computations. Based on the assumption that the thermo-mechanical processes are uncoupled, simulations were done in two steps. In the first step, a heat transfer analysis was performed to compute the thermal fields in the die. In the second step, the temperature fields obtained from the first step are combined with the cavity pressures and clamping forces to compute the strains in the die.

Simulation results show that the maximum separation occurs at the instant the maximum cavity pressure is applied (this is typically the intensification pressure). Figure 3 illustrates this phenomena. In this case, the displacement of two points on either side of the parting plane is recorded. At the moment the dies are closed these two points come in contact. Once cavity

---

* Structural Dynamics Research Corporation, Milford, OH.

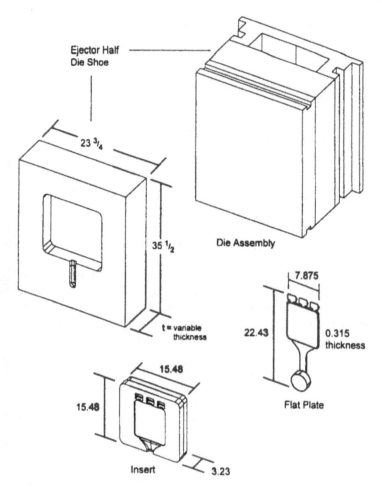

Figure 2 Schematic of the geometry used for simulations. The thickness of the plate was varied for analysis. Dimensions are given in inches.

pressure is applied, a gap is formed between the two points. After a short time, the heat load takes effect and causes the gap to collapse.

Simulation results show that thermal loads do not necessarily cancel the effects of cavity pressure. Under certain conditions deflections induced by thermal loads may actually combine with the pressure patterns to produce hazardous conditions. Figure 4 illustrates a case in which a gap that runs from the cavity to the atmosphere is created by the combined effects of all loads. If the gap is large enough, a hazardous flashing condition could occur

Table 3 Assumed Process and Environmental Conditions

	Assumed values
Casting material	Al 380
Melt temperature	600°C
Injection pressure	10,000 psi and 3000 psi
Cycle time analyzed	36 s
Clamping force	815 tons
Ambient temperature	20°C
Heat transfer coefficient steel–air	20 J/K
Heat transfer coefficient steel–steel	2000 J/K
Heat transfer coefficient steel–cooling line water	10,000 J/K
Water temperature	40°C

during operation. This figure also shows that the maximum deflection occurs behind the cavity, near the middle of the die and away from the support pillars. Additional supports in this region may reduce the magnitude of the deflections.

The behavior of the die during 4 cycles was also simulated. In general, results show that the heat load dominates the deflections even more as the operation approaches quasi-steady state. Figure 5 shows how the maximum parting plane separation keeps getting smaller as time progresses. In essence, the heat load begins to create a prestress at the parting plane as the die components try to grow due to the increased temperature.

Figure 6 shows the contributions of heat and pressure to the deflections of two points of the die. The deflections within the cavity can be used to explain the effects of each load. Before pressure is applied, the cavity is displaced from its original position by a significant amount due to the rise in temperature from room conditions. The magnitude of this displacement is not presented in the figure because the size of the cavity remains almost unchanged. As pressure is

Table 4 Sequence of Events in Simulation of the Full Die

Time (s)	Step
0–1	Dies are clamped
1–17	Heat and pressure applied to cavity
17–32	Die opens, part is ejected
32–35	Lubricant spray
35–36	Dies clamped

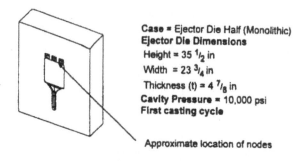

Case = Ejector Die Half (Monolithic)
Ejector Die Dimensions
Height = 35 $\frac{1}{2}$ in
Width = 23 $\frac{3}{4}$ in
Thickness (t) = 4 $\frac{7}{8}$ in
Cavity Pressure = 10,000 psi
First casting cycle

Approximate location of nodes

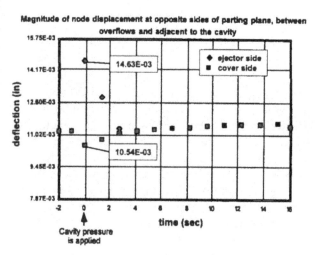

Figure 3 Chart showing the parting plane separation. The maximum occurs at the moment the pressure is applied.

applied, the cavity thickness grows instantaneously. This is illustrated in the figure by the fact that the pressure induced deflection is larger than and opposite to the heat induced deflection. After a few casting cycles, the temperature of the die increases, resulting in a larger contribution of heat to the overall deflection.

From the perspective of design, deflection models can be used to evaluate different scenarios. For example, Fig. 7 shows how the maximum separation is affected by changes in the thickness of the die shoes, or by the inclusion of inserts to manufacture the cavity. These results show that, in general, thicker plates and monolithic dies will allow for smaller separations during operation. Both measures increase the cost of the die, and consequently the designer must define the combination that results in the most economic tool while still meeting quality demands.

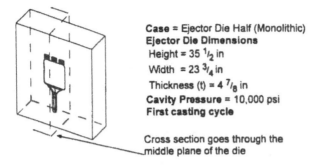

Case = Ejector Die Half (Monolithic)
Ejector Die Dimensions
Height = 35 $\frac{1}{2}$ in
Width = 23 $\frac{3}{4}$ in
Thickness (t) = 4 $\frac{7}{8}$ in
Cavity Pressure = 10,000 psi
First casting cycle

Cross section goes through the
middle plane of the die

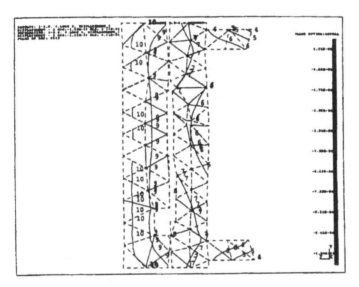

Figure 4   Cross section showing a gap that connects the cavity with the atmosphere.

V. A DESIGN EXERCISE

The analysis of the performance of a slide/lock arrangement can be used to illustrate the validity of the modeling approach introduced in the previous sections. Figure 8 presents a schematic of a slide carrier and a protruding lock horn. This type of design is commonly used in dies for transmission casings. For clarity, only a section of the ejector die half and lock are shown. At clamp up, the horn pin is inserted through the slide carrier and into the ejector die, where the tip of the horn pin engages an angled surface. During operation, the pin is intended to hold the slide in place.

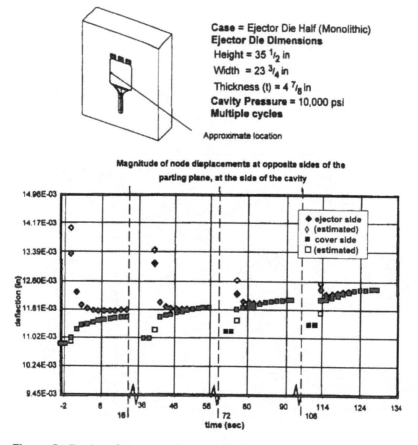

Figure 5 Parting plane separation at midpoint of plate. After the second cycle this point shows the maximum separation. For each cycle, only the deflection that occurs between the instants of metal injection and part ejection has been recorded. Estimated values were extrapolated based on the results of a single cycle because instant of maximum separation was not recorded due to large differences in time scales.

Typically, slides move back (blowback) at the instant full pressure is reached within the cavity. Figure 9 shows a blowback trace for a transmission casing slide. In this particular case, blowback due to injection pressure reaches approximately 0.028 in. Blowback after intensification reaches 0.038 in. (delay = 0.1 s).

Figure 10 presents the results of simulating blowback for this particular die. In this model, it was assumed that there was symmetry about the midplane between horn pins. For this reason, only half of the geometry was actually

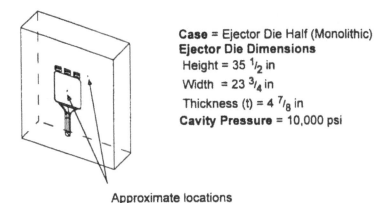

Case = Ejector Die Half (Monolithic)
Ejector Die Dimensions
Height = 35 $\frac{1}{2}$ in
Width = 23 $\frac{3}{4}$ in
Thickness (t) = 4 $\frac{7}{8}$ in
Cavity Pressure = 10,000 psi

Approximate locations

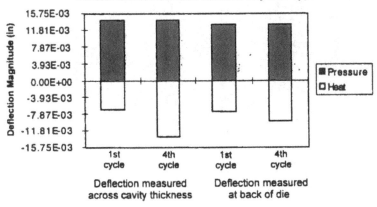

Figure 6   Contribution to overall deflection by each type of load at selected locations.

modeled. The horn pin is constrained from moving at its top. The assumption is that the horn pin is firmly attached to the cover die, and that this support is rigid. At the bottom, a small block is constructed to model the support that the inclined plane within the ejector die provides. Blowback data is read at the back of the carrier. Simulation results show that the slide will move back approximately 0.035 in. (8.8 E-04 m). The difference with respect to the field data (i.e., 0.003 in.) is in the range of error of our simulation capabilities.

Assuming that this were a new design for which field data were not available, the designer would have to make a decision as to how to deal with the deflections that the simulation reports. One possibility would be to move the slide inward (i.e., into the cavity) 0.035 in. in such a way that when the slide moves back the dimension within the cavity reaches its intended value.

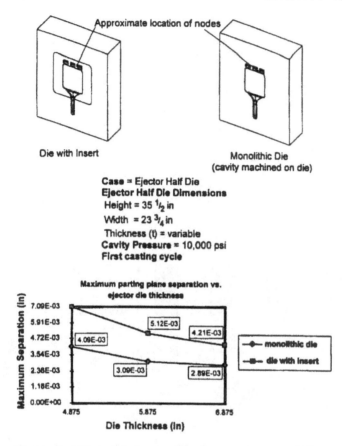

Figure 7 Effect of design modifications to structure of die on parting plane separation.

Another possibility would be to modify the design of the lock and observe how the magnitude of blowback is affected. Table 5 and Fig. 11 summarize the results of this process. In this exercise, the dimensions of the horn pin were modified and the response of the lock was observed. Table 5 presents the modifications that were analyzed: O stands for original dimension, H stands for a modification that resulted in a larger dimension (25% higher), and L in a lower dimension (25% lower). The effect of the modification is measured in terms of how the stiffness was changed (where stiffness is simply the slope of the force vs. blowback curve for the particular case—as if it were a spring). Naturally, higher stiffness results in fewer deflections. In addition to stiffness, force transmitted to the cover die was computed for each case. Larger force into the cover die requires higher clamping force to keep the die closed.

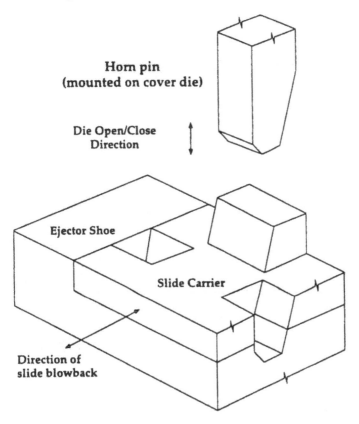

Figure 8   Schematic slide/lock arrangement showing blowback direction.

It can be seen in Fig. 11 that in only one case (5), there was a slight improvement in the stiffness of the lock. This occurred at the expense of a larger force into the cover die. The implication is that perhaps not much can be done to reduce blowback with the current lock design. This result is not surprising if one takes into account the fact that the original design (i.e., Figs. 8, 10) is a second generation (iteration) die that showed a very stable operation. Furthermore, data was collected when the die was new.

This exercise illustrates the way in which simulation results obtained with a commercially available FEM system can be used to scrutinize the performance of a given die component without the need to actually build the die. Arguably, an experienced designer armed with computer modeling tools can reduce the number of iterations needed to produce good dies.

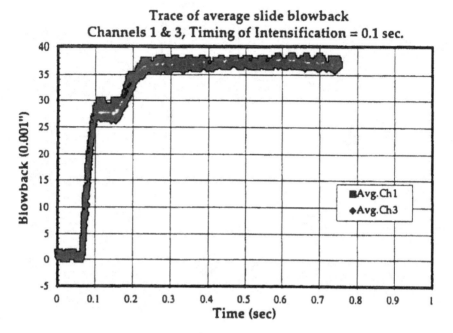

Figure 9   Field data: slide blowback, timing of intensification = 0.1 s. (Data courtesy of Exco Engineering, Newmarket, Ontario.)

VI. FUTURE TRENDS IN SIMULATION

The use of computer simulation systems has not yet been integrated fully into the design process. Perhaps the most relevant issue is the time it takes to solve a model. Clearly, before these tools can be integrated into the design process times for model solution need to be reduced drastically.

Some of these codes already offer modules that facilitate the creation of boundary conditions that are particular to die casting. Still, fundamental improvements are necessary in the models of cavity fill and heat removed by lubricant spray. In addition, the need to provide a better correlation between results and process conditions or to predict final part dimensions is also evident. Code developers are addressing these issues to different degrees.

The need to develop a more extensive and accurate database of material properties and parameters for boundary conditions (i.e., heat transfer coefficients) is also being addressed by commercial systems. While some vendors have the resources to generate some of this data by themselves (MAGMAsoft), others have developed codes that allow for the computation of some of the properties and parameters based on field data (ProCast).

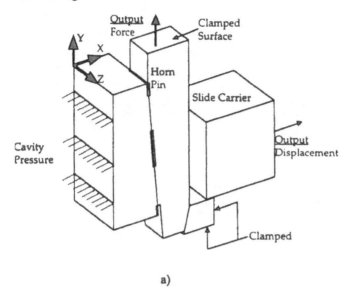

a)

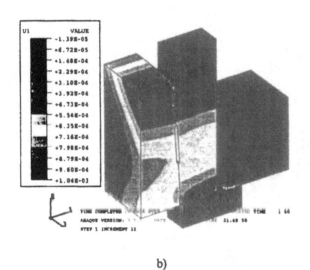

b)

Figure 10 (a) Model description for computation of slide blowback (Exco's design). (b) Typical simulation results.

Table 5   Comparison of dimensions of variations to original design

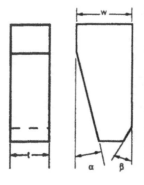

Case	t/w	α	β
Original	O	O	O
1	H	H	H
2	H	H	L
3	H	L	H
4	H	L	L
5	L	H	H
6	L	H	L
7	L	L	H
8	L	L	L
OHH	O	H	H
OHL	O	H	L

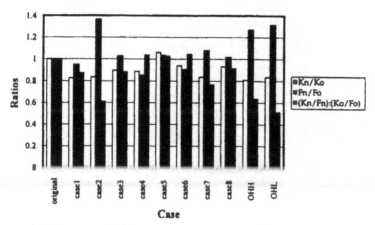

Kn/Ko = ratio of stiffnesses of pin of case n to original case (o).
Fn/Fo = ratio of force transmitted to foundation of pin of case n
        to original case (o).
n = 1, ..., 8

Figure 11   Comparison of characteristic behavior of slide/lock arrangements for small changes in the design of the lock. Desired: high stiffness and low force into the foundation.

Some of these systems are also developing models to predict die deflections. MAGMAsoft has already announced this capability for future versions of the code [19].

VII. CONCLUSIONS

Computer simulation of die casting is still in the process of gaining acceptance in the industry. To one degree or another, commercial computer codes are capable of:

- Predicting thermal fields in the die and the casting throughout the casting cycle
- Assisting the designer in the definition of size and location of cooling lines
- Assisting the designer in selecting cavity orientation, and cooling line placement to move shrink to noncritical regions
- Evaluating gate and vent designs and balance filling patterns

While the larger die casters and die manufacturers make use of computer systems, it is mainly for the purpose of troubleshooting and diagnostics. Among the obstacles for the widespread use of these codes are the high investments needed for a basic system, the lack of a proven track record, and the high cost in terms of time for performing an analysis.

Their use as tools for design is still not feasible under typical circumstances. As they currently stand, the value of these tools depends on the skill and interpretation abilities of the user. That is, these tools do not directly answer questions. Instead, they require the user to apply engineering judgment to achieve results. As users improve their ability to interpret results, these systems will find more acceptance.

REFERENCES

1. T Maier, J Kolakowski, J Wallace. Die Casting of Copper Alloys. AFS Transactions 83:279–294, 1975.
2. D Caulk. Opportunities and challenges in die casting analysis. Keynote speech presented at Flow Modeling/Thermal Simulation for Die Casting Forum. Chicago, IL, 1996.
3. Nadca Computer Modeling Task Group. Presented at Flow Modeling/Thermal Simulation for Die Casting Forum. Chicago, IL, 1996.

4.   MR Barone, DA Caulk. dieCAS—thermal analysis software for die casting: modeling approach. Presented at NADCA International Congress and Exposition, Cleveland, OH, 1993.
5.   MR Barone, DA Caulk, DE Siefker. dieCAS—thermal analysis software for die casting: results and application. Presented at NADCA International Die Casting Congress and Exposition, Cleveland, OH, 1993.
6.   MR Barone, E Kock. A method for analyzing the effect of flow on heat transfer in die casting. International Journal for Numerical Methods in Heat and Fluid Flow 3:457–472, 1993.
7.   HT Siauw, TT Nguyen. A computer-aided thermal analysis package for pressure die casting dies. Presented at NADCA 15th International Die Casting Congress and Exposition, St. Louis, MO, 1989.
8.   WT Andresen. Computer simulation and analysis of liquid metal flow and thermal conditions in die casting dies. Presented at SDCE 14th International Congress and Exposition, Toronto, Ontario, Canada, 1987.
9.   TH Siauw, AJ Davis. Flow analysis in tapered runners. Presented at 10th SDCE International Die Casting Exposition and Congress, St. Louis, MO, 1979.
10. L Kallien, JC Sturm. Simulation aided design for die casting tools. Presented at NADCA Congress and Exposition, Detroit, MI, 1991.
11. RL Smith, DE Phenicie, AK Agarwal, HL Kallien. Optimizing production of a die casting by numerical simulation of die filling and solidification. Die Casting Engineer 37:40–48, 1993.
12. D Frayce, CA Loong. Mathematical modelling of the die casting process. Presented at Light Metals, 1991.
13. GA Prince, CR Ramsey. Introduction to die movement and distortion: a case study. Presented at NADCA International Die Castung Congress and Exposition, Indianapolis, IN, 1995.
14. J Papai, C Mobley. Die thermal fields and heat fluxes during die casting of 380 aluminum alloy in H-13 steel dies. Presented at NADCA Die Casting Congress and Exposition, Detroit, MI, 1991.
15. DA Caulk. A method for analyzing heat conduction with high frequency periodic boundary conditions. Transactions of ASME, J Heat Transfer 112:280–287, 1990.
16. MR Barone, DA Caulk. A new method for thermal analysis of die casting. Transactions of the ASME, J Heat Transfer 115:284–293, 1993.
17. K Chijiiwa, K Shirahige. Behavior of molten metal with respect to the pressure in cavity of mold in aluminum die casting. J Faculty of Engineering, The University of Tokyo, 26:27–36, 1981.
18. KH Hegde. Finite element analysis of deflections in die casting dies. Masters thesis, Mechanical Engineering Department. Columbus, OH: Ohio State University, 1995, p. 148.
19. MAGMAsoft, MAGMAsoft Advanced User's Workshop II, Schaumburg, IL, 1997.

14
Semi-Solid Metalworking

Michael L. Tims
Concurrent Technologies Corporation, Johnstown, Pennsylvania

I. INTRODUCTION

In the early 1970s as Massachusetts Institute of Technology graduate student David Spenser was conducting experiments on hot tearing of metallic alloys, he observed that vigorously stirred Sn · 15wt%Pb had lower viscosity in the partially solidified (i.e., semi-solid) state than did the same material when it was allowed to cool without stirring [1]. Subsequent work showed that the shear stress in the semi-solid state can be orders of magnitude lower when continuous stirring is applied during solidification. Furthermore, Spenser noted that spherical grains developed in those specimens that were stirred during solidification, while an interlocking network of dendritic grains developed in the unstirred specimens. The spherical grains were found to be enriched alpha particles and were surrounded by a eutectic matrix. This structure is to be expected since the solid particles first formed during solidification would naturally solidify with an alpha-rich composition leaving behind a solute-rich liquid that upon further solidification would eventually reach the eutectic composition. Uniform distribution of the solid alpha particles results in the aforementioned billet microstructure. Subsequent work by a wide variety of researchers has shown similar behavior with other alloy systems [2–11]. Figure 1 shows typical microstructures for billets solidified in stirred and unstirred conditions. The term "semi-solid metal" (SSM*) has

---

* SSM also refers to "semi-solid metalworking." The abbreviation will be used throughout this chapter for both semi-solid metal and semi-solid metalworking. The context in which SSM is used in the text should be clear to the reader, however.

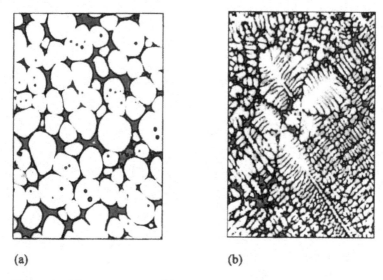

(a)   (b)

Figure 1 Microstructure of SSM and typical dendritic materials. (a) Typical SSM microstructure; (b) typical dendritic microstructure. (From Ref. 3.)

been used to describe those partially molten metallic systems that have been intentionally produced with this globular microstructure.

Flemings and coworkers [9] later reasoned that one could take advantage of the unique behavior of these materials since the particles maintained their shape during subsequent reheating to the semi-solid state. The eutectic matrix melts first, leaving behind a suspension of solid alpha particles. It is this structure that accounts for the low viscosity of SSM materials. Slugs with as much as 50% liquid can support their own weight for several minutes without appreciable sagging or slumping. These slugs can therefore be handled like a solid. Upon application of a finite shear stress, however, the semi-solid material flows like a viscous fluid—see Fig. 2. Others [2] have demonstrated this effect by easily cutting a heated SSM billet with a butter knife and then pushing the material around like peanut butter. As a result of these characteristics, SSM billets can be easily molded into a wide variety of complex shapes. Flemings [2, 12] and others [13–17] point out that a wide variety of metallic alloys and composites qualitatively behave in a similar manner. These materials include Al · Cu, Al · Si, hypereutectic Al · Si alloys (for wear resistant applications), Wood's metal, Al · Pb, Al · Be, Al · Ni, wrought aluminum alloys 6262, A6082, and A7075, Bi · Sn, Zn · Al, Al · Zn, copper-based alloys (377 and 642), cast iron, ductile iron, steel, nickel-based and cobalt-based superalloys, stainless steel, low alloy steel, hypoeutectic cast iron, and SiC in Al · 6.5wt%Si.

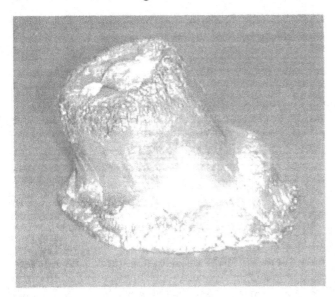

Figure 2 Demonstration of material behavior via a dropped SSM billet. 75-mm diameter × 120-mm length A356 billet at $f_s = 0.5$ dropped from a height of 1.0-m. (Courtesy of Concurrent Technologies Corporation, Johnstown, PA.)

In addition, magnesium and tin, along with titanium [18] alloys have been formed in the semi-solid state.

SSM materials exhibit a time dependent viscosity; hence the term "thixotropic" is often used to describe their behavior. As these materials are deformed, their flow resistance decreases. This time dependency is due in large part to disagglomeration of particle clusters during processing [2]. The particles have a natural affinity for each other and therefore agglomerate during periods of rest. However, particle bonds are broken during deformation as a result of shear stresses, which allows the material to flow with greater ease.

In addition to thixotropic behavior, the flow resistance of SSM materials is strongly dependent upon processing history. The resulting microstructure (average grain size and shape and the distribution of grain size and shape) in the SSM state and the degree of particle agglomeration appear to be the most significant material characteristics contributing to this process history dependent flow behavior. Additional factors that contribute to property variations during processing of SSM materials include (1) solid/liquid segregation, which leads to local variation in solid fraction, flow resistance, macroscopic chemistry, and final mechanical properties; (2) solidification due to contact with cold forming dies or the billet handling equipment; (3) nonuniform billet heating;

(4) percolation of liquid through the solid structure to the bottom of the heated billet due to gravitational effects; and (5) surface oxides.

Commercial interest in the SSM process currently stems from the automotive, aerospace, electrical connector, plumbing, and military supply industries. Sizes range from a few grams to over 9 kg [19]. The greatest interest is in production of high quality aluminum parts. Magnesium-based SSM production is gaining popularity, however. Commercial interest remains active in copper, steel, titanium, zinc, and some composite alloys [1, 13, 20]. Typical automotive applications include housings for air conditioners, steering components, suspension parts, pulleys, air bag canisters, brake disks and drums, engine brackets, master brake cylinders, bumpers, and seat frames. Other applications include housings for small motors and compressors, plumbing fixtures, seawater valves, space frames, heat sinks for electronics applications, golf putter heads, computer hardware components, ergonomic pens, golf club sole plates, and small ductile iron gears. Tolerances similar to those of machined parts have been reported [3]. In an increasing number of instances, SSM produced parts are successfully replacing machined components. For example, electrical connectors made from SSM processing have reduced machine chips to 8% of the starting billet mass compared with 81% using conventional manufacturing methods. Figure 3 shows typical SSM parts.

The primary reasons for the commercial interest in SSM produced parts are cost and performance. Since die casting methods are typically used to manufacture SSM-based components, the processes are of similar economy. Porosity defects, common to die castings as a result of liquid metal atomization and turbulence, are less likely from the SSM process. Lower porosity levels are present in finished SSM components since (1) material flow is typically laminar (resulting from the higher viscosity of the SSM slurry), (2) little or no jetting of material occurs during die filling, (3) the die is filled more evenly with a blunted flow front, and (4) less solidification shrinkage occurs as a result of the partially solidified feedstock. The laminar fill also helps to minimize the amount of entrapped gas and surface oxidation as a result of less splashing and minimal exposed surface area of the workpiece. Consequently, SSM parts can be successfully heat treated to obtain more favorable mechanical properties. In addition, the process has been shown to produce components that are pressure tight and structurally sound. Hot tears are also less likely with SSM processing owing to the lower overall shrinkage compared to liquid processing. Care must be exercised to avoid hot cracks when designing the process for typical wrought alloy applications, however. The process also results in improved flatness and dimensional repeatability [21]. Figure 4 shows typical properties from the SSM process relative to those from other competing processes.

Other benefits have been identified with SSM processing. Macrosegregation, which can be a significant problem on larger castings, can

Figure 3 Typical SSM parts. 1—Mg heat exchanger; 2—Mg metal matrix composite (MMC) space frame component; 3—Al component; 4—Cu component; 5—Al analogue of a Ti valve body (gating and vents attached); 6—Mg chain saw cover; 7—Mg exercise bicycle component. (Courtesy of Thixomat, Inc., Ann Arbor, MI and Concurrent Technologies Corporation, Johnstown, PA.)

be minimized with SSM processing. Final material properties of finished parts are less sensitive to section thickness due to the consistent spheroidal micro-structure of SSM parts. Less energy is required for heating the material to the semi-solid state than to the fully liquid state. For example, heating an alumi-num billet to a solid fraction of 0.5 requires only 71% of the energy needed to heat the material to a typical casting temperature (see Fig. 5). The reduced heat content also improves mold life (up to 20% longer than die casting [23]) and contributes to reduced cycle time (up to 20% over die casting [24]) between consecutive shots. SSM processing can also be more environmentally friendly since liquid metal handling is eliminated. Finally, it has been proposed that the SSM process may more easily accommodate materials that are difficult to process with conventional processes. These materials include tool steels, wear resistant alloys, and alloys such as titanium and magnesium that are very reactive in the liquid state [25].

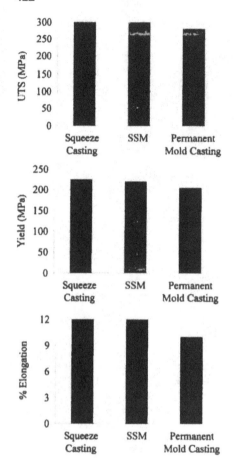

Figure 4 Typical properties from SSM and competing processes for aluminum alloy A356.0 in the T6 heat-treated condition. (From Ref. 8.)

Several limitations inhibit wider application of SSM processing. First, the scientific understanding of SSM material behavior during metalworking is insufficient to properly guide process design and to develop remediation methods to common production problems. Attempts to characterize prior processing effects on current material behavior have not been adequately quantitatively described. Some of the more fruitful work in this regard is described later in this chapter. Most of this work, however, is still in the research stage. Material properties for typical commercial processing conditions (i.e., material *heated* to the semi-solid state and then processed at suddenly applied high shear strain rates with no prior shearing) are limited. As a

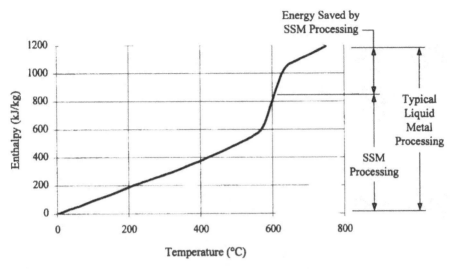

Figure 5 Energy needs for SSM and liquid metal processing of aluminum alloy A356. (Enthalpy data from Ref. 22.)

result, numerical analysts have used available material models, which typically rely upon steady state behavior (i.e., extended time at temperature and continuous shearing) of semi-solid materials. A second limitation of the process is segregation of the solid and liquid phases. Segregation occurs when the pressure gradient within the workpiece forces relative flow between the solid and liquid phases. The majority of this segregation occurs during the metalworking operation. However, significant segregation may also occur during heating when gravity induced pressure gradients cause the liquid to slowly flow through the solid structure and collect at the base of the heated billet. When severe, segregation results in significant temporal and spatial variations in material behavior during mold filling. In addition, severe segregation leads to local chemistry variations in finished parts, which leads to variations in mechanical properties throughout these parts. Process control is another production-related limitation of SSM processing. Material behavior is highly sensitive to solid fraction. Currently, no production-ready instrument has been developed that accurately measures the solid fraction within a workpiece and, therefore, commercial production relies upon temperature measurements for thermal control of the process. Uncertainty in temperature readings (and the corresponding uncertainty in solid fraction) necessitates the use of wide freezing range alloys. Until recently, SSM processing has relied upon alloys commonly used for casting. For example, aluminum alloys 356, A356, and 357 have been widely used in production of aluminum SSM parts. Alloys tailored to the unique behavior of SSM processing would greatly increase application of this advanced processing technique. A final limitation of SSM processing is the limited supplier base for the specially processed feedstock materials.

Defects common to the SSM process [3, 26] are, in general, similar to those in the die casting industry. Blisters may form on some heat-treated parts. Usually, blisters are caused by hydrogen gas formed from entrapped die lubricants. When segregation of the solid and liquid phases is severe, the liquid-rich region may atomize ahead of the main flow of material, resulting in entrapped gas in the final part. This gas can then cause blistering during subsequent heat treatment. Blisters can be minimized by careful selection of die lubricants, ensuring clean shot sleeve and die surfaces (i.e., eliminate lubricant buildup), or reducing fill velocities. Cold shuts, folds, laps, and nonfill are common production problems. Often these defects can be eliminated by heating the billets to a lower solid fraction (i.e., higher temperature) prior to forming, using higher die temperatures, or by increasing fill rate. Improved gating designs also contribute to reductions in these defects. Nonfill regions may also be caused by inadequate vent size or location, nonuniformly heated billets, inadequate injection pressure, or improper microstructure in the starting billet. On those few occasions when jetting does develop, Sigworth [26] recommends making the ingate as thick as that of the section to which it is attached. As with traditional casting processes, shrinkage porosity may result if (1) solidification does not progress toward exterior feeders and (2) fluid pressure does not allow adequate flow from the exterior feeders to the die cavity. Feeding of solidification shrinkage during SSM processing typically comes from the biscuit, through the runner and gate, and into the part (see Fig. 6). Intensification pressure from the ram provides the force necessary to drive semi-solid material

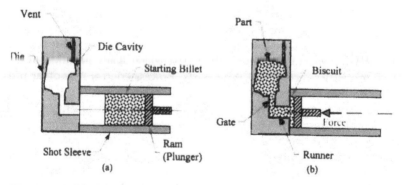

Figure 6 Thixocasting of SSM component via a modified die casting machine. (a) Heated billet in shot sleeve; (b) filled die with intensification load applied during part solidification.

into the die cavity during the feeding stage, assuming that a molten metal path is maintained from the biscuit to the die cavity during solidification of the part. Occasionally, gas porosity is present in the starting workpiece. This gas may then find its way into the finished part. Gas porosity may also result from very high fill rates when gas within the die becomes entrained in the metal stream. Oxide films on the heated billet may be entrapped in the finished part if precautions are not taken to strip this film off prior to entering the die cavity. Use of overflow cavities to trap oxide films at the flow front is also in common usage. Finally, cracking of the free surface during die filling has been reported [27].

Solidification of SSM material can be pronounced during die filling, especially during filling of long narrow parts. Consequently, computation of heat loss during die filling should be included in an SSM die filling analysis. Nonuniform temperatures of the die due to cooling channels and local hot spots from previously cast parts have been found to affect the numerically predicted solidification patterns of these parts. Uncertainty in the die/workpiece interface, which is a common simulation problem, limits the accuracy of the predictions, especially as the die lubricant is washed away from some areas during filling and deposited at other locations.

II. TYPICAL PROCESS ROUTES FOR MANUFACTURE OF SSM PARTS

Figure 7 shows the typical process route for manufacture of SSM parts. First, feedstock material is prepared. The feedstock is cut to length and is reheated—usually with induction heaters. After transfer to a modified die casting machine, the semi-solid billet is then injected into a metal die where it cools and takes the shape of the finished part. Heat treatment, trimming, final machining, and surfacing operations may then follow to yield the desired finished part.

SSM feedstock materials can be made from several processes. The most economically favorable for many alloys is electromagnetic stirring during continuous casting. Materials made from this process have a microstructure like that shown in Fig. 1a. The melt is slowly cooled to minimize temperature gradients, which enhances development of spherical grains. The stirring action also tends to break down columnar dendritic grains, which further ensures an equiaxed microstructure. Smaller grains form with vigorous stirring and with increased cooling rate [1, 2]. Cooling rate during the early stages of cooling tends to be the more important of these two effects [2]. The equiaxed microstructure is maintained during subsequent reheating of the billets into the semi-solid temperature range. These spherical grains provide many of the desirable

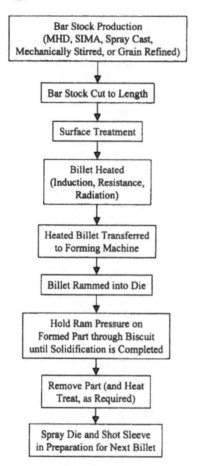

Figure 7   Typical process route during thixocasting of shaped parts.

process and final part properties unique to the SSM process. Eutectic liquid is generally trapped in individual grains as noted by the dark islands in individual grains as seen in Fig. 1a. The quantity of trapped eutectic liquid is reduced with increased shear strain rate and shearing time upon solidification [2]. The motion of the particles and the apparent viscosity of the material are affected by the percentage of eutectic liquid surrounding the particles. Therefore, the amount of trapped eutectic influences the effective solid fraction, which impacts the apparent viscosity of the semi-solid material. Probably the second most popular method of producing SSM feedstock material is the SIMA (stress induced, melt activated) method. This material is cast using conventional methods, cold-rolled to induce residual stresses in the bars, and upon remelting to the semi-solid state, the dendrite grains easily break up into many smaller, nearly spherical solid grains due in part to the residual stresses. Spheroidization occurs in less than 1 min for a variety of SSM alloys when held at typical forming temperatures [28]. Small diameter (25 mm or less) SSM bars of aluminum alloys are economically made using the SIMA process. Early SSM feedstock materials were made with the aid of mechanical stirring. While still a popular method of producing experimental lots of material, no commercial suppliers manufacture feedstock material in this manner. Spray cast material, powder preforms, and use of grain refiners in the billet production stage have also shown some promise as methods for producing alternate feedstock materials.

In commercial applications, the billets are usually heated from room temperature to the semi-solid state by induction heaters. Other methods have been suggested, such as radiant heating and electric resistance heating, however, induction heating has provided the best consistency in commercial applications. The induction heating coils and controls are generally tuned for one material and a single billet size. Billets typically range in diameter from 12 to 150 mm with length-to-diameter ratios of 1.0 to 2.0. Typically in commercial applications, a rotating, indexing table is used (see Fig. 8). Billets rest on insulated pedestals and are indexed into successive induction heating coils to rapidly heat the billets to the desired solid fraction. The last several heating coils are designed to generate just enough heat to allow the billets to soak to a uniform thermal condition prior to forming. A rule of thumb for the desired heating time t (in minutes) is $t = D_B^2 \pm 20\%$, where D_B is billet diameter (in inches) [29]. The heated billets are then transferred to the forming station by a robot.

The most common forming station is a modified die casting machine (see Fig. 6) in a process called thixocasting.* The heated slug is dropped through a hole into the shot sleeve. A ram then pushes the billet down to the end of the shot sleeve where the "casting" die is located. As the leading edge of the billet touches the die, the ram is accelerated and the material quickly fills the die cavity. Fill times of less than 1 s are typical. In addition to a blank for the

---

* The term thixoforming is commonly used to describe those processes that rely upon on reheated SSM feedstock and that use typical forming methods, such as closed die forging. Other common terms for variants of the SSM process include semi-solid metalworking, semi-solid metalforming, mashy forming, SeSoF (semi-solid forming), rheocasting (casting of SSM materials that have been cooled from the liquid state), compocasting (rheocasting techniques applied to metal matrix composites), and rheoforming (forming of SSM materials that have been cooled from the liquid state). The following treatment will focus primarily on the application of traditional casting methods to SSM processing.

Induction Coils
(surrounded by
insulation)

Pedestals

Rotary Indexing Table

Figure 8 Rotary indexing induction heating station. (Courtesy of Concurrent Technologies Corporation, Johnstown, PA.)

finished part, the die cavity usually contains a runner, gate(s), and vent(s). Vents are designed to minimize back pressure during die filling. They are typically placed where the mold last fills to avoid isolated gas pockets and to ensure complete die filling. The vents may also be designed with integral over-flow cavities to catch the material located at the free surface of the incoming material, which is prone to oxide films, excess lubricant, and other defect particles. Vacuum is sometimes applied to the vents as the ram passes the slug entry hole in the shot sleeve (to ensure a seal within the die and shot sleeve). Application of vacuum prevents certain gas and mold-fill related defects. The small portion of billet remaining in the shot sleeve is called a "biscuit." Intensification pressure from the ram on the biscuit enables feeding of solidification shrinkage after mold filling is complete. Intensification also minimizes the size of gas pores that may be present.

The dies are typically made from H13 or H21 tool steel. H13 is commonly used for aluminum parts, while H21 is typically used for copper alloys [3]. Table 1 contains some thermophysical and mechanical properties for these die materials. The dies usually contain cooling passages to enhance the solidi-fication rate of the part. A well-designed cooling system also helps to provide a favorable solidification sequence within the part (progressive solidification toward the biscuit) to ensure good liquid metal feeding to minimize solidifica-tion shrinkage defects. After solidification is complete, the dies open and the part is removed. Prior to closing for the next injection, the die segments are sprayed with a release agent to avoid welding between the die and the next part to be produced.

Table 1 Properties of H13 and H21 Tool Steels

Temperature (°C)	Thermal conductivity (W/m · K)	Thermal expansion (μm/m · K)	Modulus of elasticity (GPa)	Yield strength (MPa)	UTS (MPa)
			H13		
20			207	1196	1407
100		10.4			
200		11.5			
215	28.6				
316			197		
350	28.4				
425		12.2		972	1158
475	28.4				
482				900	1079
540		12.4	159	858	965
593				752	858
605	28.7				
650		13.1		565	638
			H21		
21				1482	1675
100	27.0	12.4			
200		12.6		1310	1579
260	29.8			1241	1531
400	29.8				
425		12.9		1207	1517
540	29.4	13.5		1124	1448
650		13.9		586	896
675	29.1				

Density at 20° C
 H13: 7.76 Mg/m^3
 H21: 8.28 Mg/m^3
Source: Refs. 30 and 31.

Alternate processing routes have been proposed. One of the original production concepts [2] involved direct injection of the SSM slurry from the initial cooling and stirring process. The corresponding processes are known as rheocasting if die casting machines are employed or rheoforming if common forming methods are employed. Subsequent billet production, reheating, and material handling are eliminated with this approach. Some difficulties in matching the capacity of the SSM caster to the forming operation were noted, however. Consequently, no known commercial entities have used this concept in production. Laboratory-scale development for this process using Sn · 15%Pb SSM alloys [32] has shown some potential promise for future commercialization, however. Another alternative processing route, known as thixomolding, is currently used to make a wide variety of production parts, most notably those from magnesium alloys [21]. In this process, the feedstock is made using traditional casting methods [33]. After machining the incoming material into small chips, the material is fed into a hopper. The hopper feeds these chips into an auger, which is surrounded by heaters. An argon gas environment eliminates oxide development during heating. As the material travels the length of the auger, it is sheared, mixed, and gradually heated into the semi-solid state and readied for die injection. A chamber at the end of the screw accumulates heated material for the next shot. The auger rapidly moves linearly forward (without rotating) to push SSM material into the die. The auger is then moved back to its starting position. Fresh material is added from the hopper, while the auger is rotating and indexing material forward through the heating stations. This method offers several advantages over the traditional processing route outlined above. First, the gates, runners, flash, and machine chips are easily recycled. The method also eliminates the need for transfer of hot billets from the induction heating station to the forming station and it offers a more compact operation. Drawbacks include increased surface area for oxide formation in the incoming material, potential contamination during the chip making process, potential safety concerns with metal chips, introduction of air (or inert gas) with the metal chips, the need for metal chipping equipment, and heavy wear on the screw mechanism [34].

III. MATERIAL MODELS

The fields of solid and fluid mechanics have been well investigated for a few hundred years. Some effort has been devoted to the study of suspensions and slurries for civil engineering and other applications. As a relatively new materials processing technique, understanding of the behavior of SSM materials is still evolving. Often the background of the researcher dictates how they approach the fundamental description of the material response. Concepts from fluid mechanics, solid mechanics, powder metal processing, colloidal suspensions, and other disciplines have been applied to describe the behavior of SSM materials. As one reviews the behavior of these materials under various conditions, it becomes clear that SSM materials exhibit significantly different behavior depending upon, among other things, solid fraction (the portion of the slurry in the solid state), shear strain rate, solid grain structure, and prior processing history. Sigworth [35] has summarized these effects for an aluminum-silicon alloy over the full range of solid fraction and at various shear strain rates—

see Fig. 9. Note especially the large variation in shear stress under the indicated process conditions. This large variation has significant implications to the SSM forming industry. Within the semi-solid state of many alloys, small variations in temperature can result in large variations in solid fraction, which in turn significantly impact material response and process loads. Consequently, undesirable material flow behavior may result from inconsistencies in the heating process within individual billets and from one billet to the next. Process control, therefore, becomes a critical issue in production. Additional complications arise in the characterization of these materials from large variations (i.e., several orders of magnitude) in properties due to processing history, including   (1) whether the current semi-solid state was arrived at upon cooling from a pure liquid state or by heating from a solid state, (2) the amount of prior shearing the material has been subjected to while in the semi-solid state, and (3) the morphology of the grains present. In summary, the rheological behavior of semi-solid materials is not yet fully characterized. The following treatment focuses on the behavior of semi-solid materials from a fluid mechanics approach as this is the most widely used approach to define SSM material behavior. In addition, a brief description is given of numerical models used to simulate industrial processes used for producing SSM parts.

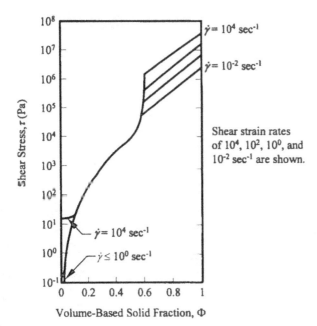

Figure 9   Shear stress for Al · Si-based SSM alloys at various solid fractions and shear strain rates. (From Ref. 35.)

Conservation laws of fluid mechanics are necessary to describe the flow of SSM materials. The continuity, momentum, and energy conservation equations must be satisfied. Viscosity models are used to characterize the behavior of SSM material. These models can be used for a wide variety of purposes, including:

- To establish the optimum location of vents to ensure complete die fill
- To determine the optimum size, number, and location of gates and runners to ensure good mold filling characteristics
- To evaluate premature freezing of the material during die filling
- To design cooling and heating passages that permit adequate die filling and favorable progression of solidification
- To determine the optimum speed profile of the ram and the forces required to do so
- To establish location of oxides (common at the surface of heated billets) in filled cavities using particle tracking

A. General Liquid Suspensions

Up to a solid fraction f_s of about 10%, no significant particle interaction occurs in liquid suspensions. Einstein analytically evaluated this phenomenon in his Ph.D. dissertation [36] and derived the following relationship for spherical particles:

$$\mu_{sus} = \mu_0(1 + 2.5f_s) \quad f_s \leq 0.1 \tag{1}$$

where μ_0 = viscosity of the pure liquid (assume melting point value for semi-solid materials)

μ_{sus} = viscosity of the suspension

Einstein also showed that this relationship is independent of particle size. Subsequent work by Thomas [37] at higher solid fractions lead to the following relationship (plotted in Fig. 10) based upon experimental data:

$$\mu^* = \mu_{sus}/\mu_0 = 1 + 2.5f_s + 10.05f_s^2 + 0.00273 \exp{(16.6f_s)} \tag{2}$$

The f_s^2 term results from agglomerates of two particles and adequately accounts for viscous effects up to a solid fraction of about 0.3. The exponential term accounts for higher order agglomerates and extends the usefulness of the equation to a solid fraction of about 0.6. In his review, Sigworth [35] describes several other models that have been proposed to describe the viscosity of liquid suspensions. Sigworth also advises against application of these relationships since they do not include the effects of shear strain rate and the Bingham fluid behavior commonly found in semi-solid materials. In addition, wall slip and the associated rejection of solid particles near the wall are ignored in these general relationships for liquid suspensions.

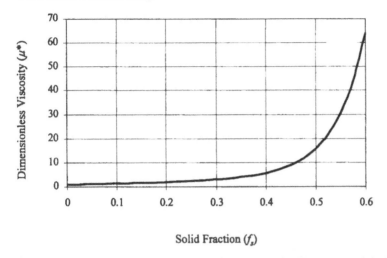

Solid Fraction (f_s)

Figure 10 Viscosity of liquid suspensions from the Thomas model. (From model found in Ref. 37.)

B. SSM Slurries

Researchers at MIT, led by Flemings, have long noted several characteristics of SSM slurries. The majority of the original work was completed under steady state shear strain rate conditions (i.e., after prolonged exposure to a fixed shear strain rate) after cooling from the liquid state to the desired solid fraction. Apparent viscosity is commonly used to characterize the fluid mechanical behavior of SSM materials. Apparent viscosity can best be described by use of Figs. 11 and 12. For a simple one-dimensional case, viscosity μ is defined as (see Fig. 11)

$$\tau = \mu \frac{du}{dy} \tag{3}$$

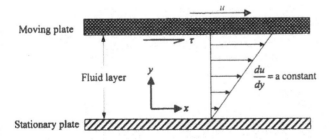

Figure 11 Definition of Newtonian viscosity.

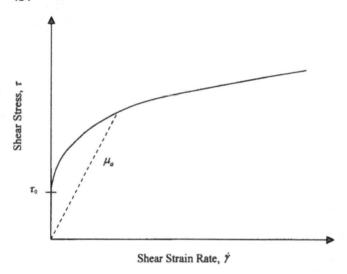

Figure 12   Typical shear stress relationship for SSM materials.

where τ = local shear stress in the fluid
          u = fluid velocity
          y = spatial coordinate cutting across streamlines

This relationship is equivalent to:

$$\tau = \mu\dot{\gamma} \tag{4}$$

where $\dot{\gamma}$ = local fluid shear strain rate. For a Newtonian fluid, μ is constant with $\dot{\gamma}$, which results in a very simple viscosity model (μ = a constant). SSM materials, however, do not behave in this manner. More typical behavior is shown schematically in Fig. 12, where the relationship between τ and $\dot{\gamma}$ is nonlinear and has a finite shear yield stress τ_0. A Newtonian fluid has a linear relationship between τ and $\dot{\gamma}$ with $\tau_0 = 0$. To accommodate the simplicity of the Newtonian model, SSM materials can be characterized by a Newtonian viscosity analogue known as the apparent viscosity μ_a, as shown in Fig. 12. In other words, the apparent viscosity is defined as the slope of the line through the origin to the point on the τ versus $\dot{\gamma}$ curve at a given shear strain rate. Although not an exact description of the fluid behavior, it is conveniently applied to standard, elementary fluid mechanics theory. Results for AISI 440C stainless steel and AISI 4340 low carbon steel are shown in Fig. 13.

Several characteristics are apparent in these *steady state* results. First, the apparent viscosity of the SSM slurry is well characterized by a power-law relationship with solid fraction. Note that the apparent viscosity is strongly

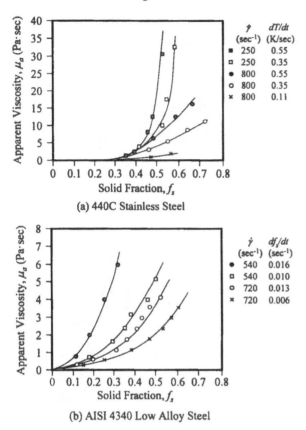

(a) 440C Stainless Steel

(b) AISI 4340 Low Alloy Steel

Figure 13 Fluid mechanical behavior of SSM steel alloy slurries over prolonged shearing at a constant shear strain rate. (From Ref. 2.)

dependent upon shear strain rate. This strong dependence is, however, partially due to the definition of apparent viscosity. As depicted in Fig. 12, one would expect larger viscosity values at lower shear strain rates due to the finite shear yield stress of the material. Various levels of sophistication have been applied to mathematically describe this behavior as noted in the models that follow. Another notable feature from the Fig. 13 results is the effect of initial cooling rate—in $K \cdot s^{-1}$ in Fig. 13a and df_s/dt in Fig. 13b. Clearly, more rapid cooling results in a more viscous fluid. This phenomenon occurs due to the formation of smaller grains in the SSM slurry during more rapid cooling. While most researchers have focused on the effects of the alloy system, shear strain rate, and solid fraction, only minimal work has been reported on viscosity models that include effects of grain size (or cooling rate) and shape.

1. Simple Shear Strain Rate Dependent Models Applied to SSM Slurries Cooled from the Liquid State

Several researchers have attempted to characterize the viscosity of SSM slurries. Early attempts accounted for effects of shear strain rate (and solid fraction) under prolonged shear deformation. This work served to identify many of the important process conditions that impact the behavior of SSM materials. The data are also useful in describing material response during casting of SSM billets. Data from those experiments that rely upon *cooling* to the SSM state are also useful in modeling rheocasting operations. Typically, power-law behavior was assumed in the corresponding viscosity models. The simplest form that has been proposed is [2, 38, 39]

$$\mu_a = K\dot{\gamma}^{n-1} \tag{5}$$

where n = power-law exponent
        K = consistency power-law index

When $n < 1$, shear thinning occurs; when $n > 1$, shear thickening occurs; and when $n = 1$, Newtonian flow (i.e., constant viscosity) occurs. For an aluminum alloy, Flemings [2] gives values for K and n under several conditions: (1) at steady state, $K = 30\,\mathrm{Pa \cdot s}^{0.1}$ and $n = 0.1$; (2) during continuous cooling and constant shear strain rate, $K = 2300\,\mathrm{Pa \cdot s}^{-0.5}$ and $n = -0.5$; and (3) for solidified and partially remelted material at low shear strain rates, $K = 39,80\,0\,\mathrm{Pa \cdot s}^{0.32}$ and $n = 0.32$. Ghosh et al. [38] give the following values under steady state, isothermal conditions and f_s less than 0.8 for magnesium alloys AZ91D and AM50, respectively, where μ_a is in Pa·s and $\dot{\gamma}$ is in s^{-1}: $K = 16.0 \times 10^{1.66 f_s}$ and $n = 0.85$ and $K = 4.58 \times 10^{2.15 f_s}$ and $n = 0.98$.

A more general form of the power-law model for a lead-tin alloy can be described as [40, 41]

$$\mu_a = -\left[K\left|0.5(\Delta:\Delta)^{1/2}\right|^{n-1}\right] \tag{6}$$

where

Δ = rate of deformation tensor

$(\Delta : \Delta)$ = dyadic product of Δ (i.e., a more complete definition of the three-dimensional effects of shear strain rate $\dot{\gamma}$)

$K = \exp(9.783 f_s + 1.4345)$

$$n = \begin{cases} 0.1055 + 0.41 f_s, & f_s < 0.30 \\ -0.308 + 1.78 f_s, & f_s \geq 0.30 \end{cases}$$

One of the few studies on defining viscosity for material reheated to the semi-solid state was completed by Loué et al. [42] using Al · 6Si · 0.3Mg and Al · 7Si · 0.3Mg alloys. In addition to evaluating heating strategies that are more consistent with typical commercial practice of producing shaped SSM parts, their work covered shear strain rates more typical of industrial processing of SSM parts. Using a backward extrusion apparatus and accepting an average shear strain rate and viscosity, they found that the materials have a pseudoplastic behavior (i.e., apparent viscosity decreases with increasing shear strain rate). That is, reheated and freshly deformed semi-solid Al · 6Si · 0.3Mg and Al · 7Si · 0.3Mg material responds to shear strain rate in a manner similar to SSM materials deformed under steady state conditions (i.e., long deformation times). Shear strain rates between $10 \, s^{-1}$ and $1000 \, s^{-1}$ were evaluated. Low shear strain rate tests (between $0.001 \, s^{-1}$ and $0.1 \, s^{-1}$) were also completed using standard compression test methods. They found that a simple power-law expression—Eq. (5)—fit all of the experimental data quite well with $n = 0.05$ and -0.06 for Al · 6Si · 0.3Mg and Al · 7Si · 0.3g, respectively. Values for K were found to be 44,000 $Pa \cdot s^{0.05}$ for Al · 6Si · 0.3Mg at a solid fraction of 0.55 and 3500 $Pa \cdot s^{-0.06}$ for Al · 7Si · 0.3Mg at a solid fraction of 0.45. Values for the material properties of the power-law relationship are summarized in Table 2.

Efforts to characterize SSM materials over a wide range of solid fractions and even higher shear strain rates can be conducted using the experimental technique outlined by Loué et al. [42]. (Independent validation of the model is recommended prior to widespread application in process design. One of the appealing aspects of the method is the ability to characterize freshly deformed and reheated material that represents the bulk of current industrial practice.) The uniform diameter plunger (see Fig. 14) is displaced at a constant speed to ensure consistent shear strain rate for material entering the annulus region between the plunger and container. The slope of the resulting force vs. extrusion time curve—shown schematically as Fig. 14c—remains approximately linear until the ram approaches the base of the container. The shear force increases linearly due to the linear growth rate of extruded material. The apparent viscosity can then be computed as follows:

$$\mu_a = \frac{1}{2\pi\lambda c_1 v_r} \frac{dF}{dt} \tag{7}$$

where
$$\lambda = \text{extrusion ratio} = r_c^2/(r_c^2 - r_p^2)$$
$$c_1 = \text{a constant} = [c_2(r_c^2 - r_p^2) - v_r]/\ln(r_p/r_c)$$
$$v_r = \text{ram speed}$$
$$dF/dt = \text{time rate of change of extrusion force}$$
$$r_c = \text{radius of container}$$
$$r_p = \text{radius of plunger}$$
$$c_2 = \text{a constant} = v_r/[(r_c^2 - r_p^2) - (r_c^2 + r_p^2)\ln(r_c/r_p)]$$

Table 2 Summary of Power-Law Parameters for SSM Materials

Material	Consistency power-law index K	Power-law exponent n	References	Comments
Al	$30\,\text{Pa}\cdot\text{s}^{0.1}$	0.1	2	Steady state
Al	$2300\,\text{Pa}\cdot\text{s}^{-0.5}$	-0.5	2	Continuous cooling; constant shear strain rate
Al	$39,800\,\text{Pa}\cdot\text{s}^{0.32}$	0.32	2	Solidified and partially remelted; low shear strain rates
AZ91D	$16.0\times10^{1.66f_s}$	0.85	38	Steady state and isothermal; μ_a in $\text{Pa}\cdot\text{s}$; $\dot{\gamma}$ in s^{-1}
AM50	$4.58\times10^{2.15f_s}$	0.98	38	Steady state and isothermal; μ_a in $\text{Pa}\cdot\text{s}$; $\dot{\gamma}$ in s^{-1}
Pb·Sn	$e^{9.783f_s+1.4345}$	$0.1055+0.41f_s$	40, 41	$f_s<0.30$; μ_a in $\text{Pa}\cdot\text{s}$; $\dot{\gamma}$ in s^{-1}
Pb·Sn	$e^{9.783f_s+1.4345}$	$-0.308+1.78f_s$	40, 41	$f_s\geq0.30$; μ_a in $\text{Pa}\cdot\text{s}$; $\dot{\gamma}$ in s^{-1}
Al·6Si·0.3Mg	$44,000\,\text{Pa}\cdot\text{s}^{0.05}$	0.05	42	K at $f_s=0.55$
Al·7Si·0.3Mg	$3500\,\text{Pa}\cdot\text{s}^{-0.06}$	-0.06	42	K at $f_s=0.45$
A356	$130,000e^{7.0(f_s-0.5)}\,\text{Pa}\cdot\text{s}^{0.1}$	0.1	49	Material heated to the semi-solid state
Sn·15%Pb	$10.5e^{20.6f_s}$	$1.78f_s-0.39$	10	$0.30<f_s<0.60$; μ_a in $\text{Pa}\cdot\text{s}$; $\dot{\gamma}$ in s^{-1}

The basic form of the power-law model is $\mu_a = K\dot{\gamma}^{n-1}$

The mean shear strain rate $\dot{\gamma}_{AV}$ is computed as

$$\dot{\gamma}_{AV} = \frac{c_1\left[\ln(-c_1/2c_2r_cr_p)-1\right]-c_2\left(r_c^2+r_p^2\right)}{r_c-r_p}. \tag{8}$$

Loué et al. [42] also give expressions for the apparent viscosity and average shear rate during compression of a cylindrical specimen as

$$\mu_a = \frac{2\pi h_\varepsilon^4 F_\varepsilon}{3V^2\dot{\varepsilon}} \tag{9}$$

and

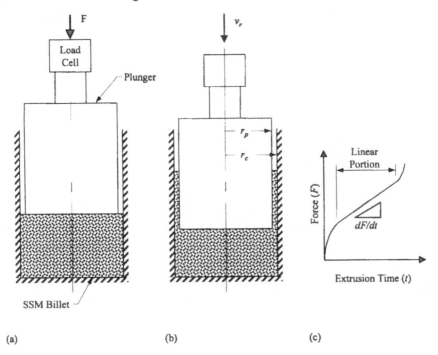

Figure 14 Back extrusion method to measure apparent viscosity. (a) Prior to Back Extrusion; (b) During Back extrusion; (c) Resulting Load Curve. (Based upon Ref. 42.)

$$\dot{\gamma}_{AV} = \frac{1}{2}\sqrt{\frac{V}{\pi h_\varepsilon^3}}\dot{\varepsilon} \tag{10}$$

where the tests are generally conducted under constant true normal strain rate $\dot{\varepsilon}$ and

h_ε = instantaneous specimen height
F_ε = instantaneous compressive force
V = sample volume (assumed to be constant)

Wang and coworkers [39] completed a curve fit to experimental data on Sn · 15%Pb. They noted that the apparent viscosity decreased with an increase in shear strain rate and approached an asymptotic value μ_∞ for large shear strain rates and prolonged shearing. Based upon these results, they proposed the following empirical relationship:

$$\mu_a = \mu_\infty(f_s)\left\{1 + \left[\frac{\dot{\gamma}^*(f_s)}{\dot{\gamma}}\right]^a\right\}^{(1-n_W)/a} \tag{11}$$

where $\quad \mu_\infty(f_s) \quad = \quad 0.0091 \exp(4.32 f_s)$
$\qquad\quad \dot{\gamma}^*(f_s) \quad = \quad$ shear strain rate where the asymptotic apparent viscosity is reached $= 525 \exp(2.35 f_s)$

Values for n_W and a need to be determined experimentally. Wang et al. [39] recommend values of 0.07 and 2, respectively, for Sn · 15%Pb under conditions of isothermal, steady state shear rate. Units for μ_a and μ_∞ in the Wang model are Pa · s, while $\dot{\gamma}$ and $\dot{\gamma}^*$ are in units of s^{-1}. An additional factor in Eq. (11) is recommended by Wang et al. [39] for material that is continuously sheared and continuously cooled:

$$\mu_a = \mu_\infty(f_s)\left(\frac{1-f_s}{f_s^*}\right)^{-m_W(\dot{\gamma})}\left\{1 + \left[\frac{\dot{\gamma}^*(f_s)}{\dot{\gamma}}\right]^a\right\}^{(1-n_W)/a} \tag{12}$$

where $\quad m_W(\dot{\gamma}) \quad = \quad 360/\dot{\gamma}$ ($\dot{\gamma}$ in s^{-1})
$\qquad\quad f_s^* \qquad\; = \quad$ solid fraction where the apparent viscosity approaches an infinite value ≈ 0.7

Okano [43] has recommended the following expression for continuously cooled (from the liquid state) and sheared SSM materials:

$$\mu_a = \mu_0\left[1 + \frac{\alpha \rho_m \dot{f_s}^{1/3} \dot{\gamma}^{-4/3}}{2\left(\frac{1}{f_s} - \frac{1}{0.72 - \beta f_s^{1/3} \dot{\gamma}^{-1/3}}\right)}\right] \tag{13}$$

where $\quad \rho_m \quad = \quad$ density of the alloy at the liquidus temperature
$\qquad\quad \dot{f_s} \quad\; = \quad$ average solidification rate (df_s/dt) up to $f_s = 0.4$

α and β are functions of the main solute concentration of the alloy x_K (in mass percentage), as follows:

$$\alpha = 203\left(\frac{x_K}{100}\right)^{1/3} \tag{14}$$

$$\beta = 19.0\left(\frac{x_K}{100}\right)^{1/3} \tag{15}$$

The material constants α and β were determined through regression analysis of apparent viscosity and various process conditions (including varying chemical compositions) for Al · Cu, Al · Si, Cu · Sn, and Fe · C alloys.

2. Kattamis Viscosity Model

A more advanced viscosity model has been proposed by Kattamis and co-workers [44, 45] for Al · 4.5Cu · 1.5Mg (alloy 2024) under steady state shear strain rate conditions. The Kattamis viscosity model is based upon an experimental curve fit to a general model that includes the relative motion of the particles and the energy associated with breaking agglomerate bonds. The model is expressed in terms of relative viscosity μ_r as follows:

$$\mu_r = \frac{\mu_a}{\mu_0} = 1 + \frac{kf_s(s_v/s_{v0})(\bar{d}/\bar{d}_0)}{k - f_s} + 2.0333 \times 10^4 f_s \dot{\gamma}^{-0.727} \tag{16}$$

where k =   limiting volume fraction for a suspension of isolated solid particles (0.5016 from experiments; $0.523 = \pi/6$ for packed uniform-diameter spheres)

s_v/s_{v0} =   normalized specific particle surface area; s_v = area of the solid/liquid interface per unit volume of solid, s_{v0} = value of s_v at the start of isothermal holding [44]

$\bar{d}/\bar{d}_0$ =   relative size of the particles ($\bar{d}$ = agglomerate diameter and $\bar{d}_0$ = average individual particle diameter)

and $\dot{\gamma}$ is in s^{-1}.

This model considers the effects of particle size and shape and degree of agglomeration through the factors s_v/s_{v0} and $\bar{d}/\bar{d}_0$ in the second term on the right-hand side of Eq. (16). In addition, agglomeration/disagglomeration effects due to shear strain rate are included in the third term on the right-hand side of the equation.

3. The BBH Model

Another class of material models has emerged and is based upon extensive experimental work. A model proposed by Barkhudarov, Bronisz, and Hirt [46] (defined here as the BBH model) assumes that the apparent viscosity decays exponentially toward the steady state value after shearing is initiated. Furthermore, they propose using the following transport equation for viscosity:

$$\frac{\partial \mu}{\partial t} + (\mathbf{u}\nabla)\mu = \frac{1}{\lambda_R}(\mu_{ss} - \mu) \tag{17}$$

where $\mathbf{u}$ =   material velocity vector

λ_R =   relaxation time constant

μ_{ss} =   steady state viscosity, which is a funciton of $\dot{\gamma}$ and f_s

The necessary experimental data include a relationship between apparent viscosity and shear strain rate under steady state conditions. This relationship is then used to compute μ_{ss} in Eq. (17). Barkhudarov et al. [46] recommend the following expression for μ_{ss}, based upon the work of Flemings [2] on Sn · 15%Pb at a solid fraction of 0.45:

$$\mu_{ss}(\dot{\gamma}) = \mu_{\infty}\left[1 + \left(\frac{\dot{\gamma}^*}{\dot{\gamma}}\right)^2\right]^{0.465} \tag{18}$$

where $\mu_{\infty} = 0.05\,\text{Pa}\cdot\text{s}$
        $\dot{\gamma}^* = 1880\,\text{s}^{-1}$

Extension of the BBH model through use of the Wang expression—Eq. (11)—above would allow for wider application of this model. An experimentally based relationship between shear stress τ and shear strain rate $\dot{\gamma}$ is also needed for the BBH model. This can be accomplished by using a single specimen to which a uniformly increasing shear strain rate is applied. The shearing should start at $\dot{\gamma} = 0$ and end at an upper bound consistent with the maximum shear strain rate expected during part production. This relationship between τ and $\dot{\gamma}$ is useful in determining the relaxation time constant. A plot of λ_R versus $\dot{\gamma}$ for Sn $\cdot$ 15%Pb is shown in Fig. 15. Using the definition of viscosity (i.e., $\tau = \mu\dot{\gamma}$) and evaluating an expression for $d\tau/d\dot{\gamma}$, a relationship for the relaxation time constant can be found from Eq. (17) (assuming uniform shear strain rate throughout the specimen):

$$\lambda_R = \frac{\dot{\gamma}(\mu_{ss} - \mu)}{d\dot{\gamma}/dt(d\tau/d\dot{\gamma} - \mu)} \tag{19}$$

Note that the relaxation time constant is dependent upon several variables including the temporal viscosity μ, solid fraction f_s, current shear strain rate

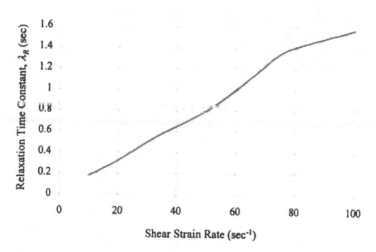

Figure 15 Relaxation time constant λ_R for Sn $\cdot$ 15%Pb at $f_s = 0.45$. Results for heated and unstrained material. (Data from Ref. 46.)

$\dot{\gamma}$, and the alloy. It is not clear if significant variations in $d\dot{\gamma}/dt$ will have a large effect on the relaxation time constant.

The above procedure requires experimental data at solid fractions that span the process to be simulated. This empirical approach can be useful in determining material behavior. Appropriate integration schemes should be employed to determine the apparent viscosity for SSM material that is (1) slowly deformed (on the order of 10–20 s or more) and (2) experiences significant variations in shear strain rate during processing. However, since industrial practice results in very rapid die filling (on the order of a few milliseconds to approximately 0.5 s), the material is not expected to progress very far toward a steady state behavior. Therefore, updating λ_R according to the local and temporal values of $\dot{\gamma}$, without time integration, should provide acceptable engineering predictions for material flow behavior during die filling.

4. Kapranos Model

A similar viscosity model to the BBH model was used by Kapranos et al. [47]. Using rapid compression testing, they measured the thixotropic behavior of aluminum alloy A357 near $f_s = 0.5$ and shear strain rates in the 3 to 60 s^{-1} range. These data were then used to fit a transport equation for the apparent dynamic viscosity given by

$$\frac{d\mu}{dt} = \left(\frac{1}{\lambda_R}\right)(\mu_{ss} - \mu) \tag{20}$$

The expression from Quaak et al. [48] for steady state viscosity was assumed:

$$\mu_{ss} = 9.0 \exp(9.5 f_s)\dot{\gamma}^{-1.3} \tag{21}$$

where, μ_{ss} has units of Pa · s and $\dot{\gamma}$ has units of s^{-1}. The initial viscosity μ_i was fit to the initial peak in the load curve. The fitted parameters are summarized in Table 3. Kapranos et al. [47] found the time dependent form given by Eq. (20) to provide a significantly better fit to the experimental data than a constant viscosity or a non-Newtonian model.

Table 3 Material Constants in Kapranos Model for Aluminum Alloy A357

Temperature (°C)	f_s	μ_i (Pa · s)	λ_R (s)
572	0.538	3.2×10^4	0.05
574	0.525	1.6×10^4	0.05
576	0.513	0.2×10^4	0.05

Source: Ref. 47.

5. Note About Models for Cooled Materials

The above models (simple shear strain rate and Kattamis models) accurately represent behavior of SSM materials under many conditions including steady state behavior and continuously cooled (from the pure liquid state) conditions. Such models are useful for evaluation and design of raw material production equipment. However, the conditions used to develop these models are rarely present during commercial production of shaped parts using the SSM process. In commercial conditions, the material is subjected to temporal and spatial variations in process conditions that require complex relationships for K and n. Process history must also be included in the expressions for K and n for the above material models to have wide spread utility in industrial applications.

6. Simple Shear Strain-Rate Dependent Models Applied to SSM Slurries Heated from the Solid State

Flemings [2] recommends $K = 39,800 \, \mathrm{Pa \cdot s}^{0.32}$ and $n = 0.32$ for an aluminum alloy that has been reheated to the semi-solid state. These values, however, apply only at low shear strain rates and for material that has been sheared for a prolonged time. Nickodemus et al. [49] have reported preliminary results of work aimed at predicting the behavior of heated SSM slugs having no prior shearing. A356 billets were heated to various solid fractions ($0.30 \le f_s \le 0.70$) and then forward extruded through conical dies at average shear strain rates between 600 and $6000 \, \mathrm{s}^{-1}$. These conditions are typical of commercial SSM production. Their results were then fit to the following viscosity model:

$$\mu_a = K e^{B(f_s - 0.5)} \dot{\gamma}^{n-1} \tag{22}$$

where B = material constant whose value was chosen to be 7.0.

Since the experimental process involved nonuniform shear stress, numerical simulations were made using Eq. (22) as the viscosity model. Trial and error was then used to determine values of K and n that would match measured pressure curves within 10%. The findings indicated that within the processing conditions evaluated, a wide range of values for K and n satisfy the viscosity relationship given by Eq. (22). The following pairs of values were found to predict pressure within 10% with μ_a in Pa·s and $\dot{\gamma}$ in s^{-1}: (1) $K = 20,000 \, \mathrm{Pa \cdot s}^{0.4}$, $n = 0.4$, (2) $K = 80,000 \, \mathrm{Pa \cdot s}^{0.2}$, $n = 0.2$, and (3) $K = 130,000 \, \mathrm{Pa \cdot s}^{0.1}$, $n = 0.1$. Obviously, other combinations of values will also satisfy the desired experimental accuracy. Nickodemus et al. [49] recommend extending the range of shear strain rates over several orders of magnitude to tighten the uncertainty associated with values for K and n.

Although much uncertainty remains in the Nickodemus results (note that further work is expected in the near future), their findings are useful in further understanding the behavior of SSM materials. Under conditions of steady state shear strain rate, Flemings [2] gives values for K and n as $30\,\text{Pa} \cdot \text{s}^{0.1}$ and 0.1, respectively for an unspecified aluminum alloy. The curve fit with $n = 0.1$ of Nickodemus (with initially unsheared material) has a K value of $130,000\,\text{Pa} \cdot \text{s}^{0.1}$, which is 4300 times larger than the Flemings value. Although Flemings doesn't indicate any restrictions on solid fraction in his values, the solid fraction effect in Eq. (22) would account for, at most, a factor of 4.0 (at a solid fraction of 0.3). Therefore, within the experimental conditions of the Nickodemus model, a factor of approximately 1000 exists between the viscosity of heated, unsheared aluminum SSM material and similar material that has been sheared for a long period of time (i.e., under steady state conditions). This points out the need on the part of the numerical analyst to carefully select material models that have been developed under SSM processing conditions similar to those present in the process being modeled.

Laxmanan and Flemings [10] evaluated viscous behavior for thixocast (i.e., solidified and reheated) $Sn \cdot 15\%Pb$ alloys by compressing cylindrical specimens between parallel plates. Their experimental methods only allowed for low shear strain rates ($1 \times 10^{-5}\,\text{s}^{-1} < \dot{\gamma} < 0.1\,\text{s}^{-1}$), which is below typical production values by several orders of magnitude. Of particular importance in their work, however, is the study of thixocast material, which is consistent with industrial production of SSM parts. Their findings resulted in several important conclusions:

1. SSM material having columnar solid particles require deformation loads that are roughly 2 orders of magnitude higher than that for equiaxed solid particles.
2. Thixocast SSM materials require deformation loads that are about 3 orders of magnitude higher than those for rheocast (i.e., cooled to the SSM state from pure liquid and under constant shearing) material.
3. The viscosity of thixocast $Sn \cdot 15\%Pb$ can be expressed as

$$\mu_a = Ke^{Bf_s}\dot{\gamma}^{c_L f_s + d_L} \tag{23}$$

where c_L and d_L are material constants that take on the following values: $c_L = 1.78$ and $d_L = -1.39$ for $0.30 < f_s < 0.60$.

Other material constants recommended by Laxmanan and Flemings [10] in this model are: $K = 10.5$ and $B = 20.6$ for $0.15 < f_s < 0.60$. Units for viscosity are $\text{Pa} \cdot \text{s}$, while units for $\dot{\gamma}$ are s^{-1}. Extrapolation of Eq. (23) beyond the shear strain rate limits of the experiments ($1 \times 10^{-5}\,\text{s}^{-1} < \dot{\gamma} < 0.1\,\text{s}^{-1}$) must be done with caution. Laxmanan and Flemings [10] thoroughly described their experimental methods. The interested reader is referred to this work in preparing similar experimental evaluation of material behavior for determination of material constitutive models.

7. Bingham Fluid Models

Figure 16 shows the relationship between shear strain rate and shear stress for a Bingham fluid. This model assumes that the fluid is free to flow only after the shear stress exceeds a yield value. After yielding, a linearly increasing relationship is assumed between shear strain rate and shear stress. The apparent viscosity can then be evaluated from

$$\mu_a = \mu_B + \frac{\tau_0}{\dot{\gamma}} \tag{24}$$

where μ_B = Bingham viscosity (i.e., the slope of the line describing the
               Bingham fluid behavior)
        τ_0 = shear yield stress of the fluid

When the local shear stress is lower than τ_0, no relative motion between particles occurs and plug flow develops. From experimental observations of SSM behavior, it is clear that μ_B and τ_0 are heavily dependent upon other processing parameters, most notably shear strain rate, local solid fraction, prior processing history, and the SSM alloy.

A phenomenological model to describe Bingham fluid behavior of SSM materials has been proposed by Alexandrou and Ahmed [50, 51]. To account for shear strain rate dependence, a power-law model is used. A mathematical relationship that provides a single-valued function for τ given $\dot{\gamma}$ was estab-

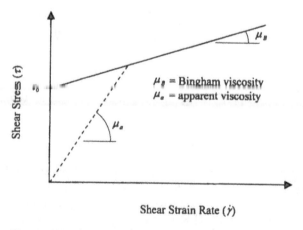

Figure 16 Comparison between Bingham viscosity model and apparent viscosity model.

lished. In addition, the discontinuity represented by a finite shear yield stress τ_0 is known to create stability and convergence difficulties during numerical simulation of the process. Hence, this behavior was approximated by a function that goes through the origin and quickly rises to τ_0 for an arbitrarily small shear strain strain rate. To this end, the following relationship for the shear stress tensor τ is proposed for Bingham fluid behavior:

$$\tau = \left\{ \mu_B + \frac{\tau_0 \left[1 - \exp\left(-m_R |\sqrt{D_{II}/2}| \right) \right]}{|\sqrt{D_{II}/2}|} \right\} D \tag{25}$$

where τ_0 = characteristic shear yield stress of the Bingham fluid

m_R = exponent that controls how quickly the model ramps from zero shear stress at a zero shear strain rate to τ_0 (Large values of m_R ramp to τ_0 at lower shear strain rates.)

D_{II} = second invariant of the rate of deformation tensor D, where D is defined as $D = [\nabla u + (\nabla u)^T]$ and u is the material velocity vector.

The second invariant of the rate of deformation tensor D_{II} represents the magnitude of the shear strain rate. In two-dimensional rectangular coordinates D_{II} is defined as:

$$D_{II} = 4 \left(\frac{\partial u}{\partial x} \right)^2 + 2 \left(\frac{\partial u}{\partial y} + \frac{\partial v}{\partial x} \right)^2 + 4 \left(\frac{\partial v}{\partial y} \right)^2 \tag{26}$$

The following equation applies to three-dimensional systems:

$$
\begin{aligned}
D_{II} = 4 &\left[\left(\frac{\partial u}{\partial x} \right)^2 + \left(\frac{\partial v}{\partial y} \right)^2 + \left(\frac{\partial w}{\partial z} \right)^2 \right] \\
+ 2 &\left[\left(\frac{\partial u}{\partial y} + \frac{\partial v}{\partial x} \right)^2 + \left(\frac{\partial u}{\partial z} + \frac{\partial w}{\partial x} \right)^2 + \left(\frac{\partial v}{\partial z} + \frac{\partial w}{\partial y} \right)^2 \right]
\end{aligned}
\tag{27}
$$

where u, v, and w are the velocity components in the x-, y-, and z-directions, respectively.

D_{II} can be used to evaluate when yielding has occurred since a zero value indicates no relative movement of local material, while a positive value requires a velocity gradient, and therefore, local yielding of the Bingham fluid. Rather than strictly enforcing this condition, Alexandrou and Ahmed [51] consider yielding to occur when D_{II} is greater than a small, positive value. This minor modification is required to accommodate the continuous functional relationship in Eq. (25) that approximates the finite shear stress of a Bingham fluid, as noted in Fig. 16. Alexandrou and Ahmed [50, 51] recommend a shear dependent viscosity model of the form:

$$\mu_B = K\left(\frac{D_{11}}{2}\right)^{(n-1)/2} \tag{28}$$

where K = consistency power-law index
        n = power-law exponent

For $n < 1$, shear thinning occurs; for $n > 1$, shear thickening occurs, and for $n = 1$, Newtonian flow (i.e., constant viscosity) occurs. The parameters τ_0, m, K, and n are obtained from best curve fits to experimental material test data. Sigworth [26] recommends the following relationship for the Bingham shear yield stress for semi-solid Al · Si alloys:

$$\tau_0 = 9615\frac{f_s^3}{\Phi_m - f_s} \tag{29}$$

where τ_0 has units of Pa and Φ_m = maximum packing fraction for uniformly sized spherical particles and is equal to 0.6.

Sigworth also recommends the following Bingham viscosity model for Al · Si alloys:

$$\mu_B = \mu_0\left(1 - \frac{f_s}{\Phi_m^\circ}\right)^{-2} \tag{30}$$

where μ_0 = viscosity of the liquid Al · Si ($= 0.0013\,\text{Pa} \cdot \text{s}$)
        Φ_m° = empirical constant equal to 0.68

Results of the continuous Bingham model of Alexandrou and Ahmed have been reported in Ref. 50. Figure 17 shows how well the model predicts the behavior of Sn · 15%Pb. In addition, the effect of the value chosen for m_R is shown schematically in Fig. 18. Note that larger values of m_R more closely behave like Bingham fluids. Large values, therefore, more accurately represent the flow behavior observed in SSM materials. However, larger values of m_R require more computational resources (finer meshes and smaller time steps) to avoid numerical convergence and stability problems so that the model can correctly capture the sharp shear stress gradients that result.

The utility of the model by Alexandrou and Ahmed is that it can be more easily incorporated into numerical schemes than can the classical discontinuous Bingham fluid model. The Alexandrou and Ahmed model has predicted that during filling of a long tube, a liquid rich region develops next to the wall. These findings are consistent with the predictions of Sigworth [35] based upon minimizing the dissipation of flow energy during filling. The phenomenological model of Alexandrou and Ahmed [50, 51] also predicted plug flow in the bulk of the flow in a long tube. Higher shear stresses develop near the surface of the tube wall. This forces the more viscous (i.e., higher solid fraction) material away from the wall, thus allowing a thin, pure liquid metal boundary layer

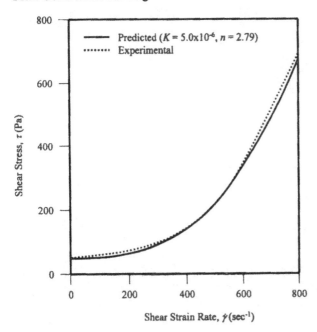

Figure 17 Comparison between the continuous Bingham model and experimental data for Sn · 15%Pb. (From Ref. 50.)

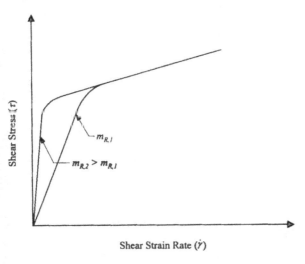

Figure 18 Shear stress vs. shear strain rate for various values of parameter m_R in the continuous Bingham model. (Based upon Refs. 50 and 51.)

to develop next to the tube wall. The material appears to slip (or slide) along over this thin liquid metal layer resulting in a flow pattern different than that of a fluid with a uniform solid fraction throughout the flow cross section. Sigworth [35] recommends the following relationship for the slip velocity V_w, that is, the particle velocity at the junction of the particles and the bounding liquid metal layer:

$$V_w \approx \frac{0.65 D_A}{\mu_0} \tau_w \tag{31}$$

where D_A = average particle diameter for uniformly sized particles or the smallest particle diameter when a wide range of particle sizes is present

   τ_w = shear stress at the wall

Note that this relationship applies to long regions of constant cross section such as Couette viscometers or long channels. Freezing of this outer layer is not considered in the above relationship, nor are transient effects or entrance effects. Nonetheless, Sigworth's cautions about the effects of wall slip are important as they may influence experimental measurements especially for materials subjected to prolonged (i.e., steady state) shearing. In addition, some consideration of wall slip may be necessary in simulation of die filling. Unfortunately, the scientific understanding of wall slip in not well developed for mold filling simulations.

8.   Internal Variable Viscosity Models

When the solid fraction of SSM materials is greater than about 10%, particle agglomeration occurs. The level of agglomeration increases with solid fraction and time at rest. Agglomeration decreases with an increase in shear strain rate. The agglomerated particles effectively form larger particles. In addition, some of the liquid becomes entrapped between particles, which increases the effective solid fraction of the mixture. The associated changes in fluid behavior can be significant as a result of these factors, which by their nature are process history related. Changes in the agglomeration structure of *continuously stirred* SSM slurries have been characterized with time constants that are generally on the order of tens of seconds. For those industrial processes involving continuously stirred material with relatively rapid part formation, the expressions provided in earlier sections of this chapter should provide sufficiently accurate information for process design. However, for processes that slowly deform the SSM material, especially those with widely varying shear strain rates in either time or space, the above expressions may be insufficient. Clearly, significant structural changes occur during the initial deformation of SSM billets during thixocasting, as discussed later in this chapter on high solid fraction processing.

Methods that account for the evolving agglomeration/particle structure provide a more accurate representation of SSM behavior. A general class of models that takes structure into account uses internal variables that correlate with the degree of particle agglomeration and the corresponding effect on material behavior.

Most of the earlier models discussed above focus on the steady state behavior of SSM slurries. Commercial processing of SSM materials is highly transient, however. Typically, a heated billet with little or no prestraining is pushed into a die cavity within a few tenths of a second. Consequently, most of these earlier material models do not adequately describe typical production processes used for the manufacture of shaped parts. A new generation of material models has been developed to describe the transient behavior of SSM materials, however. A large class of these models uses an internal variable approach [40, 41, 52–61]. In these models, the material response depends upon shear strain rate $\dot{\gamma}$, temperature T or solid fraction f_s, and internal variables s_i that describe the current material structure. The internal variables are used to describe kinetics-based effects that control the macroscopic behavior of the material. In principle, any number of these internal variables can be used. However, the only one typically considered in these models, due to its significant effect on SSM material response, is one that describes the relative amount of solid particle agglomeration. This parameter s is related to the number of actual particle bonds divided by the number of bonds that would be present if all particles were fully bonded in a cubic matrix [52]. This nondimensional internal variable is defined with arbitrary upper and lower numerical bounds. Most common are models having a value between 0 and 1 with $s = 0$ describing a fully disagglomerated condition (i.e., no particles joined) and $s = 1$ describing a completely agglomerated state (i.e., all particles joined into one large network). Other potentially useful internal variables for describing the behavior of SSM slurries include those relating to grain size (apparent viscosity decreases with an increase in particle size for the same solid fraction [27]), shape, and distribution, in addition to local chemistry variation resulting from segregation of the liquid and solid phases. Effects of process history can easily be captured and updated with internal variables allowing for a more complete description of SSM behavior.

The general form of the model proposed by Brown and coworkers [53, 55–57] is described here. Their approach was developed assuming spherical particle morphology. However, the method can be successfully applied to more general particle shapes. In the model one defines the shear stress τ as

$$\tau = f\left(\dot{\gamma}, T, s_1, s_2, s_3, \ldots, s_{k_I}\right) \tag{32}$$

where f is a flow equation and $s_i, i = 1, k_I$ represent k_I internal variables that are defined through individual rate equations as:

$$\frac{ds_i}{dt} = \dot{s}_i = \hat{g}_i\left(\dot{\gamma}, T, s_1, s_2, s_3, \ldots, s_{k_I}\right) \qquad 1 \le i \le k_I \tag{33}$$

and $\hat{g}_i$ is the evolution equation for internal variable s_i.

The flow equation is derived by considering the flow resistance associated with (1) particles bonding or debonding due to shear force and (2) hydrodynamics associated with flow around individual particles and agglomerates. The only internal variable used by Brown and coworkers [53–57], s, is one associated with the relative amount of particle agglomeration. Values for s span the range [0, 1]. Ideally, values for s would be measured directly under varying process conditions. Evaluation of the number of isolated particles or the average number of particle bonds per particle can provide a qualitative measure of s. However, as used and defined in the following formulations, s has no known observable microstructural feature [57]. The resulting general form for the flow equation is

$$\tau = f(\dot{\gamma}, T, s) = A(s)\frac{\left(c/c_{max}\right)^{1/3}}{1 - \left(c/c_{max}\right)^{1/3}}\mu_0\dot{\gamma} + (n+1)C(T)sf_s\mu_0^{n+1}\dot{\gamma}^n \tag{34}$$

where τ = shear stress

$A(s)$ = hydrodynamic coefficient that depends upon particle size, distribution, morphology, and geometric arrangement of the agglomerates

c = $f_s(1 + 0.1s)$ = effective solid fraction (includes effect of fluid trapped in agglomerates) (Note that other researchers [40, 41] use a coefficient of 0.25 for s in this expression.)

c_{max} = $0.625 - 0.1s$ = maximum effective solid fraction at a given level of agglomeration (Other sources [54, 56, 58] recommend a fixed value of 0.64 for c_{max}, while others [40, 41] use a fixed value of 0.625.)

n = power-law exponent used to describe the rate dependent deformation of the solid particles at the junction of particle bonds ($1 \le n \le 5$ [52]; $n \approx 4$ to 5 for many face-centered cubic metal alloys near the solidus temperature [56].)

$C(T)$ = collection of constants relating to particle geometry and the temperature dependent plastic deformation power-law coefficient for solid particles

Note that this model predicts shear rate thickening for constant structure (i.e., s = a constant) material. Kumar et al. successfully demonstrated this effect experimentally with a Couette rheometer [53, 55, 56].

The first term on the right-hand side of Eq. (34) accounts for hydrodynamic energy dissipation effects among particles and agglomerates including effects of entrapped fluid, particle collisions, and fluid flow around particles. The second term accounts for energy dissipation associated with plastic deformation within the agglomerate structure—mainly creation of and breaking of particle bonds. This second term, therefore, is used to introduce the time dependent, structure effect of SSM slurries. A linear estimate of variable A has been suggested [56] as

$$A(s) = \lambda_1 s + \lambda_2 \tag{35}$$

where λ_1 and λ_2 are experimentally determined material constants, as discussed below. The expression for $C(T)$ has been given as

$$C(T) = C_0 \exp\left(\frac{nQ}{R^* T}\right) \tag{36}$$

where C_0 = coefficient whose value must be determined experimentally
$\quad\quad\quad Q$ = an activation energy
$\quad\quad\quad R^*$ = universal gas constant (8.314 kJ/kgmol $\cdot$ K)

Note that as the value of c approaches c_{max}, the hydrodynamic term—the first term on the right-hand side of Eq. (34)—approaches an infinite value. In addition, if c is greater than c_{max}, the hydrodynamic term becomes negative and consequently looses its physical meaning. Therefore, this equation can only be used for $f_s < 0.5$ to 0.6.

The rate equation for s is defined as [57]

$$\frac{ds}{dt} = H(T, f_s)(1 - s) - R(T, f_s) s \dot{\gamma}^n \tag{37}$$

where $H(T, f_s)$ = agglomeration kinetics parameter
$\quad\quad\quad R(T, f_s)$ = disagglomeration kinetics parameter

Note that the agglomeration factor $H(T, f_s)$ is scaled by a factor of $1 - s$ in Eq. (37). This is reasonable since the probability of forming new agglomerates will increase with a greater number of unattached particles (i.e., lower value of s and higher values of the factor $1 - s$). Using similar reasoning, the probability of particle disagglomeration increases with larger values of both s and $\dot{\gamma}$. Therefore, factors of s and $\dot{\gamma}$ appear with the disagglomeration parameter $R(T, f_s)$ in Eq. (37). A more comprehensive expression for H may also include particle diameter, matrix fluid viscosity, and shear rate [52]. The expression for R may also include matrix fluid viscosity [52].

Values for these material models are given in Table 4 for selected materials. These data were derived from curve fits to experimental data established using a high-temperature Couette rheometer [40, 41, 53–56, 58]. An initial "steady state" structure was obtained by cooling from the liquid state to the temperature corresponding to the desired solid fraction while maintaining a

Table 4 Recommended Values for the Internal Variable Model

Property	Units	Sn · 15wt%Pb @ 460 K [54–56, 58]	Al alloys [55, 58]	Al · 5Cu with B₄C particles [40, 41]
n	—	4	5	5
Q	kJ/kgmol	74.52	167	—
μ_f	Pa · s	0.0025	0.001	—
C_0	Pa^{-n} · s^{-1}	1.971×10^{-28}	—	—
M	—	—	—	2.5×10^4
H	s^{-1}	$1.46 f_s^{5.812}$	—	0.1
R	s^{n-1}	$9 \times 10^{-11} f_s^{4.702}$	—	—
G	—	—	—	0.01
c_{max}	—	0.64	0.64	0.625
$A(s)$	—	$24s + 16$	8×10^{11}	$\frac{9}{8}$

Note that the form of the rate equation and viscosity for Al · 5Cu with B₄C particles was [40, 41]:

$$\frac{ds}{dt} = H(1 - s) - Gs^2 \dot{\gamma} \tag{38}$$

and

$$\mu^* = \frac{\mu}{\mu_0} = A\left[\frac{(c/c_{max})^{1/3}}{1 - (c/c_{max})^{1/3}}\right] + Mf_s \dot{\gamma}^{(1-n)/n} s \tag{39}$$

desired shear strain rate. After reaching the desired solid fraction, shearing was maintained until the torque became constant. The rotational speed of the outer rotating cup was designed to quickly change speeds to permit sudden jumps or drops in shear strain rate and allow determination of transient shear stress as a result of developing agglomerate structure. Brown and coworkers [53, 55, 56] found that the rate of change of structure (on the order of tens of seconds) is significantly slower during shear strain rate drops than during shear strain rate jumps. In addition, they determined that the transient response of the fluid velocity gradients in a well-designed rheometer is orders of magnitude smaller than the transient development of internal SSM agglomeration structure. Therefore, one is able to determine the viscosity of a constant structure SSM material over a range of shear strain rates using a ramp change in shear strain rate. Curve fits to the corresponding shear strain rate $\dot{\gamma}$ and shear stress τ over a range of solid fractions enable a best-fit estimate for λ_1, λ_2, C_0, Q, and s. Results from Brown and coworkers [56] are presented in Figs. 19 through 21. The model does very well at capturing the material response for Sn · 15wt%Pb at all solid fractions shown (Fig. 19). The model also does well at solid fractions less than 0.4 for the Al · 7Si · 0.6Mg alloy (Fig. 20). Above a solid fraction of about 0.4 (Fig. 21), the Bingham fluid response

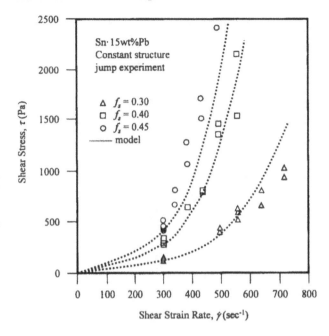

Figure 19   Constant structure material response of Sn · 15%Pb. (Reprinted from Ref. 56 with permission from Elsevier Science.)

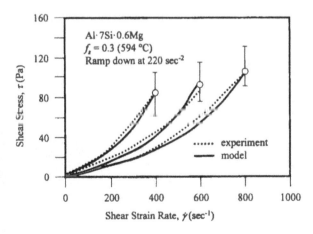

Figure 20   Comparison between the Brown internal variable model and experimental data for Al · 7Si · 0.6Mg at $f_s = 0.3$. (Reprinted from Ref. 56 with permission from Elsevier Science.)

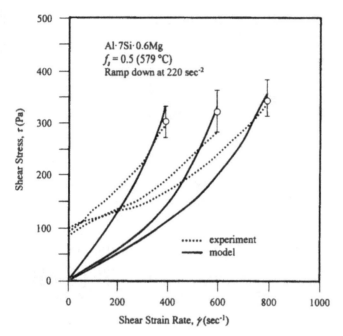

Figure 21 Comparison between the Brown internal variable model and experimental data for Al · 7Si · 0.6Mg at $f_s = 0.5$. (Reprinted from Ref. 56 with permission from Elsevier Science.)

noted by Alexandrou and Ahmed [50, 51] becomes significant, and therefore should be included for a more complete material model.

Perhaps the most limiting aspect of the internal variable approach, as defined above, is a clear definition of the initial condition for *heated* SSM billets just prior to being formed into a shaped component. A relationship between initial solid fraction and the initial value of the agglomeration factor s is necessary. Zavaliangos and Lawley [52] showed through numerical simulation that deformation loads predicted by these models vary by as much as a factor of 2, depending upon the assumed initial value of s. Additional complications may arise in defining the initial distribution of the agglomeration parameter in industrial applications due to nonuniform thermal conditions present in billets at the start of the casting or forming operation. Simple upset experiments can be used to define an appropriate starting value for the agglomeration factor at a given solid fraction. These tests should also be run to ensure that the material responds in a manner consistent with that predicted by Eqs. (34) and (37).

The material constants λ_1, λ_2, C_0, and Q, along with the agglomeration and disagglomeration kinetics parameters $H(T,f_s)$ and $R(T,f_s)$, in the above internal variable constitutive model can be systematically determined one solid fraction (i.e., temperature) value at a time as recommended by Kumar et al. [56, 57]. Notationally, we will define solid fraction i and initial shear strain rate j in the following discussion.

1. Use the ramp down shear strain rate test at several initial shear strain rates to obtain τ versus $\dot{\gamma}$ for a specified value of solid fraction. Since a steady state internal structure is obtained prior to actually ramping $\dot{\gamma}$ and assuming that the test is completed quickly (usually within 2–5 s), the structure remains constant and $ds/dt = 0$. (Note that ramp down experiments are used since the change in structure is much slower than during ramp up experiments. Flemings results [2] suggest that the difference in magnitude is about 10 to 1.) Results are shown schematically in Fig. 22.

2. For every measured τ versus $\dot{\gamma}$ curve (Fig. 22), use a suitable curve fitting procedure to determine values for parameters $A'_j(s, f_s)$ and $C'_j(s, T)$ for the ramp down shear stress results in the following modified form of Eq. (34), where μ_0, n, and $f_{s,i}$ are constant, known values for a given curve:

$$\tau = A'_j \mu_0 \dot{\gamma} + (n+1) C'_j f_{s,i} \mu_0^{n+1} \dot{\gamma}^n \tag{40}$$

where

$$A'_j = A'(s_j, f_s) = A(s_j) \frac{(c/c_{max})^{1/3}}{1 - (c/c_{max})^{1/3}} \tag{41}$$

$$C'_j = C'(s_j, T_i) = s_j C(T_i) = s_j \left[C_0 \exp \left(\frac{nQ}{R^* T_i} \right) \right] \tag{42}$$

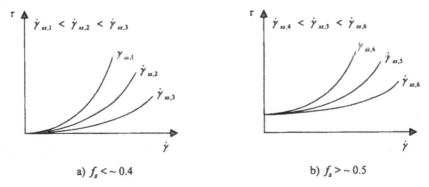

a) $f_s < \sim 0.4$ b) $f_s > \sim 0.5$

Figure 22 Schematic representation of measured shear stress τ as a function of shear strain rate $\dot{\gamma}$ at constant agglomeration structure.

Since the structure is constant during each individual ramp down experiment and the solid fraction (i.e., temperature) is held constant during each individual experiment, the coefficients A_j' and C_j' in Eq. (40) are constant for each individual ramp down experiment. Pairs of initial steady state, shear strain rate values ($\dot{\gamma}_{ss,j}$) and A_j' values can be plotted. Pairs of $\dot{\gamma}_{ss,j}$ and C_j' values can also be plotted. Curves through these points are shown schematically in Fig. 23. Note that the form of Eq. (40) is presupposed and during an individual ramp down experiment, the only independent variable is $\dot{\gamma}$.

   3. From Eq. (37) at steady state structure, H and R take on constant values H_j and R_j. Equation (37) then reduces to

$$\frac{ds_j}{dt} = 0 = H_i(1 - s_j) - R_i s_j \dot{\gamma}_{ss,j}^n \tag{43}$$

or

$$s_j = \frac{1}{1 + (R/H)_i \dot{\gamma}_{ss,j}^n} \tag{44}$$

Note that the initial steady state shear strain rate $\dot{\gamma}_{ss,j}$ is used since it corresponds to the present value of s_j (which is assumed constant during the experiment). Therefore, from Eq. (42) under steady state shear strain rate, where the subscript j has been dropped for a more general relationship,

$$C'(s, T_i) = \frac{1}{1 + (R/H)_i \dot{\gamma}_{ss}^n} C(T_i) \tag{45}$$

Pairs of C_j' and $\dot{\gamma}_{ss,j}$ values can then be used to determine the best fit for $(R/H)_i$ and $C(T_i)$ in this relationship. Note that this expression yields a shear rate-thinning behavior, which is consistent with the phenomena reported by other

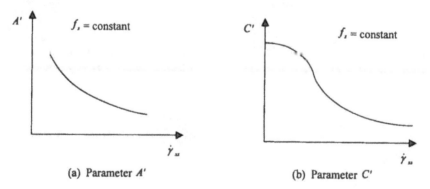

(a) Parameter A' (b) Parameter C'

Figure 23 Schematic representation of the effect of initial steady state shear strain rate $\dot{\gamma}_{ss}$. (a) Parameter A'; (b) Parameter C'.

researchers [2, 35] at steady state. Again, note that the form of Eq. (45) is presupposed, and values of $(R/H)_i$ and $C(T_i)$ that best match this form to the curve schematically represented in Fig. 23b are desired. At this point, we have values for A'_j, C'_j, $C(T_i)$, and $(R/H)_i$ for the solid fraction currently being evaluated.

4. s_j values can be computed from Eq. (44). Using these values and results from step 2, plot A'_j versus s_j and draw a curve through these points as shown schematically in Fig. 24. Combining Eqs. (35) and (41) and using the defined expression for c and c_{max}, we obtain the following equation expressed in its general form:

$$A' = (\lambda_1 s + \lambda_2)\frac{\left[\frac{f_i(1+0.1s)}{0.625-0.1s}\right]^{1/3}}{1-\left[\frac{f_i(1+0.1s)}{0.625-0.1s}\right]^{1/3}} \tag{46}$$

Using the pairs of A'_j and s_j values used to create a plot like Fig. 24, find the best fit for λ_1 and λ_2 in Eq. (46). Note that once again we have a presupposed form for the relationship in Eq. (35). At this point, we have estimates for A'_j, C'_j, s_j, $(R/H)_i$, $C(T_i)$, λ_1, and λ_2 for the solid fraction i evaluated to this point.

5. Repeat the above procedure at additional solid fraction values.

6. Using Eq. (36), values for C_0 and Q can be computed from a curve fit to the T_l, $C(T_i)$ data pairs. [Recall $C(T_i)$ values were computed in step 3.] One may be tempted to consider $C(T)$ constant over the temperature range in which metallic alloys are in the semi-solid state. Even though this temperature range is typically a small fraction of the absolute temperature within this range, the factor nQ/R^* is in the tens of thousands of Kelvins. Consequently, small

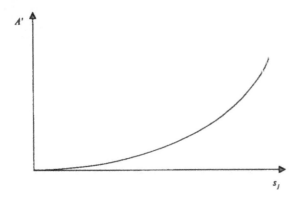

Figure 24 Schematic representation of parameter A' as a function of initial value of the agglomeration structure parameter s_j.

variations in temperature have significant effects on the value of $\exp(nQ/R^*T)$ and therefore on the computed shear stress.

7. Verify the complete model using the experimentally measured steady state shear stress values at various shear strain rates. Additional modifications to λ_1, λ_2, C_0, Q, and $(R/H)_i$ values may be necessary to provide the best overall fit of the model to the experimental data. Alternatively, one may wish to have solid fraction dependent values for λ_1, λ_2, C_0, and Q.

8. We can establish separate values for H_i and R_i by running a shear rate jump experiment. After reaching steady state for an initial shear strain rate $\dot{\gamma}_0$, an initial value of the agglomeration parameter is established as s_0. After the sudden shear rate jump to $\dot{\gamma}_f$, the agglomeration parameter changes as defined in Eq. (37). Upon prolonged shearing at $\dot{\gamma}_f$, the SSM slurry reaches a new steady state condition with an agglomeration parameter value of s_f. Mathematically, these conditions can be expressed as

$$@t = 0, \ s = s_0 \tag{47}$$

and using Eq. (44),

$$\lim_{t \to \infty}(s) = s_f\left(= \frac{1}{1 + (R_i/H_i)\dot{\gamma}_f^n}\right) \tag{48}$$

Integration of Eq. (37), use of the initial condition given by Eq. (47), and substitution of the expression in Eq. (48) yields

$$s = s_f + (s_0 - s_f)\exp\left[-(H_i + R_i\dot{\gamma}_f^n)t\right] \tag{49}$$

Values for H_i and R_i can be computed by substituting this expression into Eq. (34) and completing a curve fit to the transient τ versus t curve from the shear rate jump experiment.

This procedure should then be repeated for other values of solid fraction to obtain functional relationships for $H(T,f_s)$ and $R(T,f_s)$. Martin et al. [57] found that H_i and R_i only weakly depend upon $\dot{\gamma}_0$ and $\dot{\gamma}_f$ and therefore ignored their effect. Strong dependency was found for f_s (which depends upon T), however.

After completing this procedure for every solid fraction of interest, one may wish to establish solid fraction (i.e., temperature) and/or internal variable dependent expressions for λ_1, λ_2, C_0, Q, R, and H. Using the experimental results of Brown and coworkers [52–57], one can construct a material model for $Sn \cdot 15wt\%Pb$ as follows, where c_{max} is assumed constant at 0.625.

$$\frac{ds}{dt} = 1.46 f_s^{5.812}(1-s) - 9 \times 10^{-11} f_s^{4.702} s \dot{\gamma}^4 \tag{50}$$

$$\tau = 0.0232(3s+2)\frac{[f_s(1+0.1s)]^{1/3}}{1-1.17[f_s(1+0.1s)]^{1/3}}\dot{\gamma}$$

$$+ 9.62 \times 10^{-41} s f_s \dot{\gamma}^4 e^{35.9/T} \tag{51}$$

where $\dot{\gamma}$ has units of s^{-1}, τ has units of Pa, and T has units of K. This is equivalent to an apparent viscosity of

$$\mu_a = 0.0232(3s+2)\frac{[f_s(1+0.1s)]^{1/3}}{1-1.17[f_s(1+0.1s)]^{1/3}}$$

$$+ 9.62 \times 10^{-41} s f_s \dot{\gamma}^3 e^{35.9/T} \tag{52}$$

Brown and coworkers [57] tested these models by conducting experiments where steady state shear strain rate was obtained, followed by a slow reduction

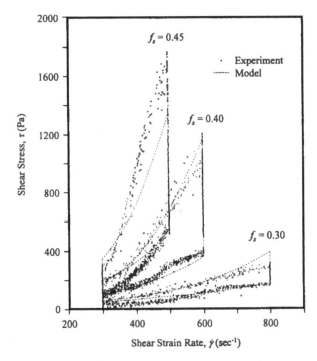

Figure 25 Hysteresis loops for the Brown internal variable model and experiments for Sn · 15%Pb. (Reprinted from Ref. 57 with permission from Elsevier Science.)

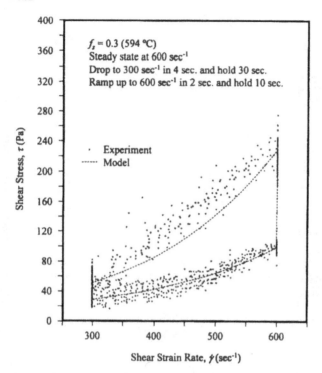

Figure 26 Hysteresis loops for the Brown internal variable model and experiments for Al · 7Si · 0.6Mg. (Reprinted from Ref. 57 with permission from Elsevier Science.)

in shear strain rate to a lower value that was maintained for a brief time, followed by a slow increase in shear strain rate back to the starting value. Shear stress was continuously monitored during this hysteresis loop and compared with predictions from the model—see Figs. 25 and 26. The results are sufficiently accurate for engineering purposes.

To test the concept of structure dominate flow, Kumar et al. [53] conducted an experiment where the structure for a Sn · 15%Pb alloy was stabilized at a steady state shear strain rate of $764 s^{-1}$. The shear strain rate was then reduced at a constant rate to $0.77 s^{-1}$ in 3.47 s. Immediately, the shear strain rate was increased at a constant rate back to $764 s^{-1}$ in 1 s. The times chosen for the ramp down and ramp up segments were small enough to ensure uniform structure. Figure 27 shows the results of this experiment. Note that the material, at constant structure, is shear rate thickening. In addition, the ramp down curve closely matches that of the ramp up results. These results are very relevant for modeling industrial processing of SSM materials where the deformation time is typically less than 1 s. If the

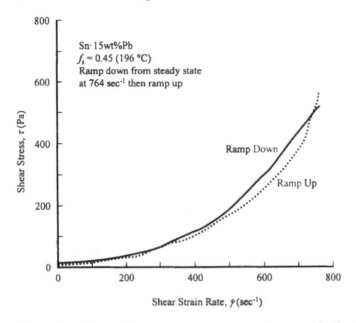

Figure 27 Effect of shear strain rate on material response for Sn · 15%Pb at constant structure. (From Ref. 53.)

Brown model is applicable to heated and unsheared SSM material, then no significant change in structure is accomplished during typical industrial processing for shape making. Furthermore, a shear rate-thickening model should be used to simulate the local shear stress from which die filling and other phenomena are calculated.

9. Exponential Decay Models

Using the internal variable approach, Mada and Ajersch [59, 60] developed some simple relationships that define the variation in shear stress over time when SSM alloys are subjected to a sudden change in shear strain rate. The material models are based upon results from experiments involving sudden changes in shear strain rate. Mada and Ajersch [59, 60] recommend the following expression, based upon disagglomeration of SiC particles in A356:

$$\tau(t) = \tau_e + (\tau_a - \tau_e)\exp\left(-\frac{\dot{\gamma}_f}{a_1 + b_1\dot{\gamma}_f}t\right) \tag{53}$$

where τ =   transient shear stress

$$t \quad = \quad \text{time}$$
$$\dot{\gamma}_f \quad = \quad \text{final shear strain rate}$$
$$\tau_a \quad = \quad \text{shear stress just after the jump in shear strain rate}$$
$$\tau_e \quad = \quad \text{shear stress at steady state at the new shear strain rate}$$

while a_1 and b_1 are material constants that can be determined through experimentation.

Quaak et al. [63] proposed a double exponential relationship in the case of a shear strain rate jump (i.e., disagglomeration event), as follows:

$$\tau(t) = \tau_e + (\tau_a - \tau_e) \left[\alpha_Q \exp\left(-\frac{t}{\lambda_1^*}\right) + (1 - \alpha_Q) \exp\left(-\frac{t}{\lambda_2^*}\right) \right] \tag{54}$$

where λ_1^* and λ_2^* are time constants that depend upon f_s and both the initial and final shear strain rates and α_Q is a constant.

Effects of reagglomeration at rest can be determined from the following expression [64]:

$$\tau(t_r) = \tau_\infty - (\tau_\infty - \tau_e) \exp\left(-\frac{\dot{\gamma}_0}{a_2 + b_2 \dot{\gamma}_0} t_r\right) \tag{55}$$

where $\quad t_r \quad = \quad$ time at rest

$\tau_\infty \quad = \quad$ shear stress in the fully agglomerated condition (i.e., at a long rest time)

$\tau_e \quad = \quad$ initial shear stress (before resting the material)

while a_2 and b_2 are material constants that must be determined from experimentation.

10. Xu Model

Solid/liquid segregation is a common problem during production of SSM parts. The two phases have different chemical compositions—the solid phase is typically composed of few alloying elements, while the liquid phase is typically rich in alloying elements and is often a eutectic mixture. As a consequence of chemical gradients induced by solid/liquid segregation, the mechanical properties in finished SSM parts can vary significantly from location to location. Usually, this is undesirable for finished components. Semi-solid materials that are mixtures of low melting point alloys and high melting point solid particles of a different alloy system can be even more susceptible to segregation (due to greater density variations between the two materials), resulting in wide variations in mechanical properties throughout finished parts.

To account for solid/liquid segregation, Xu et al. [65] developed a two-phase material model that relies upon two calculation steps for description of material deformation and segregation. The first step in the analysis assumes that the semi-solid material deforms macroscopically with no microscopic solid/liquid segregation. In the second step, the macroscopically fixed geometry from the first step is used. Here, the microscopic solid/liquid segregation is computed using a generalized Darcy's potential flow of liquid metal through the solid particle structure. Darcy's original law assumes that the solid structure is stationary. In the present application, both the liquid and solid phases are moving. Therefore, the volumetric flow rate in Darcy's original law is replaced in the present work by the difference in the velocities between the liquid and solid phases. In addition, a more generalized flow potential is required since the solid structure is deforming and fluid pressure gradients are not easily evaluated experimentally. A constitutive model of the following form is proposed:

$$\mathbf{u}_l - \mathbf{u}_s = -\nabla\Psi \tag{56}$$

where $\mathbf{u}_l$ = liquid velocity
$\mathbf{u}_s$ = solid velocity
Ψ = generalized Darcy potential

Xu et al. [65] reasoned that Ψ is in general a function of (1) the permeability of the liquid through the solid network, (2) the viscosity of the liquid, (3) the solid fraction of the SSM slurry, (4) the normal strain E, (5) the normal strain rate $\dot{E}$, and (6) the macroscopic normal stress Σ. For the Sn $\cdot$ Pb alloy system, the following functional form for Ψ is suggested:

$$\Psi = \omega\left|\Sigma_m\right|^{a'}(f_s)^{b'}\left(\dot{E}_{II}\right)^{c'} \tag{57}$$

where Σ_m = macroscopic mean normal stress
$\dot{E}_{II}$ = second invariant of normal strain rate

and ω, a', b', and c' are material constants.

The macroscopic single-phase velocity u from the first step can be expressed as the sum of the individual phase velocities times their phase fraction.

$$\mathbf{u} = f_s\mathbf{u}_s + f_l\mathbf{u}_l \tag{58}$$

The velocity fields can then be solved for both the solid and liquid phases from Eqs. (56) and (57) along with the fact that $f_s + f_l = 1$ to yield

$$\mathbf{u}_s = \mathbf{u} + f_l\nabla\Psi \tag{59}$$

$$\mathbf{u}_l = \mathbf{u} - f_s\nabla\Psi \tag{60}$$

Material constants are determined in the above models through a series of uniaxial compression tests (upset tests) under adiabatic conditions. First, upsetting should be conducted under conditions yielding zero friction between the die and test piece to eliminate solid/liquid segregation. The corresponding normal stress–normal strain curves at various solid fractions and normal strain rates can be used to develop a constitutive relationship for the SSM material. This relationship is used in the first step of the two-step approach outlined above. Additional uniaxial compression tests with friction between the die and workpiece need to be conducted to calibrate the generalized Darcy's law in Eq. (57). Image analysis can be used to determine the radial variation in solid fraction at varying amounts of reduction. Varying solid fraction and normal strain rate and using the measured uniaxial normal stress permits a determination of the material constants in Eq. (57). Some trial-and-error using numerical models may be necessary to find the best material constants. Friction factors of 0.015 [66] to 0.3 [67] have been recommended for simulation of SSM forming at a solid fraction of 0.5. If necessary, ring compression tests [68–71] can be conducted to determine the friction factor more accurately for SSM materials.

Xu et al. [65] used the above approach to model the upsetting of cylindrical Sn·Pb alloy billets. The best values for material constants a', b', and c', based upon the experimental work of Suéry and Flemings [72] and Pinsky et al. [73], were 1.0, −3.0, and −0.23, respectively. An acceptable correlation of solid fraction with radial position was observed, as noted in Fig. 28. The constitutive model was reasoned to take the following form:

$$\Sigma_0 = \sigma_0 (f_s)^{\alpha_x} (\dot{E}_{\mathrm{II}})^{\beta_x} \tag{61}$$

where Σ_0 = the normal flow stress (i.e., stress after yielding)

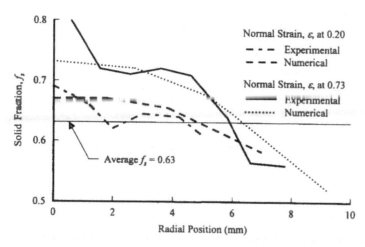

Figure 28 Radial variation of solid fraction in a compressed SSM billet. (From Ref. 65.)

$$\sigma_0 = \text{a material constant}$$

Xu et al. [65] showed that:

$$\alpha_x = -\frac{b'}{a'} \tag{62}$$

$$\beta_x = -\frac{c'}{a'} \tag{63}$$

Therefore, for the Sn · Pb alloy:

$$\Sigma_0 = \sigma_0 (f_s)^{3.0} (\dot{E}_{\mathrm{II}})^{0.23} \tag{64}$$

A similar analysis of an Al · 4.5Cu alloy, based upon the results of Yoshida et al. [74] gave values of 2.1, −6.3, and 0 for material constants a', b', and c', respectively. The material model for the latter alloy was

$$\Sigma_0 = \sigma_0 (f_s)^{3.0} \tag{65}$$

The above model of Xu was applied in a finite difference, Volume of Fluid (VOF), control volume code [75] to predict the segregation during filling of shaped parts. Segregation is computed during die filling, as shown in Fig. 29, while shear strain rate, velocity, and pressure are shown in Fig. 30. Also included in these results are thermal and solidification effects during die filling, which affects the local solid fraction and hence the local viscosity of the material.

11. Modigell Model

More recently, an interesting model was proposed by Modigell et al. [61]. This model uses a fluid mechanics approach (i.e., conservation of momentum and continuity) to describe both the liquid phase and the solid structure. A structural internal variable s_M indicates the degree of agglomeration (fully disagglomerated at $s_M = 0$ and fully agglomerated as $s_M \to \infty$). By tracking the velocities of both the liquid and solid phases, segregation effects can be computed with this model. First, conservation of momentum is applied to the solid phase to obtain

$$\rho \frac{\partial}{\partial t}(f_s \mathbf{u}_s) = -f_s \nabla p + \nabla \cdot \left[f_s \mu_s (\nabla \mathbf{u}_s + \nabla^T \mathbf{u}_s) \right] + C_{\mathrm{LS}}(\mathbf{u}_l - \mathbf{u}_s) \tag{66}$$

where p = pressure in the semi-solid
C_{LS} = drag coefficient
μ_s = equivalent viscosity of solid particle network

Conservation of momentum is applied to the liquid phase with the realization that the two most significant terms are those related to the pressure

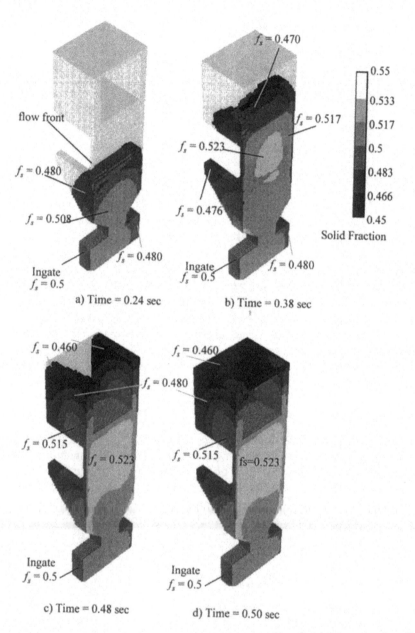

Figure 29 Solid fraction distribution as a result of solid/liquid segregation during filling of a shaped part. (a) Time = 0.24s; (b) time = 0.38s; (c) time = 0.48s; (d) time = 0.50s.(From Ref. 75.)

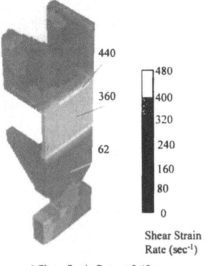

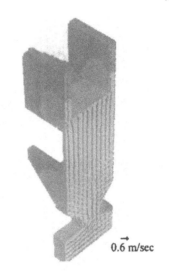

a) Shear Strain Rate at 0.48 sec

b) Velocity Vector Field at 0.48 sec

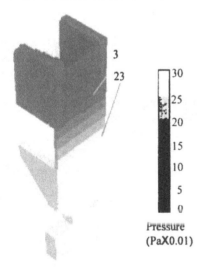

c) Pressure at 0.48 sec

Figure 30 Process parameters at 0.48 seconds during filling of a shaped part. (a) strain rate (b) velocity vector field (c) pressure. (From Ref. 75.)

gradient and drag. As a result, the momentum equation for the liquid phase reduces to Darcy's law:

$$-f_l \nabla p + C_{LS}(\mathbf{u}_s - \mathbf{u}_l) = 0 \tag{67}$$

where

$$C_{LS} = \frac{f_l \mu_l}{k_S} = \frac{k_0 \mu_l (1 - f_l)^2}{d_p^2 f_l^2} \tag{68}$$

or

$$f_l(\mathbf{u}_l - \mathbf{u}_s) = -\frac{k_S}{\mu_l} f_l \nabla p \tag{69}$$

and

$$k_S = \frac{1}{k_0} d_p^2 \frac{f_l^3}{(1 - f_l)^2} \tag{70}$$

where k_0 = shape constant
μ_l = liquid viscosity
d_p = mean diameter of globules in the alloy
k_S = permeability of the solid structure

Note that this model predicts that segregation increases with particle size [76]. Adding Eqs. (66) and (67) eliminates the liquid velocity term $\mathbf{u}_l$. This is also where a major, unstated assumption is made in the Modigell model. By assuming that the pressure fields are equal in both the liquid and solid phases*, then the addition of Eqs. (66) and (67) yields

$$\rho \frac{\partial}{\partial t}(f_s \mathbf{u}_s) = -\nabla p + \nabla \cdot \left[f_s \mu_s (\nabla \mathbf{u}_s + \nabla^T \mathbf{u}_s) \right] \tag{71}$$

Continuity is applied independently to both the two-phase material and the solid phase. Assuming incompressibility, applying the Darcy law, Eq. (69), and defining the mean velocity of the mixture $\mathbf{u}$ as

$$\mathbf{u} = f_s \mathbf{u}_s + f_l \mathbf{u}_l \tag{72}$$

the continuity of the two-phase material model takes on the following form:

$$\nabla \cdot \mathbf{u}_s - \nabla \cdot \left(\frac{k_S}{\mu_l} f_l \nabla p \right) = 0 \tag{73}$$

Application of continuity to the solid phase yields

---

* This is a significant assumption within the model. One would expect the loads supported by the solid phase to be several times that of the liquid phase for a structure of solid particles that are in contact with each other. Hence, the pressure gradients in each phase are also likely to be very different from each other.

$$\frac{\partial f_s}{\partial t} + \nabla \cdot (f_s \mathbf{u}_s) = 0 \tag{74}$$

Now, the viscosity of the network of solid particles must be defined and be related to the structure of the solid particles. Modigell et al. [61] have assumed the following form:

$$\mu_s = \delta_M e^{Bf_s} s_M \dot{\gamma}^{m_M - 1} \tag{75}$$

where δ_M, B, and m_M are material constants, which must be determined through experimentation. The functional form for s_M under equilibrium conditions is given by

$$s_{M,e} = \frac{1}{(\psi \dot{\gamma})^{m_M - n_M}} \tag{76}$$

where ψ = constant to ensure unit consistency = 1 s
      n_M = strain hardening coefficient ($\sigma \propto \varepsilon^{n_M}$, σ = flow stress, and ε = true normal strain) for plastic flow in the pure solid material.

The rate of change of s_M is defined as

$$\frac{ds_M}{dt} = g(\dot{\gamma})[s_{M,e}(\dot{\gamma}) - s_M] \tag{77}$$

where $g(\dot{\gamma})$ = shear rate dependent function that controls how quickly s_M approaches its equilibrium value.

Modigell et al. [61] assumed the following functional form:

$$g(\dot{\gamma}) = a_M \exp(b_M \dot{\gamma}) \tag{78}$$

where a_M and b_M are material constants that can be determined experimentally using shear rate jump tests on a Couette rheometer.

Equations (71), (73), (74), (75), and (77) are then used to describe the material flow during SSM processing including segregation. If necessary, Eq. (69) can then be used to directly solve for the liquid velocity. Based upon a least squared curve fit to the experimental data presented by Modigell et al. [61], the following values were obtained for Sn · 15%Pb: $\delta_M = 0.0766$, $B = 11.25$, and $n_M = 0.387$ with $\dot{\gamma}$ in units of s^{-1} and μ in units of Pa · s.

12. Grain Growth and Coalescence Effects

Some interesting characterization work was completed by Salvo et al. [77] on semi-solid slurries of aluminum alloy AS7U3G (Al · 7Si · 3Cu · Mg). Their efforts focused on upsetting cylindrical specimens between two flat, parallel dies to obtain classical normal stress-normal strain relationships. In addition to characterizing this alloy for its utility in SSM processing, they also characterized the effect of hold time (after reaching the desired processing temperature) on rheological behavior. Grain growth was found to follow the classical Ostwald ripening law[*]:

$$\bar{D}^3 - \bar{D}_0^3 = k_W t \qquad\qquad (79)$$

where $\bar{D}$ = mean particle diameter
        $\bar{D}_0$ = initial mean particle diameter
        t = time at temperature
        k_W = 650 μm^3/s at a temperature of 573°C ($f_s \approx 0.45$)

Han et al. [80] give experimentally determined values of k_W for 2014 aluminum alloy as 210 μm^3/s and 320,000 μm^3/s at 560°C ($f_s = 0.90$) and 600°C ($f_s = 0.78$), respectively. Obviously, k_W is highly dependent upon temperature. The grains were found to quickly globularize at first and then maintain a significantly slower globularization after approximately 6 min at processing temperature. Material having rounded particles will deform more easily, which will have some impact on flow resistance during forming of SSM alloys. However, Salvo et al. [77] found that of more significance is the amount of entrapped liquid in the solid alpha particles. At the start of the hold temperature, 5.5% of the eutectic liquid was entrapped in the solid particles. This level decreased to 3.5% in 30 s, then increased to 5.0% after holding for 2 min at temperature. The entrapped volume fraction then decreased in an exponential fashion to 2.0% at 50 min. The explanation for this behavior is the competition between Ostwald ripening, which reduces the quantity of entrapped liquid, and grain coalescence, which increases the amount of entrapped liquid. The true stress (at a true normal strain of 0.2) was found to correlate to the amount of entrapped liquid. The flow stress quickly increased from 0.006 MPa at a 1-min hold time to 0.010 MPa at a hold time of 2.0 min. After 2 min, the flow stress decreased with hold time to 0.0055 MPa at 30 min of hold time.

This work clearly points to the time dependency of SSM properties. Based upon this work, viscosity models should also be sensitive to time at temperature prior to casting the material. A simple approximation for aluminum alloy AS7U3G would be to adjust the shear yield stress of a Bingham fluid model by a factor equivalent to that of the change in flow stress noted above in

---

[*] While most reports on grain growth in SSM slurries also mention a kinetic growth rate of $t^{-1/3}$, others have calculated the growth rate kinetics with $t^{-3/7}$ [28] during *rheocasting*, while Poirier et al. [78] have computed that grains grow at a kinetic rate of $t^{-\theta}$, where $1/6 < \theta < 1/5$, when the growth is dominated by coalescence of particles. Poirier and coworkers conclude that Ostwald ripening dominates at low solid fractions or high solid/liquid interface energies, while high volume fraction and/or systems with low solid/liquid interface energies coarsen by coalescence. Experimentally determined relationships for k_W and θ for various SSM starting conditions are given in Ref. 79.

the study by Salvo et al. [77]. These authors also point out that the rheological behavior of aluminum alloy AS7U3G is similar to that of aluminum alloy A356.

13. Particle Diffusing Motivated Solid/Liquid Segregation

In a review by Sigworth [81], a paper by Leighton and Acrivos [82] is summarized. The most significant and interesting aspect of the paper is a method for computing solid/liquid segregation in concentrated suspensions. A particle diffusion term is defined for this purpose as

$$D_p = k' \dot{\gamma} a_p^2 \left(\frac{f_s^2}{\mu} \right) \frac{\partial \mu}{\partial f_s} \tag{80}$$

where D_p =   particle diffusion coefficient
       k' =   constant having a value between 0.6 and 1.0
      a_p =   particle radius

The model was seen to correctly predict the migration of particles to the center of the annulus region in a Couette rheometer. This segregation then results in a liquid-rich layer near the walls that is responsible for wall slip. This effect can be seen in macrographs of material sheared in Couette rheometers [9]. One would expect a larger velocity gradient in the liquid-rich region near the walls than that in the more particle-rich region near the midplane of the material. As such, one must expect that the measurements made on these segregated SSM materials in these tests lead to an underestimation of viscosity for SSM materials.

C. High Solid Fraction Processing

Nearly all the models developed for high solid fraction processing are based upon deformation of porous solids. July and Mehrabian [11] first suggested an empirical relationship of the form

$$\sigma = A_J e^{B_J f_s} \dot{\varepsilon}^m \tag{81}$$

where $\dot{\varepsilon}$ =   normal strain rate
      σ =   normal flow stress
     m =   strain rate sensitivity

and A_J, B_J, and m are determined through experimentation.
    Reported values for B_J are 20 for Sn · 15wt%Pb [10] and Al · Si alloys [62], while for magnesium alloys, the value of B_J was found to be closer to 15 [62]. Values for m were found to be 0.5 at $f_s = 0.5$ [10], while a near zero or

negative value was observed for Al · Si alloys [62]. Laxmanan and Flemings [10] later proposed a linear relationship for m based upon solid fraction: $m = c_J f_s + d_J$, where c_J and d_J are experimentally determined material constants. Other researchers have focused on a more comprehensive approach that relies upon more sophisticated solid mechanics approaches that use tensor mathematics to separate the stress field into hydrostatic and devortoric terms. One of the more important reasons for this description of stress is to predict solid/liquid segregation.

Han et al. [80] evaluated wrought 2014 aluminum alloy in the range: $0.70 < f_s < 0.86$. They found that the following constitutive law best fits the limited experimental data from constant load compression tests:

$$\dot{\varepsilon}^m = \varepsilon^{-n_H}[\sinh(b\sigma)] \qquad 0.70 < f_s < 0.86 \tag{82}$$

where $\dot{\varepsilon}$ = true normal strain rate
              ε = true normal strain
              b = compliance (i.e., inverse stiffness)

Variables n_H, b, and m are material parameters that depend upon f_s. For 2014 aluminum alloy, the following relationships were found (based upon limited experimental results and $0.70 < f_s < 0.86$):

$$n_H = 0.106 + 0.438 f_s$$

$$b = 4.89 \times 10^{-4} - 5.21 \times 10^{-5} f_s$$

$$m = 1.947 - 1.263 f_s \tag{83}$$

The most common models of high solid fraction i.e., (f_s approximately greater than 0.60) processing assume a solid structure of particles with isolated liquid islands. In this case, flow models from metal powder compaction or soil mechanics are often used to describe the behavior of SSM alloys. Viscoplastic behavior describes the deformation of the solid structure (network of connected solid grains), while Darcy flow is used to describe the flow of liquid metal through the deforming solid skeleton. The Darcy equation is defined as

$$\mathbf{u}_{l/s} = \frac{k_P}{\mu_l f_s} \operatorname{grad}\left(p_{\text{liquid}}\right) \tag{84}$$

where $\mathbf{u}_{l/s}$ = velocity of liquid phase relative to the solid phase
              p_{liquid} = liquid pressure
              k_P = permeability of the porous medium (i.e., the structure of solid particles)

Note that for porous materials, k_P is proportional to the square of a characteristic length—typically associated with the distance between particle centers, which for SSM materials is approximately the particle diameter [58, 83].

Zavaliangos and Lawley [58] recommend the following expression for determination of permeability:

$$k_P = \frac{(1 - f_s)^2}{24\pi N \vartheta^3} \tag{85}$$

where N = number of channels per unit area ($N \propto 1/\bar{D}^2$)
ϑ = tortuosity factor
$\bar{D}$ = mean particle diameter

Consequently, fine grain materials are less prone to solid/liquid segregation during processing. Sigworth [35] points out that for high solid fraction processing, the normalized flow stress (δ^* = ratio of the flow stress of the mushy material to that in the solid state) is closely approximated by the following relationship when volumetric changes and solid/liquid segregation are ignored:

$$\delta^* \approx f_s^{7.5} \tag{86}$$

A more elaborate model was proposed by Gunasekera [84] for high (0.80–1.00) values of solid fraction:

$$\sigma = K_G \dot{\varepsilon}^m \exp\left(\frac{Q_D}{R^* T_H}\right) [1 - (\beta_G f_l)^{2/3}] \tag{87}$$

where σ = flow stress
$\dot{\varepsilon}$ = normal strain rate
Q_D = activation energy for deformation in the solid state
R^* = universal gas constant
T_H = homologous temperature (using an absolute temperature scale)
f_l = liquid volume fraction (i.e., $f_l = 1 - f_s$)

The values of K_G, m, and β_G must be experimentally determined for each alloy. The value for β_G is approximately 1.4–1.5 [84].

Based upon unconstrained compression tests on spray cast aluminum alloys 2014 and Al·4Cu, Tzimas et al. [25, 85] showed that SSM material with a high solid fraction behaves like highly packed granular solids. Little or no deformation occurred in the grains. Rather, grains slid or rotated past each other to accommodate shear strains and material deformation. As seen in Fig. 31, the flow stress increased rapidly with normal strain (up to a value of approximately 0.05–0.1), then generally fell quickly to a much lower stress (at a normal strain of approximately 0.1–0.3). This lower stress value was then maintained for large normal strain values. As expected, both the peak and sustained stresses increase with an increase in solid fraction (i.e., lower temperature). Tzimas et al. [25] also show experimental evidence that the yielding of SSM materials is pressure dependent. The peak stress was shown to have a

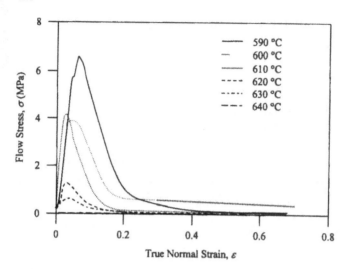

Figure 31 True normal stress–true normal strain behavior of Al · 4Cu SSM alloy at various temperatures in the semi-solid state. (From Ref. 25.)

strain rate sensitivity m of approximately 0.1. The ratio of peak stress to sustained stress varied greatly depending upon solid fraction and grain size. Small grains (60–90 μm) at a solid fraction of 0.8 resulted in a stress ratio of 4.0, while large grains (140–200 μm) at the same solid fraction resulted in a stress ratio of 11.5—roughly a 3 to 1 difference associated with grain size. This same relative effect of grain size was found at solid fractions of 0.65 and 0.5.

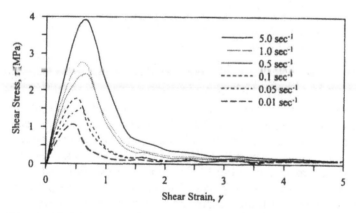

Figure 32 Shear stress–shear strain relationship of a Sn · Pb alloy at a solid fraction of 0.7. (From Ref. 86.)

However, at these lower solid fractions, no peak was found for the small grain material, while the peak was about 3 times the sustained stress for the large grain material. These results indicate the need for a more elaborate material model than that proposed by Gunasekera [84].

Experimental work by Martin et al. [86]—see Fig. 32—showed a similar behavior for dendritic semi-solid Sn · Pb alloys at high solid fraction ($f_s > 0.6$). The stresses for the dendritic structure were found to be 2–3 orders of magnitude higher than for a globular structure, however. At a solid fraction of 0.7, the peak shear stress (= 1.05 MPa) occurred at a shear strain of 0.45 for a shear strain rate of $0.01 \, s^{-1}$. As shear strain rate increased, the peak stress shifted to higher shear strains. At a shear strain rate of $5.0 \, s^{-1}$, the peak shear stress of 3.9 MPa occurred at a shear strain of 0.65. The shear stress vs. shear strain curves appeared roughly parabolic about the peak. In all cases, the shear stress fell to low values (roughly 9% of peak value) at shear strains above 2.5. Sannes et al. [87] evaluated several magnesium alloys and found a similar drop in shear stress with increased deformation in a vane deformation test rig. For the magnesium alloys, however, the steady state shear stress was approximately 20% of the peak value.

A material model by Nguyen et al. [88] for high solid fraction simulation is based upon solid mechanics principles. A viscoplastic constitutive model is used to describe the deformation of the solid phase. The Darcy law is applied to the liquid phase to determine solid/liquid segregation. The interested reader should consult this work and an interesting application of the model [89] as applied to (1) upsetting cylindrical test pieces (both with and without friction) and (2) slumping and solid/liquid segregation in a freestanding SSM billet. Experimental measurements of liquid pressure during deformation of SSM materials have been made by Martin et al. [90]. Their methodology is useful in establishing how the macrostress components should be distributed to the solid deforming state and the liquid phase. Fluid pressure gradients necessary for prediction of solid/liquid segregation in the Darcy porous model can then be computed. Deformation of the solid structure can also be determined from the stress components acting on the solid structure. Loué et al. [42] found that as deformation rate is increased, the amount of solid/liquid segregation is reduced until a critical deformation speed is reached above which no segregation was observed. Other researchers [62, 74] have drawn similar conclusions.

Tzimas et al. [85] noted significant strain localization during upsetting of very high solid fraction ($f_s > 0.8$) spray cast aluminum alloys. This results in variations in material response that lead to poor process control. In addition, strain localization may also lead to an increase in certain internal defects such as porosity.

D. Intermediate Solid Fraction

Material behavior of both low solid fraction and high solid fraction SSM slurries have been characterized as outlined above. Description of the transition from low solid fraction behavior to high solid fraction behavior can be problematic, however. The material makeup transitions from a suspension of solid particles (individual particles or clusters of agglomerated particles) to a structure consisting of an extensive network of joined particles. Transition for rounded particles generally occurs between a solid fraction of 0.5 and 0.7 [27] and depends upon shear strain rate and particle morphology (size and shape). More rounded particles delay the transition to high solid fraction behavior. Materials consisting of dendrites have a transition that is much lower ($0.2 \leq f_s \leq 0.3$). For well grain-refined alloys, load carrying capacity does not develop until a solid fraction of about 0.4 is reached [2]. Sigworth [35] has summarized the relationship between shear stress, shear strain rate, and solid fraction for aluminum-silicon alloys, as shown in Fig. 9.

E. Boundary Conditions

Shirai et al. [91] conducted some experiments to determine the effective convection heat transfer coefficient h between SSM material and the die. They investigated Sn · 15%Pb, Al · 10%Cu, 0.4%C steel, and AISI 304 stainless steel at solid fractions less than 0.5. They also evaluated superheated material to compare the difference between the SSM and liquid metal states. The experimental method consisted of plunging a thin chill plate into a stirred SSM slurry (or liquid metal bath) and measuring the following for various immersion times: temperature of chill plate, thickness of solidified SSM material adhering to the chill plate, and the secondary dendrite arm spacing (SDAS) throughout the solidified layer. Analyses were conducted for immersion times of up to two seconds. Based upon the SDAS, the cooling rate was determined and h was computed using results of a finite difference model of the test setup. The findings of Shirai et al [91]. indicated little difference in the effective heat transfer coefficient between the SSM state and the pure liquid state. Results in the liquid phase compared well with those from other researchers. There were significant differences from one alloy to another, however, in both the effective heat transfer coefficient and the rate of solidified shell buildup on the chill plate.

   The ratio of maximum to minimum SDAS during freezing of the first 2 mm of shell resulted in high uncertainty in h. The scatter in h during this time in Shirai et al.'s results [91] was 4 to 1 (maximum to minimum) for Sn · 15%Pb and 3 to 1 for 0.4%C steel and AISI 304 stainless steel. Beyond the first 2 mm, the scatter was reduced to about 2 to 1 for each alloy. Table 5 summarizes their

Table 5 Recommended Values of Effective Convective
Heat Transfer Coefficient Between the Die and SSM
Workpiece Based upon Material and Immersion Time

SSM material	Immersion time (s)	h $(W/m^2 \cdot K)$
Sn · 15%Pb	0.5	8700
	1.0	6900
	2.0	5700
Al · 10%Cu	0.5	8600
	1.0	6800
	2.0	5400
AISI304 stainless steel	0.5	5000
	1.0	4000
	2.0	3000

Source: Ref. 91.

results. Their values are approximately 2–3 times the constant value of 2300
$W/m^2 \cdot K$ recommended by Sigworth [35].

F. Induction Heating of SSM Billets

As discussed in the introduction to this chapter, the typical process route for
producing SSM parts includes induction heating of cylindrical SSM billets.
An optimally designed induction heating process will produce billets that are
uniform in solid fraction. Poor heating practices, on the other hand, will lead
to wide variations in solid fraction throughout the heated billet. Solid fraction differences of 10% or more within a billet can result in significantly
different material behavior from one part of the billet to another. Many
researchers [2, 35] show that a 10% difference in solid fraction can result
in large (as much as 100 times) differences in apparent viscosity from $f_s = 0.5$
to $f_s = 0.6$. Furthermore, the die fill characteristics (e.g., required plunger
loads, die fill sequence, and ability to fill fine die details) are altered as a
result of the magnitude and distribution of the nonuniformity in viscosity and
consequent variations in flow behavior. These variations usually lead to
decreases in quality, mechanical properties, and part yield. Effective heating
practices must balance two competing needs: (1) quick heating for improved
productivity and (2) uniform and consistent heating for optimum die filling
characteristics. Prolonged heating times or overheating of billets can result in
(1) "elephant foot" defects where the liquid metal percolates through the
solid mesh and pools at the base of billets (see Fig. 33) and (2) slumping

Figure 33 Elephant's foot on heated SSM billet. (Courtesy of Concurrent Technologies Corporation, Johnstown, PA.)

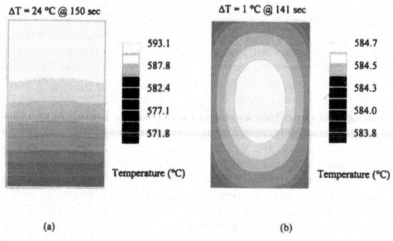

ΔT = 24 °C @ 150 sec

593.1
587.8
582.4
577.1
571.8

Temperature (°C)

ΔT = 1 °C @ 141 sec

584.7
584.5
584.3
584.0
583.8

Temperature (°C)

(a)   (b)

Figure 34 Solid fraction distribution within induction heated aluminum alloy A357 billets using differing process equipment. (a) Glass pedestal and open coils; (b) insulation surrounding billet. (Courtesy of Concurrent Technologies Corporation, Johnstown, PA.)

of low solid fraction billets due to their inability to withstand their own weight. Larger billets are more susceptible to these heating problems. Analysis of induction heating design [75, 92] has led to several strategies for improving temperature (i.e., solid fraction) distribution in heated billets. These strategies include [75] the following.

1. Use of an insulator on the top and bottom of the billet results in more uniform temperature throughout the billet, as depicted in Fig. 34. (Note that insulation is most effective when placed above, around the inside, and between the turns of the induction coil along with a layer of insulation between the billet and the pedestal upon which it stands.)
2. The total energy required to heat a billet is a strong function of the billet weight but the relationship results in a slightly concave down curve when energy is plotted against billet weight.
3. Heating time is a strong function of the initial current but is not as dependent upon frequency.
4. Billet heating is due to eddy current heating on and near the surface and requires the effect of conduction to heat the center of the billet.
5. Use of a properly designed flux concentrator will reduce the time required to heat the billet.
6. The type of flux concentrator influences the heating time.
7. There is an optimum thickness for the flux concentrator material.

The rate of heat generated decays approximately exponentially from the billet surface inward. Approximately 86% of the heat is generated in a thin layer at the surface of the billet [93, 94] in a region known as the skin depth δ. To accurately capture this effect in numerical models, Dantzig and Midson [95] recommend a minimum of three elements through the skin depth on the outer layer of the billet. The skin depth can be computed from the following relationship, where uniform billet properties are assumed [96]:

$$\delta = \sqrt{\frac{\rho_E}{2\pi v \mu_{0f} \mu_R}} \qquad (88)$$

where δ = depth of penetration (i.e., skin depth)
v = induction (i.e., electrical or line) frequency
μ_{0f} = permeability of free space (= $4\pi \times 10^{-7}$ webers/Amp · m)
μ_R = relative permeability
ρ_E = electrical resistivity

Typically, the temperature distribution in heated aluminum billets decays from the outside surface inward. Top and/or bottom outside corners may be hotter or colder than the midlength surface temperature depending upon coil and billet geometry, material properties, and the frequency and current in the coil [92, 97]. Temperature variations throughout the heated billet of less than a few degrees Celsius are possible when heating aluminum alloys to the semi-solid state. This uniformity is due in large part to the high thermal conductivity* of aluminum. Heating of high temperature, low thermally conductive semi-solid materials (such as iron-, nickel-, and titanium-based alloys) often results in different temperature distributions and larger temperature variations throughout heated billets. Conventional induction coils packed in insulation and operated under constant frequency resulted in a maximum billet temperature below the surface of heated billets for both Ti · 20Ni [97] and 310 stainless steel [98]. Furthermore, temperature variations of 50°C were common in these heated billets with the cold spot located at either the top or bottom outside corner of the heated billet [97, 98]. Surface heat losses, especially radiation, are much more significant in these high temperature alloys. Larger temperature variations are also caused by the lower thermal conductivity of Ti · 20Ni and 310 stainless steel. Kapranos et al. [98] have recommended a high frequency, high power induction field at the end of soaking to reduce this temperature variation in 310 stainless steel. This technique reduced the temperature variation from 40 to 18°C in heated billets from their analyses.

The electrical resistivity of most metals is typically highly temperature dependent over the range of SSM processing. Its value for aluminum increases by a factor of 3 from room temperature to SSM processing temperatures [89]. Such variations can be easily accounted for in numerical simulations, which can be used to optimize the induction heating process.

Additional complicating factors in the prediction of billet temperature distribution include end effects, which are influenced by the frequency of the line current, the coil and billet geometry (air gap size and relative lengths), properties of the slug, surface heat transfer conditions (emissivity and convective heat transfer coefficient), coil refractory, power density, and cycle time [88]. These end effects have been summarized as (see Fig. 35) (1) r_{billet}/δ, the skin effect (r_{billet} = billet radius), (2) ξ, coil overhang length (i.e., the length of "excess" coil beyond the end of the billet), and (3)

---

* Thermal diffusivity ($\alpha_t = k_t/\rho c_p$) is defined as the ratio of thermal conductivity k_t to the product of density ρ and specific heat c_p. This material property arises in the transient heat conduction equation and establishes how quickly and uniformly a body reaches thermal steady state. High values of thermal diffusivity result in a more rapid development of steady state conditions. The final thermal steady state condition (i.e., temperature distribution) is, however, dictated by the thermal conductivity of the material (and the boundary conditions of the part). It becomes appropriate, therefore, to focus on thermal diffusivity during transient events such as during billet heating, while focusing on thermal conductivity in a billet that has reached its final, steady state condition.

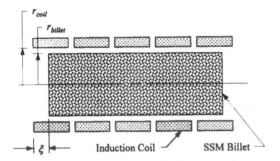

Figure 35 Schematic of induction heating coil.

r_{coil}/r_{billet}, coil tightness ratio (r_{coil} = inside radius of the coil) [89]. Either overheating or underheating can occur at the billet ends as a result of the value of these parameters. Overheating is possible, for example, when ξ is a large positive value.

Sebus and Henneberger [99] demonstrated that coils having windings of various diameters result in a more uniformly heated billet. Nearly uniform diameter windings were used near the midlength of the billet, while larger diameter windings near the ends of the billet resulted in a more uniform heat generation and axial temperature distribution within the billets.

Readers interested in modeling the induction heating process are referred to general purpose, commercial codes that have the ability to compute electromagnetic effects. As an alternative, Dantzig and Midson [95] outline the mathematics needed to define the electrical current distribution in a computationally meshed cylindrical billet coaxially located in an induction coil. In addition,

Table 6 Freezing Rate of Aluminum Alloy A356 Billets of Varying Diameters and Infinite Length

	Grip Holding Time (s)					
Billet Diameter (mm)	0.5	1.0	2.0	5.0	10.0	25.0
	Solid Fraction (%)					
25	40	49	68	100	100	100
50	35	40	49	77	100	100
100	32	35	40	54	77	100
200	31	32	35	42	54	89

Grip temperature = 175°C

Initial fraction solid of billets = 30% (i.e., initial temperature = 595°C)

they describe equations for current distribution in a cylindrical billet in the annulus region between two induction coils of differing radii. These expressions include several terms involving series solutions.

G. Design of Robot Transfer Grips

Heated SSM billets are typically transferred to the metalworking station by mechanical grips attached to the end of a robot arm. Proper thermal management of these grips is necessary to avoid inadvertent freezing of the heated billets. Table 6 shows how quickly small SSM billets of an aluminum alloy can completely freeze by contact with a thin steel clamshell, which was typical of first generation grip designs. For excessively long transfer times (i.e., greater than 10.0 s), billets up to 50 mm in diameter can be completely frozen. With this in mind, Tims et al. [75, 100] designed improved robot grips through use of computer models. This work showed that minimal heat loss is accomplished by (1) using grips constructed of insulating materials and (2) resting the grip against a heated post during the time between consecutive billet transfers. Figure 36 shows the effect of such a grip (i.e., the one using, a heater, having a closed end, and using insulation) compared to that of a steel clamshell design. Clearly, the engineered grip (i.e., the one using a heater, having a closed end, and using insulation) provides a more uniformly heated billet. In addition, the

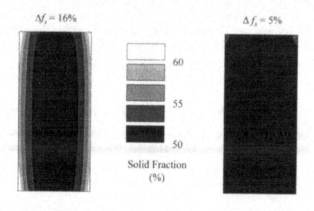

a) No Grip Heating, Open Grip Ends, No Insulation

b) Grip Heating Used, Closed Grip End, Use of Insulation

Figure 36 Uniformity of solid fraction in aluminum alloy A357 billets handled with grips made from a variety of designs. (Courtesy of Concurrent Technologies Corporation, Johnstown, PA.)

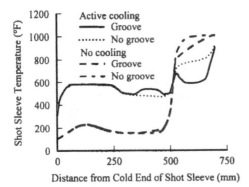

Figure 37 Temperature variation along the length of shot sleeves of various thermal management designs. (From Ref. 101.)

engineered grip also achieves this performance after only one or two billet transfers in a typical production run. The steel clamshell design took considerably longer to achieve billet-to-billet thermal consistency at the start of a production run. Another benefit to the use of insulation is the reduced tendency for actuator seal damage due to overheating in the robot grip assembly.

H. Design of Shot Sleeves

Horizontal die casting machines have been designed to allow liquid metal to lie along the bottom surface of the shot sleeve along its entire length. As a result, the temperature variation along the length of the shot sleeve is minimal. However, when these same horizontal die casting machines are employed as

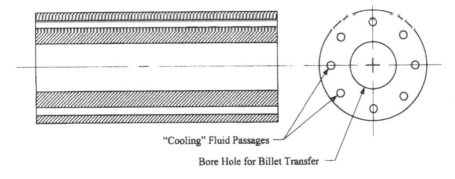

Figure 38 Cross section of a thermally well-managed shot sleeve.

SSM forming machines, significant temperature variations may exist along the length of the shot sleeve. As a result, the bore diameter can vary along the length of the shot sleeve resulting in significant differences in the clearance between the bore of the shot sleeve and the plunger used to push the heated slug. Some simple work by Tims and Creeden [101] showed that active thermal management of the shot sleeve is necessary to ensure a consistent clearance. Figure 37 shows the variation of temperature along the length of the shot sleeve with various shot sleeve designs. The major active thermal management designs evaluated here are (1) use of a thermal break between the outside of the shot sleeve and the die and (2) use of fluid circulating channels placed in the wall of the shot sleeve (see Fig. 38) for active thermal management of the shot sleeve temperature. Clearly, fluid circulating channels have the most pronounced effect on temperature uniformity. Using this design, along with the thermal break, resulted in a reduction in the bore diameter variation from 0.017 in without thermal management to 0.004 in with active thermal management. Upon subsequent evaluation in production equipment, the actively thermally managed shot sleeve eliminated a severe galling problem and minimized the amount of flash developed as material was squeezed behind the plunger at the locations of greatest clearance between the shot sleeve bore and plunger [102].

IV. SUMMARY

The scientific understanding of how metallic semi-solid materials behave is still evolving. Currently, most efforts to describe their behavior centers around a fluid mechanics approach with a constitutive material model to describe the viscosity of the material. Early work on SSM characterization focused on steady state behavior, most notably with material cooled to the semi-solid state. Although useful for understanding much about the material behavior, these models are unsuitable for application in modeling the production of shaped parts made from heated SSM billets. The basic power-law model offers the simplest mathematical form. When used, the consistency power-law index K and power-law exponent n are strong functions of alloy and solid fraction. These parameters may also be sensitive to prior processing history and grain size and shape. The internal variable models are the only ones to adequately account for processing history, although such detail may not be necessary for shape part manufacture given the typical short processing time compared with the time constants for particle agglomeration and disagglomeration. The continuous Bingham fluid model appears to offer the best description of actual material response to shear loading. For simulation of solid/liquid segregation, the generalized Darcy model is recommended. Experimental calibration is required for most of these models to establish the various material constants, especially when dealing with new alloys, differing shear strain rates, or extended solid fraction regimes.

Other processing steps must also be properly controlled if high quality SSM components are to be produced. The heating system must deliver billets that are uniform in solid fraction to ensure consistent material response during die filling. Transfer mechanisms must minimize thermal disturbances in the heated billets. Die casting shot sleeves must be redesigned to accommodate the unique thermally induced strains that accompany the handling of semi-solid billets.

Discrete part production through semi-solid processing offers many opportunities for improved quality, economics, and performance. Development and application of accurate, robust process models will allow further advancement of this relatively new production process.

SYMBOLS

A	hydrodynamic coefficient
A'	collection of terms involving internal variable s and solid fraction
A'_i	ith value of A'
A_J	material constant in Joly and Mehrabian model
A'_j	jth value of A'
a	material constant (power-law index modifier) for Wang model
a'	material constant in Xu model
a_M	material constant in Modigell model
a_p	particle radius
a_1	material constant in Mada and Ajersch disagglomeration model
a_2	material constant in Mada and Ajersch reagglomeration model
B	material constant—exponential coefficient associated with solid fraction
B_J	material constant in Joly and Mehrabian model
b	compliance
b'	material constant in Xu model
b_M	material constant in Modigell model
b_1	material constant in Mada and Ajersch disagglomeration model
b_2	material constant in Mada and Ajersch reagglomeration model
C	collection of constants relating to particle geometry and the temperature-dependent plastic deformation power-law coefficient for solid particles
C'	collection of terms involving internal variable s and absolute temperature
C_i	ith value of C
C_j	jth value of C
C'_j	jth value of C'
C_{LS}	drag coefficient
C_0	coefficient on the deformation power-law expression for the internal variable model of Brown

c	effective solid fraction
c'	material constant in Xu model
c_J	material constant in Joly and Mehrabian model
c_j	jth value of c
c_L	material constant in Laxmanan and Flemings model
c_{max}	maximum effective solid fraction
$c_{max,j}$	jth value of c_{max}
c_p	specific heat
c_1	constant $= \left[c_2(r_c^2 - r_p^2), - v_r\right]/\ln(r_p/r_c)$
c_2	constant $= \dfrac{v_r}{(r_c^2 - r_p^2) - (r_c^2 + r_p^2)\ln(r_c/r_p)}$
$\mathbf{D}$	rate of deformation tensor
$\bar{D}$	mean particle diameter
D_A	average particle diameter; smallest particle diameter
D_B	billet diameter
D_{II}	second invariant of the rate of deformation tensor
$\bar{D}_0$	initial mean particle diameter
D_p	particle diffusion coefficient
d	agglomerate diameter
d_J	material constant in Joly and Mehrabian model
d_L	material constant in Laxmanan and Flemings model
d_0	average individual particle diameter
d_p	mean diameter of globules
E	normal strain
$\dot{E}$	normal strain rate
$\dot{E}_{II}$	second invariant of normal strain rate
F	extrusion force
F_e	instantaneous compressive force
f	flow equation
f_l	liquid fraction
f_s	solid fraction
f_s^*	fraction solid (≈ 0.7) where the apparent viscosity approaches an infinite value
$\bar{f}_s$	average solidification rate
$f_{s,i}$	ith solid fraction value
$f_{s,j}$	jth solid fraction value
G	disagglomeration kinetics factor
g	shear strain rate dependent function in Modigell model
$\hat{g}_i$	evolution equation for internal variable number i
H	agglomeration kinetics parameter
H_i	ith value of H
H_j	jth value of H
h	convective heat transfer coefficient
h_e	instantaneous specimen height
i	dummy variable; solid fraction value index
j	initial shear strain rate value index

K	consistency power-law index
K_G	material constant in Gunasekera model
k	limiting volume fraction for a suspension of isolated solid particles
k'	constant with a value in the range [0.6, 1.0] used for estimating particle diffusion
k_I	number of internal variables
k_0	shape constant
k_P	permeability of a porous medium
k_S	permeability of the solid structure
k_t	thermal conductivity
k_W	proportionality constant in Ostwald ripening law
M	material constant for use in internal variable model of Al $\cdot$ 5Cu with B_4C particles
m	strain rate sensitivity
m_M	material constant in Modigell model
m_R	exponent that controls how quickly the continuous Bingham fluid model goes to τ_0
$m_{R,1}$	first value of m_R
$m_{R,2}$	second value of m_R
m_W	exponent on solid fraction ratio in Wang model
N	number of channels per unit area
n	power-law exponent
n_H	strain exponent in Han model
n_M	strain hardening coefficient in Modigell model
n_W	shear strain rate exponent in Wang model
p	pressure
p_{liquid}	liquid pressure
Q	an activation energy
Q_D	activation energy for deformation in the solid state (as used in the Gunasekera model)
R	disagglomeration kinetics parameter
R^*	universal gas constant
R_i	ith value of R
R_j	jth value of R
r_{billet}	radius of the billet
r_c	radius of container
r_{coil}	inside radius of induction coil
r_p	radius of plunger
s	internal variable to account for some material characteristic (the most notable one being the relative amount of solid particle agglomeration)
s_f	value of s at steady state
s_i	ith value of s
$\dot{s}_i$	rate of change of internal variable number i
s_j	jth value of s_{k_I}
s_{k_I}	k_Ith value of s

s_M	particle structure internal variable in Modigell model
$s_{M,e}$	value of s_M at prolonged, constant shear strain rate
s_0	initial value of s (i.e., at initiation of shear strain rate jump experiment)
s_v	area of the solid/liquid interface per unit volume of solid
s_{v0}	value of s_v at the start of isothermal holding
T	absolute temperature
T_H	homologous temperature
T_i	ith value of T
t	time
t_r	time at rest
u	material velocity component in the x-direction
$\mathbf{u}$	material velocity vector
$\mathbf{u}_l$	liquid phase velocity vector
$\mathbf{u}_{l/s}$	velocity vector of liquid phase relative to the solid phase
$\mathbf{u}_s$	solid phase velocity vector
V	sample volume
V_w	slip velocity
v	material velocity component in the y-direction
v_r	ram speed
w	material velocity component in the z-direction
x	first Cartesian coordinate
x_K	main solute concentration (in mass percentage) of alloy in Okano model
y	second Cartesian coordinate
z	third Cartesian coordinate
α	first solute concentration factor in Okano model
α_Q	constant in Quaak model
α_t	thermal diffusivity
α_x	solid fraction exponent in normal flow stress model of Xu
β	second solute concentration factor in Okano model
β_G	multiplier on liquid fraction in Gunasekera model
β_x	exponent on $\dot{E}_{II}$ in normal flow stress model of Xu
γ	shear strain
$\dot{\gamma}$	shear strain rate
$\dot{\gamma}^*$	shear strain rate where the asymptotic apparent viscosity is reached
$\dot{\gamma}_{AV}$	mean shear strain rate
$\dot{\gamma}_f$	final shear strain rate
$\dot{\gamma}_0$	initial shear strain rate
$\dot{\gamma}_{ss}$	initial shear strain rate prior to initiation of ramp-down experiment
$\dot{\gamma}_{ss,j}$	jth value of $\dot{\gamma}_{ss}$
$\boldsymbol{\Delta}$	rate of deformation tensor
$\boldsymbol{\Delta} : \boldsymbol{\Delta}$	dyadic product of $\boldsymbol{\Delta}$
δ	skin depth in induction heating
δ^*	normalized flow stress
δ_M	material constant in Modigell model
ε	true normal strain

$\dot{\varepsilon}$	true normal strain rate
ϑ	tortuosity factor
θ	Ostwald growth rate exponent
λ	extrusion ratio $= \dfrac{r_c^2}{r_c^2 - r_p^2}$
λ_R	relaxation time constant
λ_1	material constant in the internal variable model of Brown
λ_1^*	time constant in Quaak model
$\lambda_{1,i}$	ith value of λ_1
λ_2	material constant in the internal variable model of Brown
λ_2^*	time constant in Quaak model
$\lambda_{2,i}$	ith value of λ_2
μ	viscosity
μ^*	dimensionless viscosity $(= \mu_{sus}/\mu_0)$
μ_a	apparent viscosity
μ_B	Bingham viscosity
μ_i	initial viscosity
μ_l	liquid viscosity
μ_0	viscosity of pure liquid at its melting point
$\mu_{0,f}$	permeability of free space $(= 4\pi \times 10^{-7}$ webers/Amp $\cdot$ m$)$
μ_R	relative permeability
μ_r	relative viscosity $(= \mu_a/\mu_0)$
μ_s	equivalent viscosity of solid particle network
μ_{ss}	steady state viscosity
μ_{sus}	viscosity of suspension
μ_∞	asymptotic apparent viscosity value
ξ	coil overhang length
ρ	density
ρ_E	electrical resistivity
ρ_m	density of alloy at the liquidus temperature
v	induction frequency
Σ	macroscopic normal stress
Σ_m	macroscopic mean normal stress
Σ_0	normal flow stress
σ	flow stress
σ_0	material constant in Xu model
τ	shear stress
$\boldsymbol{\tau}$	shear stress tensor
τ_a	shear stress just after a shear strain rate jump
τ_e	steady state shear stress at a fixed shear strain rate
τ_0	shear yield stress for a Bingham fluid
τ_w	shear stress at the wall
τ_∞	shear stress for long rest times
Φ	volume-based solid fraction
Φ_m	maximum packing fraction (≈ 0.6)

Φ_m° empirical constant (=0.68)
Ψ generalized Darcy potential
ψ constant to ensure unit consistency in Modigell model
ω material constant in Xu model

REFERENCES

1. SB Brown, MC Flemings. Net-shape forming via semi-solid processing. Adv Matls and Procs 36–40, January 1993.
2. MC Flemings. Behavior of metal alloys in the semi-solid state. Met Trans B 22B:269–293, June 1991.
3. MP Kenney, JA Courtois, RD Evans, GM Farrior, CP Kyonka, AA Koch, KP Young. Semi-solid metal casting and forging. In: ASM Metals Handbook, Vol. 15: Casting. Metals Park, OH: ASM International, 1988, pp. 327–338.
4. SB Brown and MC Flemings, eds. Proc 2nd Int Conf on Semi-solid Processing of Alloys and Composites, Cambridge, MA, June 10–12, 1992.
5. M Kiuchi, ed. Proc of 3rd Int Conf on Processing of Semi-solid Alloys and Composites, Tokyo, June 13–15, 1994.
6. DH Kirkwood, P. Kapranos, eds. Proc 4th Int Conf on Semi-solid Processing of Alloys and Composites, Sheffield, June 19–21, Exeter, UK: SRP Ltd.
7. AK Bhasin, JJ Moore, KP Young, S Midson, eds. Proc 5th Int Conf on Semi-solid Processing of Alloys and Composites, Golden, CO, June 23–25, 1998.
8. NADCA Product Specification Standards for Semi-solid and Squeeze Casting Processes. Rosemont, IL: North American Die Casting Association, 1997.
9. DB Spencer, R Mehrabian, MC Flemings. Rheological behavior of Sn-15 pct Pb in the crystallization range. Met Trans 3:1925–1932, July 1972.
10. V Laxmanan, MC Flemings. Deformation of semi-solid Sn-15 pct Pb alloy. Met Trans A 11A:1927–1937, Dec 1980.
11. PA Joly, R Mehrabian. The rheology of a partially solid alloy. J Mat Sci 11:1393–1418, 1976.
12. I Diewwanit, MC Flemings. Semi-solid forming of hypereutectic Al-Si alloys. In: W Hale, ed. Light Metals 1996, Warrendale, PA: TMS, pp. 787–793.
13. NH Nicholas, MR Trichka, KP Young. Application of semi-solid metal forming to the production of small components. Proc 5th Int Conf on Semi-solid Processing of Alloys and Composites, Golden, CO, June 23–25, 1998, pp. 79–86.
14. XP Niu, BH Hu, SW Hao, FC Yee, I Pinwill. Semi-solid forming of cast and wrought aluminum alloys. Proc 5th Int Conf on Semi-solid Processing of Alloys and Composites, Golden, CO, June 23–25, 1998, pp. 141–148.
15. CM Wang, GH Nickodemus, TP Creeden. Determining optimal semi-solid forming process parameters by simulation technique. Proc 5th Int Conf on Semi-solid Processing of Alloys and Composites, Golden, CO, June 23–25, 1998, pp. 327–334.
16. LS Turng, KK Wang. Modelling the flow and solidification for semi-solid Sn-Pb alloy. Trans 15th Die Casting Congress, St. Louis, MO, Oct 16–19, 1989, Paper No. G-T89-043.

17. P Kapranos, DH Kirkwood, PH Mani. Semi-solid metal processing of ductile iron. Proc 5th Int Conf on Semi-solid Processing of Alloys and Composites, Golden, CO, June 23–25, 1998, pp. 431–438.
18. FR Dax. Semi-solid metalworking technology for titanium fluid handling components. TR No. 95-109D, Concurrent Technologies Corp, Johnstown, PA, Jan 7, 1996.
19. J Boylan. Semi-solid formed aluminum. Adv Matls and Procs 27–28, Oct 1997.
20. G Chiarmetta. Thixoforming and weight reduction—industrial application of SeSoF. Proc 5th Int Conf on Semi-solid Processing of Alloys and Composites, Golden, CO, June 23–25, 1998, pp. 87–95.
21. RF Decker, RD Carnahan, E Babij, J Mihelich, G. Spalding, L Thompson. Magnesium semi-solid metal forming. Adv Matls and Procs 41–42, February 1996.
22. RE Taylor, H Groot, J Ferrier. Thermophysical properties of A356-MHD. Purdue Univ, West Lafayette, IN, Report to Concurrent Technologies Corp, Feb 1994.
23. KP Young. Semi-solid metal cast automotive components: new markets for die casting. Trans 17th Int Die Casting Congress, Cleveland, OH, Oct 18–21, 1993, pp. 387–393.
24. KP Young, R Fitze. Semi-solid metal cast aluminium automotive components. Proc 3rd Int Conf on Processing of Semi-solid Alloys and Composites, Tokyo, June 13–15, 1994, pp. 155–177.
25. E Tzimas, A Zavaliangos, A Lawley, C Pumberger. Physical mechanisms of the flow resistance of semi-solid materials at a high volume fraction of solid. Proc 4th Int Conf on Semi-solid Processing of Alloys and Composites, Sheffield, June 19–21, 1996, pp. 40–46.
26. GK Sigworth. Defect formation during semi-solid casting. Int J Cast Metals Res 9:113–123, 1996.
27. SB Brown. Semi-solid processing: new advances in net shape forming. In: W Hale, ed. Light Metals 1996, Warrendale, PA: TMS, pp. 763–766.
28. G Wan, PR Sahm. Particle characteristics and coarsening mechanisms in semi-solid processing. Proc 2nd Int Conf on Semi-solid Processing of Alloys and Composites, Cambridge, MA, June 10–12, 1992, pp. 328–335.
29. SP Midson, K Brissing. Semi-solid casting of aluminum alloys: a status report. Modern Casting 41–43, February 1997.
30. MF Rothman, ed. High-Temperature Property Data: Ferrous Alloys. Metals Park, OH: ASM International, 1989.
31. ASM Metals Handbook, Vol. 3: Properties and Selection: Stainless Steels, Tool Materials and Special-Purpose Metals. Metals Park, OH: ASM International, 1980.
32. N Wang, H Peng, KK Wang. Rheomolding—a one-step process for producing semi-solid metal castings with lowest porosity. In: W Hale, ed. Light Metals 1996, Warrendale, PA: TMS, pp. 781–786.
33. NL Bradley, PS Frederick, DF Pawlowski, WJ Schafer. Injection molding of thixotropic magnesium: machine development. Trans 15th Die Casting Congress, St. Louis, MO, Oct 16–19, 1989, Paper No G-T89-112.

34. H Peng, SP Wang, N Wang, KK Wang. Rheomolding—injection molding of semi-solid metals. Proc 3rd Int Conf on Processing of Semi-solid Alloys and Composites, Tokyo, June 13–15, 1994, pp. 191–200.
35. GK Sigworth. Rheological properties of metal alloys in the semi-solid state. Canadian Met Qtr 35:101–122, April–June 1996.
36. A Einstein. Inaugural Dissertation, Bern, 1905.
37. DG Thomas. J Colloid Sci. 20:267, 1965.
38. D Ghosh, R Fan, C VanSchilt. Thixotropic properties of semi-solid magnesium alloys AZ91D and AM50. Proc 3rd Int Conf on Processing of Semi-solid Alloys and Composites, Tokyo, June 13–15, 1994, pp. 85–94.
39. SP Wang, KK Wang, LS Turng. Die-casting of semi-solid metals. Proc 2nd Int Conf on Semi-solid Processing of Alloys and Composites, Cambridge, MA, June 10–12, 1992, pp. 336–345.
40. OJ Ilegbusi. Application of a time-dependent constitutive model to rheocast systems. J Matls Engr and Perf 5:117–123, February 1996.
41. OJ Ilegbusi, J Szekely. The role of constitutive relationships in predicting the behavior of rheocast systems. Proc 2nd Int Conf on Semi-solid Processing of Alloys and Composites, Cambridge, MA, June 10–12, 1992, pp. 364–375.
42. WR Loué, M Suéry, JL Querbes. Microstructure and rheology of partially remelted AlSi-alloys. Proc 2nd Int Conf on Semi-solid Processing of Alloys and Composites, Cambridge, MA, June 10–12, 1992, pp. 266–275.
43. S Okano. Research activities in Rheo-Technology Ltd. Proc 3rd Int Conf on Processing of Semi-solid Alloys and Composites, Tokyo, June 13–15, 1994, pp. 7–18.
44. TZ Kattamis, AI Nakhla. Rheological, microstructural and constitutional studies of semi-solid Al-4.5%Cu-1.5%Mg alloy. Proc 2nd Int Conf on Semi-solid Processing of Alloys and Composites, Cambridge, MA, June 10–12, 1992, pp. 237–247.
45. TZ Kattamis, TJ Piccone. Rheology of semisolid Al-4.5%Cu-1.5%Mg alloy. Matls Sci and Engr A A131:265–272, 1991.
46. MR Barkhudarov, CL Bronisz, CW Hirt. Three-dimensional thixotropic flow model. Proc 4th Int Conf on Semi-solid Processing of Alloys and Composites, Sheffield, June 19–21, 1996, pp. 110–114.
47. P Kapranos, DH Kirkwood, MR Barkhudarov. Modeling of structural breakdown during rapid compression of semi-solid alloy slugs. Proc 5th Int Conf on Semi-solid Processing of Alloys and Composites, Golden, CO, June 23–25, 1998, pp. 11–19.
48. CJ Quaak, MG Horsten, WH Kool. Rheological behaviour of partially solidified aluminum matrix composites. Matls Sci and Engr A A183:247–256, 1994.
49. GH Nickodemus, CM Wang, ML Tims, JJ Fisher, JJ Cardarella. Rheology of materials for semi-solid metalworking applications. Proc 5th Int Conf on Semi-solid Processing of Alloys and Composites, Golden, CO, June 23–25, 1998, pp. 29–34.

50. AN Alexandrou. Constitutive modeling of semi-solid materials. Proc 4th Int Conf on Semi-solid Processing of Alloys and Composites, Sheffield, June 19–21, 1996, pp. 132–136.
51. AN Alexandrou, A Ahmed. Processing of semi-solid materials: final report. Worcester Polytechnic Institute, Worcester, MA, Report to Concurrent Technologies Corp, Jan 10, 1996.
52. SB Brown. An internal variable constitutive model for semi-solid slurries. Modeling of Casting, Welding and Advanced Solidification Processes V, Warrendale, PA: TMS, 1991, pp. 31–38.
53. P Kumar, CL Martin, S Brown. Flow behavior of semi-solid alloy slurries. Proc 2nd Int Conf on Semi-solid Processing of Alloys and Composites, Cambridge, MA, June 10–12, 1992, pp. 248–262.
54. P Kumar, CL Martin, S Brown. Predicting the constitutive flow behavior of semi-solid metal alloy slurries. Proc 3rd Int Conf on Processing of Semi-solid Alloys and Composites, Tokyo, June 13–15, 1994, pp. 37–46.
55. P Kumar, CL Martin, S Brown. Shear rate thickening flow behavior of semi-solid slurries. Met Trans A 24A:1107–1116, May 1993.
56. P Kumar, CL Martin, S Brown. Constitutive modeling and characterization of the flow behavior of semi-solid metal alloy slurries—I. The flow response. Acta Met 42:3595–3602, 1994.
57. CL Martin, P Kumar, S Brown. Constitutive modeling and characterization of the flow behavior of semi-solid metal alloy slurries—II. Structural evolution under shear deformation. Acta Met 42:3603–3614, 1994.
58. A Zavaliangos, A Lawley. Numerical simulation of thixoforming. J Matls Engr and Perf 4:40–47, February 1995.
59. M Mada, F Ajersch. Rheological model of semi-solid A356-SiC composite alloys—Part I: Dissociation of agglomerate structures during shear. Matls Sci and Engr A A212:157–170, July 15, 1996.
60. M Mada, F Ajersch. Rheological model of semi-solid A356-SiC composite alloys—Part II: Reconstitution of agglomerate structures at rest. Matls Sci and Engr A A212:171–177, July 15, 1996.
61. M Modigell, J Koke, J Petera. Two-phase model for metal alloys in the semi-solid state. Proc 5th Int Conf on Semi-solid Processing of Alloys and Composites, Golden, CO, June 23–25, 1998, pp. 317–326.
62. M Suéry, CL Martin, L Salvo. Overview of the rheological behaviour of globular and dendritic slurries. Proc 4th Int Conf on Semi-solid Processing of Alloys and Composites, Sheffield, June 19–21, 1996, pp. 21–29.
63. CJ Quaak, L Katgerman, WH Kool. Viscosity evolution of partially solidified aluminum slurries after a shear rate jump. Proc 4th Int Conf on Semi-solid Processing of Alloys and Composites, Sheffield, June 19–21, 1996, pp. 35–39.
64. M Mada, F Ajersch. Viscosity measurements of A356–15% Si semi-solid alloys using a squeezing flow viscometer. Proc 2nd Int Conf on Semi-solid Processing of Alloys and Composites, Cambridge, MA, June 10–12, 1992, pp. 276–289.

65. J Xu, S Cheng, S Hsu, ML Tims, FR Dax. A two-step approach for the simulation of semi-solid metalworking. Proc Int Symp on Recent Advs in Constitutive Laws for Engr Matls, Mauna Lani, HI, July 30–Aug 3, 1995.

66. W Yunhua, G Zhiqiang, Z Mingfang, S Huaqin. An improved net inflow FEM simulation on the squeezing process of semi-solid ZA12 alloy. Proc 5th Int Conf on Semi-solid Processing of Alloys and Composites, Golden, CO, June 23–25, 1998, pp. 699–704.

67. CG Kang, YH Moon. Finite element simulation of the semi-solid forming based on the deformation models of the spheroidal structure. Proc 5th Int Conf on Semi-solid Processing of Alloys and Composites, Golden, CO, June 23–25, 1998, pp. 573–580.

68. V DePierre, F Gurney, AT Male. Mathematical calibration of the ring test with bulge formation, Wright-Patterson AFB, OH, AFML-TR-72-37, March 1972.

69. AT Male, MG Cockcroft. A method for the determination of the coefficient of friction of metals under conditions of bulk plastic deformation. J Inst Metals 93:38–46, 1964–1965.

70. AT Male, V DePierre. The use of the ring compression test for defining realistic metal processing parameters, Wright-Patterson AFB, OH, AFML-TR-70-129, June 1970.

71. G Garmong, NE Paton, JC Chesnutt, LF Nevarez. An evaluation of the ring test for strain-rate-sensitive materials. Met Trans A 8A:2026–2027, December 1977.

72. M Suéry, MC Flemings. Effect of strain rate on deformation behavior of semi-solid dendritic alloys. Met Trans A 13A:1809–1819, October 1982.

73. DA Pinsky, PO Charreyron, MC Flemings. Compression of semi-solid dendritic Sn-Pb alloys at low strain rates. Met Trans B 15B:173–181, March 1984.

74. C Yoshida, M Moritaka, S Shinya, S Yahata, K Takebayashi, A Nanba. Semi-solid forging of aluminum alloy. Proc 2nd Int Conf on Semi-solid Processing of Alloys and Composites, Cambridge, MA, June 10–12, 1992, pp. 95–102.

75. ML Tims, J Xu, G Nickodemus, FR Dax. Computer based numerical analysis of semi-solid metalworking. Proc 4th Int Conf on Semi-solid Processing of Alloys and Composites, Sheffield, June 19–21, 1996, pp. 120–125.

76. A Zavaliangos, E Tzimas. A two-phase model of the mechanical behavior of semi-solid metallic alloys at high volume fractions of solid. Proc 5th Int Conf on Semi-solid Processing of Alloys and Composites, Golden, CO, June 23–25, 1998, pp. 705–712.

77. L Salvo, M Suéry, Y DeCharentenay, W Loué. Microstructural evolution and rheological behaviour in the semi-solid state of a new Al-Si based alloy. Proc 4th Int Conf on Semi-solid Processing of Alloys and Composites, Sheffield, June 19–21, 1996, pp. 10–15.

78. DR Poirier, S Ganesan, M Andres, P Ocansey. Isothermal coarsening of dendritic equiaxial grains in Al-15.6wt.%Cu alloy. Matls Sci and Engr A A148:289–297, Dec 14, 1991.

79. GK Sigworth, Grain growth in SSM materials, Concurrent Technologies Corp, Johnstown, PA, 1994.

80. DS Han, G Durrant, B Cantor. Semi-solid deformation of 2014 Al alloys. Proc 5th Int Conf on Semi-solid Processing of Alloys and Composites, Golden, CO, June 23–25, 1998, pp. 43–50.
81. GK Sigworth. Technical articles on the rheology of suspensions. CTC/GKS-M0052-94, Concurrent Technologies Corp, Johnstown, PA, Jan 9, 1994.
82. D Leighton, A Acrivos. The shear induced migration of particles in concentrated suspensions. J Fluid Mechs 181:415–439, 1987.
83. S Toyoshima. A FEM simulation of densification in forming processes for semi-solid materials. Proc 3rd Int Conf on Processing of Semi-solid Alloys and Composites, Tokyo, June 13–15, 1994, pp. 47–62.
84. JS Gunasekera. Development of a constitutive model for mushy (semi-solid) materials. Proc 2nd Int Conf on Semi-solid Processing of Alloys and Composites, Cambridge, MA, June 10–12, 1992, pp. 211–222.
85. E Tzimas, A Zavaliangos, A Lawley. Mechanical behavior of spray cast alloys in the semi-solid regime under unconstrained compression. In: W Hale, ed. Light Metals 1996, Warrendale, PA: TMS, pp. 799–806.
86. CL Martin, SB Brown, D Favier, M Suéry. Mechanical behavior of coarse dendritic semi-solid Sn-Pb alloys under various stress states. Proc 3rd Int Conf on Processing of Semi-solid Alloys and Composites, Tokyo, June 13–15, 1994, pp. 27–36.
87. S Sannes, H Gjestland, L Arnberg, JK Solberg. Yield point behaviour of semi-solid Mg alloys. Proc 3rd Int Conf on Processing of Semi-solid Alloys and Composites, Tokyo, June 13–15, 1994, pp. 271–280.
88. TG Nguyen, D Favier, M Suéry. Theoretical and experimental study of the iso-thermal mechanical behaviour of alloys in the semi-solid state. Int J Plasticity 10:663–693, 1994.
89. JC Gebelin, D Favier, M Suéry, C Guarneri. A FEM simulation of semi-solid materials behaviour. Proc 4th Int Conf on Semi-solid Processing of Alloys and Composites, Sheffield, June 19–21, 1996, pp. 126–131.
90. CL Martin, D Favier, M Suéry. Experimental measure of the bulk deformation and liquid pressure of a semi-solid specimen under drained and undrained conditions. Proc 4th Int Conf on Semi-solid Processing of Alloys and Composites, Sheffield, June 19–21, 1996, pp. 51–57.
91. Y Shirai, T Moriya, C Yoshida, S Okano. Heat transfer properties and solidification structure in the initial solidification of semi-solid metals. Proc 4th Int Conf on Semi-solid Processing of Alloys and Composites, Sheffield, June 19–21, 1996, pp. 97–102.
92. KC Bearden. Computer simulation of the induction heating process used in the semi-solid metalworking process. TR No. 00496, Concurrent Technologies Corp, Johnstown, PA, March 29, 1995.
93. CA Tudbury. Basics of Induction Heating, vol. 1, New York: John Rider, 1960.
94. S Zinn, SL Semiatin. Elements of Induction Heating: Design, Control, and Applications. Metals Park, OH: ASM International, 1988.

95. JA Dantzig, SP Midson. Billet heating for semi-solid forming. Proc 2nd Int Conf on Semi-solid Processing of Alloys and Composites, Cambridge, MA, June 10–12, 1992, pp. 105–118.

96. RC Gibson. Computer model SCEDDY ensures uniform partial melting of slugs for SSM applications. Proc 4th Int Conf on Semi-solid Processing of Alloys and Composites, Sheffield, June 19–21, 1996, pp. 137–141.

97. DK Moyer. Semi-solid metalworking technology for titanium fluid handling components—report on induction heating analysis: induction heating development and analysis for titanium alloys. TR No. 96-175, Concurrent Technologies Corp, Johnstown, PA, March 31, 1997.

98. P Kapranos, RC Gibson, DH Kirkwood, CM Sellars, Induction heating and partial melting of high melting point thixoformable alloys. Proc 4th Int Conf on Semi-solid processing of Alloys and Composites, Sheffield, June 19–21, 1996, pp. 148–152.

99. R Sebus, G Henneberger. Optimisation of coil-design for inductive heating in the semi-solid state. Proc 5th Int Conf on Semi-solid Processing of Alloys and Composites, Golden, CO, June 23–25, 1998, pp. 481–487.

100. ML Tims. Semi-solid metalworking—billet handling thermal response model: final report. TR No. 95-082D, Concurrent Technologies Corp, Johnstown, PA, Dec 14, 1995.

101. ML Tims, TP Creeden. Semi-solid metalworking—application of SSM process analysis models: comparison with part manufacturing experience. TR No. 95-161, Concurrent Technologies Corp, Johnstown, PA, Dec 31, 1995.

102. TP Creeden, RS Corrente, ML Tims, FR Dax. Tooling design for semi-solid metalworking. Trans 18th Int Die Casting Congress, Indianapolis, IN, Nov 1995, p. 373.

15
Continuous Casting

Brian G. Thomas
University of Illinois at Urbana-Champaign, Urbana, Illinois

I. INTRODUCTION

Continuous casting is used to solidify most of the 750 million tons of steel produced in the world every year. Like most commercial processes, continuous casting involves many complex interacting phenomena. Most previous advances have been based on empirical knowledge gained from experimentation with the process. To further optimize the design and improve the continuous casting process, mathematical models are becoming increasingly powerful tools to gain additional quantitative insight. The best models for this purpose are mechanistic models based on the fundamental laws and phenomena which govern the process, because they are more reliably extended beyond the range of data used to calibrate them.

This chapter first presents an overview of the many interacting phenomena that occur during the continuous casting of steel. It then reviews some of the advanced mechanistic models of these phenomena and provides a few examples of the information and insights gained from them. These model applications focus on the mold region, where many continuous casting defects are generated.

II. PROCESS DESCRIPTION

In the continuous casting process, pictured in Fig.1, molten steel flows from a ladle, through a tundish into the mold. It should be protected from exposure to air by a slag cover over each vessel and by ceramic nozzles between vessels. Once in the mold, the molten steel freezes against the water-cooled copper

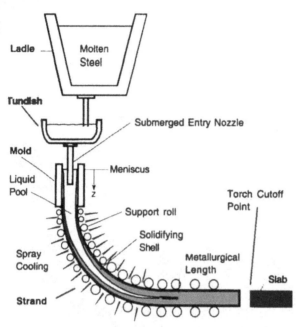

Figure 1 Schematic of steel continuous casting process.

mold walls to form a solid shell. Drive rolls lower in the machine continuously withdraw the shell from the mold at a rate or "casting speed" that matches the flow of incoming metal, so the process ideally runs in steady state. Below mold exit, the solidifying steel shell acts as a container to support the remaining liquid. Rolls support the steel to minimize bulging due to the ferrostatic pressure. Water and air mist sprays cool the surface of the strand between rolls to maintain its surface temperature until the molten core is solid. After the center is completely solid (at the "metallurgical length") the strand can be torch cut into slabs.

III. BASIC PHENOMENA

Some of the important phenomena which govern the continuous casting process and determine the quality of the product are illustrated in Fig. 2. Steel flows into the mold through ports in the submerged entry nozzle, which is usually bifurcated. The high velocities produce Reynolds numbers exceeding 100,000 and fully turbulent behavior.

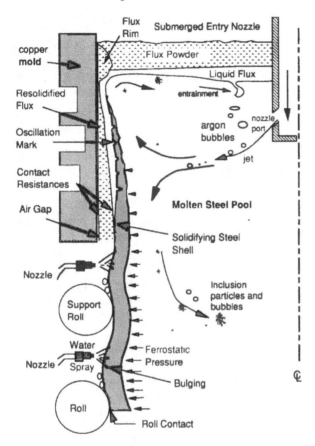

Figure 2   Schematic of phenomena in the mold region.

Argon gas is injected into the nozzle to prevent clogging. The resulting bubbles provide buoyancy that greatly affects the flow pattern, both in the nozzle and in the mold. They also collect inclusions and may become entrapped in the solidifying shell, leading to serious surface defects in the final product.

The jet leaving the nozzle flows across the mold and impinges against the shell solidifying at the narrow face. The jet carries superheat, which can erode the shell where it impinges on locally thin regions. In the extreme, this may cause a costly breakout, where molten steel bursts through the shell.

Typically, the jet impinging on the narrow face splits to flow upwards towards the top free surface and downwards toward the interior of the strand. Flow recirculation zones are created above and below each jet. This flow pattern changes radically with increasing argon injection rate or with the application of electromagnetic forces, which can either brake or stir the liquid. The flow pattern can fluctuate with time, leading to defects, so transient behavior is important.

Liquid flow along the top free surface of the mold is very important to steel quality. The horizontal velocity along the interface induces flow and controls heat transfer in the liquid and solid flux layers, which float on the top free surface. Inadequate liquid flux coverage leads to nonuniform initial solidification and a variety of surface defects.

If the horizontal surface velocity is too large, the shear flow and possible accompanying vortices may entrain liquid flux into the steel. This phenomenon depends greatly on the composition-dependent surface tension of the interface and possible presence of gas bubbles, which collect at the interface and may even create a foam [1]. The flux globules then circulate with the steel flow and may later be entrapped into the solidifying shell lower in the caster to form internal solid inclusions.

The vertical momentum of the steel jet lifts up the interface where it impinges the top free surface. This typically raises the narrow face meniscus, and creates a variation in interface level, or *standing wave*, across the mold width. The liquid flux layer tends to become thinner at the high points, with detrimental consequences.

Transient fluctuations in the flow cause time variations in the interface level which lead to surface defects such as entrapped mold powder. These level fluctuations may be caused by random turbulent motion, or changes in operating conditions, such as the sudden release of a nozzle clog or large gas bubbles.

The molten steel contains solid inclusions, such as alumina. These particles have various shapes and sizes and move through the flow field while colliding to form larger clusters and may attach to bubbles. They either circulate up into the mold flux at the top surface, or are entrapped in the solidifying shell to form embrittling internal defects in the final product.

Mold powder is added to the top surface to provide thermal and chemical insulation for the molten steel. This oxide-based powder sinters and melts into the top liquid layer that floats on the top free interface of the steel. The melting rate of the powder and the ability of the molten flux to flow and to absorb detrimental alumina inclusions from the steel depends on its composition, governed by time dependent thermodynamics. Some liquid flux resolidifies against the cold mold wall, creating a solid flux rim which inhibits heat transfer at the meniscus. Other flux is consumed into the gap between the shell and mold by the downward motion of the steel shell, where it encourages uniform heat transfer and helps to prevent sticking.

Periodic oscillation of the mold is needed to prevent sticking of the solidifying shell to the mold walls, and to encourage uniform infiltration of the mold flux into the gap. This oscillation affects the level fluctuations and associated defects. It also creates periodic depressions in the shell surface, called *oscillation marks*, which affect heat transfer and act as initiation sites for cracks.

Initial solidification occurs at the meniscus and is responsible for the surface quality of the final product. It depends on the time dependent shape of the meniscus, liquid flux infiltration into the gap, local superheat contained in the flowing steel, conduction of heat through the mold, liquid mold flux and resolidified flux rim, and latent heat evolution. Heat flow is complicated by thermal stresses which bend the shell to create contact resistance, and nucleation undercooling, which accompanies the rapid solidification and controls the initial microstructure.

Further solidification is governed mainly by conduction and radiation across the interfacial gap between the solidifying steel shell and the mold. This gap consists mainly of mold flux layers, which move down the mold at different speeds. It is greatly affected by contact resistances, which depend on the flux properties and shrinkage and bending of the steel shell, which may create an air gap. The gap size is controlled by the amount of taper of the mold walls, which is altered by thermal distortion. In addition to controlling shell growth, these phenomena are important to crack formation in the mold due to thermal stress and mold friction, which increases below the point where the flux becomes totally solid.

As solidification progresses, microsegregation of alloying elements occurs between the dendrites as they grow outward to form columnar grains. The rejected solute lowers the local solidification temperature, leaving a thin layer of liquid steel along the grain boundaries, which may later form embrittling precipitates. When liquid feeding cannot compensate for the shrinkage due to solidification, thermal contraction, phase transformations, and mechanical forces, then tensile stresses are generated. When the tensile stresses concentrated on the liquid films are high enough to nucleate an interface from the dissolved gases, then a crack will form.

After the shell exits the mold and moves between successive rolls in the spray zones, it is subject to large surface temperature fluctuations, which cause phase transformations and other microstructural changes that affect its strength and ductility. It also experiences thermal strain and mechanical forces due to ferrostatic pressure, withdrawal, friction against rolls, bending and unbending. These lead to complex internal stress profiles which cause creep and deformation of the shell. This may lead to further depressions on the strand surface, crack formation, and propagation.

Lower in the caster, fluid flow is driven by thermal and solutal buoyancy effects, caused by density differences between the different compositions created by the microsegregation. This flow leads to macrosegregation and associated defects, such as centerline porosity, cracks, and undesired property variations. Macrosegregation is complicated by the nucleation of relatively pure crystals, which move in the melt and form equiaxed grains that collect near the centerline.

Large composition differences through the thickness and along the length of the final product can also arise due to intermixing after a change in steel grade. This is governed by transient mass transport in the tundish and liquid portion of the strand.

IV. MODEL FORMULATION

Mathematical models are being applied to quantify and investigate interactions between these phenomena as a function of the controllable process parameters. Mechanistic models are based on satisfying the laws of conservation of heat, mass, force, and momentum in an appropriate domain with appropriate boundary conditions. Each phenomenon considered is represented by term(s) in these governing equations. The equations are discretized using finite difference or finite element methods and are solved numerically with computers, which are becoming increasingly fast and affordable. Because of the overwhelming complexity, no model can include all of the phenomena at once. An essential aspect of successful model development is the selection of the key phenomena and the making of reasonable assumptions.

V. FLOW THROUGH THE SUBMERGED ENTRY NOZZLE

The geometry and position of the submerged entry nozzle (SEN) are easy and inexpensive to change. These design variables have a critical influence on steel quality through their effect on the flow pattern in the mold. Fluid velocities in the nozzle have been calculated by solving the three-dimensional Navier Stokes equations for mass and momentum balance [2]. Turbulence is modeled by solving two additional partial differential equations for the turbulent kinetic energy K (m^2/s^2) and the rate of turbulence dissipation ε (m^2/s^3) and focusing on the time-averaged flow pattern.

Flow through the SEN is gravity driven by the pressure difference between the liquid levels of the tundish and the mold top free surfaces. This is generally not modeled, as the inlet velocity to the SEN is simply imposed based on the casting speed of interest. In practice, the flow rate is further controlled by other means which strive to maintain a constant liquid level in the mold. In one method, a "stopper rod" extends down through the tundish to partially plug the exit. In another method, a "slide gate" blocks off a portion of the SEN pipe section by moving a disk-shaped plate through a horizontal slit across the entire SEN. These flow-control devices strongly influence the flow pattern in the nozzle and beyond, so should be modeled.

Figure 3 shows a typical time-averaged flow pattern calculated in a 50% open slide gate system [3] using a commercial finite difference program [4]. Multiphase flow effects, caused by the injection of argon gas in the upper tundish well just above the slide gate, were modeled by solving additional transport equations for the gas phase. Velocity vectors on the left and corre-

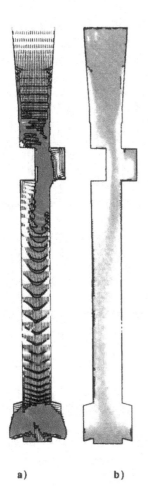

a)   b)

Figure 3 Flow pattern in nozzle with slide gate. (a) Velocities with 10% volume fraction argon injection; (b) gas fraction (lighter areas have more argon). (From Ref. 3.)

sponding gas bubble fraction on the right show that gas collects in at least five different recirculation zones. The largest regions form in the cavity created by the slide gate, and just below the slide gate. Gas also collects in the corners above the slide gate and in the upper portion of oversized nozzle exit ports. In each of these zones, steel flow is minimal so gas bubbles tend to collect, leading to large bubble fractions in these regions. These bubbles might collide to form large pockets of gas. If large gas pockets are entrained into the downward flowing steel, they may cause detrimental sudden changes in flow pattern, such as *annular flow* [5]. The slide gate also creates significant asymmetry. In single phase flow with a 75% closed slide gate, twice as much fluid exits the port opposite the gate opening and with a shallower jet angle [6]. The random nature of gas bubbles diminishes this asymmetry (assuming annular flow does not occur).

Figure 4 shows a close-up of steady flow near a nozzle port [7]. One quarter of the nozzle is modeled using a commercial finite element program [8] by assuming two-fold symmetry. Flow exits only the bottom portion of the nozzle port, due to the oversized area of the port (90 × 60 mm) relative to the nozzle bore (76 mm diameter). This creates stagnant recirculating flow in the upper portion of the ports, where gas collects and alumina particles can attach to form clogs. Figure 3 also shows that the jet's momentum causes it to exit at a steeper downward angle than machined into the bottom edge of the ceramic nozzle port. This particular jet exits at a downward angle of 10°, even though its nozzle has ports angled 15° upward.

The view looking into the nozzle (4b) reveals swirling flow with two recirculation zones spiraling outward from each port. Due to flow instabilities, one of these zones usually grows to dominate the entire port, leading to swirl in a particular direction. The swirl is stronger for larger, upward angled nozzles, and with a 90° aligned slide gate (which moves perpendicular to the direction pictured in Fig. 3). After a jet with a single swirl direction enters the mold cavity, it deflects toward one of the wide faces, leading to asymmetric flow.

This model has been validated with experimental measurements [2] and applied to investigate the effects of nozzle design parameters, such as the shape, height, width, thickness, and angles of the ports, on the jet leaving the nozzle [6]. This jet is characterized by its average speed, direction, spread angle, swirl, turbulence intensity, dissipation rate ε, and degree of symmetry. These conditions can be used as input to a model of flow in the mold.

VI.   FLUID FLOW IN THE MOLD

Due to its essentially turbulent nature, many important aspects of flow in continuous casting are transient and difficult to control. However, the time-

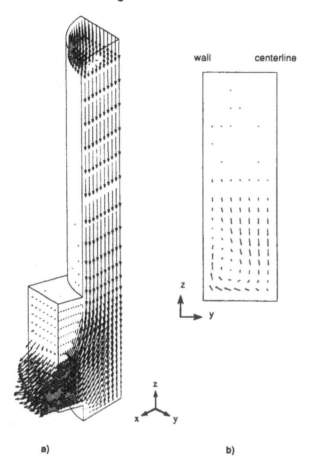

wall centerline

Figure 4 Flow pattern in SEN with 15° up-angled nozzle ports. (a) Closeup of three-dimensional velocities near port; (b) view looking into port showing swirl. (From Ref. 7.)

averaged flow pattern in the mold is greatly influenced by the nozzle geometry, submergence depth, mold dimensions, argon injection rate, and electromagnetic forces.

A. Effect of Argon Gas Injection

One of the important factors controlling flow in the mold is the amount of argon injected into the nozzle to control clogging [9]. Because the injected gas heats quickly to steel temperature and expands, the volume fraction of gas bubbles becomes significant. Those bubbles which are swept down the nozzle into the mold cavity create a strong upward force on the steel jet flowing from the nozzle, owing to their buoyancy. A few models have been applied to simulate this complex flow behavior [9, 10].

Figure 5 shows two flow patterns in the upper region of a 220 × 1320 mm mold for 1 m/min casting speed, calculated using a three-dimensional finite difference turbulent-flow model [8]. Bubble dispersion is modeled by solving a transport equation for the continuum gas bubble concentration [9], assuming that turbulent diffusivity of the gas bubble mixture is the same as that of the fluid eddies [9]. Bubble momentum and drag are ignored, so each grid point is assumed to have only a single "mixture" velocity.

Without gas injection (Fig. 5a) the jet typically hits the narrow face and is directed upward and back along the top surface toward the SEN. Maximum velocities near the center of the top surface reach almost 0.2 m/s. With optimal argon, Fig. 5b, top surface velocities are greatly reduced. With too much argon, the jet may bend upward to impinge first on the top surface, and then flow along this interface toward the narrow face. Recirculation in the upper mold is then reversed, and there are no longer separate recirculation zones above and

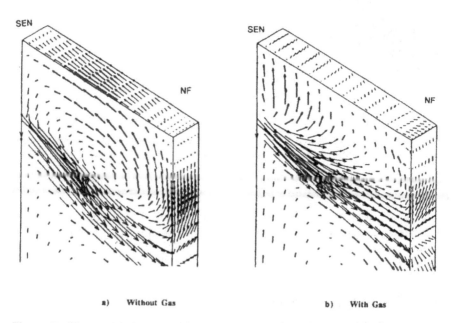

a)   Without Gas   b)   With Gas

Figure 5 Flow pattern in slab casting mold showing effect of argon gas on top surface velocities. (From Ref. 28.)

below the jet. These changes in flow pattern may have important consequences for steel quality, discussed later in this chapter.

B. Effect of Electromagnetic Forces

Electromagnetic forces can be applied to alter the flow in continuous casting in several different ways. A rotating magnetic field can be induced by passing electrical current through coils positioned around the mold. This forces electromagnetic "stirring" of the liquid in the horizontal plane of the strand. Alternatively, a strong DC magnetic field can be imposed through the mold thickness, which induces eddy currents in the metal. The resulting interaction creates a "braking" force which slows down the fluid in the flow direction perpendicular to the imposed field. Slower flow has several potential benefits: slower, more uniform fluid velocities along the top surface; more uniform temperature [11]; less inclusion entrapment in the solidifying shell below the mold [12]; and the ability to separate two different liquids to cast clad steel, where the surface has a different composition than the interior [13].

Electromagnetic phenomena are modeled by solving Maxwell's equations and then applying the calculated electromagnetic force field as a body force per unit volume in the steel flow equations [14]. Significant coupling between the electromagnetic field and the flow field may occur for DC braking, which then requires iteration between the magnetic field and flow calculations. Idogawa and coworkers applied a decoupled model to suggest that the optimal braking strategy was to impose a field across the entire width of the mold in two regions: above and below the nozzle inlet [14]. Care must be taken not to slow down the flow too much, or the result is the same as angling the ports to direct the jet too steeply downward: defects associated with freezing the meniscus. In addition, the field also increases some velocity components, which has been modeled to increase free surface motion in some circumstances [15].

Others have examined the application of electromagnetic fields near the meniscus to change the surface microstructure [16]. Finally, electromagnetic stirring both in and below the mold is reported to reduce centerline macrosegregation [17], presumably due to the flow effects on heat transfer and nucleation.

C. Transient Flow Behavior

Transient surges in the steel jets leaving the nozzle parts may cause asymmetric flow, leading to sloshing or waves in the molten pool [18]. Jet oscillations are periodic and increase in violence with casting speed, making them a particular concern for thin slab casting [19]. Huang has shown that a sudden change in inlet velocity creates a large transient flow structure, that appears to be a large vortex shed into the lower region of the liquid cavity [20]. Recent transient models have reproduced periodic oscillations of the jet, even with constant inlet conditions [21]. The consequences of nonoptimal flow, such as top surface level fluctuations are discussed next.

VII.   CONSEQUENCES OF FLUID FLOW IN THE MOLD

The steady flow pattern in the mold is not of interest directly. However, it influences many important phenomena, which have far-reaching consequences on strand quality. These effects include controlling the dissipation of superheat (and temperature at the meniscus), the flow and entrainment of the top surface powder layers, top-surface contour and level fluctuations, the motion and entrapment of subsurface inclusions and gas bubbles. Each of these phenomena associated with flow in the mold can lead to costly defects in the continuous-cast product [22]. Design compromises are needed to simultaneously satisfy the contradictory requirements for avoiding each of these defects.

A.   Superheat Dissipation

An important task of the flow pattern is to deliver molten steel to the meniscus region that has enough superheat during the critical first stages of solidification. Superheat is the sensible heat contained in the liquid represented by the difference between the steel temperature entering the mold and the liquidus temperature.

The dissipation of superheat has been modeled by extending the three-dimensional finite difference flow model to include heat transfer in the liquid [23]. The effective thermal conductivity of the liquid is proportional to the effective viscosity, which depends on the turbulence parameters. The solidification front, which forms the boundary to the liquid domain, has been treated in different ways. Some researchers model flow and solidification as a coupled problem on a fixed grid. However, this approach is subject to convergence difficulties and requires a fine grid to resolve the thin porous mushy zone. In addition, properties such as permeability of this mush are uncertain and care must be taken to avoid any improper advection of the latent heat. An alternative approach for columnar solidification of a thin shell, such as found in the continuous casting of steel, is to treat the boundary as a rough wall fixed at the liquidus temperature [24]. This approach has been shown to match plant measurements in the mold region, where the shell is too thin to affect the flow significantly [23, 25, 26].

Figure 6 compares the measured temperature distribution in the mold [27] with recent calculations using a three-dimensional finite difference flow model [28]. Incorporating the effects of argon on the flow pattern were very important in achieving the reasonable agreement observed. This figure shows that the temperature drops almost to the liquidus by mold exit, indicating that

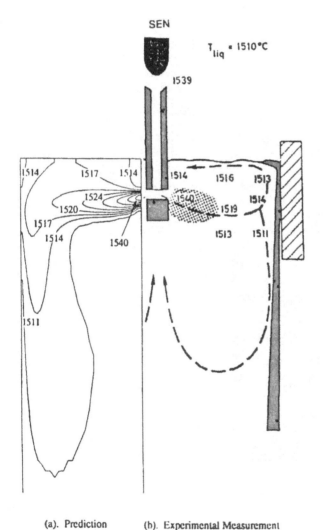

(a). Prediction   (b). Experimental Measurement

Figure 6 Superheat temperature distribution in mold [28] compared with measurements [27].

most of the superheat is dissipated in the mold. The hottest region along the top surface is found midway between the SEN and narrow face. This location is directly related to the flow pattern. The coldest regions are found at the meniscus at the top corners near the narrow face and near the SEN.

The cold region near the meniscus is a concern because it could lead to freezing of the meniscus, and encourage solidification of a thick slag rim. This could lead to quality problems such as deep oscillation marks, which later initiate transverse cracks. It can also disrupt the infiltration of liquid mold flux into the gap, which can induce longitudinal cracks and other surface defects. The cold region near the SEN is also a concern because, in the extreme, the steel surface can solidify to form a solid bridge between the SEN and the shell against the mold wall, which often causes a breakout. To avoid meniscus freezing problems, flow must reach the surface quickly. This is why flow from the nozzle should not be directed too deep.

The jet of molten steel exiting the nozzle delivers most of its superheat to the inside of the shell solidifying against the narrow face [23]. The large temperature gradients found part way down the domain indicate that the maximum heat flux delivery to the inside of the solidifying shell is at this location near mold exit. If there is good contact between the shell and the mold, then this heat flux is inconsequential. If a gap forms between the shell and the mold, however, then the reduction in heat extraction can make this superheat flux sufficient to slow shell growth and even melt it back. In the extreme, this can cause a breakout. Breakouts are most common at mold exit just off the corners, where contact is the poorest. This problem is worse with higher flow rates and non-uniform flow from the nozzle [23].

B. Top-Surface Shape and Level Fluctuations

The condition of the meniscus during solidification has a tremendous impact on the final quality of the steel product. Meniscus behavior is greatly affected by the shape of the top "free" surface of the liquid steel, and in particular, the fluctuations in its level with time. This surface actually represents the interface between the steel and the lowest powder layer, which is molten.

If the surface waves remain stable, then the interface shape can be estimated from the pressure distribution along the interface calculated from a simulation with a fixed boundary [29].

$$\text{Standing wave height} = \frac{\text{surface pressure} - 1 \text{ atm}}{(\text{steel density} - \text{flux density})g} \qquad (1)$$

When casting with low argon and without electromagnetics in a wide mold, the interface is usually raised about 25 mm near the narrow face meniscus, relative to the lower interface found near the SEN. This rise is caused by the vertical momentum of the jet traveling up the narrow face and depends greatly on the flow pattern and flow rate. The rise in level increases as the density difference between the fluids decreases, so water/oil models of the steel/flux system tend to exaggerate this phenomenon. Recent models are being developed to predict the free surface shape coupled with the fluid flow [29]. Additional equations must be solved to satisfy the force balance, at the interface, involving the pressure in the two phases, shear forces from the moving fluids, and the surface tension. Numerical procedures such as the *volume of fluid method* (VOF) are used to track the arbitrary interface position.

The transient simulation of level fluctuations above a turbulent flowing liquid is difficult to model. However, a correlation has been found between the steady kinetic energy (turbulence) profile across the top surface and level fluctuations [20]. Figure 7 compares calculated turbulence levels [20] along the top surface of the moving steel with measured level fluctuations [30]. The flow calculations were performed using a steady state, three-dimensional, turbulent K-ε finite difference model of fluid flow in the mold. The observed agreement is expected because the kinetic energy contained in the moving fluid corresponds to the time-averaged velocity fluctuations, which may be converted temporarily to potential energy in the form of a rise or fall in level.

For typical conditions, Fig. 7a, the most severe level fluctuations are found near the narrow face, where turbulence and interface level are highest. These level fluctuations can be reduced by directing the jet deeper and changing the flow pattern. It is interesting to note that increasing argon injection moves the location of maximum level fluctuations (and accompanying turbulence) toward the SEN at the central region of the mold (0.5 m away from the narrow faces in Fig. 7b. Casting at a lower speed with a smaller mold width means that steel throughput is less. This increases the volume fraction of argon, if the argon injection rate (liters/second) is kept constant. It is likely that the higher argon fraction increases bubble concentration near the SEN, where it lifts the jet, increases the interface level near the SEN, and produces the highest level fluctuations.

Sudden fluctuations in the level of the free surface are very detrimental because they disrupt initial solidification and can entrap mold flux in the solidifying steel, leading to surface defects in the final product. Level fluctuations can deflect the meniscus and upset the infiltration of the mold flux into the gap, building up a thick flux rim and leaving air gaps between the shell and the mold. Together, this can lead to deep, nonuniform oscillation marks, surface depressions, laps, bleeds, and other defects. The thermal stress created in the tip of the solidifying shell from a severe level fluctuation has been predicted to cause distortion of the shell, which further contributes to surface depressions [31].

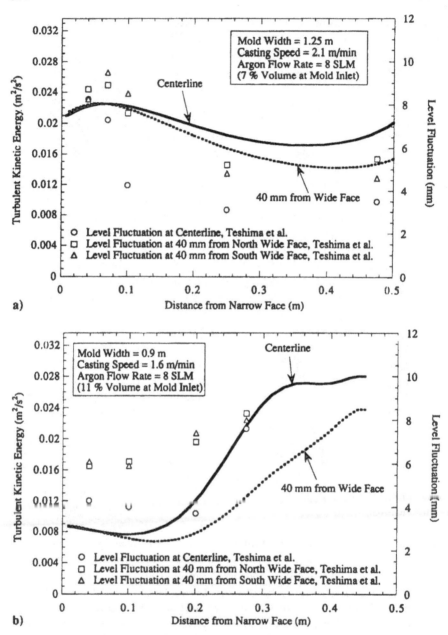

Figure 7 Comparison between calculated turbulent kinetic energy and measured level fluctuations. (a) Low argon fraction (high throughput); (b) high argon fraction (low throughput). (From Ref. 20.)

C. Top Surface Powder/Flux Layer Behavior

The flow of the steel in the upper mold greatly influences the top surface powder layers. Mold powder is added periodically to the top surface of the steel. It sinters and melts to form a protective liquid flux layer, which helps to trap impurities and inclusions. This liquid is drawn into the gap between the shell and mold, where it acts as a lubricant and helps to make heat transfer more uniform. The behavior of the powder layers is very important to steel quality and powder composition is easy to change. It is difficult to measure or to accurately simulate with a physical model, so is worthy of mathematical modeling.

A three-dimensional finite element model of heat transfer and fluid flow has been developed of the top surface of the continuous caster to predict the thickness and behavior of the powder and flux layers [32]. The bottom of the model domain is the steel/flux interface. Its shape is imposed based on measurements in an operating caster, and the shear stress along the interface is determined through coupled calculations using the three-dimensional finite difference model of flow in the steel. Uniform consumption of flux was imposed into the gap at the bottom edges of the domain along the wide and narrow faces. The model features temperature dependent properties of the flux, with enhanced viscosity in the temperature range where sintering occurs and especially in the resolidified liquid flux, which forms the rim. Temperature throughout the flux was calculated, including the interface between the powder and liquid flux layers.

When molten steel flows rapidly along the steel/flux interface, it induces motion in the flux layer. If the interface velocity becomes too high, then the liquid flux can be sheared away from the interface, become entrained in the steel jet, and be sent deep into the liquid pool to become trapped in the solidifying shell as a harmful inclusion [1]. If the interface velocity increases further, then the interface standing wave becomes unstable, and huge fluctuations contribute to further problems.

The model results in Fig. 8 show the calculated flux layer thickness relative to the assumed steel profile [32]. Input flow conditions were similar to those in Fig. 5a; maximum total powder and liquid flux thickness was 35 mm, and flux consumption was $0.6 \, kg/m^2$. The flux layer was found to greatly reduce the steel velocity at the interface, due to its generally high viscosity relative to the steel. However, the rapid flow of steel along the interface was calculated to drag the liquid flux, in this case toward the SEN. This induces recirculation in the liquid flux layer, which carries powder slowly toward the narrowface walls, opposite to the direction of steel flow. The internal recirculation also increases convection heat transfer and melts a deeper liquid flux layer near the SEN. The result is a steel flux interface that is almost flat.

The thickness of the beneficial liquid flux layer is observed to be highly non-uniform. It is generally thin near the narrow face because the steel flow

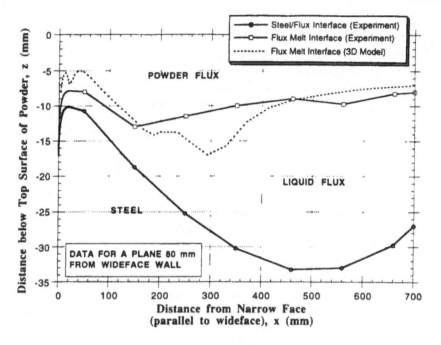

Figure 8 Comparison of measured and predicted melt-interface positions. (From Ref. 32.)

drags most of the liquid toward the center. The thinnest liquid flux layer is found about 150 mm from the narrow face. In this region, a flow separation exists where liquid flux is pulled away in both directions: toward the narrow face gap, where it is consumed, and toward the SEN. This shortage of flux, compounded by the higher, fluctuating steel surface contour in this region, makes defects more likely near the narrow face for this flow pattern. This is because it is easier for a level fluctuation to allow molten steel to touch and entrain solid powder particles. In addition, it is more difficult for liquid flux to feed continuously into the gap. Poor flux feeding leads to air gaps, reduced, nonuniform heat flow, thinning of the shell, and longitudinal surface cracks [30]. It is important to note that changing the flow pattern can change the flow in the flux layers, with corresponding changes in defect incidence.

D. Motion and Entrapment of Inclusions and Gas Bubbles

The jets of molten steel exiting the nozzle carry inclusions and argon bubbles into the mold cavity. If these particles circulate deep into the liquid pool and become trapped in the solidifying shell, they lead to internal defects. Inclusions lead to cracks, slivers, and blisters in the final rolled product [22]. Trapped argon bubbles elongate during rolling and in low-strength steel, may expand during subsequent annealing processes to create surface blisters and "pencil pipes" [1].

Particle trajectories can be calculated using the Langrangian particle tracking method, which solves a transport equation for each inclusion as it travels through a previously calculated molten steel velocity field. The force balance on each particle can include buoyancy and drag force relative to the steel. The effects of turbulence can be included using a "random walk" approach [4]. A random velocity fluctuation is generated at each step, whose magnitude varies with the local turbulent kinetic energy level. To obtain significant statistics, the trajectories of several hundred individual particles should be calculated, using different starting points. Inclusion size and shape distributions evolve with time, which affects their drag and flotation velocities. Models are being developed to predict this, including the effects of collisions between inclusions [33] and the attachment of inclusions to bubbles [3].

Particle trajectories were calculated in the above manner for continuous casting at 1 m/min through a 10 m radius curved-mold caster [3] using a finite difference program [4]. Most of the argon bubbles circulate in the upper mold area and float out to the slag layer. This is because the flotation velocities are very large for 0.3–1.0 mm bubbles. Many inclusions behave similarly, as shown by three of the inclusions in Fig. 9. Bubbles circulating in the upper recirculation region without random turbulent motion were predicted to always eventually float out. Bubbles with turbulent motion often touch against the solidification front. This occurs on both the inside and outside radius with almost equal frequency, particularly near the narrow faces high in the mold, where the shell is thin (<20 mm). It is likely that only some of these bubbles stick, when there is a solidification hook, or other feature on the solidification front to catch them. This entrapment location does not correspond to pencil pipe defects.

A few bubbles manage to penetrate into the lower recirculation zone. Here, a very small argon bubble likely behaves in a similar manner to a large inclusion cluster. Specifically, a 0.15 mm bubble has a similar terminal velocity to a 208 μm solid alumina inclusion, whose trajectories are plotted in Fig. 16. These particles tend to move in large spirals within the slow lower recirculation zones, while they float toward the inner radius of the slab. At the same time, the bubbles are collecting inclusions, and the inclusions are colliding. When they eventually touch the solidifying shell in this depth range, entrapment is more likely on the inside radius. This could occur anywhere along their spiral path, which extends from roughly 1–3 m below the meniscus. This distance corresponds exactly with the observed location of pencil pipe defects [3]. There is a slight trend that smaller bubbles are more likely to

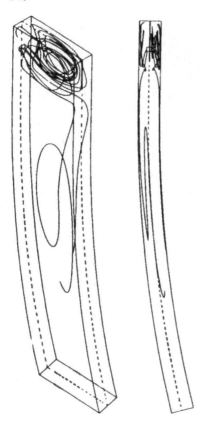

Figure 9 Sample trajectories of five 208 μm solid alumina inclusion particles. (From Ref. 3.)

enter the lower zone to be entrapped. However, the event is relatively rare for any bubble size, so the lack of sufficient statistics and random turbulent motion mask this effect.

Recent work with a transient K-ε flow model [20] suggests that sudden changes to the inlet velocity may produce transient structures in the flow that contribute to particle entrapment. After an inlet velocity change, the large zone of recirculating fluid below the jet was modeled to migrate downwards like a shedding vortex. The particles carried in this vortex are then transported very deep into the liquid pool. Thus, they are more likely to become entrapped in the solidifying shell before they float out.

To minimize the problem of particle entrapment on the inside radius, some companies have changed their machine design from the usual curved mold design to a costly straight mold design. Through better understanding of the fluid dynamics using models, it is hoped that alternative, less expensive modifications to the nozzle geometry and casting conditions might be found to solve problems such as this one.

VIII. COMPOSITION VARIATION DURING GRADE CHANGES

The intermixing of dissimilar grades during the continuous casting of steel is a problem of growing concern, as the demand for longer casting sequences increases at the same time as the range of products widens. Steel producers need to optimize casting conditions and grade sequences to minimize the amount of steel downgraded or scrapped. In addition, the unintentional sale of intermixed product must be avoided. To do this requires knowledge of the location and extent of the intermixed region and how it is affected by grade specifications and casting conditions.

A one-dimensional transient mathematical model, MIX1D, has been developed to simulate the final composition distributions produced within continuous-cast slabs and blooms during an arbitrary grade transition [34]. The model is fully transient and consists of three submodels, that account for mixing in the tundish, mixing in the liquid core of the strand, and solidification. The three submodels have been incorporated into a user-friendly FORTRAN program that runs on a personal computer in a few seconds.

Figure 10 shows example composition distributions calculated using the model. All compositions range between the old grade concentration of 0 and the new grade concentration of 1. This dimensionless concept is useful because alloying elements intermix essentially equally [35]. This is because the turbulent mass diffusion coefficient, which does not depend on the element, is about 5 orders of magnitude larger than the laminar diffusion coefficient.

To ensure that the MIX1D model can accurately predict composition distribution during intermixing, extensive verification and calibration has been undertaken for each submodel [36]. The tundish submodel must be calibrated to match chemical analysis of steel samples taken from the mold in the nozzle port exit streams, or with tracer studies using full-scale water models. Figure 10a shows an example result for a small (10 tonne) tundish operation. The strand submodel was first calibrated to match results from a full three-dimensional finite difference flow model, where an additional three-dimensional transport equation was solved for transient species diffusion [35]. Figure 10b shows reasonable agreement between the three-dimensional model and the simplified one-dimensional diffusion model. Figure 10b also compares the MIX1D model predictions with composition measurements in

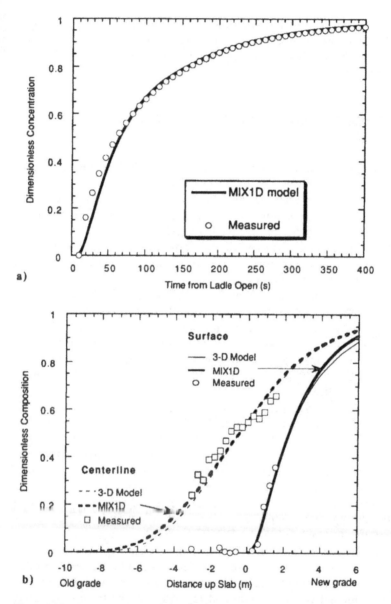

Figure 10 Predicted intermixing during a grade change compared with experiments. (a) Composition history exiting tundish into mold; (b) composition distribution in final solid slab. (From Ref. 34.)

the final slab, corresponding to the tundish conditions in Fig. 10a. The predictions are seen to match reasonably well for all cases. This is significant because there are no parameters remaining to adjust in the strand submodel.

The results in Fig. 10b clearly show the important difference between centerline and surface composition. New grade penetrates deeply into the liquid cavity and contaminates the old grade along the centerline. Old grade lingers in the tundish and mold cavity to contaminate the surface composition of the new grade. This difference is particularly evident in small tundish, thick-mold operations, where mixing in the strand is dominant.

The MIX1D model has been enhanced to act as an on-line tool by outputting the critical distances which define the length of intermixed steel product which falls outside the given composition specifications for the old and new grades. This model is currently in use at several steel companies. In addition, it has been applied to perform parametric studies to evaluate the relative effects on the amount of intermixed steel for different intermixing operations [34] and for different operating conditions using a standard ladle-exchange operation [36].

IX. SOLIDIFICATION AND MOLD HEAT TRANSFER

Heat transfer in the mold region of a continuous caster is controlled primarily by heat conduction across the interface between the mold the solidifying steel shell. Over most of the wide face in the mold, the interface consists of thin solid and liquid layers of mold flux of varying thicknesses. Heat transfer is determined greatly by the factors that control these thicknesses (such as feeding events at the meniscus, flux viscosity, strength, and shell surface uniformity), the thermal conduction and radiation properties of the mold flux layers, and the contact resistances.

Many models have been developed to predict shell solidification in continuous casting. One such model, CON1D, couples a one-dimensional transient solidification model of the shell with a two-dimensional steady state heat conduction model of the mold. It features a detailed treatment of the interface, including heat, mass, and momentum balances on the flux in the gap and the effect of oscillation marks on heat flow and flux consumption [37, 38]. Axial heat conduction can be ignored in models of steel continuous casting because it is small relative to axial advection, as indicated by the small Peclet number (casting speed multiplied by shell thickness divided by thermal diffusivity).

Typical boundary conditions acting on the shell in the mold region are presented in Fig. 11. Heat transfer to the gap is greatest at the meniscus, where the shell is thin and its thermal resistance is small. Delivery of superheat from the liquid is greatest near mold exit, where the hot turbulent jet of molten steel

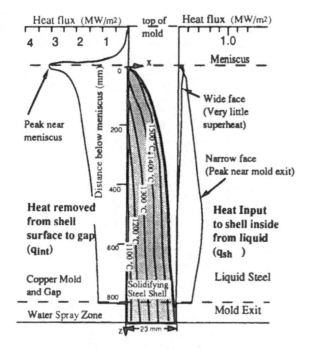

Figure 11 Typical isotherms and boundary conditions calculated on shell solidifying in mold. (From Ref. 65.)

impinges on the solidification front. The model incorporates the superheat effect using a data base of heat flux results calculated from the turbulent fluid flow model of the liquid pool [23].

Figure 12 schematically shows the temperature and velocity distributions that arise across the interface [37]. The steep temperature gradients are roughly linear across each layer. Liquid flux attached to the strand, including flux trapped in the oscillation marks, moves downward at the casting speed. Solid flux is attached to the mold wall at the top. Lower down, it creeps along intermittently. For simplicity, the model assumes a constant average downward velocity of the solid flux, calibrated to be 10–20% of the casting speed.

This modeling tool can predict thermal behavior in the mold region of a caster as a function of design and operating variables, after calibration with mold thermocouple and breakout shell measurements from a given caster [39]. In addition to temperatures, the CON1D model also predicts important phenomena, such as the thicknesses of the solid and liquid flux layers in the gap. The complete freezing of the liquid powder in the interfacial gap generates

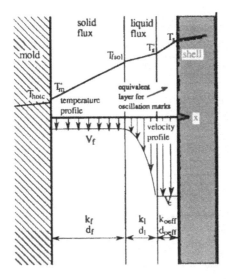

Figure 12   Velocity and temperature profiles across interfacial gap layers (no air gap). (From Ref. 38.)

friction and increases the likelihood of transverse crack formation. So long as a liquid slag film is present in contact with the steel shell, hydrodynamic friction forces remain fairly low, and a sound shell is more likely.

One example application of this model is the interpretation of thermo-couple signals which are routinely monitored in the mold walls. Measured temperature fluctuations in the mold walls are already useful in the prediction and prevention of breakouts [40]. If better understood, patterns in the tem-perature signals might also be used to predict surface quality problems, which would ease the burden of surface inspection [41].

Toward this goal, the model was calibrated with thermocouple data and applied to predict shell thickness [42] Then, the effect of local variations in oscillation mark depth were modeled using the same interface parameters [26]. Predictions for a section of shell with deep surface depressions are shown in Fig. 13. The four oscillation marks in this section of shell are seen to be very wide and deep, with areas of 1.1–2.5 mm^2. Even filled with flux, they signifi-cantly reduce the local shell growth. Temperatures at the oscillation mark roots are predicted to increase significantly. The resulting decrease in shell thickness, relative to that predicted with no oscillation marks, is predicted to range from 0.5 to 0.9 mm over this section (6% maximum loss). These predictions almost exactly match measurements from the breakout shell, as shown in Fig. 17. This match is significant because the model calibration was performed for a differ-

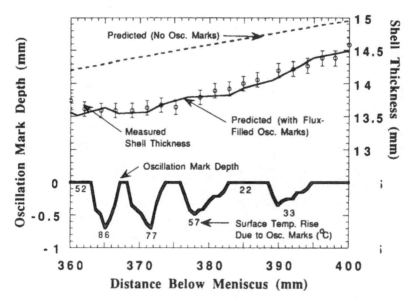

Figure 13 Comparison of predicted and measured shell thickness with oscillation mark profile and surface temperatures for large, flux-filled oscillation marks. (From Ref. 26.)

ent average oscillation mark depth and did not consider local variations. It suggests that oscillation marks are indeed filled with flux, at least in this case. The drop in heat flux that accompanies each depression causes a corresponding drop in mold temperature as it passes by a given thermocouple location. Such a temperature disturbance displaced in time at several different thermocouples indicates the presence of a surface depression moving down the mold at the casting speed [43].

X. THERMAL MECHANICAL BEHAVIOR OF THE MOLD

Thermal distortion of the mold during operation is important to residual stress, residual distortion, fatigue cracks, and mold life. By affecting the internal geometry of the mold cavity, it is also important to heat transfer to the solidifying shell. To study thermal distortion of the mold, a three-dimensional finite element model with an elastic-plastic-creep constitutive model has been developed [44]. In order to match the measured distortion [44], the model had to incorporate all of the complex geometric features of the mold, including the four copper plates with their water slots, reinforced steel water box assemblies, and tightened bolts. Ferrostatic pressure pushes against the inside of the copper, but was found to have a negligible influence on mold distortion [44]. This pressure is balanced by mold clamping forces on the eight points joining the water chambers to the rigid steel frame. Its four-piece construction makes the slab mold behave very differently from single-piece bloom or billet molds, which have also been studied using thermal stress models [45].

Figure 14 illustrates typical temperature contours and the displaced shape calculated in one quarter of the mold under steady operating conditions [44]. The hot exterior of each copper plate attempts to expand, but is constrained by its colder interior and the cold, stiff, steel water jacket. This makes each plate bend in towards the solidifying steel. Maximum inward distortions of more than one millimeter are predicted just above the center of the mold faces, and below the location of highest temperature, which is found just below the meniscus.

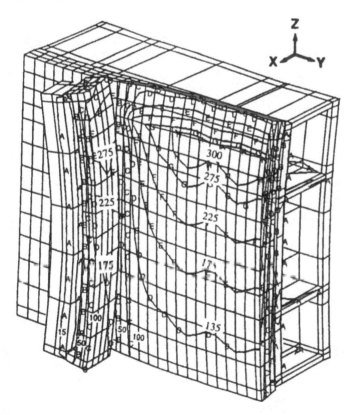

Figure 14 Distorted mold shape during operation with temperature contours (°C). (From Ref. 44.)

The narrow face is free to rotate away from the wide face and contact only along a thin vertical line at the front corner of the hot face. This hot edge must transmit all of the clamping forces, so is prone to accelerated wear and crushing, especially during automatic width changes. If steel enters the gaps formed by this mechanism, this can lead to finning defects or even a sticker breakout. In addition, the widefaces may be gouged, leading to longitudinal cracks and other surface defects [44].

The high compressive stress due to constrained thermal expansion induces creep in the hot exterior of the copper plates which face the steel. This relaxes the stresses during operation, but allows residual tensile stress to develop during cooling. Over time, these cyclic thermal stresses and creep build up significant distortion of the mold plates. This can contribute greatly to remachining requirements and reduced mold life. A copper alloy with higher creep resistance is less prone to this problem. Under adverse conditions, this stress could lead to catastrophic fatigue fracture of the copper plates, when stress concentration initiates cracks at the water slot roots. Finally, the distortion predictions are important for heat transfer and behavior of the shell in the mold. This is because the distortion is on the same order as the shell shrinkage and interfacial gap sizes. It should be taken into account when designing mold taper to avoid detrimental air gap formation.

The model has been applied to investigate these practical concerns by performing quantitative parametric studies on the effect of different process and mold design variables on mold temperature, distortion, creep and residual stress [44]. This type of stress model application will become more important in the future to optimize the design of the new molds being developed for continuous thin-slab and strip casting. For example, thermal distortion of the rolls during operation of a twin-roll strip caster is on the same order as the section thickness of the steel product.

XI THERMAL MECHANICAL BEHAVIOR OF THE SHELL

The steel shell is prone to a variety of distortion and crack problems, owing to its creep at elevated temperature, combined with metallurgical embrittlement and thermal stress. To investigate these problems, a transient, thermal-elastic-viscoplastic finite element model, CON2D [46] has been developed to simulate thermal and mechanical behavior of the solidifying steel shell during continuous casting in the mold region [42]. This model tracks the behavior of a two-dimensional slice through the strand as it moves downward at the casting speed through the mold and upper spray zones. The two-dimensional nature of this modeling procedure makes the model ideally suited to simulate phenomena such as longitudinal cracks and depressions, when simulating a horizontal section [47]. A vertical section domain can be applied to simulate transverse phenomena [31].

The model consists of separate finite element models of heat flow and stress generation that are step-wise coupled through the size of the interfacial gap. The heat flow model solves the two-dimensional transient energy equation, using a fixed Lagrangian grid of three-node triangles. The interface heat flow parameters (including properties and thickness profiles of the solid and liquid mold flux layers in the gap) are chosen by calibrating the CON1D with plant measurements down the wide face, where ferrostatic pressure prevents air gap formation. The one-dimensional results are extrapolated around the mold perimeter by adding an air gap, which depends on the amount of shrinkage of the steel shell. The gap thickness is recalculated at each location and time, knowing the position of the strand surface (calculated by the stress model at the previous time step), and the position of the mold wall at that location and time.

Starting with stress-free liquid at the meniscus, the stress model calculates the evolution of stresses, strains, and displacements, by interpolating the thermal loads onto a fixed-grid mesh of six-node triangles [48]. The loads are calculated from the heat transfer model results, and arise from both thermal and phase transformation strains. The out-of-plane z stress state is characterized by the *generalized-plane-strain* condition, which constrains the shell section to remain planar while it cools and contracts in the z direction. This allows the two-dimensional model to reasonably estimate the complete three-dimensional stress state, for the thin shells of interest in this work.

The stress calculation incorporates a temperature dependent elastic modulus and temperature, strain rate, composition, and stress state dependent plastic flow due to high-temperature plastic creep. The elastic-viscoplastic constitutive equations (model III) developed by Kozlowski et al. [49] are used to describe the unified inelastic behavior of the shell at the high temperatures, low strains, and low strain rates important for continuous casting. The equations reproduce both the tensile test data measured by Wray [50] and creep curves from Suzuki et al. [51] for austenite. They have been extended to model the enhanced creep rate in delta-ferrite, by matching measurements from Wray [52].

These constitutive equations are integrated using a new two-level algorithm [53] which alternates between solutions at the local node point and the global system equations. The effects of volume changes due to temperature changes and phase transformations are included using a temperature and grade dependent thermal-linear-expansion function.

Intermittent contact between the shell and the mold is incorporated by imposing spring elements to restrain penetration of the shell into the mold. The shape of the rigid boundary of the water-cooled copper mold is imposed from a database of results obtained from a separate three-dimensional calculation of thermal distortion of the mold [54]. To extend model simulations below the mold, the shell is allowed to bulge outward only up to a maximum displacement, whose axial z profile is specified.

The thermal-mechanical model has been validated using analytical solutions, which is reported elsewhere [53]. The model was next compared with measurements of a breakout shell from an operating slab caster. An example of the predicted temperature contours and distorted shape of a region near the corner are compared in Fig. 15. The interface heat flow parameters (including thickness profile of the solid and liquid mold flux layers) were calibrated using thermocouple measurements down the centerline of the wideface for typical conditions [42]. Thus, good agreement was expected and found in the region of good contact along the wideface, where calibration was done. Around the perimeter, significant variation in shell thickness was predicted, due to air gap formation and nonuniform superheat dissipation.

Near the corner along the narrow face, steel shrinkage is seen to exceed the mold taper, which was insufficient. Thus, an air gap is predicted. This air gap lowers heat extraction from the shell in the off-corner region of the narrow face, where heat flow is one-dimensional. When combined with high superheat delivery from the bifurcated nozzle directed at this location, the shell growth is greatly reduced locally. Just below the mold, this thin region along the off-corner narrow-face shell caused the breakout.

Near the center of the narrow face, creep of the shell under ferrostatic pressure from the liquid is seen to maintain contact with the mold, so much less

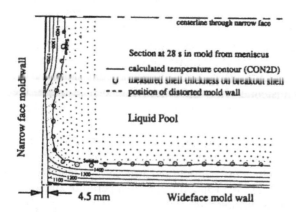

Figure 15 Comparison between predicted and measured shell thickness in a horizontal xy section through the corner of a continuous-cast steel breakout shell. (From Ref. 42.)

thinning is observed. The surprisingly close match with measurements all around the mold perimeter tends to validate the features and assumptions of the model. It also illustrates the tremendous effect that superheat has on slowing shell growth, if (and only if!) there is a gap present, which lowers heat flow.

This model has been applied to predict ideal mold taper [55] to prevent breakouts such as the one discussed here [56] and to understand the cause of other problems such as off-corner surface depressions and longitudinal cracks in slabs [25].

A. Longitudinal Surface Depressions

One of the quality problems which can affect any steel grade is the formation of longitudinal surface depressions, 2–8 mm deep, just off the corners along the wide faces of conventionally cast slabs. These "gutters" are usually accompanied by longitudinal subsurface cracks and by bulging along the narrow face. Gutters are a costly problem for 304 stainless steel slabs, because the surface must be ground flat in order to avoid slivers and other defects after subsequent reheating and rolling operations.

Simulations were performed for a 203 × 914 mm stainless steel strand cast through a 810 mm long mold at 16.9 mm/s with 25°C superheat, and no wide face taper [25]. Results in the mold revealed that away from the corners, ferrostatic pressure maintains good contact between the shell and the mold walls. However, "hot spots" are predicted to develop on the off-corner regions of the shell surface in the mold, as shown in both Fig. 15 (narrow face) and in Fig. 16 (wide face).

The hot spot on the wide face is predicted in Fig. 16 to reach several hundred degrees in extreme cases, and has been reported previously [57, 58]. It may arise in several different ways. The typical lack of taper on the wide face creates an inherent tendency to form a gap between the shell and mold wide face in the off-corner region. Thermal distortion tends to bend the mold walls away from the shell [44]. Insufficient taper of the narrow face mold walls may cause the solidified corner to rotate inside the mold, lifting the shell slightly away from the off-corner mold wall [59]. Finally, excessive taper of the narrow face mold walls may compress the wide face shell, causing it to buckle at its weakest point: the thin shell at the off-corner region [59]. With large enough taper, this mechanism alone could create gutters, entirely within the mold [59, 60].

Although a thin (about 0.2 mm) gap may form in several different ways, it becomes critical only if it is filled with air. This can happen if the mold flux solidifies completely part way down the mold. It is even more likely when there are problems with flux feeding at the meniscus. Thus, mold flow issues, discussed earlier, can contribute to the hot spots. The accompanying lack of heat

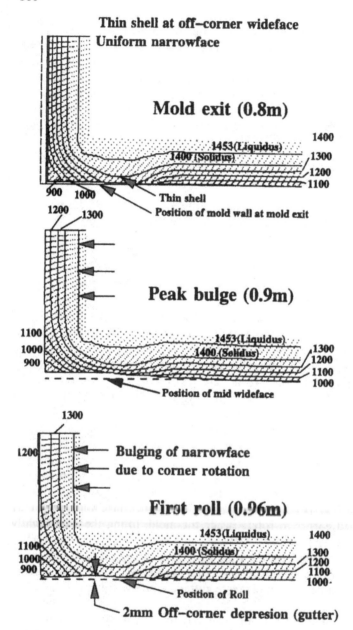

Figure 16 Calculated evolution of shape of shell below mold (to scale) with isotherms (°C). (From Ref. 25.)

flow causes thinning of the shell, which is compounded by the high superheat delivered to the inside of the shell by the impinging jet in this region. The thinner shell in the off-corner region is the critical step needed to cause bending of the shell below the mold, leading to off-corner depressions.

Continuing the model simulations below the mold revealed how longitudinal off-corner depressions form when the shell at the off-corner region was thin at mold exit. Figure 16 shows the steps that occur below the mold, assuming uniform narrow-face growth, achieved from a high 1.35%/m narrow-face taper. No depression is predicted at mold exit (0.8 m below meniscus). In order to observe the effect of the bulging over only two roll spacings, the specified bulging displacements (3 mm maximum) were much larger than those normally encountered (less than 0.5 mm).

Bulging between mold exit and the first support roll (reaching a first maximum peak at 0.90 m below the meniscus) causes the weak off-corner region to bend and curve away from the cold, rigid corner. Then, passing beneath the first support roll (0.96 m) bends the wide face back into alignment with the corner. Due to permanent creep strain, the curvature remains, in the form of a longitudinal depression. At the same time, the rotation of the rigid corner causes the narrow face to bend outward, which is observed in practice [25]. High subsurface tension is generated beneath the depression, leading to subsurface cracks in crack-sensitive steel grades.

The maximum depth of the calculated gutter after the first roll is 2 mm, growing to 3 mm at the second roll. The calculated shape corresponds to the location of the thin spot. It roughly matches the observed surface profile, as seen in Fig. 17. Passing beneath each subsequent support roll causes the depression to deepen, but by lessening amounts. This continues for several more rolls, until solidification evens out the shell thickness, and the shell becomes strong enough to avoid further permanent curvature.

Other model simulations that assumed uniform heat transfer and shell growth along the wide face prior to mold exit never produced any longitudinal depressions. This suggests that solutions to the problem should focus on achieving uniform heat transfer in the mold, even though the depressions mainly form below the mold. Ensuring adequate spray cooling and proper alignment and minimum spacing of rolls to minimize the bulging below the mold is also beneficial. The effect of important casting variables on longitudinal depressions and solutions to the problem that have been achieved in practice are explained in the light of this mechanism [25].

B. Crack Prediction

Continuous-cast steel is subject to a variety of different surface and subsurface cracks. Using a thermal stress model to investigate crack formation requires a

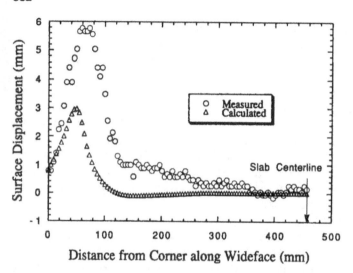

Figure 17 Comparison of calculated and measured surface shape profile across wide-face, showing gutter. (From Ref. 25.)

valid fracture criterion, linking the calculated mechanical behavior with the microstructural phenomena that control crack initiation and propagation. The task is difficult because crack formation depends on phenomena not included in a stress model, such as microstructure, grain size and segregation. Thus, experiments are needed to determine the fracture criteria empirically.

Many of the cracks that occur during solidification are hot tears. These cracks grow between dendrites when tensile stress is imposed across barely solidified grains while a thin film of liquid metal still coats them. Very small strains can start hot tears when the liquid film is not thick enough to permit feeding of surrounding liquid through the secondary dendrite arms to fill the gaps. Long, thin, columnar grains and alloys with a wide freezing range are naturally most susceptible. The phenomenon is worsened by microsegregation of impurities to the grain boundaries, which lowers the solidus temperature locally and complicates the calculations.

Crack formation in the shell is being investigated using the CON2D thermal stress model [61]. Figure 18 presents sample distributions of temperature, stress, and strain through the thickness of the shell, for a 0.1%C steel after 5 s of solidification below the meniscus. To minimize numerical errors, a very fine, graded mesh was required, including 201 nodes per row across the 20 mm thick domain. The time step increases from 0.001 s initially to 0.005 s at 1 s.

The temperature profile is almost linear across the 6 mm thick shell in Fig. 18a, with a mushy zone (liquid and delta-ferrite) that is 1.4 mm thick

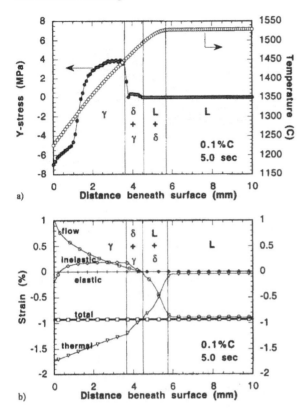

Figure 18 Typical strain and stress distributions across shell (0.1%C steel, 5 s below meniscus). (From Ref. 61.)

(33°C). The stress profile shows that the slab surface is in compression. This is because, in the absence of friction with the mold, the surface layer solidifies and cools stress free. As each inner layer solidifies, it cools and tries to shrink, while the surface temperature remains relatively constant. The slab is constrained to remain planar, so complementary subsurface tension and surface compression stresses are produced. To maintain force equilibrium, note that the average stress through the shell thickness is zero. It is significant that the maximum tensile stress of about 4 MPa is found near the solidification front. This generic subsurface tensile stress is responsible for hot tear cracks, when accompanied by metallurgical embrittlement.

Although it has no effect on stress, the model tracks strain in the liquid. When liquid is present, fluid flow is assumed to occur in order to exactly match the shrinkage. A finite element is treated as liquid in this model when any node

534   Thomas

in the element is above the specified coherency temperature (set to the solidus). These liquid elements are set to have no elastic strain, and consequently develop no stress. The difference between the total strain and thermal strain in liquid elements is assumed to be made up by a *flow strain*. This method allows easy tracking of various fracture criteria. For example, a large flow strain when the solid fraction is high indicates high cracking potential. It is also needed for future macrosegregation calculations.

Typical strain distributions are shown in Fig. 18b for 0.1%C steel. At the surface, the inelastic and elastic strains are relatively small, so the total strain matches the thermal strain accumulated in the solid. Ideal taper calculations can therefore be crudely estimated based solely on the temperature of the surface. At the solidification front, liquid contraction exceeds solid shell shrinkage, causing flow into the mushy zone. It may be significant that 0.1% inelastic strain is accumulated in the solid during the δ to γ transformation. This occurs within the critical temperature range 20–60°C below the solidus, where segregation can embrittle the grain boundaries, liquid feeding is difficult, and strain can concentrate in the thin liquid films. However, the model predictions are less than the measured values needed to form cracks using either critical stress [62] or critical strain [63] criteria. Thus, other unmodeled phenomena are surely important.

Longitudinal surface cracks might initiate within 1 s of the meniscus if the shell sticks to the mold. After this time, the shell is able to withstand the ferrostatic pressure and should contract to break away any sticking forces. This creates nonuniform surface roughness, however, which leads to local variations in heat transfer and shell growth rate. Strain concentrates in the hotter, thinner shell at the low heat flux regions. Strain localization may occur on both the small scale (when residual elements segregate to the grain boundaries) and on a larger scale (within surface depressions or hot spots). Later sources of tensile stress (including constraint due to sticking, unsteady cooling below the mold, bulging, and withdrawal) worsen strain concentration and promote crack growth. Further work is needed to investigate these phenomena.

XII. FUTURE FRONTIERS

This paper has shown several examples of the application of mathematical models to quantifying and understanding the mechanisms of defect formation during the continuous casting of steel slabs. Significant gains have been made in recent years. Despite the sophistication of existing models, much more work is needed to quantify, incorporate, and simplify the key phenomena into better models before the use of mathematical modeling tools to design process improvements reaches its full potential and usefulness.

The emerging continuous casting processes, such as thin slab and strip casting, present several special challenges, which offer opportunities for the application of modeling tools. Reducing the strand thickness from 8 in. to 2–4 in. requires higher casting speeds, if productivity is to be matched. When combined with the thinner mold, these processes are more prone to the fluid flow problems discussed in this chapter. Thinner product (with more surface) puts more demand on solving surface quality problems. Mold distortion is a greater concern, both because it is relatively larger for the thinner product, and because mold life is more critical for the expensive molds needed in these processes. The integration of casting and deformation processing together requires better understanding of the interaction between microstructure, properties, and processing. Conventional slab casting has had three decades of development and is still not perfected. For these new processes to thrive, the learning curve toward process optimization to produce quality steel needs to be shorter [64].

As computer hardware and software continue to improve, mathematical models will grow in importance as tools to meet these challenges. By striving toward more comprehensive mechanistic models, better understanding and process improvements may be possible.

ACKNOWLEDGMENTS

The author wishes to thank the member companies of the Continuous Casting Consortium at UIUC for the many years of support which has made this work possible and the National Center for Supercomputing Applications (NCSA) at UIUC for computing time. Special thanks are extended to the many excellent graduate students and research associates supported by the CCC who performed the years of hard work referenced here.

REFERENCES

1. WH Emling, TA Waugaman, SL Feldbauer, AW Cramb. Subsurface mold slag entrainment in ultra-low carbon steels. In: Steelmaking Conference Proceedings, 77, Chicago, IL. Warrendale, PA: ISS, 1994, pp 371–379.
2. DE Hershey, BG Thomas, FM Najjar. Turbulent flow through bifurcated nozzles. International Journal for Numerical Methods in Fluids 17:23–47, 1993.
3. BG Thomas, A Dennisov, H Bai. Behavior of argon bubbles during continuous casting of steel. In: Steelmaking Conference Proceedings, 80, Chicago, IL. Warrendale, PA: ISS, 1997, pp. 375–384.
4. FLUENT. Report, Fluent, Inc., Lebanon, NH, 1996.

5. M Burty. Experimental and theoretical analysis of gas and metal flows in submerged entry nozzles in continuous casting. In: PTD Conference Proceedings, 13, Nashville, TN. Warrendale, PA: ISS, 1995, pp. 287–292.

6. FM Najjar, BG Thomas, DE Hershey. Turbulent flow simulations in bifurcated nozzles: effects of design and casting operation. Metall Trans B, 26B(4):749–765, 1995.

7. D Hershey. Turbulent flow of molten steel through submerged bifurcated nozzles in the continuous casting process. Masters thesis, University of Illinois, 1992.

8. MS Engleman. FIDAP. Fluid Dynamics International, Inc., 500 Davis Ave., Suite 400, Evanston, IL 60201, 1994.

9. BG Thomas, X Huang, RC Sussman. Simulation of argon gas flow effects in a continuous slab caster. Metall Trans B, 25B(4):527–547, 1994.

10. N Bessho, R Yoda, H Yamasaki, T Fujii, T Nozaki, S Takatori. Numerical analysis of fluid flow in the continuous casting mold by a bubble dispersion model. Iron Steelmaker 18(4):39–44, 1991.

11. K Takatani, K Nakai, N Kasai, T Watanabe, H Nakajima. Analysis of heat transfer and fluid flow in the continuous casting mold with electromagnetic brake. ISIJ International 29(12):1063–1068, 1989.

12. P Gardin, J Galpin, M Regnier, J Radot. Liquid steel flow control inside continuous casting mold using a static magnetic field. IEEE Trans on Magnetics 31(3):2088–2091, 1995.

13. M Zeze, H Tanaka, E Takeuchi, S Mizoguchi. Continuous casting of clad steel slab with level magnetic field brake. In: Steelmaking Conference Proceedings, 79, Pittsburgh, PA. Warrendale, PA: ISS, 1996, pp. 225–230.

14. A Idogawa, M Sugizawa, S Takeuchi, K Sorimachi, T Fujii. Control of molten steel flow in continuous casting mold by two static magnetic fields imposed on whole width. Mat Sci & Eng A173:293–297, 1993.

15. N Saluja, OJ Ilegbusi, J Szekely. Three-dimensional flow and free surface phenomena in electromagnetically stirred molds in continuous casting. In: Proc Sixth Internat. Iron Steel Congress, 4, Nagoya, JP. Tokyo: ISIJ, 1990, pp. 338–346.

16. Y Kishida, K Takeda, I Miyoshino, E Takeuchi. Anisotropic effect of magnetohydrodynamics on metal solidification. ISIJ International 30(1):34–40, 1990.

17. JP Birat and J Chone. Electromagnetic stirring on billet, bloom, and slab continuous casters: state of the art in 1982. Ironmaking and Steelmaking 10(6):269–281, 1983.

18. D Gupta, AK Lahiri. A water model study of the flow asymmetry Inside a continuous slab casting mold. Metall and Materials Trans 27B(5):757–764, 1996.

19. T Honeyands, J Herbertson. Flow dynamics in thin slab caster moulds. Steel Research 66(7):287–293, 1995.

20. X Huang, BG Thomas. Modeling transient flow phenomena in continuous casting of steel. In: 35th Conference of Metallurgists, 23B, C. Twigge-Molecey, ed., Montreal, Canada: CIM, 1996, pp. 339–356.

21. BM Gebert, MR Davidson, MJ Rudman. Calculated oscillations of a confined submerged liquid jet. In: Inter Conf on CFD in Mineral & Metal Processing and Power Generation, MP Schwarz, ed., Melbourne, Australia: CSIRO, 1997, pp. 411–417.

22. J Herbertson, OL He, PJ Flint, RB Mahapatra. In: 74th Steelmaking Conference Proceedings, 74. Warrendale, PA: ISS, 1991, pp. 171–185.

23. X Huang, BG Thomas, FM Najjar. Modeling superheat removal during continuous casting of steel slabs. Metall Trans B, 23B(6):339–356, 1992.

24. BG Thomas, FM Najjar. Finite-element modeling of turbulent fluid flow and heat transfer in continuous casting. Applied Mathematical Modeling 15:226–243, 1991.

25. BG Thomas, A Moitra, R McDavid. Simulation of longitudinal off-corner depressions in continuously-cast steel slabs. ISS Trans, 23(4):57–70, 1996.

26. BG Thomas, D Lui, B Ho. Effect of transverse and oscillation marks on heat transfer in the continuous casting mold. In: Applications of Sensors in Materials Processing, V Viswanathan, ed., Orlando, FL. Warrendale, PA: TMS, 1997, pp. 117–142.

27. C Offerman. Internal structure in continuously cast slabs by the metal flow in the mould. Scandanavian J Metall 10:25–28, 1981.

28. X Huang, BG Thomas. Unpublished research report, University of Illinois, 1996.

29. GA Panaras, A Theodorakakos, G Bergeles. Numerical investigation of the free surface in a continuous steel casting mold model. Metall Mater Trans B. 29B(5):1117–1126,1998.

30. T Teshima, M Osame, K Okimoto, Y Nimura. Improvements of surface property of steel at high casting speed. In: Steelmaking Conference Proceedings, 71. Warrendale, PA: Iron and Steel Society, 1988, pp. 111–118.

31. BG Thomas, H Zhu. Thermal distortion of solidifying shell in continuous casting of steel. In: Proceedings of Internat. Symposia on Advanced Materials and Tech. for 21st Century, I. Ohnaka and D. Stefanescu, eds., Honolulu, HI. Warrendale, PA: TMS, 1996, pp. 197–208.

32. R McDavid, BG Thomas. Flow and thermal behavior of the top-surface flux/powder layers in continuous casting molds. Metall Trans B, 27B(4):672–685, 1996.

33. Y Miki, BG Thomas, A Denissov, Y Shimada. Model of inclusion removal during RH degassing of steel. Iron and Steelmaker 24(8):31–38, 1997.

34. X Huang, BG Thomas. Intermixing model of continuous casting during a grade transition. Metall Trans B, 27B(4):617–632, 1996.

35. X Huang, BG Thomas. Modeling of steel grade transition in continuous slab casting processes. Metall Trans 24B:379–393, 1993.

36. BG Thomas. Modeling study of intermixing in tundish and strand during a continuous-casting grade transition. ISS Trans 24(12):83–96, 1996.

37. B Ho. Characterization of interfacial heat transfer in the continuous slab casting process. Masters thesis, University of Illinois at Urbana-Champaign, 1992.

38. BG Thomas, B Ho, G Li. CON1D User's Manual. Report, University of Illinois, 1994.

39. BG Thomas, B Ho. Spread sheet model of continuous casting. J. Engineering Industry 118(1):37–44, 1996.

40. WH Emling, S Dawson. Mold instrumentation for breakout detection and control. In: Steelmaking Conference Proceedings, 74. Warrendale, PA: Iron and Steel Society, 1991, pp. 197–217.

41. JK Brimacombe. Empowerment with knowledge—toward the intelligent mold for the continuous casting of steel billets. Metall Trans B, 24B:917–935, 1993.

42. A Moitra, BG Thomas. Application of a thermo-mechanical finite element model of steel shell behavior in the continuous slab casting mold. In: Steelmaking Proceedings, 76. Dallas, TX: Iron and Steel Society, 1993, pp. 657–667.

43. MS Jenkins, BG Thomas, WC Chen, RB Mahapatra. Investigation of strand surface defects using mold instrumentation and modelling. In: Steelmaking Conference Proceedings, 77. Warrendale, PA: Iron and Steel Society, 1994, pp. 337–345.

44. BG Thomas, G Li, A Moitra, D Habing. Analysis of thermal and mechanical behavior of copper molds during continuous casting of steel slabs. Iron and Steelmaker (ISS Transactions) 25(10):125–143, 1998.

45. IV Samarasekera, DL Anderson, JK Brimacombe. The thermal distortion of continuous casting billet molds. Metall Trans B, 13B(March):91–104, 1982.

46. A Moitra, BG Thomas, W Storkman. Thermo-mechanical model of steel shell behavior in the continuous casting mold. Proceedings of TMS Annual Meeting, San Diego, CA, 1992. Warrendale, PA: The Minerals, Metals, and Materials Society.

47. A Moitra. Thermo-mechanical model of steel shell behavior in continuous casting. PhD thesis, University of Illinois at Urbana-Champaign, 1993.

48. H Zhu. Coupled thermal-mechanical fixed-grid finite-element model with application to initial solidification. PhD thesis, University of Illinois, 1997.

49. P Kozlowski, BG Thomas, J Azzi, H Wang. Simple constitutive equations for steel at high temperature. Metall Trans A, 23A(3):903–918, 1992.

50. PJ Wray. Effect of carbon content on the plastic flow of plain carbon steels at elevated temperatures. Metall Trans A, 13A(1):125–134, 1982.

51. T Suzuki, KH Tacke, K Wunnenberg, K Schwerdtfeger. Creep properties of steel at continuous casting temperatures. Ironmaking and Steelmaking, 15(2):90–100, 1988.

52. PJ Wray. Plastic deformation of delta-ferritic iron at intermediate strain rates. 7A(Nov.):1621–1627, 1976.

53. H Zhu, BG Thomas. Evaluation of finite element methods for simulation of stresses during solidification. Report, University of Illinois, 1994.

54. BG Thomas, A Moitra, DJ Habing, JA Azzi. A finite element model for thermal distortion of continuous slab casting molds. Proceedings of the 1st European Conference on Continuous Casting, Florence, Italy, 1991, Associazione Italiana di Metallurgia 2, pp. 2.417–2.426.

55. BG Thomas, A Moitra, WR Storkman. Optimizing taper in continuous slab casting molds using mathematical models. In: Proceedings, 6th International Iron and Steel Congress, 3, Nagoya, Japan. Tokyo: Iron and Steel Inst. Japan, 1990, pp. 348–355.

56. GD Lawson, SC Sander, WH Emling, A Moitra, BG Thomas. Prevention of shell thinning breakouts associated with widening width changes. In: Steelmaking Conference Proceedings, 77. Warrendale, PA: Iron and Steel Society, 1994, pp. 329–336.

57. K Kinoshita, H Kitaoka, T Emi. Influence of casting conditions on the solidification of steel melt in continuous casting mold. Tetsu-to-Hagane 67(1):93–102, 1981.

58. K Sorimachi, M Shiraishi, K Kinoshita. Continuous casting of high carbon steel slabs at Chiba Works. 2nd Process Tech. Div. Conference, Chicago, IL, 1981, pp. 188–193.

59. WR Storkman, BG Thomas. Mathematical models of continuous slab casting to optimize mold taper. Modeling of Casting and Welding Processes, Palm Coast, FL, 1988, Engineering Foundation, 4, pp. 287–297.

60. RB Mahapatra, JK Brimacombe, IV Samarasekera. Mold behavior and its influence on product quality in the continuous casting of slabs: Part II. Mold heat transfer, mold flux behavior, formation of oscillation marks, longitudinal off-corner depressions, and subsurface cracks. Metall Trans B, 22B(December):875–888, 1991.

61. BG Thomas, JT Parkman. Simulation of thermal mechanical behavior during initial solidification. Thermec 97 Internat. Conf. on Thermomechanical Processing of Steel and Other Materials, Wollongong, Australia, 1997. Warrendale, PA: TMS, pp. 2279–2285.

62. C Bernhard, H Hiebler, M Wolf. Simulation of shell strength properties by the SSCT test. ISIJ International 36:S163–S166, 1996.

63. A Yamanaka, K Nakajima, K Okamura. Critical strain for internal crack formation in continuous casting. Ironmaking and Steelmaking 22(6):508–512, 1995.

64. IV Samarasekera, BG Thomas, JK Brimacombe. The frontiers of continuous casting. Julian Szekely Memorial Symposium on Materials Processing, MIT, Boston, MA, 1997. Warrendale, PA: TMS.

65. BG Thomas. Mathematical modeling of the continuous slab casting mold, a state of the art review. In: Mold Operation for Quality and Productivity, A. Cramb, ed., Warrendale, PA: Iron and Steel Society, 1991, pp. 69–82.

16
Direct Chill Casting

Hallvard G. Fjær and Dag Mortensen
Institute for Energy Technology, Kjeller, Norway

I. INTRODUCTION

The direct-chill (DC) casting process is an important method for the production of a number of nonferrous metal products. Among these are the aluminium extrusion ingots and rolling sheet ingots. A schematic drawing of the DC casting process for sheet ingots is depicted in Fig. 1. A starting block is initially positioned inside a bottomless, water-cooled mold. During casting superheated liquid metal is fed into the mold through a nozzle or launder. When the metal starts to solidify, the starting block, placed on top of a casting table, is lowered. From holes or slits in the lower part of the mold, a water jet impinges on the ingot surface as this is lowered from the mold. Heat transfer from the cast metal to the mold is called primary cooling, whereas the heat transfer to the direct water cooling is called secondary cooling.

Compared to continuous casting of steel, the higher thermal conductivity of aluminium makes curved continuous casting impossible. Vertical semicontinuous casting is the standard procedure. This normally limits the length of the ingots to 8–10 m. However, the major part of this length is cast under quasi-steady state conditions, and therefore it may from a modeler's viewpoint be considered as a continuous casting process. Extrusion ingots are usually cylindrical with diameters typically within the range of 100–400 mm. Rolling sheets ingots typically have thickness of 400–600 mm and widths of 1500–2200 mm. For extrusion ingot casting up to 200 molds may be situated on a single casting table, depending on the billet diameter. Depending on the alloying elements, aluminium may obtain a wide range of physical and mechanical properties [1]. Wrought aluminium alloys from DC casting are classified as heat-treatable or as non-heat-treatable alloys.

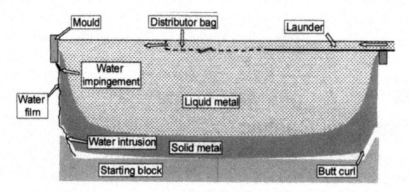

Figure 1 Illustration of the sheet ingot DC casting process.

The DC casting process was developed in the 1930s, and it has in principle not been changed since. The technology has, however, been continuously refined in order to satisfy the increasing demands for quality of the cast products. This quality is defined by a number of criteria. The geometric flatness or straightness of the ingots is important in order to minimize cutting and scalping. It is furthermore essential to avoid cracks in the products, and demands on grain structure and uniform distribution of particles should be satisfied. Chemical homogeneity is also required on the scale of the ingot (small macrosegregations).

The typical means of improving the ingot quality and cast house productivity are better melt treatments to remove inclusions, better mold designs to improve surface appearance, shell zone structure and ingot geometry, controlled water cooling during start-up to avoid butt curl, and better starting block designs to avoid cracking.

It has for a long time been focused on making the mold itself or the mold's effective cooling height shorter in order to improve the structure of the surface shell zone. A long primary cooling zone leads to a coarse grain structure close to the surface, and may result in severe macrosegregation at the surface. The early spout and float method for extrusion ingot castings has been replaced, to a large extent, by the use of hot-top molds. The surface quality has been further improved by continuously feeding of gas and lubricant, either through a porous graphite ring or through distributors placed between the mold and the hot-top. Hot-top molds are also being applied for casting of sheet ingots. The ultimate answer to the request for the perfect surface has been to completely remove the mold, and instead control the molten metal by an electromagnetic shield. Electromagnetic casting (EMC) of sheet ingots has proven to be the superior technology with respect to surface quality [2]. This technology is, however, only used to a limited extent due to its complexity and high investment costs.

II. MATHEMATICAL MODELING

Several characteristics of the DC casting process make the modeling of the process challenging. In DC casting, the dimensions of the casting are large and the thermal gradients are locally very high due to the water cooling. This combination of large dimensions and high thermal gradients may induce severe distortions. The direct water cooling itself is a rather complicated mechanism where the impingement angle, the velocity, the quality and the quantity of the water flow, the cast surface roughness, and the temperatures of both the water and the cast surface are of importance. Fluid flow is important for the distribution of superheated liquid, especially for large ingot dimensions, and for the microstructure evolution.

Analytical tools generate general solutions providing quantitative and qualitative insight, but with the loss of accuracy for real-world problems that are both nonlinear and have irregular time dependence. Numerical solutions, on the other hand, may give accurate solutions to a particular problem, but they lack the generality. Dimensional analyses combined with numerical simulations may enhance the value of the simulation results, and it may provide a more general understanding of mechanisms such as for the general thermal field [3] and for the pull-in phenomenon in sheet ingots [4]. Insight obtained by dimensional analyses may quite directly be applied in practical solutions, such as for mold design [5].

In order to produce reliable results from simulations, one needs accurate input of both boundary conditions and material data. One method to provide such data is to carry out experiments, e.g., thermocouple measurements, and to subsequently apply inverse modeling techniques to extract the desired information. For DC casting, inverse modeling has mostly been applied to find the boundary conditions for the water cooling and to determine the temperature dependent thermal conductivity [6, 7].

No commercial software packet has been specifically developed for DC casting. However, in the literature one can find numerous examples of adaption of general purpose commercial FEM (finite element method) or FDM (finite difference method) software. Generally, different codes have been applied for addressing problems associated with deformations or stresses and heat distribution in the liquid. For thermomechanical analyses some authors [8, 9] have applied ABAQUS; others, i.e., Ref. 10, have used MARC. For fluid flow analyses there are examples with use of PHOENICS [11, 12] and FIDAP

[13, 14]. Some special purpose codes have also been written in university and research institutes [15–18].

An example of a process improvement based on using modeling tools is the development of the conical shaped starting block in order to eliminate the problem of center cracks generated by hot tearing during start-up for billet casting [19]. Based on the idea that the cracking was related to an excessive sump depth, the effects on varying a number of parameters, in particular the shape of the starting block, were investigated [20]. In a later thermomechanical analysis [21] a local maximum in tensile stresses was computed for the traditional bowl-shaped starting block at the position where cracks were observed (Fig. 2). When simulating casting with the new type of conical shaped starting block, this stress maximum vanished.

The casting speed is limited by the hot-tearing tendency. Special arrangements of the cooling conditions may overcome these limitations [22, 23]. A general hot-tearing criterion has for a long time been a goal. The best macro-scale type of criteria available is perhaps the idea that a casting is susceptible for hot tearing if the induced viscoplastic strains overshoot a limit during a critical part of the solidification interval [24, 25]. It is, however, difficult to achieve an understanding of this phenomenon from macro-scale analyses. Recently, there has been a revival interest for the fundamental aspects of

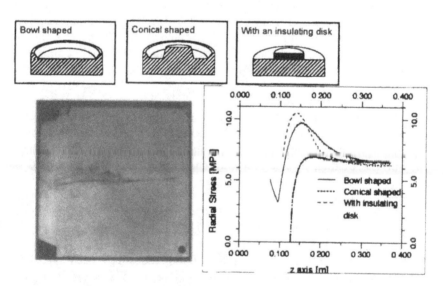

Figure 2 Bottom left: A start crack in an AA6063 extrusion ingot. Bottom right: Computed stresses along center axis at temperature 550°C for the three different starting block shapes schematically drawn above.

deformation of the solid network, melt flow, and morphology of the two-phase mushy zone [26, 27].

The start-up is the most critical phase of the sheet ingot casting. Many problems (see Fig. 1) can be related to the butt curl, which is a deformation of the ingot butt characterized by a bowing-up of the shell formed against the starting block. A more general discussion of such defects and difficulties is given in [28]. The most common way to limit the butt curl, is to reduce the water cooling during start-up, inducing a slower and more homogeneous cooling of the ingot shell. This can be achieved by using cooling water with dissolved CO_2 [29] or by applying pulsed water cooling [30]. In addition, the shape of the starting block has been shown to have significant influence on the butt curl [31]. This has also been illustrated by accompanying modeling work [32, 33]. Computed butt curl results have also been reported in Refs. 8–10.

Before rolling, the sheet ingot must have a rectangular shape. Due to the pull-in effect, which is a distortion characterized by a significant inward bending of the solidifying shell, the actual ingot cross section may deviate significantly from the mold opening. Due to nonoptimal mold shapes and the butt-swell phenomenon (i.e., lack of pull-in during start-up) excessive scalping of the ingot may be required. The pull-in effect has lately been investigated by numerical models [4, 8].

An important topic, especially for sheet ingot casting, is the design of the melt distribution systems that redirect the inflow of metal. A general goal is to achieve an even and narrow shell zone along the mold wall. Down in the sump thermal convection produces a nearly isothermal volume of liquid, which affects the extent and the shape of the solidification zone, as well as the transport of solute, free grains and the resulting segregation. Free convection generates typical flow patterns with downward flow along the cold solidification front and upward flow in the warm center of the pool. This is seen in numerical experiments, and it has also been reported from analytical methods [34, 35].

Turbulence will generally be induced from the feeding system or from the buoyancy affected flow. Typical Rayleigh numbers for a commercial size sheet ingot are in the order of 10^9. In the extrusion ingot it is more likely that the flow may be laminar, but also in this case there may be turbulence generated in the distribution pan. However, in both the sheet ingot and the billet the damping of turbulence down into the sump is a process that is difficult to predict. Vertical damping caused by buoyancy in a thermally stratified flow is one of the mechanisms in this process. The level of turbulence is important, not only in the calculation of the flow field, but also from the enhanced heat transfer due to turbulent mixing. Remark also that the widely used k-ε model is developed on the basis of isotropic eddy viscosity and eddy diffusivity model, which is a weak point in the common modeling of turbulence during solidification, since the condition of warm liquid above cold liquid suppresses the vertical turbulent flux of momentum.

Several models for the heat and fluid flow in DC casting have been described in the literature. The modeling of the flows in the steady state period has been investigated by various authors, also included the effects of macrosegregation, surface segregation and grain transport [36–41]. Dynamic heat and fluid flow have been studied in Refs. 42–44 for two-dimensional start-up calculations and for three-dimensional flow start-up calculations [16].

III. BOUNDARY HEAT TRANSFER

In DC casting, only a small fraction of the total heat is extracted through the mold and the starting block. The direct water chill is the dominating cooling mechanism. This is in particular the case for large sheet ingot dimensions. However, a number of defects and problems are associated with the ingot surface, and the mold and bottom block heat transfer is essential for the solidification of the cast metal surface. Additionally, the structural stiffness of the initially solidified shell may influence the butt curl development, and the surface temperature prior to the water impingement may be decisive to whether film boiling will occur.

A. Heat Transfer to the Water Film

The heat transfer to the water film may be represented by different mathematical correlations for the different cooling regimes, as outlined in Fig. 3. For temperatures below and to some extent above the boiling temperature T_{SAT} the water film cools the surface by convection calculated by a correlation for q_{Con}. When the surface temperature of the ingot is well above the boiling temperature, nucleate boiling q_{NUB} is the dominating heat transfer mechanism. The heat transfer increases very rapidly with the temperature until the vapor close to the surface restricts the bubble nucleation. The heat transfer rate in this region is represented by the critical heat flux q_{CHF}. The amount of vapor at the surface increases and the transition heat flux q_{TRA} decreases as the surface is nearly covered by vapor, which acts as an insulator. At the end of the transition region, a minimum of the heat flux at the minimum film boiling temperature T_{MFB}, or the Leidenfrost temperature is reached. For higher surface temperatures the film boiling heat flux q_{FB} will be larger, due to radiation and conduction in the vapor film. The vapor film is not stable and the correlations of the critical heat flux, transition boiling, and film boiling depend on the water film undercooling and also on the water film velocity.

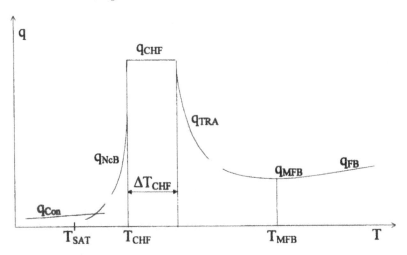

Figure 3 Heat flow density as a function of the boundary temperature of the ingot, against the falling water film.

It is advantageous to distinguish the heat transfer rates between an "impingement zone" close to the water hit point and a "streaming zone" below the impingement zone [45]. In general, the correlations depend on the water quality and the alloy (or the surface roughness which is dependent on the alloy and the mold) as well as the water spraying system. However, because of the complex phenomena involved, one has to develop simple correlations between surface temperature, water temperature, water amount and measurements in the cast house. This has been done in Ref. 16, where a number of heat transfer correlations are suggested for the impingement zone and the streaming zone. An overview of water cooling in DC casting is given in Refs 46, 47.

Under steady state conditions the water film operates in the stable nucleate boiling regime, where an increase in surface temperature causes an increase in heat transfer rate from the surface. However, in the beginning of the start-up period, the process may operate in the film boiling regime because of the high initial temperatures. If the process is not properly designed, this may cause severe bleed out problems. However, the film boiling regime may also be deliberately used to help reduce ingot butt deformations. This is achieved by adding gas (usually CO_2) or other additives to the cooling media, or by using a very small amount of water. This allows a controlled transition to the nucleate boiling regime after the formation of the very first solid shell.

B. Heat Transfer Between the Ingot and the Mold

A meniscus will form between the liquid metal surface and the mold wall. Below the meniscus there will be a contact zone. The heat transfer in the contact zone is crucial for the surface quality of the ingot and for the formation of segregated bands or Bergmann zones [48, 49]. Typical values of the heat transfer coefficient in the contact zone are about 1000–5000 W/m^2 K, depending on the amount of lubrication and other factors such as the casting speed. The phenomenon of air gap formation has generally been considered as too complicated to be calculated directly by a thermomechanical approach. A pure thermal criteria has been applied in Ref. 16 which results in contact lengths of the order of 10–20 mm for sheet ingots. A similar length of a high heat transfer zone has been measured and reported in Ref. 50. Below the high heat transfer zone, a semisolid material appears. The heat transfer rate in this region will depend on the mechanical properties of the semisolid alloy and the amount of exudation as well as the hydrostatic pressure. Typical values are in the range of 200–600 W/m^2 K.

C. Heat Transfer to the Starting Block

The heat transfer rate between the ingot and the bottom block is influenced by the contact resistance in the ingot/bottom block interface, as well as the development of the butt curl gap and the water flowing into this gap. To handle these phenomena, it is necessary to specify local variations of the heat transfer rate with position and time. This may be done by a division between a contact zone and a heat sink zone (water intrusion zone between ingot and bottom block). The water intrusion will typically appear shortly after the water hits the ingot in the beginning of the start-up period. However, if film boiling is present, the water intrusion may be delayed. In the following section, a coupled thermomechanical investigation of these phenomena is outlined.

IV. A COUPLED THERMAL, FLUID FLOW, AND STRESS SIMULATION

During DC casting, the development of temperatures, melt flow, and solid deformations is interrelated. A model incorporating such coupled effects for sheet ingot casting is presented in this section, based on Ref. 51. Special attention is given to the heat transfer at the bottom of the ingot which is severely affected by water intrusion following the thermally induced deformation (butt curl) of the solidified shell.

In a previous work on the modeling of laboratory scale sheet ingot casting with the heat and fluid flow model ALSIM [16], elaborate adjustments of heat transfer coefficients (as function of position and time) were necessary in order to obtain an agreement with measured temperatures of the starting block. These simulations were the basis for the subsequent mechanical analysis with the stress and deformation model ALSPEN [17, 18], resulting in a good agreement with measured butt curl values. A work was then carried out to use a coupled thermomechanical modeling approach to substitute this procedure of specifying boundary conditions.

A. Modeling Concepts

The finite element model ALSIM has been developed to analyze the time dependent heat and fluid flows during DC casting of aluminium ingots. This model is based on a continuum mixture model for the solid-liquid material, following Ref. 52. The velocity in the solid regions is set equal to the casting velocity. This model assumes that a dendritic network exists in the whole mushy zone, neglecting the effect of free solid grains. A Darcy force is used in the mixture momentum equation which accounts for the interfacial friction due to the different velocities of the solid and the liquid, following Ref. 53. The mixture model is simplified by neglecting the solidification shrinkage and the solute transport caused by gradients in the alloy composition. Thermal convection is included by the Boussinesq approximation. The turbulence is modeled by using a low Reynolds number (LRN) k-ε model [54]. In the LRN model, the damping of the turbulence in the mushy zone is done by the usage of a Darcian term in the turbulent transport equations.

During the start-up period, the computational grid has to expand to account for the increase of ingot height. There are also different needs on the kinematic description (i.e. the relationship between the moving media and the computational grid). In the solid part a Lagrangian description is preferred, so that the computational grid in the thermal analysis would coincide with the grid in the stress analysis. On the other hand, a Euler description is used in the fluid flow analysis.

In the stress analysis model ALSPEN, the compatibility equations, the momentum equations and the constitutive equations are solved by a finite element technique. The material is treated as an elastic-viscoplastic material for which plasticity and creep are treated in a uniform manner. The solution domain is the part of the ingot that is considered to be solid, i.e., where the temperature is predicted to be below a given coherency temperature. During a casting simulation this solution domain increases in size accordingly. Symmetry boundary conditions is applied when a half or a quarter of an ingot is considered. At the solidification front the metallostatic pressure is taken into account. At the mold inner surface no-penetration constraints are imposed. A pressure distribution is applied at the ingot underside.

B. Thermomechanical Coupling

In the treatment of the mechanical problem in the DC casting process, the temperature development represents the major load, or the cause of stresses, due to inhomogeneous thermal contractions. The temperature field also influences the local response of this load through the temperature dependency of the elastic and viscoplastic mechanical properties. These effects can be accounted for in sequentially coupled thermal and mechanical models. However, the effect of distortions on the thermal problem is then neglected. Thermal effects associated with strains and distortions are generally:

- Heat dissipation due to deformation work
- Influence of distortions on thermal boundary conditions
- Change of geometry due to thermally induced distortions

As pointed out by Du et al. [9], the heat dissipation work associated with the viscoplastic strains is negligible in this process.

The thermal boundary conditions involve heat transfer from the ingot to the bottom block as well as to the cooling water and to the mold. For the heat transfer to the bottom block, there are two major effects associated with the development of an air gap between the bottom block and the ingot. The first effect is the reduction of the rate of heat extraction from the ingot due to reduced thermal contact. This might lead to re-melting of the ingot sole. The other effect is the enhanced cooling of the ingot due to water intrusion into the gap. It is assumed that both the extension and the intensity of this water cooling are related to the size of the air gap. In the remaining contact area between the ingot and the bottom block, the heat transfer is assumed to depend on the contact pressure, which can be derived from the size of the contact area and the weight of the ingot. The pull in phenomenon also results in an air gap between the ingot and the mold. The extent and the size of this air gap depends on the metal level, the primary cooling conditions, the strength of the solid/mushy ingot shell, and surface segregation effects.

For the water cooling, both the vertical displacements (i.e., the butt curl) and the horizontal contractions (i.e., the pull-in) affect the water hit point position. The butt curl results in a temporarily lower local casting speed, especially close to the narrow ingot sides. The pull-in increases the vertical distance from the mold to the water hit point. This distance will, however, also depend on the mold water outlet design and the water flow rate. The butt curl effect is significant during the early start-up phase, whereas the pull-in effect will be fully evolved at steady state casting conditions, when the cast length becomes larger than the butt swell length.

As the pull-in phenomenon [5] may involve an ingot thickness reduction of about 10%, the thermal problem should be solved on a computational domain corresponding to the actual ingot geometry. Such a procedure requires input of the total displacement field from the mechanical model to the thermal model, and the application of an updated Lagrangian formulation.

C. Coupled Model Implementation

Algorithms of thermomechanical problems follow two kinds of strategies. One is to solve the mechanical and the thermal problem simultaneously. For the current process this will inevitably lead to a very computationally expensive task. What is often done in the case of casting is to apply a staggered time-stepping algorithm where a time increment of the coupled problem is split into a thermal problem, solved for a given displacement field, followed by a mechanical problem solved for a given evolution of the temperature field. In this way one may benefit from the use of different time steps in the thermal and the mechanical problems. This is in particular the case with the models ALSPEN and ALSIM. With ALSIM the energy and momentum conservation equations are solved by a projection method where temperature, pressure, and each of the velocity components are evaluated separately. The time step, limited by the Courant number or the nonlinear thermal boundary conditions, is usually of the order of one tenth of a second. With ALSPEN each time step is more costly because a larger system of equations is solved several times for each time increment, due to iterations with respect to the nonlinear material properties. However, it was found that the size of the ALSPEN time steps usually can be increased to a number of seconds without seriously affecting the calculated results.

ALSIM and ALSPEN are separate computer codes. However, by using the public domain PVM (parallel virtual machines) software [55], runtime communication between these two programs is achieved. PVM is an integrated set of software tools and libraries that emulate a computing framework on interconnected computers with varied architecture. In the implementation of thermomechanical coupling, ALSIM sends out a geometry and temperature data package to ALSPEN before each ALSPEN time step. After each calculation time step, ALSPEN sends back a displacement and pressure data package to ALSIM.

D. Coupled Thermal Boundary Conditions

The coupling mechanisms that have been implemented so far, are the influence of calculated displacements and pressure on the thermal boundary conditions between the ingot and the bottom block, and the influence of the displacements on the water impingement position.

A simulation time step for the deformation analyses might not be successful, or give accurate results, unless a solid shell of some thickness has been predicted. It is therefore questionable whether it is feasible to predict the air gap development during the mold filling and the initial solidification accurately. It was therefore chosen to start the stress simulations a short time before the water hits the ingot. During the very first period of casting, a heat transfer coefficient depending on the ingot surface temperature is applied for the bottom of the ingot. When displacement and surface pressure fields calculated from ALSPEN become accessible in ALSIM, the heat transfer coefficient becomes dependent on both the local gap distance and the estimated contact pressure. In any area where the calculated gap is bigger than 0.2 mm, an air gap thermal boundary condition is applied. Following Ref. 56 the total heat transfer coefficient h is calculated from terms representing conduction through the air gap h_{cond}, radiation in the air gap h_{rad}, and an initial value h_{init}, which may be associated with the insulating effect of the lubrication.

$$h = \frac{1}{1/(h_{cond} + h_{rad}) + 1/h_{init}} \tag{1}$$

Where the gap distance is calculated to be less than 0.2 mm, the heat transfer coefficient h is assumed to depend on the local ingot surface temperature T_b, the normal pressure p_n, and three constants k_1, k_2, and k_3.

$$h = k_1 \cdot (T_{liq} - T_b)^{-k_2} \cdot p_n^{k_3} \tag{2}$$

With ALSPEN, vertical forces, counterbalancing the total weight of the ingot, are distributed underneath of the ingot. Emulating some elastic response from the starting block, the local contact pressure p_n is estimated from the calculated gap distance. Although not believed to be accurate, this formulation enables us to incorporate some effect of temperature and pressure dependence in the contact heat transfer. In any area where the gap distance is larger than a prescribed value, 3 mm, water intrusion is assumed. In this water intrusion zone both the starting block and the ingot have thermal contact with the water inside the gap. Water will seek the lowest level of the starting block bowl, and eventually flow down the drainage holes. It is also expected that the water, due to boiling, is easily rejected from the ingot surface above the gap at higher temperatures. From these considerations, different temperature dependent heat transfer coefficient functions $h(T)$ have been applied. In the case study presented later, these functions were adjusted in order to obtain agreement with temperature measurements.

E. Experimental Description

The case study presented here is based on an experimental series carried out at Hydro Aluminium R&D Materials Technology [57]. AA5005.03 aluminium ingots, with dimension 2150 × 600 mm were cast. This alloy contains 0.71wt% Mg, 0.25wt% Fe and 0.09wt% Si. A bowl-shaped, 150 mm deep starting block, and a hot-top mold with a total height of 200 mm and a cooling height of 35 mm, was applied. Liquid metal was fed through one of the narrow sides of the mold in a horizontal U-shaped steel launder. As an extension of this launder, a permeable distributor bag, with a length corresponding to one third of the total ingot width, was situated in the center of the mold (see Fig. 4). During mold filling, the liquid metal flowed vertically out through the bottom of this bag. When the mold was filled, the flow out of the bag became mainly horizontal.

The starting block was instrumented with groups of thermocouples along the longitudinal symmetry axis at positions 0, 200, 400, 600, 800, 1000, and 1060 mm measured from the center of the ingot in the direction of the metal inlet flow. Each group (with a few exceptions) had thermocouples placed 5 mm inside the ingot, at the starting block surface, and at positions 10 and 30 mm inside the starting block, as shown in Fig. 4. In this work, one particular casting was simulated. Here, after a filling time of 105 s, the starting block was lowered at a rate of 72 mm/min. The cooling water rate was kept constant at 38 m^3/h until a cast length of 700 mm was reached. Then it was rapidly increased to 64 m^3/h. The total cast length was about 1 m. In this casting series, the casting parameters were varied outside the normal process window. Therefore, many defects such as run-out and cracking occurred. The simulated casting had an unusual short holding time and an extraordinary long period with a small amount of cooling water, but it was nevertheless reported to be successful.

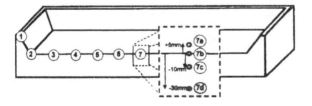

Figure 4 Positions of thermocouples placed inside and above the starting block.

Some temperatures measured at the surface of the starting block are shown in Fig. 5. It is evident that the water must have entered into the gap and reached the thermocouple group 3 as early as after 155 s, i.e., 30 s after the water hits the ingot. Unfortunately, no thermocouples were situated inside the ingot in groups 1 and 2, and others placed inside the ingot failed. They may have been torn apart, or pulled out of the ingot due to the butt curl.

F. Simulations

Three simulations of different cases have been carried out:

- *Case A*: A thermal simulation on one quarter of the ingot, with a temperature dependent heat transfer coefficient from the ingot to the bottom block, as in Ref. 44 (no water intrusion)
- *Case B*: A coupled thermomechanical simulation on one quarter of the ingot
- *Case C*: A coupled thermomechanical simulation including fluid flow on one half of the ingot

Usually, due to symmetry, computations can be carried out on a domain corresponding to one quarter of the ingot. The asymmetric flow conditions in case C necessitate a calculation on one half of the ingot. It was, however, essential to impose symmetry conditions in the stress calculation, i.e., zero longitudinal displacement in the central plane at the middle of the wide ingot, in order to prevent "spurious rocking." Currently, it is not possible to

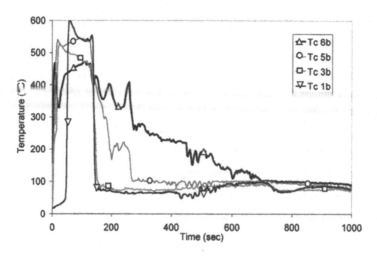

Figure 5 Temperatures measured at surface of starting block.

simulate filling of the starting block bowl with ALSIM. In order to incorporate some effect of the metal inflow from the start, the metal level was, in the simulation, continuously raised from the top of the starting block to the top of the mold, but at a slower rate than in the actual casting.

Thermal properties were estimated by use of the program ALSTRUC [58]. For the AA5005 alloy, it was calculated a liquidus temperature of 654.1°C, a solidus temperature of 610.5°C, and thermal conductivity values for the solid phase in the range 185–202 W/m K. Thermal conductivity values in the range 110–160 W/m K were applied for the (Al-Mn) starting block. The measurements on the as-cast mechanical properties for the AA5005 alloy were not available. Parameters in an elastic-viscoplastic constitutive law [17] were adjusted until a reasonable accordance was obtained with flow stress values found by interpolation into data for related alloys. The coherency temperature, defining the upper boundary of the ALSPEN solution domain, was given a value 639°C, corresponding to a solid fraction of approximately 0.85.

G. Results and Discussion

In Fig. 6 calculated temperatures are compared with measured temperatures 10 mm inside the center of the starting block. It is noticeable that the high central peak temperature (Tc 7c) is only reproduced in the simulation including fluid flow (case C).

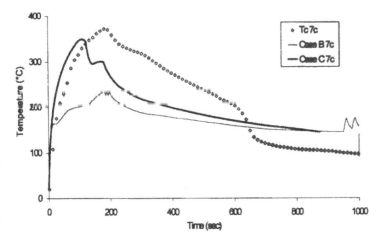

Figure 6 Comparison of calculated and measured temperatures 10 mm inside the center of the starting block.

Figure 7 shows the comparison between calculated and measured temperatures inside the starting block at 0.5 m from the narrow side. The rapid decline for temperatures in the water intrusion zone (Tc 4c) is successfully reproduced by the coupled thermomechanical simulation (case B). The simulation without water intrusion (case A) clearly overestimates the starting block temperatures for high time values.

Figures 8 and 9 compare calculated and measured ingot temperatures. Temperatures calculated in case A is seen to decrease rather continuously. Results from case B, where heat transfer to water in the gap is accounted for, show a better agreement with the measured values. It is noticed that even if water is present inside the gap at a given position, the cooling of the ingot is initially very slow. In contrast, the cooling rate is much higher at lower temperatures. This effect has been reproduced in case B by the use of low values of the heat transfer coefficient (200 W/m^2 K) for high surface temperatures and much higher values (2000–3500 W/m^2 K) for lower temperatures. This indicates that the cooling of the ingot underside depends not mainly on whether water gets in between the ingot and starting block, but perhaps rather on whether the water is able to wet the ingot surface and flow along the ingot underside. This seems, in this case, not to have happened before the water rate was increased at 700 s. The low cooling rate for Tc 2a, together with the starting block surface temperatures close to 100°C (seen in Fig. 5), indicates that poor water cooling of the short ingot side between 500 and 700 s may be due to an uneven water distribution.

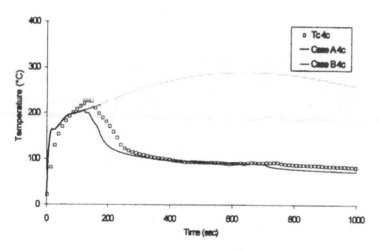

Figure 7 Comparison of calculated and measured temperatures 10 mm inside the starting block about 0.5 m from the narrow side.

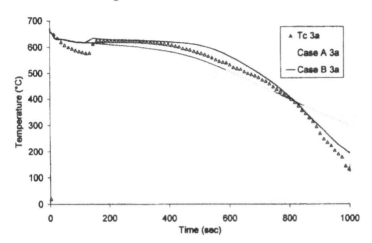

Figure 8 Comparison of calculated and measured temperatures 5 mm above the bottom of the ingot at thermocouple position 3.

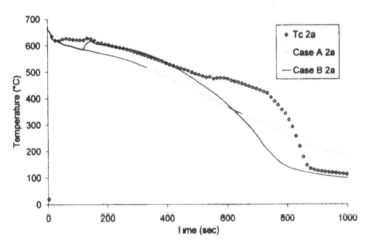

Figure 9 Comparison of calculated and measured temperatures 5 mm above the bottom of the ingot at thermocouple position 2.

From Fig. 10, it can be seen that the calculated butt curl (i.e., the calculated vertical displacement at the center of the narrow ingot side) is not significantly influenced by the incorporation of fluid flow in the thermal simulation. On the other hand, it has experienced a rather strong effect on the butt curl development of the heat transfer coefficient in the water impingement zone. In current simulations, relatively high values were used in order to

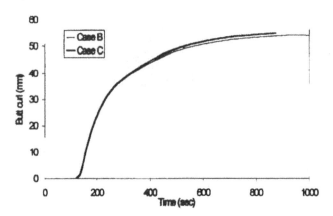

Figure 10 Calculated butt curl development in case B and case C.

suppress film boiling and to achieve a rapid developing butt curl in accordance with the measured water intrusion. Figure 11 shows the calculated (case C) flow field and the temperature distribution in a central longitudinal cross section. It is observed that the isotherms in the melt pool are somewhat asymmetric due to the inflow conditions. In Fig. 12, the calculated deformations in the same cross section as a distorted mesh is displayed. It is noticed that the thickness of the solid shell is comparable with the typical finite element size, and this clearly expresses a limited numerical accuracy. Incorporating only elements where the metal is solid reduces the problem size in ALSPEN, but the solid shell possesses little stiffness to bulk deformations, which is reflected in a slow convergence for the equation solver. Numerical problems associated with the prediction of remelting were also experienced. In simulations with a

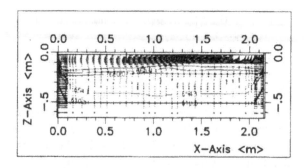

Figure 11 Calculated melt flow velocities and isotherms after 250 s in central symmetry plane in case C.

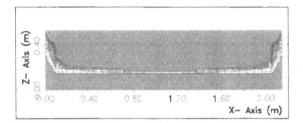

Figure 12 Calculated distortions (magnified by a factor 2) after 250 s in case C

small ingot [33], the choice of constitutive model did not affect the predicted butt curl notably. The important factor was found to be the relative strength of the cast metal at different temperatures. However, for a large ingot, the effect of gravity becomes more pronounced, and the absolute level of flow stress becomes important. In this case, where solid mechanical properties have been extrapolated into the mushy zone, the stiffness of the solid shell has probably been overestimated.

H. Concluding Remarks

A combined thermal, melt flow, and stress simulation has been carried out for the start-up period of a full scale DC casting. A concept for a coupled thermo-mechanical modeling has been discussed and compared against measurements of temperatures in the ingot and the bottom block. In this concept new phenomena connected to the concurrent evolution of the thermal field, the melt flow, and the deformations were addressed. In the work with the water intrusion criteria it became evident that this phenomenon is closely related to the level of the water film cooling, which is the most important parameter for the development of the butt curl. In the future there is a need for a systematic examination of the experimental data available, in order to improve the mathematical model presented here.

ACKNOWLEDGMENTS

The development of DC casting models at Institute for Energy Technology has been funded by Hydro Aluminium, Elkem Aluminium, Hydro Raufoss Automotive, and The Research Council of Norway. The inspiring cooperation with individuals within these companies and SINTEF is deeply appreciated.

REFERENCES

1. ASM Specialty Handbook: Aluminium and Aluminium Alloys. ASM International, 1993, pp. 60–87.
2. JW Evans, R Kageyama, Deepak, DP Cook, DC Prasso, S Nishioka. Electromagnetic casting. Proceedings of the seventh conference on Modeling of Casting, Welding and Advanced Solidification Processes, London. Warrendale, PA: TMS, 1995, pp. 779–792.
3. A Håkonsen, OR Myhr. Dimensionless diagrams for the temperature distribution in direct chill continuous casting. Cast Metals 8(3):147–157, 1995.
4. HG Fjær, A Håkonsen. The mechanism of pull-in during DC casting of aluminium sheet ingots. In: Light Metals. Warrendale, PA: TMS, 1997, pp. 683–690.
5. A Håkonsen. A model to predict the steady state pull-in during DC-casting of aluminium sheet ingots. In: Light Metals. Warrendale, PA: TMS, 1997, pp. 675–682.
6. JB Wiskel, SL Cockcroft. Heat-flow-based analysis of surface crack formation during start-up of the direct chill casting process: Part I. Development of the inverse heat-transfer model. Metall and Mater Trans 27B:119–128, 1996.
7. J-M Drezet. Direct chill and electromagnetic casting of aluminium alloys: thermo-mechanical effects and solidification aspects. PhD thesis, EPLF, Lausanne, Switzerland, 1996, pp. 107–112.
8. J-M Drezet, M Rappaz. Modeling of ingot distortions during direct chill casting of aluminium alloys. Metall and Mater Trans 27A:3214–3225, 1996.
9. J Du, BS-J. Kang, K-M. Chang, J Harrisl. Computational modeling of DC casting of aluminum alloy using finite element method. In: Light Metals. Warrendale, PA: TMS, 1998, pp. 1025–1030.
10. B Hannart, F Cialti, R Schalkwijk. Laboratory thermal stresses in DC casting of aluminum slabs: application of a finite element method. In: Light Metals. Warrendale, PA: TMS, 1994, pp. 879–887.
11. SC Flood, L Katgerman, AH Langille, S Rogers, CM Read. Modeling of fluid flow and stress phenomena during DC casting of aluminium alloys. In: Light Metals. Warrendale, PA: TMS, 1989, pp. 943–947.
12. Ch Raffourt, Y Fautrelle, JL Meyer, B Hannart. Thermal and fluid flow calculations in an aluminium slab DC casting. Modeling of Casting, Welding and Advanced Solidification Processes-V, ed. M Rappaz, MR Özgü, KW Mahin. Warrendale, PA: TMS, 1991, pp. 691–698.
13. C Devadas, JF Grandfield. Experiences with modeling DC casting of aluminium. In: Light Metals. Warrendale, PA: TMS, 1991, pp. 883–892.
14. GU Grün, I Eick, D Vogelsang. 3D-modeling of flow & heat transfer for DC-casting of rolling ingots. In: Light Metals. Warrendale, PA: TMS, 1994, pp. 863–869.
15. JB Wiskel. Thermal analysis of the startup phase for DC casting of an AA5182 aluminium ingot. PhD thesis, The University Of British Columbia, Canada, 1995.
16. D Mortensen. A mathematical model of the heat and fluid flows in direct-chill casting of aluminium sheet ingots and billets. Metall and Mater Trans 30B:119–133, 1999.

17. HG Fjær, A Mo. ALSPEN—a mathematical model for thermal stresses in direct chill casting of aluminium billets. Metall Trans 21B:1049–1061, 1990.

18. HG Fjær, EK Jensen. Mathematical modeling of butt curl deformation of sheet ingots. Comparison with experimental results for different starting block shapes. In: Light Metals. Warrendale, PA: TMS, 1995, pp. 951–959.

19. W Schneider, EK Jensen. Investigations about starting cracks in DC-casting of 6063-type billets. Part I: experimental results. In: Light Metals. Warrendale, PA: TMS-AIME, 1990, pp. 931–936.

20. EK Jensen, W Schneider. Investigations about starting cracks in DC-casting of 6063-type billets. Part II: modeling results. In: Light Metals. Warrendale, PA: TMS-AIME, 1990, pp. 937–943.

21. HG Fjær, A Mo. Influence of starter block shape on center crack formation in DC casting of aluminium billets. Mathematical Predictions. Stranggießen, Deutsche Gesellshaft für Materialkunde, Oberursel, Germany, 1991, pp. 127–134.

22. NB Bryson. Increasing the productivity of aluminium DC casting. In: Light Metals. Warrendale, PA: TMS-AIME, 1972, pp. 429–435.

23. HG Fjær, EK Jensen, A Mo. Mathematical modeling of heat transfer and thermal stresses in aluminium billet casting. Influence of the direct water cooling conditions. Proceedings of the 5th International Aluminum Extrusion Technology Seminar, The Aluminum Association,1992, 1:113–120.

24. ML Nedreberg. Thermal stress and the hot tearing during the DC casting of AlMgSi billets. PhD thesis, University of Oslo, Dept of Physics, 1991.

25. JWB Magnin, L Katgerman, B Hannart. Physical and numerical modeling of thermal stress generation during DC casting of aluminium alloys. Proceedings of the seventh conference on Modeling of Casting, Welding and Advanced Solidification Processes, London. Warrendale, PA: TMS, 1995, pp. 303–310.

26. M Rappaz, J-M Drezet, M Gremaud. A new hot-tearing criterion. Metall and Mater Trans 30A:449–455, 1999.

27. I Farup, A Mo. Two-phase modeling of mushy zone parameters associated with hot tearing. Accepted for publication in Metall and Mater Trans 1999.

28. W Droste, W Schneider. Laboratory investigation about the influence of starting conditions on butt curl and butt swell of DC cast sheet ingots. In: Light Metals. Warrendale, PA: TMS, 1991, pp. 945–951.

29. H Yu. A process to reduce DC ingot butt curl and swell. In: Light Metals. Warrendale, PA: TMS, 1980, pp. 613–628.

30. FA Sergerie, NB Bryson. Reduction of ingot bottom "bowing and bumping" in large sheet ingot casting. In: Light Metals. Warrendale, PA: TMS-AIME, 1974, pp. 587–590.

31. W Schneider, EK Jensen, B Carrupt. Development of a new starting block shape for DC casting of aluminium sheet ingots. Part I: Experimental results. In: Light Metals. Warrendale, PA: TMS, 1995, pp. 961–967.

32. EK Jensen, W Schneider. Development of a new starting block shape for the DC casting of aluminium sheet ingots. Part II: Modeling results. In: Light Metals. Warrendale, PA: TMS, 1995, pp. 969–978.

33. HG Fjær, EK Jensen. Mathematical modeling of butt curl deformation of sheet ingots. Comparison with experimental results for different starting block shapes. In: Light Metals. Warrendale, PA: TMS, 1995, pp. 951–959.

34. PA Davidson, SC Flood. Natural convection in an aluminium ingot: a mathematical model. Metall Trans 25B:293–302, 1994.

35. JM Reese. Characterization of the flow in the molten metal sump during direct chill aluminium casting. Metall Trans 28B:491–499, 1997.

36. SC Flood, L Katgerman, VR Voller. The calculation of macrosegregation and heat and fluid flows in the DC casting of aluminium alloys. Modeling of Casting, Welding and Advanced Solidification Processes V, M Rappaz, MR Özgü, KW Mahin, eds. Warrendale, PA: TMS, 1991, pp. 683–690.

37. AV Reddy, C Beckermann. Simulation of the effects of thermosolutal convection, shrinkage induced flow and solid transport on macrosegregation and equiaxed grain size distribution in a DC continuous cast Al-Cu round ingot. Materials Processing in the Computer Age II, VR Voller, SP Marsh, N El-Kaddah, eds. Warrendale, PA: TMS, 1994, pp. 89–102.

38. A Håkonsen, D Mortensen. A FEM model for the calculation of heat and fluid flows in DC casting of aluminium slabs. Modeling of Casting, Welding and Advanced Solidification Processes VII, M Cross, J Campbell, eds. Warrendale, PA: TMS, 1995, pp. 763–770.

39. GU Grün, W Schneider. Influence of fluid flow field and pouring temperature on thermal gradients in the mushy zone during level pour casting of billets. In: Light Metals. Warrendale, PA: TMS, 1997, pp. 1059–1064.

40. A Mo, T Rusten, HJ Thevik. Modeling of surface segregation development during DC casting of rolling slab ingots. In: Light Metals. Warrendale, PA: TMS, 1997, pp. 667–674.

41. AV Reddy, C Beckermann. Modeling of macrosegregation due to thermosolutal convection and contraction-driven flow in direct chill continuous casting of an Al-Cu round ingot. Metall Trans 28B:479–489, 1997.

42. BQ Li, JC Liu, JA Brock. Numerical simulation of transient fluid flow and solidification phenomena during continuous casting of aluminium. EPD Congress 1993, JP Hager, ed. Warrendale, PA: TMS, 1992, pp. 841–857.

43. BQ Li, PN Anyalebechi. A micro/macro model for fluid flow evolution and microstructure formation in solidification processes. Int J Heat Mass Transfer 38:2367–2381, 1995.

44. BR Henriksen, EK Jensen, D Mortensen. The modeling of transient heat and fluid flows in the start-up phase of the DC casting process. Modeling of Casting, Welding and Advanced Solidification Processes VIII, BG Thomas, C Beckermann, eds. Warrendale, PA: TMS, 1998, pp. 623–630.

45. L Maenner, B Magnin, Y Caratini. A comprehensive approach to water cooling in DC casting. In: Light Metals. Warrendale, PA: TMS, 1997, pp. 701–707.

46. JF Grandfield, A Hoadley, S Instone. Water cooling in direct chill casting. Part 1, Boiling theory and control. In: Light Metals. Warrendale, PA: TMS, 1997, pp. 691–699.

47. JF Grandfield, K Goodall, P Misic, X Zhang. Water cooling in direct chill casting: Part 2, Effect on billet heat flow and solidification. In: Light Metals. Warrendale, PA: TMS, 1997, pp. 1081–1090.
48. WJ Bergmann. Solidification in continuous casting of aluminium. Metall Trans 1:3361–3364, 1970.
49. S Benum, A Håkonsen, JE Hafsås, J Sivertsen. Mechanisms of surface formation during direct chill (DC) casting of extrusion ingots. In: Light Metals. Warrendale, PA: TMS, 1999, pp. 737–742.
50. JM Drezet, G-U Grün, M Gremaud. Determination of thermal properties and boundary conditions in the DC casting process using inverse stationary methods. In: Light Metals. Warrendale, PA: TMS, 2000.
51. HG Fjær, D Mortensen, A Håkonsen, EA Sørheim. Coupled stress, thermal and fluid flow modeling of the start-up phase of aluminium sheet ingot casting. In: Light Metals. Warrendale, PA: TMS, 1999, pp. 743–748.
52. WD Bennon, FP Incropera. A continuum model for momentum, heat and species transport in binary solid-liquid phase change systems—I. Model formulation. Int J Heat Mass Transfer 30:2161–2170, 1987.
53. VR Voller, C Prakash. A fixed grid numerical modeling methodology for convection-diffusion mushy region phase-change problems. Int J Heat Mass Transfer 30:1709–1719, 1987.
54. BE Launder, BI Sharma. Application of the energy-dissipation model of turbulence to the calculation of flow near a spinning disc. Letters in Heat and Mass Transfer 1:131–138, 1974.
55. A Geist, et al. PVM: parallel virtual machine, a users guide and tutorial for networked parallel computing. MIT Press, 1994.
56. TG Kim, Y-S Choi, Z-H Lee. Heat transfer coefficients between a hollow cylinder casting and metal mold. Modeling of casting, Welding and Advanced Solidification Processes VIII, BG Thomas, C Beckermann, eds. Warrendale, PA: TMS, 1998, pp. 1023–1030.
57. S Grådahl, ST Johansen. Måleprogram for Valseblokk, Del III. SINTEF-report F93198, Trondheim, 1993.
58. AL Dons, EK Jensen, Y Langsrud, E Trømborg, S Brusethaug. The alstruc microstructure solidification model for industrial aluminium alloy. Metall and Mater Trans 30A: 2135–2146, 1999.

17

Vacuum Arc Remelting and Electroslag Remelting

Lee A. Bertram
Sandia National Laboratories, Livermore, California

Ramesh S. Minisandram
Allvac, an Allegheny Technologies Company, Monroe, North Carolina

Kuang-O (Oscar) Yu
RMI Titanium Company, Niles, Ohio

I. INTRODUCTION

Vacuum arc remelting (VAR) [1, 2] and electroslag remelting (ESR) [3, 4] are two secondary melting processes used for achieving high metallurgical quality ingot structure necessary for high performance metal components. Both processes are commonly used to produce nickel-base alloys and specialty steels. High quality titanium and its alloys are not currently produced using the ESR process. When compared with conventional static ingot casting processes, both VAR and ESR provide the advantages of chemical composition refining, non-metallic inclusion removal, and ingot structure enhancement.

The performance of VAR and ESR processed materials depends largely on ingot structure and chemical uniformity [5–12]. VAR and ESR are semi-continuous casting processes which typically produce ingot structures with columnar dendritic grains near the ingot surface and (in large diameter ingots) equiaxed dendritic grains in the center of the ingot (Figs. 1 and 2). The final ingot grain structure is strongly influenced by the molten metal pool profile, which in turn depends on the power input and ingot heat transfer characteristics. The grain structure is similar to that in direct chill casting (DC casting) and, to a lesser extent, structures obtained in continuous casting processes, but

Figure 1   Grain structures and pool profiles of three top sections of 508 mm diameter, Inconel 718, VAR ingots. From top to bottom, melting rates are 182, 322 (without hot top), and 355 kg/h (with hot top). (From Ref. 6.)

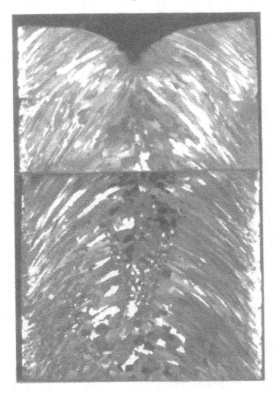

Figure 2 Grain structure and pool profile of a 432 mm diameter Inconel 718 ingot ESR processed (without hot top) at 591 kg/h melting rate. (From Ref. 6.)

much more controlled. This chapter emphasizes process characteristics and modeling techniques that are unique to VAR and ESR.

II. PROCESS DESCRIPTION

Figures 3 and 4 show schematics of VAR and ESR processes [3]. The two processes are similar in that both involve the remelting of a consumable electrode into a water-cooled copper mold or crucible. The consumable electrodes are typically made by conventional static ingot-casting processes VIM (vacuum induction melting) and EAF/AOD (electric arc furnace/argon oxygen decarburization) for nickel-base alloys and steels. For titanium alloys, electrodes can be either welded compacts (mechanically compacted titanium sponge

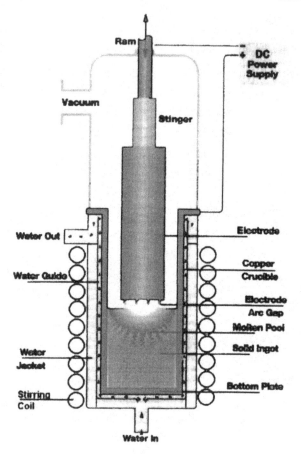

Figure 3 Schematic of VAR furnace. (Courtesy of Allvac, an Allegheny Technologies Company.)

briquettes welded together with bulk weldable scrap pieces in an inert gas atmosphere) or previously melted ingots.

The major difference between VAR and ESR is the means by which the energy for melting is obtained. In VAR, the energy is supplied by a direct current (dc) arc that is struck between the electrode and the surface of the ingot pool. In ESR, the energy is provided by a layer of liquid slag into which the tip of the electrode is immersed. The electrically resistive slag is kept molten by Ohmic heating due to the passage of alternating current (ac) through it. In VAR, the falling of metal droplets from the electrode tip to ingot top in a vacuum environment provides a favorable condition for VAR processing to remove entrapped gases and high vapor pressure elements. In ESR, the

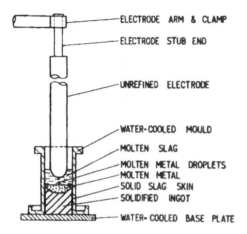

Figure labels:
ELECTRODE ARM & CLAMP
ELECTRODE STUB END
UNREFINED ELECTRODE
WATER-COOLED MOULD
MOLTEN SLAG
MOLTEN METAL DROPLETS
MOLTEN METAL
SOLID SLAG SKIN
SOLIDIFIED INGOT
WATER-COOLED BASE PLATE

Figure 4   Schematic of ESR furnace. (From Ref. 3.)

slag layer permits melting of nickel-base alloys and steels in air by serving as a barrier to keep oxygen of the air from the molten metal, which falls as droplets through the slag to form an ingot or slab in a moving mold or fixed crucible. The passage of metal droplets through the slag offers an opportunity for chemical refining reaction such as desulfurization and for removal of ceramic oxide inclusions. Due to the strong chemical reaction between the slag and molten titanium, ESR is not used by Western countries to melt titanium. However, Russians have used ESR to melt nonpremium titanium. The fundamental aspects of the different energy sources in the VAR and ESR are reflected in the heat distribution in the ingot and the shape of the resultant liquid metal pool.

In DC casting and continuous casting processes, the only heat input to the ingot is the molten metal with a predetermined superheat. The melting and casting steps are decoupled, and the molten metal superheat and ingot casting rate can be controlled independently. In contrast, the melting and casting steps are closely related in VAR and ESR processes. The power input (arc heating for VAR or slag Joule heating for ESR) not only results in electrode melting to generate molten metal droplets but also provides additional heat input to the top surface of the ingot. An increase (or decrease) of the power input will result in an increase (or decrease) of electrode melting rate (and, hence, ingot casting rate), molten metal droplet temperature, and ingot top surface heat input simultaneously. Consequently, electrode melting (molten metal droplet temperature) and ingot casting (casting rate and top surface heat input) cannot be controlled independently in VAR and ESR processes.

Another major difference between VAR and ESR versus DC casting and continuous casting is the presence of a strong electromagnetic body force in VAR and ESR ingots. This Lorentz force has a significant influence on the fluid flow behavior in the molten metal pool, and therefore affects the resultant ingot structure.

A complete VAR or ESR melting cycle includes three separate regions: start-up, steady state melting, and hot-top (Fig. 5). In the start-up region, the depth of the liquid metal pool gradually increases from zero to the pool depth of the steady state melting region. The power input level of this region is usually higher than that of the steady state melting region. Depending on the actual power input, the maximum depth of the molten metal pool in the start-up region can briefly be larger than that in the steady state melting region [7]. The start-up region typically ends when the height of the ingot is between 1 and $1\frac{1}{2}$ times the ingot diameter.

The steady state melting region usually has the deepest liquid metal pool, the largest mushy zone size, and the longest local solidification time. As discussed later, this means it has the strongest tendency to form freckle-type defects.

The power input level in the hot-top region is reduced from the steady state melting level, in order to gradually decrease the depth of the liquid metal pool from the steady state pool depth to zero. The purpose of the hot top is to

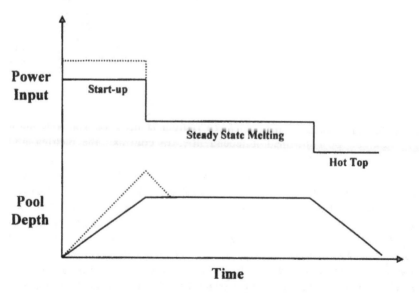

Figure 5 Complete cycle of VAR and ESR processing.

minimize the material loss due to the formation of a shrinkage pipe on the top of the ingot (Fig. 1). A good hot-top practice can eliminate the formation of shrinkage pipe completely, without undue increase in total process time.

For nickel-base superalloys and tool steels, the ingot diameter is relatively small and the height of the ingot is typically about 5 to 6 times the ingot diameter. During VAR and ESR processing, start-up and hot-top regions each accounts for about 1 and $1\frac{1}{2}$ times the ingot diameter height and the steady state melting region accounts for 3 to 4 times the ingot diameter height. For titanium melting, these proportions can differ depending on the alloy chemical composition. Ingots of Ti-6Al-4V in VAR can have much larger diameters than those of superalloy and tool steel. The final ingot height to diameter ratio is only around 2 or 3. Consequently, the steady state melting region shown in Fig. 5 is practically nonexistent for VAR processing of this titanium alloy.

A useful process model for the simulation of VAR and ESR processes should have the capability to simulate all three (start-up, steady state melting, and hot-top) regions.

III. APPLICATIONS

Secondary melting processes such as VAR and ESR are primarily used to produce high performance wrought materials of titanium alloys, superalloys, nickel-base high temperature and corrosion resistant alloys, and specialty steels (tool steels, stainless steels, and low alloy steels). The shapes of the products are usually cylindrical ingots (VAR and ESR) and rectangular slabs (ESR). In Russia and Europe, ESR is also used to produce shaped components such as rolls and crank shafts.

Except titanium and zirconium alloys, which are only produced by VAR process in Western countries, all other aforementioned alloys can be produced by either VAR or ESR. For one particular class of application, namely, superalloy turbine disks for jet engines and industrial gas turbines, material is produced by the so-called double-melt and triple-melt processes [11]. Alloys (e.g., Inconel 718, Waspaloy, and Udimet 720) for turbine disk applications are required by equipment manufacturers to be produced by vacuum induction melting (VIM) to make static cast electrodes followed by VAR processing of the electrodes to make the final ingots. This double vacuum melting procedure is commonly called VIM/VAR technology. Recently, an ESR step has been added between the VIM and VAR steps to reduce the material ceramic inclusion content and to produce a pipe free, fully dense ESR ingot as input electrode for the subsequent VAR processing. This process is called VIM/ESR/VAR triple melt [11].

For titanium alloys, ingots are typically melted two or three times. For the first VAR, many electrodes are made by the welded titanium sponge compacts and bulk scraps. Other melting and consolidation methods such as EBM (electron beam melting), plasma arc consolidation, and Rototrode consolidation are also being used to make the first electrodes. The resultant first VAR ingot is then used as the electrode for the second VAR processing. If the third VAR processing is required, the second VAR ingot will be used as the electrode for remelting. Two or three times melted VAR ingots are commonly referred as Two VAR (2X VAR) or Three VAR (3X VAR) ingots.

Several types of defects have been found in billet materials which were forged from VAR and ESR ingots [8–14]. For nickel-base alloys, the most frequent defects are freckles, white spots, and sonic defects. Beta flecks, type II segregation, type I hard alpha, and high density inclusions (HDI) are corresponding defects for titanium alloys. Details regarding the formation mechanisms and modeling approaches for these defects will be discussed in Sec. V.E.

Improving VAR and ESR furnace performance means increasing the production rate of defect-free ingots. To accomplish this, the solidification conditions which lead to defect formation must first be established by a combination of modeling and experiment [15, 16]. This combined approach is illustrated here, starting with the description of electrode melting .

IV. HEAT GENERATION AND ELECTRODE MELTING

A. Heat Generation

For VAR, the heat source for melting is the DC power deposited on the (negative "straight" polarity) electrode. The fundamental theory of vacuum arc is very complex, involving cathode spot structure and dynamics and arc plasma dynamics [17, 18]. At present, the vacuum arc physics of the VAR furnace is only partially characterized. However, by combining theory and experiment as described below, furnace control parameters such as current can be related to electrode heating [19, 20].

In ESR, the electrode is melted by immersing it in a slag bath whose temperature is sustained by Joule heating. The molten slag not only provides heat for electrode melting but also is the source to provide heat input to ingot top surface and to cause a significant heat loss to water-cooled copper crucible in slag cap region. Choudhary and Szekely [21] simulated the molten slag behavior. This model included:

1. Electromagnetic equations to quantify current paths, voltage field, body forces, and Joule heating

2. Incompressible Navier-Stokes equations with the body forces to calculate fluid flow
3. An energy equation including heat conduction with the Joule heating effect and radiative and convective boundary conditions

Based on the above discussion, it can be envisioned that developing a comprehensive simulation for the heat generation phenomena in VAR and ESR processes is quite demanding. From the practical application point of view, the characterization of the energy partition between electrode melting and ingot/crucible surfaces, based on macro heat balance, provides a valuable starting approach.

For both VAR and ESR, the total electrical power P_t received by the remelting furnace can be expressed as

$$P_t = IV = P_e + P_i + P_l \tag{1}$$

where I = furnace current
V = furnace voltage
P_e = power of electrode melting
P_i = power of ingot heat input
P_l = power loss through arc gap (VAR) or molten slag cap (ESR)

The partition of P_e, P_i, and P_l depends on furnace dimensions and operating conditions. For example, the bigger the arc gap between the electrode tip and the ingot top surface in VAR and the thickness of the molten slag cap in ESR, the higher the power loss P_l. On the other hand, a bigger ingot diameter will result in a lower percentage of total power loss through the arc gap or molten slag cap.

During steady state operation, the partition of P_e, P_i, and P_l is also steady and the steady state electrode melting rate dm/dt is related to P_e:

$$P_e = \Delta H_f (1 + \text{St}) \frac{dm}{dt} \tag{2}$$

$$\text{St} = \frac{C_p(T_d - T_0)}{\Delta H_f} \tag{3}$$

where St = Stefan number
ΔH_f = latent heat of fusion of the material
C_p = specific heat of the material
T_d = temperature of the molten metal droplet dripping from the electrode face
T_0 = electrode initial temperature

Many VAR and ESR furnaces are equipped with load cells which measure electrode weight changes. Thus, the electrode's average melting rate can be measured to around 5% precision by load cells over, typically, a 20 min interval. Assuming molten metal droplets are dripping from the electrode face at material liquidus temperature plus some superheat, the power for electrode melting P_e can be estimated with 10% accuracy or better [22, 23].

The experimental estimation of the sum $P_i + P_l$ is somewhat more difficult than the measurement of P_e, but separating P_i and P_l proves very difficult [15, 16, 24]. To capture the sum (the power loss through arc gap or molten slag cap), one possible way is to use copper crucible temperature/history data obtained from thermocouples. Based on the measured temperature difference, temperature gradient and, hence, heat flux across the thickness of the copper crucible wall can be calculated. Then, by assuming a steady state and one-dimensional heat transfer across the crucible wall thickness, part of this calculated heat flux can be used as an estimation for P_l. Once P_l is known, the heat input to the ingot P_i can be calculated from $P_i = P_t - P_e - P_l$.

B. Electrode Melting

Considering a concentric cylindrical electrode, and assuming axial symmetry about the x axis, the electrode temperature is governed by the heat equation in the form of Eq. (21) of Chapter 2, in cylindrical coordinates (x, r):

$$\rho C_p \frac{\partial T}{\partial t} + \nabla \cdot (\rho \mathbf{v} \mathbf{H}) = \nabla \cdot (k \nabla T) + \frac{dQ}{dt} \tag{4}$$

where ρ is density, C_p is heat capacity, $\mathbf{v}$ is velocity, H is enthalpy, t is time, T is temperature, k is thermal conductivity, and Q is heat generation term. The axial coordinate x is positive upward, with origin at the original electrode bottom; the radial coordinate is r. The electrode fills

$0 < x < L_e$ and $0 < r < R_e$ at time $t = 0$

where L_e is the electrode original length and R_e is the radius of the electrode.

The boundary condition at the electrode melting face is the "Stefan condition" [25, 26], which balances melting at a constant temperature (liquidus temperature) with the applied heat flux estimated from the power for electrode melting P_e. The side wall of the electrode is subject to radiation (VAR and ESR) and natural convection (ESR) heat loss.

The electrode has some initial temperature T_0 at time $t = 0$; in general, this can be a function of position:

$T(x, r, 0) = T_0(x, r)$ at time $t = 0$

The rate at which the electrode melts, dm/dt, can be computed by solving the problem posed above, using one of several numerical techniques available for "moving boundary problems" [25, 26].

The basis for predicting dynamic electrode melting rate is to capture the relevant arc physics for a given furnace geometry and alloy by measuring P_e. For example, remelting of Inconel 718 from 432 mm (17 in.) diameter electrode to 508 mm (20 in.) diameter ingot under quasisteady conditions, Zanner et al. [27, 29] related the independent furnace parameters I (furnace current), g_e (arc gap), p_g (chamber pressure) to the dependent values melting rate dm/dt and voltage V:

$$\frac{dm}{dt} = Z_{mr}(I, g_e, p_g) = C_{mr1}(g_e, p_g)I + C_{mr12}(g_e, p_g)I^2 \tag{5}$$

$$V = Z_V(I, g_e, p_g) = V_c + C_{V1}(g_e, p_g)I + C_{V12}(g_e, p_g)I^2 \tag{6}$$

where Z's and C's are experimentally established empirical functions and V_c is the voltage drop in the cathode spots, also experimentally determined.

These statistical regressions are alloy specific, and must be developed from a factor space experiment on a furnace of interest. The effect of ingot size is also included in the extended melting rate correlation developed by Schved [30].

Despite its derivation for quasisteady conditions, the empirical melting rate from Eqs. (5) and (6) can be substituted into Eq. (2) to give the instantaneous cathode heating power:

$$P_e(t; I; g_e, p_g) = h_m \frac{dm}{dt} = h_m Z_{mr}(I, g_e, p_g) \tag{7}$$

during transient arc conditions. This provides a basis for dynamic simulation of melting rate of the VAR electrode [22, 23].

V. INGOT SOLIDIFICATION

Modeling of ingot solidification involves heat transfer, electromagnetic field, fluid flow, micro/grain structure evolution, and macrosegregation formation. The solidifying ingot receives molten metal droplets and heat flux from its top surface and loses heat to the water cooled copper crucible from the side wall and bottom surface. The heat extraction mechanisms are thus the most important physical phenomenon for ingot solidification. The sizes and shapes of the resultant molten pool and mushy zone are strong indicators of ingot structure and quality. The electromagnetic fields influence macro fluid flow patterns in the molten pool by exerting the Lorentz body force, which is typically as important as buoyancy forces in creating the pool shapes. In VAR and ESR ingot solidification, besides the macro fluid flow in the molten pool which dominates heat transfer, there is a much smaller interdendritic fluid flow in the mushy zone responsible for transport of solute, which can cause the formation of macrosegregation defects such as freckles and beta flecks.

A. Heat Transfer

The heat transfer for a cylindrical ingot is also described by Eq. (4). Details of boundary conditions are discussed in the following sections.

1. Top Boundary Conditions

In cylindrical coordinates with the axial coordinate origin on the metal pool surface of the ingot, positive pointing downward, a heat flux boundary condition can be written as

$$-k\frac{\partial T}{\partial x} = q_{tot} \qquad \text{on } x = 0, \text{ for } 0 < r < R_0 \tag{8}$$

where q_{tot} is the net equivalent heat flux from arc heating, slag heating, evaporation of metal vapor, and radiative cooling.

For VAR, heat input at the ingot top surface is due to arc heating:

$$q_{tot} = q_{arc} - H_{vap}\dot{m}_{evap} - q_{rad} \tag{9}$$

$$P_i = \int q_{arc}dA \tag{10}$$

where A is the ingot cross-sectional area, $\dot{m}_{evap}$ is the mass rate of evaporation, H_{vap} is the enthalpy difference between the vapor and reference temperature, including the latent heat of evaporation, and q_{rad} is the net radiation flux for heat loss at the annulus area.

In normal VAR processing (straight polarity), the electrode is the cathode and ingot is the anode. The energy partition for this melting mode is such that less energy is used for electrode melting whereas more energy is deposited on the ingot top surface, resulting in a deeper molten pool and higher macro segregation tendency in the solidifying ingot. On the other hand, the electrode is the anode and ingot is the cathode for reverse polarity melting mode. At the same total power input level, this melting mode generally has a faster melting rate but shallower pool depth and lower segregation tendency than the straight polarity melting mode. As a result, reverse polarity melting mode is commonly used by titanium melters to melt segregation prone alloys such as Ti-10V-2Fe-3Al. It should be noted that reverse polarity melting usually results in an ingot with poorer surface quality. In addition, the water-cooled copper crucible can be damaged by cathode spots during reverse polarity melting. Thus, reverse polarity is not a preferred melting mode for titanium melters. During VAR ingot modeling, care should be taken to ensure that appropriate energy partitions (q_{arc} and P_e) for straight and reverse polarity melting modes are being used.

At the top of an ESR ingot (liquid metal-slag interface), the boundary condition specifies a convective heat flux which can be written as

$$q_{\text{tot}} = h_{\text{sm}}(T_s(r, t) - T(r, t)) \tag{11}$$

where $h_{\text{sm}}(r, t)$ is the heat transfer coefficient between the slag and molten metal and $T_s(r, t)$ is the slag temperature at radius r at time t. The superheat of the liquid metal droplets, the slag temperature, and the slag-metal interface heat transfer coefficient are functions of time or ingot height and are strongly dependent on power input and melting rate. A complete heat transfer model for ESR would include both slag and metal. Details for this approach can be found in Ref. 21. A more convenient way is to determine the metal droplet temperature, slag temperature, and slag-metal interface heat transfer coefficient based on experimental data and engineering judgment.

The heat flux q_{arc} is deposited by the metal vapor arc in VAR, and is expected to be strongest in the area underneath the electrode tip. For practical modeling, the total arc power deposited on the ingot pool P_i can be calculated from the macro heat balance between total power input, electrode melting rate, and heat loss to the copper crucible above the pool surface. Alternatively, P_i can be derived from a similar (but more conveniently measured) electric current partition for the furnace [22]. Once P_i is known, a spatial distribution can be assumed for q_{arc}, such as a Gaussian function [15, 22, 31], $q_{\text{arc}} = q_0 \exp\{-3(r/r_{2\sigma})^2\}$, where the parameter $r_{2\sigma}$ is a measure of the concentration of the heat deposition near the axis.

Each surface point also receives an inflow of metal droplets at a rate determined by the melting rate dm/dt with temperature T_d. This temperature T_d is taken to be some superheat above the liquidus temperature T_l of the electrode material:

$$T_d = T_l + \Delta T_{\text{sup}} \tag{12}$$

where superheat ΔT_{sup} is 50–150 K for VAR, based on both experimental evidence and physical arguments. In ESR, the droplets have undergone longer exposure to the slag, and experiment suggests that $\Delta T_{\text{sup}} = 100$–150 K.

2. Side and Bottom Boundary Conditions

The flow of heat from VAR and ESR ingot surfaces through copper crucibles to cooling water is a complex process and depends on several factors such as slag skin thickness (w_{sk}, ESR), helium pressure (p_{He}, VAR), shrinkage gap width w_{gap}, cooling water flow rate and velocity, and crucible surface condition. Yu [5] compared VAR and ESR heat flux distribution profiles along the height of copper crucible (Fig. 6).

For ESR, there are three heat transfer zones in copper crucible. The zone above the free surface of the slag has little effect on the heat transfer to the crucible. In the region of the molten slag cap and liquid metal head, the molten slag film has good contact with the crucible. The effect of convective motion in the slag and molten metal results in high heat transfer rates in this region. When the ingot surface temperature is lower than the metal liquidus temperature, the ingot starts to pull away from the crucible and a shrinkage gap forms between the slag skin and crucible. Heat is lost by radiation and air conduction across this gap.

In contrast, VAR crucibles have only two heat transfer zones. During VAR melting, splattering and condensed metal vapor produce crucible wall deposits which form a collar or crown on the top edge of the ingot. With increasing ingot height, this crown forms a shelf or layer at the ingot surface. Hence, good contact between the ingot surface and the crucible can only exist if this crown is partially melted by the pool. Far enough down from the ingot top, the ingot surface is cool enough that it has shrunk away from the crucible and formed a shrinkage gap. Heat transfer in the shrinkage gap region between the

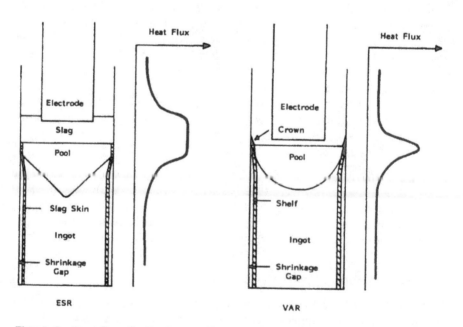

Figure 6 Heat flux distribution profiles along the height of VAR and ESR copper crucibles. (From Ref. 5.)

ingot and crucible is primarily by radiation if this gap is a vacuum. This gap heat transfer is greatly improved by admitting helium gas into the gap to allow thermal conduction by the rarefied helium gas [5, 32, 33]. The conduction rate depends primarily on gas pressure and gap width. The partial contact between the top crown and crucible forms a seal which prevents the leakage of helium gas. Higher power input results in better contact and hence, allows higher helium pressure. Higher helium input flow rate also increases helium pressure in the shrinkage gap.

For both the ESR and VAR processes, the zone where the shrinkage gap is formed has the greatest effect on the ingot heat transfer rate. Hence, the following discussion will concentrate upon the heat transfer characteristics of this zone [5].

During operation, both axial and radial heat conduction occur in the copper crucible. Due to the small thickness of the crucible wall (typically about 19 mm), however, the effect of axial conduction is small and hence only radial conduction needs to be considered. Figure 7 schematically shows the heat transfer resistance from the surface of the ESR and VAR ingots to the crucible cooling water [5]. Heat transfer from ingot surface to the crucible cooling water includes:

1. Conduction across the slag skin of thickness (ESR) w_{sk}
2. Radiation and helium (VAR)/air (ESR) conduction in the gap between the slag skin (ESR) or ingot (VAR) surface and crucible
3. Conduction across the crucible wall of thickness w_{cu}
4. Conduction through the scale deposit on the outside surface of the crucible, with film coefficient h_{sc}
5. The crucible-water heat transfer with film coefficient h_{H2O}

The overall heat transfer coefficient h_{iw} can be obtained by the summation of the individual resistances of these mechanisms [5, 33]:

$$h_{iw} = \frac{1}{\dfrac{w_{sk}}{k_s} + \dfrac{1}{h_{rad} + h_{gas}} + \dfrac{w_{cu}}{k_{cu}} + \dfrac{1}{h_{sc}} + \dfrac{1}{h_{H2O}} + R_w} \qquad (13)$$

where h_{rad} is the heat transfer coefficient due to ingot surface radiation heat loss, h_{gas} is the gas conduction heat transfer coefficient by rarefied helium gas in VAR [5, 32, 33] or air in ESR, and R_w is a temperature-sensitive interfacial resistance characterizing the partial contact [24, 31]. Figure 8 shows calculated heat transfer coefficient at VAR and ESR ingot surface as functions of helium pressure and gap width. The overall heat transfer boundary condition then takes the form:

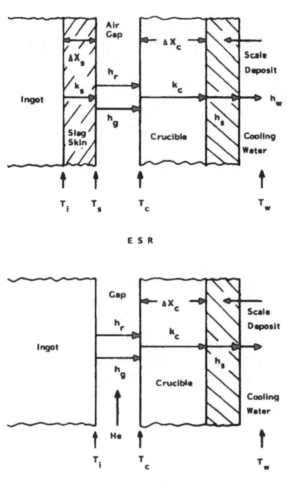

Figure 7 Schematics of heat transfer resistance from the surface of VAR and ESR ingots to the crucible cooling water. (From Ref. 5.)

$$-k\frac{\partial T}{\partial r} = h_{iw}(T - T_w)$$ (14)

where T_w is the temperature of crucible cooling water.

For steady state model, the bottom of the computational mesh is placed in the solidified ingot at a distance L_c from the pool surface. L_c may be chosen to be large enough to make the effect of the boundary condition minor. Under

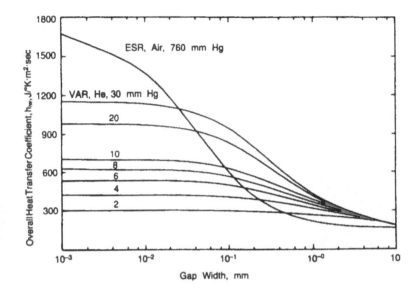

Figure 8 Calculated heat transfer coefficient at VAR and ESR ingot surfaces as functions of gas pressure (helium for VAR and air for ESR) and gap width. (From Ref. 5.)

these conditions, rigid body motion (with a velocity equal to the ingot growth rate) and thermally insulated surface conditions may be imposed on the outflow boundary.

B. Electromagnetic Field

For industrial scale VAR and ESR furnaces, electrical currents are in the range of 5000–40,000 A. In general, nickel-base alloys are melted at the low end of this scale whereas titanium alloys are melted at the higher current levels. When the current passes through the melting ingot, it produces a significant electromagnetic field which, in turn, generates Lorentz forces which have important effects on the fluid flow pattern of the molten metal pool.

The Lorentz forces $\mathbf{F}_L$ are computed from electrical current density $\mathbf{j}$ and the magnetic induction vector $\mathbf{B}$:

$$\mathbf{F}_L = \mathbf{j} \times \mathbf{B} \tag{15}$$

Determining the fields of $\mathbf{j}$ and $\mathbf{B}$ in complete detail requires the solution of the Maxwell equations of electrodynamics [35, 36]. This is a challenging task and will not be discussed in detail here. Instead, a general discussion regarding the characteristics of VAR and ESR will be presented.

How the current passes through electrode, ingot, crucible, and ground has a significant effect on the distribution of current density and the resultant Lorentz force. In VAR, the only path for current flow between the ingot and crucible is assumed to be metal-metal contact. Thus, the shrinkage gap is treated as a perfect electrical insulator, and all the current entering through the liquid metal pool surface must find its way to the crucible through the top, bottom, or crown areas. The actual partition of the current going through the top, bottom, or crown areas depends on the furnace design. Some furnaces are grounded through the bottom of the crucible. In this case, most of the current flows vertically, through the ingot top to bottom and thence to ground. Some other furnaces are designed to encourage current passage to the crucible wall, by grounding the crucible flange. In ESR, both slag and ingot have a direct contact with a copper crucible. Thus, current may be relatively evenly distributed through the crucible wall and ingot bottom to ground.

During the melting of titanium alloys in VAR furnaces, the arc is steered by excitation of the solenoidal windings to provide additional heat to the edge of ingot top surface to improve the ingot surface quality. This excitation creates an independent axial magnetic field which interacts with the furnace current, resulting in a more complex overall magnetic field [24, 35]. In the case of ESR, since ac power supplies are commonly used, the period-averaged values of j and B are typically used in calculations.

In VAR, the molten metal pool is located right next to the electrode tip, thus the influence of electromagnetic field on the pool fluid flow pattern is relatively strong. For ESR, the electrode is immersed in the molten slag which is on top of the molten metal pool. Consequently, the effect of the electromagnetic field on the fluid flow of the molten slag is stronger than that of the molten metal pool. For both the VAR and ESR, because the mushy zone is far away from the electrode tip, the effects of the electromagnetic body forces on the interdendritic fluid flow is considered to be much weaker than Lorentz effects in the molten pool.

C. Fluid Flow

During melting, there are two fluid flow phenomena in VAR and ESR ingots: macro fluid flow in the liquid metal pool and interdendritic fluid flow in the mushy zone. Interdendritic fluid flow is the primary cause for the formation of freckles and beta flecks type macrosegregation; its modeling approach will be discussed in Sec. V.E.. In the liquid metal pool, fluid flow involves both natural convection that is caused by the buoyancy force and forced convection that is caused by the Lorentz forces from the electromagnetic field. The modeling approach for the fluid flow in liquid metal pool will be briefly discussed here.

Consider cylindrical coordinates (x, r, θ), with the x axis pointing vertically downward along the symmetry axis of the crucible. In these coordinates, an axisymmetric flow depends only on the x and r coordinates, and time t. If this is generalized slightly by allowing a nonzero component of velocity in the azimuthal θ direction, then the Navier-Stokes equations for incompressible unsteady flow of a viscous fluid which conducts an electrical current density vector $\mathbf{j}(x, r, t)$ in the presence of a magnetic induction $\mathbf{B}(x, r, t)$ are given by

$$\rho_m \frac{D\mathbf{u}}{Dt} = -\nabla p + \nabla \cdot \tau + \mathbf{j} \times \mathbf{B} + \rho_m \mathbf{g} \tag{16}$$

where ρ_m is the mass density of the fluid, $\mathbf{u}$ is the velocity of a particle and D/Dt is the Lagrangian derivative following the particle motion:

$$\frac{D\mathbf{u}}{Dt} = \frac{\partial \mathbf{u}}{\partial t} + \frac{1}{2}\nabla|\mathbf{u}|^2 - \mathbf{u} \times \nabla \times \mathbf{u} \tag{17}$$

Pressure is denoted by p and deviatoric stress by (tensor) τ. The velocity vector $\mathbf{u}$ has components $\{u, v, w\}$ all dependent on (x, r, t) but not θ. The use of k-ε model to take into account the turbulence of the three-dimensional movement in VAR can be found in reference [36].

These equations of motion contain source terms involving the temperature T, so it is necessary to solve a coupled energy equation. The most convenient form of the energy equation [37, 38] contains enthalpy $H(x, r, t)$ as unknown:

$$\frac{\partial H}{\partial t} + \frac{\partial(uH)}{\partial x} + \frac{\partial(rvH)}{r\,\partial r} = \frac{\partial}{\partial x}\left(\alpha\frac{\partial H}{\partial x}\right) + \frac{\partial}{r\,\partial r}\left(r\alpha\frac{\partial H}{\partial r}\right) + \frac{1}{2}\rho_e|\mathbf{j}|^2 \tag{18}$$

where α is thermal diffusivity and ρ_e is the electrical resistivity of the alloy, so that the term containing this factor becomes the Joule heating of the ingot (usually negligible in VAR) The temperature T is found from an equation of state $H = H(T)$, in which the phase change is rendered by a piecewise linear version of the Gulliver-Scheil solidification path for a pseudobinary alloy [24, 39].

In VAR, the fluid flow pattern and the volume of the molten metal pool are the resultant of the interaction between the thermal buoyancy forces (natural convection) and the electromagnetic Lorentz forces. The thermal buoyancy forces tend to drive hot metal up the axis of the ingot and down the side walls. As a result, they bring relatively cool metal to the bottom of the pool. In contrast, the electromagnetic Lorentz forces drive the molten metal down the axis and up along the walls. Thus, they bring hot metal to the bottom of the liquid pool. This difference of flow pattern between the thermal buoyancy forces and electromagnetic Lorentz forces has important effects on the volume of the molten metal pool and macrosegregation formation tendency.

From a pure heat transfer point of view, a higher furnace power input results in a higher melting rate, heat input to the ingot top surface and larger volume of the liquid metal pool. Additionally, the fluid flow in the liquid metal pool can also influence the pool volume. Because they bring relatively cool metal to the bottom of the pool, the effects of thermal buoyancy forces on the increase of pool volume will be minimal. On the other hand, electromagnetic Lorentz forces bring hot metal to the bottom of the liquid pool and hence, have a more profound effect on the increase of the liquid metal pool volume.

In general, buoyancy flows dominate at a low current operation. As the current is gradually increased, the influence of the electromagnetic Lorentz forces begin increasing. Nickel-base superalloys are typically melted at a relatively low current level. For example, the current level for melting a 508 mm (20 in.) diameter Inconel 718 ingot is about 6000 A. At this power level, the effects of Lorentz forces on pool volume are small. As the power input increases, the Lorentz forces become bigger and the total pool volume increases rapidly. Figure 9 shows the rapid increase of pool depth

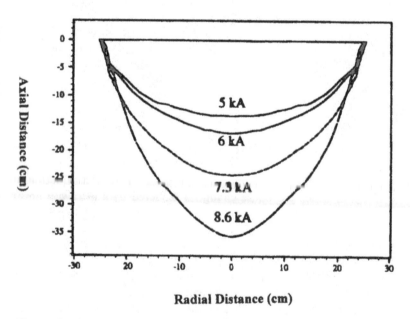

Figure 9 Experimental pool shapes for a 508 mm diameter Inconel 718 VAR ingot melted at four different power levels. The measured pool corresponds to the liquidus isotherm. (From Ref. 40.) See Fig. 34 below.

with power input for a 508 mm diameter Inconel 718 VAR ingot [40]. This rapid increase of pool volume is probably one of the reasons that the tendency for freckles formation in nickel-base superalloys is very sensitive to the increase of furnace power input. At very large current levels [15], the Lorentz forces become dominant and result in a very deep pool (Fig. 10). At this current level, the effects of the increase of furnace power input on the increases of pool volume and macrosegregation formation tendency becomes less significant. The melting of titanium alloy Ti-6Al-4V (done at 30,000–35,000 A for 863–914 mm (34–36 in.) diameter ingots) is typical of this situation.

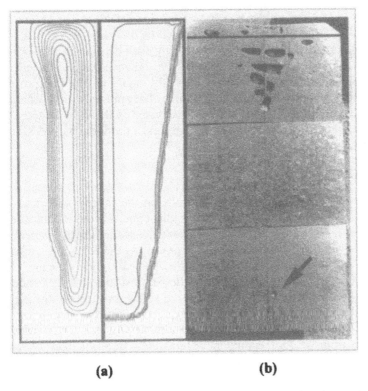

(a)   (b)

Figure 10 (a) Simulated flow streamlines (left) and pool profile (right) for a 910 mm diameter Ti-6Al-4V VAR ingot melted at 33 kA, 44.7 V, and 0.45 kg/s. (b) Grain structure of the ingot. Near the end of the melting, a tantalum ball dropped into the pool. The ball came to rest on the edge of the columnar grain region growing normal to the baseplate surface (arrow). This columnar grain frontier is presumed to mark the end-of-melt pool along the bottom and ingot lateral surface, indicating a "soda can" containment of liquid metal. (From Ref. 15.)

D. Micro/Grain Structure

Ingots made by the conventional ingot casting process can have severe micro-segregation, macrosegregation, and hot tearing which pose a limit on ingot forgeability during the subsequent open die forging operations. VAR and ESR, as previously described, provide a way to increase the ingot cooling rate and hence reduce the levels of microsegregation, macrosegregation, and hot tearing, resulting in a vastly improved ingot forgeability. This improved forgeability of VAR and ESR ingots is primarily due to a reduced microse-gregation, refined microstructure, and more controllable grain structure. The approach for modeling these features will be discussed here. Although they have a reduced degree of macrosegregation than the conventionally cast ingots, VAR and ESR ingots can still have macrosegregation defects. An approach to modeling the formation of macrosegregation will be discussed in Sec. V.E.

Microstructure of an as-cast ingot is strongly alloy dependent. For exam-ple, the as-cast microstructure for the alloy Inconel 718 includes γ-γ''eutectic, δ phase, Laves phase, and various carbides, e.g., NbC and TiC. The grain struc-ture includes both columnar and equiaxed grains. The approach for stochastic modeling of micro/grain structure has been discussed in detail in Chapter 5. Here, only a summary of how to apply that approach to model the VAR and ESR ingot structures is presented.

1. Grain Structure

The morphology of the grain structure depends on the casting solidification parameter G/R, where G is the temperature gradient and R is the solidification rate [41]. A low G/R value will result in equiaxed dendritic grains whereas a high G/R value will form columnar dendritic grains. At the edge of the ingot, where molten metal directly contacts the water-cooled copper crucible, the quench effect results in a very fast solidification rate R and low G/R value. As a result, a very thin layer of very fine equiaxed chill grains form at the ingot surface. Once getting away from the chill grain layer, the high temperature gradient in the ingot radial direction results in a high G/R value. Thus, colum-nar grains start to grow. As the columnar grains grow from the ingot surface toward the center, temperature gradient also starts to decrease. For nickel-base superalloys, when ingot diameter is equal to or less than 508 mm (20 in.), these columnar grains can continue to grow until they reach the center of the ingot (Figs. 1 and 2). Consequently, from a practical point of view, these ingots have a 100% columnar grains structure. For large diameter ingots, at some point the temperature gradient will decrease to a level that results in a low G/R value and hence, equiaxed grains structure at the ingot center portion. Because the center of the ingot always has a low cooling rate, these equiaxed grains have relatively large grain sizes and a high degree of microsegregation. Thus, for nickel-base alloys, the formation of equiaxed grains at the center of the ingot indicates a low ingot cooling rate, high degree of microsegregation, and high tendency to form macrosegregation defects.

The growth pattern for the columnar grains depends on the shape of the molten metal pool. Yu and Flanders [6] showed that VAR ingots have a U-shaped pool whereas ESR ingots have a V-shaped pool. This difference in pool shape results in a very different grain structure between VAR and ESR ingots (Figs. 1, 2 and 11). Primary dendrite and grain growth direction is directly opposite to the heat flow direction and perpendicular to the liquidus isotherm. As a result, VAR ingots have columnar grains that are perpendicular to the crucible wall at the ingot surface and parallel to the ingot vertical axis at the ingot center. In ESR, grain growth direction is parallel to ingot axis at the ingot surface and forms an approximately 45° angle with the ingot axis at the center of the ingot.

The grain structure for titanium VAR ingots [15] is quite different from those of superalloy ingots. In general, titanium alloys are melted with much larger ingot diameters and power input than those of superalloys. As shown in

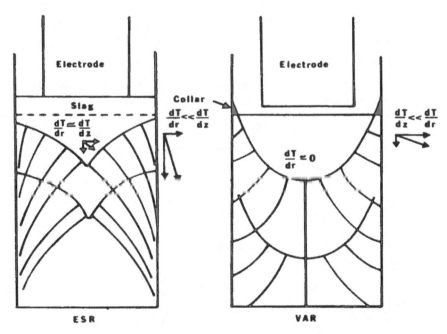

Figure 11 Schematics of pool profile and grain growth pattern of VAR and ESR superalloy ingots. (From Ref. 6.)

Fig. 10, this high power input results in a strong electromagnetic field and fluid flow in the ingot center portion. This strong fluid flow can break the dendrites that grow from columnar grains and mix those broken dendrite tips with the bulk liquid. At the end of the solidification, those broken dendrite tips can serve as nuclei and promote the formation of equiaxed grains in the center of the ingot. Because these equiaxed grains are formed due to a heterogeneous nucleation effect, their grain sizes are actually smaller than those outside columnar grains.

A comprehensive grain structure model needs to have capabilities to predict grain morphology (columnar grains, equiaxed grains, and columnar-equiaxed transition), grain size, and grain growth pattern. Nastac et al. [42, 43] developed a stochastic model to simulate grain structures of VAR and ESR processed superalloy ingots. Their model includes nucleation and growth kinetics, as well as the growth anisotropy and grain selection mechanisms. They use macro thermal history as the input data for the stochastic calculation, which includes (1) local cooling rates at the liquidus temperature to compute the grain nucleation kinetics, (2) average cooling rate or local solidification time in the mushy zone to compute grain growth kinetics, (3) time-dependent temperature gradients in the mushy zone to decide grain growth direction, and (4) G/R ratio to decide columnar-equiaxed grains transition. Figures 12 to 14 show their calculated results.

2. Microstructure

For alloy Inconel 718, the relative volume fractions of both Laves phase and NbC carbides depend on the alloy C/Nb ratio. Alloys with a higher C/Nb ratio will have a higher volume fraction of carbides than alloys with a lower C/Nb ratio. It is known that the Laves phase can decrease the ingot incipient melting temperature, promote intergranular liquation cracking during homogenization heat treatment, and reduce the ingot forgeability for the open die forging operation. Laves phase is also closely related to the formation of freckle defects. In addition, the size, amount, and distribution of NbC carbides affect the alloy mechanical properties. Thus, a capability to model the evolution of Laves phase and NbC carbides during the solidification of Inconel 718 ingot can provide valuable insights regarding how to control and enhance the ingot as-cast structure.

Nastac et al. [44–46] developed a stochastic model to simulate the grain growth as well as the evolution of NbC and Laves phase during ingot solidification. The model accounts for (1) nucleation and growth of columnar or equiaxed dendritic grains, (2) coarsening/remelting of spherical instabilities through the evolution of the fraction of solid, (3) nucleation and growth of NbC carbides are governed by the carbon diffusion from the liquid to the

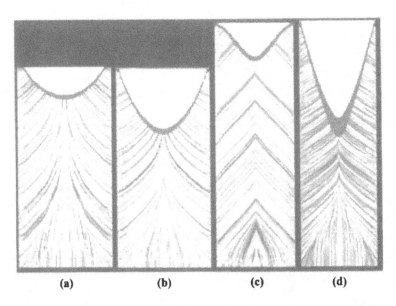

(a)   (b)   (c)   (d)

Figure 12 Simulated pool profiles and grain growth patterns of VAR and ESR Inconel 718 ingots. (a) and (b): 508 mm diameter, VAR, 172 and 327 kg/h melting rate; (c) and (d): 432 mm diameter, ESR, 272 and 591 kg/h melting rate. (From Ref. 43.)

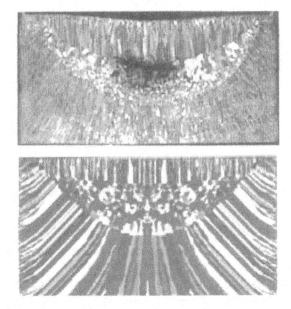

Figure 13 Comparison of model predicted pool profile and grain growth pattern with experimental results of a 508 mm diameter Inconel 718 VAR ingot. (From Ref. 43.)

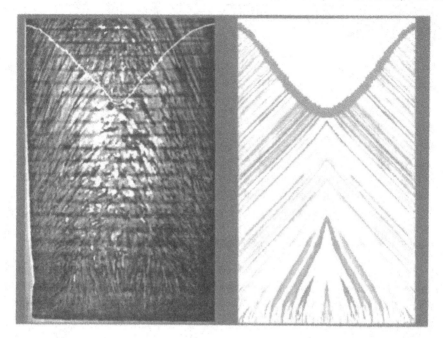

Figure 14 Comparison of model predicted pool profile and grain growth pattern with experimental results of a 432 mm diameter Inconel 718 ESR ingot. (From Ref. 42.)

NbC/liquid interface and by the reaction kinetics between Nb and C, (4) redistribution of Nb and C concentrations is calculated in both solid and liquid for solute transport, (5) redistribution of NbC particles between the solid and liquid phases by incorporating pushing/engulfment effects, and (6) nucleation and growth of Laves phase when the Nb concentration reaches the eutectic (γ + Laves) composition which is 19.1wt.% Nb. Figures 15 and 16 show the experimental and model calculated amount of NbC and Laves phase as a function of cooling rate [44]. For titanium alloys, the calculation of primary and secondary dendrite arm spacing as functions of solidification-kinetics parameters is shown in Ref. 47.

E. Macrosegregation Defects

Both endogenous and exogenous defects are found in VAR and ESR ingots [8–14, 48, 49]. Endogenous defects are intrinsic macrosegregations that resulted from an undesirable solidification condition. Exogenous defects are due to exogenous (drop-in) materials, which remain unmelted after falling into the

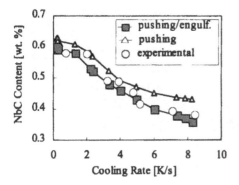

Figure 15 Experimental and model calculated amount of NbC as a function of cooling rate for the alloy Inconel 718. (From Ref. 44.)

molten pool. The chemical composition of these macrosegregation defects is strongly alloy dependent.

1. Endogenous Macrosegregation

Flemings and coworkers [50–61] have shown that the flow of solute-rich/lean interdendritic liquid in the mushy zone during solidification is responsible for most types of endogenous macrosegregation. Movement of this interdendritic liquid occurs as a result of several effects such as solidification contraction, gravity acting on a fluid of variable density, penetration of bulk liquid in front of the liquidus isotherm into the mushy zone, and electromagnetic forces (Fig. 17).

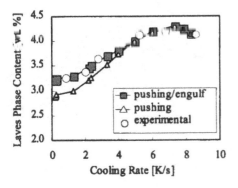

Figure 16 Experimental and model calculated amount of Laves phase as a function of cooling rate for the alloy Inconel 718. (From Ref. 44.)

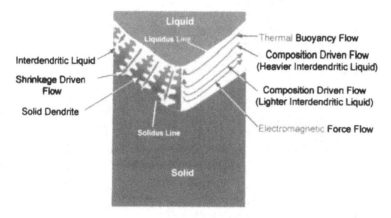

Figure 17   Schematic illustration of interdentritic fluid flow.

The magnitude of the fluid flow in the interdendritic region depends strongly on the power input level. High power input causes a deep pool and extends the mushy zone, which can enhance the intensity of the interdendritic fluid flow. Higher power input also induces a stronger electromagnetic force field. Thus, when the power input is sufficiently high to cause intensive interdendritic fluid flow, endogenous macrosegregation can result.

For nickel-base superalloys, freckles are the most common type of endogenous macrosegregation defects. Freckles are channel segregates [56, 59] and appear as long vertical linear trails of equiaxed grains and/or eutectic enriched metal [62, 63]. VAR and ESR ingots as well as directionally solidified (DS) and single crystal (SX) castings [64] are subject to freckling. In general, freckles are unremovable by subsequent thermomechanical treatments and can reduce material yield strength and ductility. As a result, they are considered as unacceptable defects for industrial applications.

Depending on the chemistry of the alloy, densities of interdendritic liquid can be either higher or lower than their corresponding bulk liquid [62, 63]. Freckles and channel-type segregation defects have been found in both types of alloys. For nickel-base alloys, in general, elements such as Al, Ti, Ta, and Nb tend to segregate toward interdendritic liquid whereas elements like W segregates toward solid dendrites [65]. Consequently, freckles for Nb containing alloys such as Inconel 718, Inconel 625, and Inconel 706 are rich in Nb. On the other hand, alloys that have fairly high Al and/or Ti content (e.g., DS and SX alloys) usually have interdendritic liquid densities lower than the densities of their corresponding bulk liquid. The variation of interdendritic liquid density can have significant effect on the interdendritic fluid flow pattern and freckle formation characteristics. For alloy Inconel 718, because its interden-

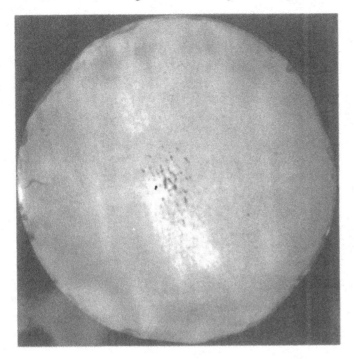

Figure 18   Freckles in a forged Inconel 718 billet.

dritic liquid is heavier than its bulk liquid [62, 63], freckles usually are located near the center of the ingot (Fig. 18). However, for the extreme case, freckles are channels that follow the ingot pool profile (Fig. 19). In this case, freckles may look like they are located at the midradius of the ingot (Figs. 20 and 21).

   Ingots processed at a relatively high melting rate are more often affected by freckles, and as ingot diameter increases, the phenomenon is more likely to occur [8]. Both VAR and ESR can generate freckles, but the frequency of the incidence in ESR is higher than VAR. Yu and Flanders [6] demonstrated that the difference of mushy zone shape in ESR and VAR ingots results in different macrosegregation formation phenomena (Fig. 22). The typical pool in an ESR ingot is V-shaped with an extended mushy zone that can enhance the intensity of fluid flow considerably compared to a VAR ingot of the same size, which has a shallower U-shaped pool profile with a shorter and more uniformly distributed mushy zone. This implies that for a similar size ingot and melting rates, there is a greater propensity for freckles formation in ESR than in VAR. This correlation is substantiated by the fact that the typical production size of VAR-processed Inconel 718 ingots is 508 mm (20 in.), while 432 mm (17 in.) diameter

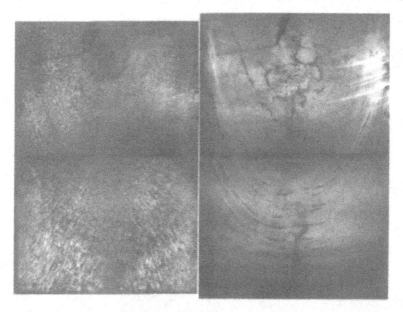

Figure 19 Freckle channels following the pool profile of a 432 mm diameter ESR Inconel 718 ingot. Left: grain etch; right: segregation etch. (From Ref. 10.)

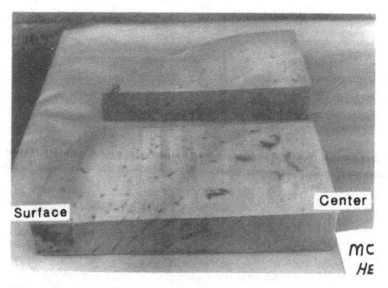

Figure 20 Large freckles outlining the pool profile near the top end of a 34 in. (864 mm) diameter ingot of VIM + ESR Alloy 706. The topmost transverse face of the plate is on the ground. (From Ref. 11.)

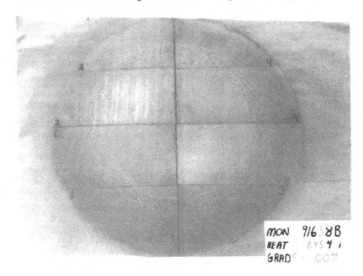

Figure 21 Pattern of freckles near the bottom end of the same 34 in. (864 mm) diameter ingot of Alloy 706 pictured in Fig. 20. (From Ref. 11.)

Ingot Type		VAR	VAR	ESR
Power Input Kw		125	200	240
Melting Rate Kg/Hr.		180	322	273
Ingot Diameter mm		508	508	432
Dendrite Arm Spacing μm	C	131	114	113
	E	101	104	74
Mushy Zone Thickness* mm	C	123	146	167 (152)**
	E	57	113	48 (40)**
Mushy Zone Shape				

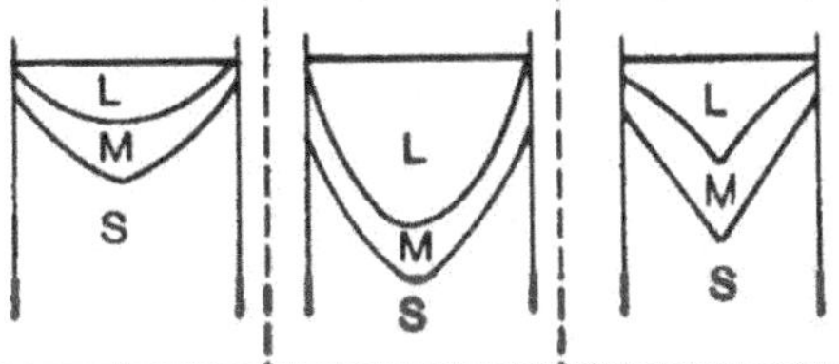

*Calculation based on the relationship of d=33.85 ℓ-0.338, where d is secondary dendrite arm spacing (μm) and ℓ is average cooling rate in mushy zone (°K/sec.)
**Obtained from previous thermal computations

Figure 22 Pool profile and estimated mushy zone shape (based on measured dendrite arm spacing) of VAR and ESR processed Inconel 718 ingots. (From Ref. 6.)

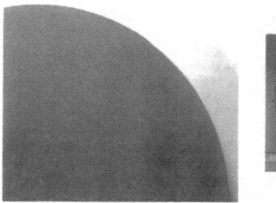

Figure 23 Macrostructure of beta flecks in Beta CEZ billet (white areas, left) and Ti-10-2-3 billet (dark areas, right). (From Ref. 65.)

ingots ESR-processed at comparable melting rate will have freckles unless an adequate solidification control strategy is employed.

Beta flecks (Figs. 23 and 24) are one type of endogenous defects found in titanium alloys and are recognized as localized volumes of large grain transformed β phase within the normal fine grain $\alpha + \beta$ microstructure of the billet product [12, 13, 65]. Beta flecks are attributed to the segregation of β stabilizing elements (e.g., Cr and Fe, etc.) which decrease the $\alpha + \beta$ to all-β transus temperature. During heat treatment, local areas with a higher content of β stabiliz-

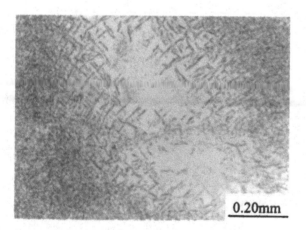

0.20mm

Figure 24 Beta fleck: alpha phase denuded zone caused by beta element stabilized beta phases. These defects have a low beta transus temperature. (From Ref. 13.)

ing elements tend to have a lower β transus temperature than the surrounding matrix and hence, form localized large β grains, i.e., beta flecks. The major segregation element for beta flecks of the alloy Ti-17 (Ti-5Al-2Sn-2Zr-4Mo-4Cr-0.1O) is Cr [12, 13, 65]. For alloy Ti-10V-2Fe-3Al, the major segregation element for beta flecks are Fe and V [14, 65]. Cr and Fe segregations occur during VAR ingot solidification and are partitioned toward the interdendritic liquid. VAR ingots that are processed at a larger diameter and higher melting rate tend to have a higher incidence of beta flecks. In order to produce beta flecks-free products, it is common practice for titanium alloy producers to reduce melting rate and ingot diameter for beta flecks prone alloys.

   Another type of endogenous defect found in titanium alloys is aluminum segregation stabilized alpha phase zone (Fig. 25). The aluminum rich area has a slightly higher hardness and β transus temperature than the matrix material. These defects are commonly called type II hard alpha because they are not as hard or brittle as the interstitially stabilized type I hard alpha particles (see details in Sec. V.E.2).

   The approach for modeling the interdendritic fluid flow has been investigated extensively by Flemings and coworkers [50–61]. For binary alloys, the "solute redistribution equation" describing the effect of interdendritic fluid flow in a representative volume element in the mushy zone is [54, 56]

$$\frac{1}{g_L}\frac{\partial g_L}{\partial t} = -\frac{\rho_L}{\rho_s(1-k)}\left(1 - \frac{\mathbf{n}\cdot\mathbf{v}}{\mathbf{n}\cdot\mathbf{u}}\right)\frac{1}{C_L}\frac{\partial C_L}{\partial t} - \frac{\mathbf{n}\cdot\mathbf{v}}{\mathbf{n}\cdot\mathbf{u}} = \frac{\mathbf{v}\cdot\nabla T}{\partial T/\partial t}. \tag{19}$$

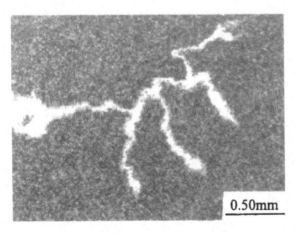

0.50mm

Figure 25   Type II hard alpha: aluminum segregation stabilized alpha phase zone, not as hard or brittle as type I. These defects have a slightly higher beta transus temperature. (From Ref. 13.)

where g = volume fraction liquid
        C_L = liquid composition
        k = equilibrium partition ratio
        $\mathbf{n}$ = unit vector normal to isotherms
        t = time
        T = temperature
        $\mathbf{u}$ = isotherm velocity
        $\mathbf{v}$ = interdendritic fluid flow velocity
        ρ_L = liquid density
        ρ_s = solid density

The interdendritic fluid flow velocity $\mathbf{v}$ can be related to the various fluid flow driving forces, including applied external forces (e.g., electromagnetic and centrifugal forces), if the mushy zone is viewed as a porous medium in which volume fraction of porosity varies with fraction solid and is continuously changing during solidification. The equation is

$$\mathbf{v} = -\frac{K}{\mu g_L}\left(\nabla P + \rho_L \mathbf{g} + \mathbf{F}_{\text{applied}}\right) \tag{20}$$

where K = permeability of the porous medium
        μ = viscosity of interdendritic liquid
        P = pressure
        $\mathbf{g}$ = acceleration due to gravity
        $\mathbf{F}_{\text{applied}}$ = externally applied force

The permeability K is a measure of the resistance of the medium to flow and is a function of pore size, geometry, orientation, and volume fraction liquid. Theoretical and experimental relationships between K and volume fraction liquid have been developed [54, 56]. Taking one simple model, K is proportional to the square of the volume fraction liquid, i.e.,

$$K = \gamma g_L^2 \tag{21}$$

where γ is a constant and is a function of dendrite arm spacings.

The dimensionless parameter, $1 - (\mathbf{n} \cdot \mathbf{v})/(\mathbf{n} \cdot \mathbf{u})$, in Eq. (19) can be viewed as *local flow velocity perpendicular to isotherm relative to isotherm velocity.* Macrosegregation criteria can then be established from the following equation [54]:

$$1 - \frac{\mathbf{n} \cdot \mathbf{v}}{\mathbf{n} \cdot \mathbf{u}} \gtrless \frac{\rho_s g_L + g_E(\rho_{sE} - \rho_s)}{\rho_L g_L} \tag{22}$$

When the left-hand side is equal to the right-hand side, Eq. (19) reduces to the simple differential form of the nonequilibrium segregation equation for a binary alloy (written in terms of volume fraction liquid). Therefore, no macrosegregation results. When the left-hand side of Eq. (22) is larger, most of the flow is going from the hotter (liquidus isotherm) to the cooler (solidus isotherm) region of the mushy zone, negative segregation results. When the right-hand side of Eq. (22) is larger, flow patterns start to change and some flow is going from the edge cooler region to the center hotter region and positive segregation is expected at the ingot centerline. In the extreme case, when

$$1 - \frac{\mathbf{n} \cdot \mathbf{v}}{\mathbf{n} \cdot \mathbf{u}} < 0 \qquad (23)$$

In this case, fluid is going from cooler to hotter regions and is flowing faster than the isotherms. This type of flow results in remelting (volume fraction liquid, g_L, increases with increasing C_L) instead of solidification. Thus, there is a driving force for the formation of expanding channels in the mushy zone through which the solute-rich/lean interdendritic liquid preferably flows. Growth of a channel can occur until it becomes optically visible as a severe, localized segregate after solidification is complete. This type of flow instability is the basic mechanism of formation of various channel-type segregates, including freckles.

The intensity of interdendritic fluid flow depends on the mushy zone size and shape (depth and slope), the time available for flow (i.e., local solidification time), the resistance of the porous interdendritic medium (surface tension and dendrite arm spacing), the material thermophysical property (viscosity), and the liquid density gradient in the mushy zone. Kou [61] showed that the interdendritic fluid flow intensity and the resultant segregation for the alloy Al-4.4% Cu increase with the mushy zone size and depth (Fig. 26). In Al-4.4% Cu, the density of the interdendritic liquid is higher than the bulk liquid density and increases with the decrease of temperature (i.e., increase of solute content) in the mushy zone as solidification proceeds from the liquidus to the solidus isotherm. With a small and shallow mushy zone, flow lines would fan outward as shown in Fig. 26b and c. At the ingot centerline, the left-hand side of Eq. (22) would become larger, whereas at the mold wall, the opposite would be true. Consequently, negative segregation tends to occur at the centerline of the ingot and positive segregation at the mold wall. When the mushy zone becomes larger and/or deeper, the flow lines of the solute-rich interdendritic liquid start to turn inward and result in positive segregation (Fig. 26e and f).

Although Flemings' theory provides satisfactory explanations for most types of endogenous macrosegregation defects, it does not show how the freckle channels are formed. As a result, it predicts that when the mushy zone profile is completely flat, there will be no macrosegregation (Fig. 26a and d). In reality, this is not exactly true. Yu et al. [9] showed that axially oriented dark streaks (looked like freckles in transverse direction) presented in the center of an ESR-processed Inconel 718 ingot with an almost completely

INCREASING DEGREE OF MACROSEGREGATION WITH
INCREASING DEPTH OF MUSHY ZONE

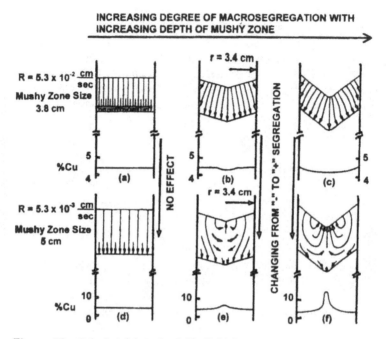

Figure 26 Calculated interdendritic fluid flow pattern and the resultant segregation for the alloy Al-4.4% Cu. (From Ref. 61.)

flat mushy zone shape which is indicated by the vertical grain growth direction (Fig. 27). Another example of flat mushy shape during casting is the DS and SX turbine airfoils. During casting, the mushy zone shapes of these turbine airfoils are typically very flat in order to maintain a primary dendrite growth direction parallel to the casting stacking axis. However, freckle defects are quite common for these castings [62–64]. Another phenomenon that cannot be explained by Flemings' theory is freckle channels following the ingot molten pool profile (Figs. 19–21).

A comprehensive freckle formation model needs to have a capability to calculate interdendritic fluid flow as well as to predict the formation of freckle channels for multicomponent alloy systems. To accomplish this, the thermo-solutal convection in the liquid pool and the interaction of the convective transport phenomena between the liquid pool and the mushy zone of a solidifying ingot needs to be modeled [66]. Continuum models have recently been developed to address this issue [67–73]. In continuum models, the dendritic mushy zone of the alloy is treated as a porous medium of variable porosity. The porosity is a function of the volume fraction of liquid, which varies from zero (all-solid region) to one (all-liquid region). A unique set of equations,

Figure 27 Longitudinal grain (top) and segregation (bottom) structures of a 432 mm Inconel 718 ingot ESR processed at a fairly low melting rate. (From Ref. 9.)

governing the conservation of mass, momentum, energy and solute, is solved in the whole domain, with no tracking of internal interfacial conditions. The equations in the mushy zone automatically reduce to the governing equations for the all-liquid or all-solid regions as the fraction of liquid varies from one to zero, respectively. These models have been used to predict the formation of liquid plumes in the form of chimney convection emanate from freckle channels within the mushy zone of directionally solidified nickel-base superalloy single crystal castings (see Figs. 25 and 26 of Chapter 11). In addition, they can be used to evaluate the effects of solidification conditions (e.g., cooling rate, temperature gradient, solidification rate, and mushy zone shape) and alloy compositions on freckle formation tendency. However, because of their intensive computational requirements, the application of these models to industrial scale VAR and ESR ingots would be a challenging task. Furthermore, whether these models can predict freckle channels following the molten pool profile still needs to be proven.

2. Exogenous Defects

The formation of exogenous defects in VAR and ESR ingots is outside the scope of Flemings' macrosegregation theory. The most plausible explanation for exogenous defects is that of drop-in materials that fall into the molten pool and remain unmelted.

In nickel-base superalloys, exogenous defects manifest themselves as sonic defects and white spots. Small ceramic inclusions originated from VIM processing may not be totally removed during subsequent ESR and/or VAR processing. During ingot to billet conversion by open die forging, small cracks may be formed around these ceramic inclusions. These small cracks can be detected by ultrasonic inspection and are named as sonic defects (Fig. 28). White spots (Figs. 29 and 30) can be classified into three types: discrete, dendritic, and solidification [48]. Depending on factors such as size, chemistry, grain size, and the presence or absence of oxide/nitride clusters, discrete and dendritic white spots may be deleterious to low-cycle fatigue (LCF) life.

Figure 28 Sonic defect in an Inconel 718 billet. (From Ref. 6.)

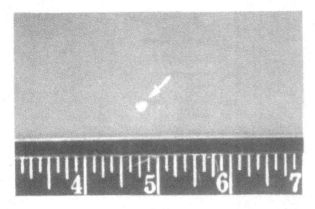

Figure 29 Discrete white spot in an Inconel 718 billet forged from a VAR ingot. (From Ref. 48.)

Solidification white spots are generally not associated with non-metallic clusters and appear to have little effect on mechanical properties. Possible sources of discrete and dendritic white spots are metallic drop-ins such as torus, crown, and shelf (Fig. 31). During VAR melting, condensed metal vapor and splattering metal droplets form a crown on the top edge of the ingot. With the increase of ingot height during melting, a shelf (chill zone) is formed at the surface

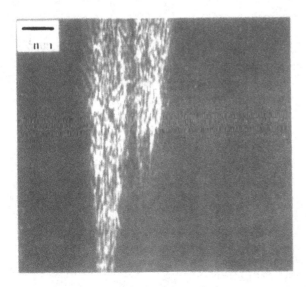

Figure 30 Dendritic white spots of alloy Inconel 718. (From Ref. 48.)

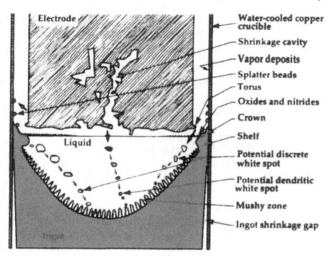

Electrode

Liquid

Water-cooled copper crucible

Shrinkage cavity

Vapor deposits

Splatter beads

Torus

Oxides and nitrides

Crown

Shelf

Potential discrete white spot

Potential dendritic white spot

Mushy zone

Ingot shrinkage gap

Figure 31 Schematics showing the torus, crown, and shelf of a VAR ingot. (From Ref. 48.)

periphery of the ingot. Pieces of shelf may be undercut by the arc due to arc instability and fall into the molten pool. The chemical composition of the shelf is usually lean in niobium and similar to that of the white spot. Only VAR-processed ingots exhibit white spot phenomenon [8]. No white spots have been found in ESR-processed materials.

For titanium alloys, exogenous defects typically appear as remnant electrodes, type I hard alpha particles, and high density inclusions (HDI). VAR electrodes are sometimes formed by welding two short electrodes into a long full size electrode. During the subsequent VAR melting, when the lower short electrode is mostly melted, the weld line between two short electrodes becomes very hot and may break. As a result, the remnants of the lower short electrode fall into the molten pool and may not be totally melted. Type I hard alpha particles (Fig. 32) are formed during sponge processing whereas the HDI (Fig. 33) may come from tungsten carbide tool bits in titanium turnings. Because both hard alpha particles [74] and HDI have higher melting points than bulk titanium alloys, these particles may not be melted during electrode melting and can fall into the liquid pool and may not be removed during VAR processing.

From the practical point of view, exogenous defects often have more detrimental effects on material quality than endogenous defects. Despite this, very little modeling work relating to the formation and elimination of exogenous defects in VAR and ESR ingots have been found in the open literature.

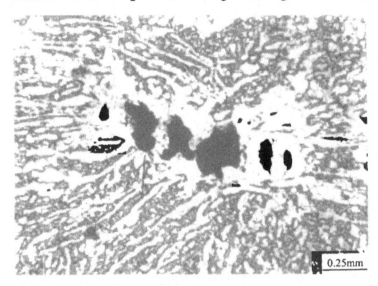

Figure 32 Type I hard alpha inclusion microstructure: an interstitially stabilized hard and brittle zone. Continuous light areas in these inclusions are alpha stabilized by nitrogen, oxygen or carbon. (From Ref. 13.)

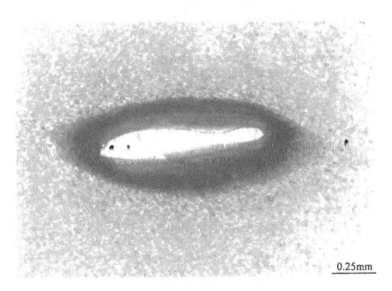

Figure 33 High density inclusion (HDI): refractory metal rich inclusions which have higher density and melting points than titanium. The core of this defect is tungsten; it is surrounded by a beta stabilized diffusion zone. (From Ref. 13.)

VI. INDUSTRIAL APPLICATIONS OF PROCESS MODELING

One of the most immediate impacts of modeling to date has been the enhanced understanding of the processes. Such understanding is necessary before any attempt can be made at process optimization. In VAR, for example, it is now well established that for low-current operations, the steady state pool is an interplay of the thermal buoyancy forces (which tend to drive hot metal up the axis of the ingot and down the side walls) and the electromagnetic Lorentz forces (that drive the molten metal down the axis and up along the walls). At the lower end of the current range, buoyancy flows dominate. As the current is gradually increased, the electromagnetic influences begin increasing until at a certain current level, there is a discontinuous increase in pool volume. This represents the stage at which electromagnetic flows become dominant. Knowledge of this interplay of pool forces in at least semi-quantitative mode allowed direct control of centerline-to-edge macrosegregation in U-6w/o Nb VAR ingots by varying melt current [20]. In high current melting, it has now been established that pool volumes are much larger than previously surmised [15]. Also, there is little, if any "steady-state" in the sense of a pool of constant volume that one observes in low-current melting.

The ability to simulate electrode melting allowed hundreds of trials by which controller architecture and parameters could be selected [15, 22]. It is expected that it will provide the basis for intelligent control in the future, with control of metallurgically important melt rate rather than arc gap being implemented.

Reference 15 outlining several applications of process modeling of both electrode melting and VAR ingot melting and solidification is indicative of an increasing application of the modeling methodology in the shop floor. Novel applications in this article include:

- The use of the electrode melting code to modify start-up and/or hot-topping procedures to increase yield.
- The use of the VAR melt simulations to demonstrate the beneficial effects of the introduction of helium cooling in the annulus between the ingot and crucible on the pool depths and ingot structures for various specialty alloys.
- The use of VAR simulations in determining the feasibility of increasing the maximum size alloy 718 ingot that can be melted by the triple melt (VIM/ESR/VAR) process from a diameter of 610–686 mm (24–27 in.) through comparison of predicted local solidification times (LST) for various size ingots.

Figure 34 illustrates this last example, wherein the predicted and measured pool depths are compared for the 686 mm (27 in.) diameter ingot. The

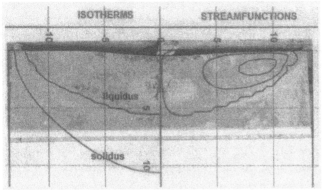

Figure 34 VAR model predictions for pool depth superimposed on an actual pool for a 686 mm diameter Alloy 718 ingot. (From Ref. 40.)

predicted depth obviously came very close to predicting the actual size. The 686 mm diameter product is now routinely melted.

With the dramatic growth in computing power and an equally dramatic drop in associated costs, it is expected that process modeling will attain increasing relevance in the design and modification of remelt processes.

REFERENCES

1. A Choudhury. Vacuum Metallurgy. Materials Park, OH: ASM, 1990.
2. O Winkler, R Bakish. Vacuum Metallurgy. New York: Elsevier, 1971.
3. G Hoyle. Electroslag Processes—Principles and Practice. London and New York: Applied Science Publishers, 1983.
4. RH Nafziger, et al. The electroslag remelting process. Bull 669, Albany Metallurgy Research Center, Albany, OR, 1974.
5. KO Yu. Comparison of ESR-VAR processes. Part I: Heat transfer characteristics of crucibles. Proceedings AVS Vacuum Metallurgy Conference, GK Bhat, M Lherbier, eds. Pittsburgh, PA: ISS, 1986, pp. 83–92.
6. KO Yu, HD Flanders. Comparison of ESR-VAR processes. Part II: Melting phenomena and ingot structure. Proceedings AVS Vacuum Metallurgy Conference, GK Bhat, M Lherbier, eds. Pittsburgh, PA: ISS, 1986, pp. 107–118.
7. KO Yu, CB Adasczik, WH Sutton. Heat transfer characteristics in an industrial scale ESR system. Proceedings of the 7th International Conference on Vacuum Metallurgy (ICVM), November 1982, Japan, pp. 1495–1501.
8. KO Yu, JA Dominque, GE Maurer, HD Flanders. Macrosegregation in ESR and VAR processes. J Metals 1:46–50, 1986.

9. KO Yu, JA Dominque, HD Flanders. Control of macrosegregation in ESR and VAR processed IN-718. Proceedings of the 8th International Conference on Vacuum Metallurgy, Linz, Austria, Sept. 30–Oct. 4, 1985.

10. JA Dominque, KO Yu, HD Flanders. Characterization of macrosegregation in ESR IN-718. Proceedings of the Symposium on Fundamentals of Alloy Solidification Applied to Industrial Processes, NASA Lewis Research Center, Cleveland, September 12–13, 1984, pp. 139–149.

11. JM Moyer, LA Jackman, CB Adasczik, RM Davis, RF Jones. Advances in triple melting superalloys 718, 706, and 720. Proceedings of the International Symposium on Superalloys 718, 625, 706 and Various Derivatives, Pittsburgh, PA. EA Loria, ed., June 26–29, 1994, pp. 39–48, Warrendale, PA: TMS, 1994.

12. CE Shamblen. Minimizing beta flecks in the Ti-17 alloy. Met Trans B, 28B:899–903, 1997.

13. WH Buttrill, GB Hunter, CE Shamblen. Manufacturing technology for premium quality titanium alloys for gas turbine engine rotating components. AFWAL, No. F33615-88-C-5418. Interim Technical Report No. 8, Sept. 1, 1992–Feb. 28, 1993.

14. JA Brooks, JS Krafcik, JA Schneider, JA Van Den Avyle, F Spadafora. Fe segregation in Ti-10V-2Fe-3Al 30 inch VAR ingot β-fleck formation. Proceedings of the 1999 International Symposium on Liquid Metal Processing and Casting, Santa Fe, New Mexico, Feb. 21–24, 1999, pp. 130–144, A. Mitchell, Ridgway, Baldwin, eds., Vacuum Metallurgy Division, American Vacuum Society, 1999.

15. LA Bertram, PR Schunk, SN Kempka, F Spadafora, R Minisandram. The macro-scale simulation of remelting processes. J Metals 50(3):18–21, 1998.

16. CB Adasczik, LA Bertram, DG Evans, RS Minisandram, PA Sackinger, DD Wegman, RL Williamson. Quantitative simulation of a superalloy VAR ingot at the macroscale. Proceedings AVS Vacuum Metallurgy Conference, A Mitchell, P Aubertin, eds. Pittsburgh, PA: AVS, 1997, p. 110.

17. RL Boxman, DM Sanders, PJ Martin, eds. Handbook of Vacuum Arc Science and Technology. Park Ridge, NJ: Noyes Publications, 1992.

18. JM Lafferty, ed. Vacuum Arcs: Theory and Application. New York: John Wiley & Sons, 1976.

19. FJ Zanner, LA Bertram. Behavior of sustained high-current arcs on molten alloy electrodes during vacuum consumable arc remelting. Trans Plas Sci PS-11(3):223, 1983.

20. LA Bertram, FJ Zanner. Plasma and magnetohydrodynamics problems in VAR. IUTAM Symposium on Metallurgical Applications of Magnetohydrodyamics, HK Moffatt, JA Shercliff, MRE Proctor, eds. Cambridge: TMS, London, 1982, p. 283.

21. M Choudhary, J Szekeley. A mathematical representation of the pool profile, the velocity and temperature fields in a laboratory scale ESR system. Proceedings AVS Vacuum Metallurgy Conference, GK Bhat, M Lherbier, eds. Pittsburgh, PA: ISS, 1986, p. 484.

22. LA Bertram, FJ Zanner. Electrode tip melting simulation during vacuum arc remelting of Inconel 718. In: Modeling and Control of Casting and Welding Processes, S Kou, R Mehrabian, eds. Warrendale, PA: TMS, 1986, p. 95.

23. EA Aronson, LA Bertram. Development of a parsimonious VAR furnace gap controller using simulated dynamics. In: Proceedings of AVS Vacuum Metallurgy Conference, A Mitchell, P Aubertin, eds. Pittsburgh, PA: AVS, 1997, p. 330.

24. FJ Zanner, LA Bertram. Interaction between computational modeling and experiments for vacuum consumable arc remelting. In: Modeling of Casting and Welding Processes, HD Brody, D Apelian, eds. Warrendale, PA: TMS, 1981, p. 333.

25. PJ Roache. Computational Fluid Dynamics. Albuquerque, NM: Hermosa Press, 1998.

26. JR Ockendon, WR Hodgkins, eds. Moving Boundary Problems in Heat Flow and Diffusion. Oxford: Clarendon Press, 1975.

27. FJ Zanner. Vacuum consumable arc remelting electrode gap control strategies based on drop short properties. Metall Trans 12B:721, 1981.

28. FJ Zanner, LA Bertram, R Harrison, H Flanders. Relationship between furnace voltage signatures and the operational parameters arc power, arc current, CO pressure and electrode gap during VAR of Inconel 718. Metall Trans 17B:357, 1986.

29. FJ Zanner, LA Bertram. Overview of VAR processing. Proceedings of the 8th International Conference on Vacuum Metallurgy, Linz, Austria, Sept. 30–Oct. 4, 1985.

30. F Schved. Vacuum arc remelting—review of the results of some Russian process investigations. Proceedings of 1994 AVS Vacuum Metallurgy Conference, Santa Fe, NM, A Mitchell, ed. Pittsburgh, PA: AVS, 1995, p. 112.

31. KH Tacke. Int J Num Meth Eng 21:543, 1985.

32. LG Hosamani, WE Wood, JH Devletian. Solidification of Alloy 718 during VAR with helium gas cooling between ingot and crucible. Proceedings of International Symposium on Metallurgy and Application of Alloy 718, Pittsburgh, PA, June 1989, EA Loria, ed., Warrendale, PA: TMS, 1989, pp. 49–57.

33. R Schlatter. Effect of helium cooling on VAR ingot quality of Alloy 718. Proceedings of the International Symposium on Superalloys 718, 625, 706 and Various Derivatives, Pittsburgh, PA, June 26–29, 1994, EA Loria, ed., Warrendale, PA: TMS, 1994, pp. 55–64.

34. M Jakob. Heat Transfer. New York: John Wiley & Sons, 1949.

35. LA Bertram, FJ Zanner, B Marder. Current paths and magnetohydrodynamics in vacuum arc remelting. In: Single- and Multi-Phase Flows in an Electromagnetic Field, AIAA Progress Series, vol. 100, H Brannover, PS Lykoudis, M Mond, eds., AIAA, 1986, p. 618.

36. S Hans, A Jardy, D Ablitzer. A numerical model for the prediction of transient turbulent fluid flow, heat transfer and solidification during vacuum arc remelting. International Symposium on Liquid Metals Processing and Casting, Vacuum Metallurgy Conference, Santa Fe, NM, 1994, pp. 143–154.

37. LA Bertram. A mathematical model for vacuum-consumable-arc-remelt casting. In Proceedings of the First International Conference on Math Model, vol. III. U of Missouri, Rolla, 1977, p. 1173.

38. G Comini, S del Guidice, RW Lewis, OC Ziendiewicz. Finite element solution of nonlinear heat conduction problems with special reference to phase change. International Journal Num Meth Eng 8:613, 1974.

39. J Ni, C Beckermann. A volume-averaged two-phase model for transport phenomena during solidification. Metall Trans 22B:349, 1991.

40. Specialty Metals Processing Consortium: the perspective of industrial members. JOM 3:26–29, 1998.

41. M McLean. Directionally solidified materials for high temperature service. The Metals Society, London, 1983.

42. JS Chou, L Nastac, S Sundarraj, Y Pang, KO Yu. Secondary remelting of titanium and superalloys. Presented at AeroMat '98, Washington, D.C., June 15–18, 1998.

43. L Nastac, S Sundarraj, KO Yu, Y Pang. Stochastic modeling of grain structure formation during solidification of superalloy and Ti alloy remelt ingots. International Symposium on Liquid Metals Processing and Casting, Vacuum Metallurgy Conference, Santa Fe, NM, 1997, pp. 145–165.

44. L Nastac, S Sundarraj, KO Yu. Stochastic modeling of solidification structure in Alloy 718 remelt ingots. Fourth International Special Emphasis Symposium on Superalloy 718, 625, 706, and Derivatives, EA Loria, ed. Pittsburgh, PA, 1997, pp. 55–66,.

45. L Nastac, S Sundarraj, KO Yu, Y Pang. The stochastic modeling of solidification structures in Alloy 718 remelt ingots: research summary. JOM 3:30–35, 1998.

46. L Nastac. Numerical modeling of solidification morphologies and segregation patterns in cast dendritic alloys. Acta Mater 47,17:4253–4262, 1999.

47. L Nastac, JS Chou, Y Pang. Assessment of solidification-kinetics parameters for titanium-base alloys. International Symposium on Liquid Metals Processing and Casting, Vacuum Metallurgy Conference, Santa Fe, NM, 1999, pp. 207–233.

48. LA Jackman, GE Maurer, S Widge. White spots in superalloys. Proceedings of the International Symposium on Superalloys 718, 625, 706 and Various Derivatives, Pittsburgh, PA, June 26–29, 1994, EA Loria, ed. Warrendale, PA: TMS, 1994, pp. 133–166.

49. JF Wadier, G Raisson, J Morlet. A mechanism for white spot formation in remelted ingots. Proceedings Vacuum Metallurgy Conference on Specialty Metals Melting and Processing, Pittsburgh, PA, June 11–13, 1984, Bhat and Lherbier, eds. Iron and Steel Society, 1985, pp. 119–126.

50. MC Flemings, GE Nereo. Macrosegregation, Part I. Trans TMS-AIME 239:1449–1461, 1967.

51. MC Flemings, R Mehrabian, GE Nereo. Macrosegregation, Part II. Trans TMS-AIME 242:41–49, 1968.

52. MC Flemings, GE Nereo. Macrosegregation, Part III. Trans TMS-AIME 242:50–55, 1968.

53. R Mehrabian, MC Flemings. Trans TMS-AIME 245:2347, 1969.

54. R Mehrabian, MA Keane, MC Flemings. Interdendritic fluid flow and macro-segregation; influence of gravity. Met Trans 1:1209, 1970.
55. R Mehrabian, MC Flemings. Macrosegregation in ternary alloys. Met Trans 1:455, 1970.
56. T Fujii, DR Poirier, MC Flemings. Macrosegregation in a multicomponent low alloy steel. Met Trans 108:331–339, 1979.
57. S Kou, DR Poirier, MC Flemings. Macrosegregation in rotated remelted ingots. Met Trans B, 9B:711–719, 1978.
58. S Kou, DR Poirier, MC Flemings. Macrosegregation in electroslag remelted ingots. Electric Furnace Proceedings, 1977, vol. 35, pp. 221–228.
59. SD Ridder, FC Reyes, R Chakravorty, R Mehrabian, JD Nauman, JH Chen, JH Klein. Steady state segregation and heat flow in ESR. Met Trans B, 9B:415–425, 1978.
60. CL Jeanfils, JH Chen, HJ Klein. Modeling of macrosegregation in electroslag remelting of superalloys. Superalloys 1980, Tien, et al., eds., ASM, 1980, pp. 119–130.
61. S Kou. Macrosegregation in electroslag remelted ingots. PhD dissertation, MIT, 1978.
62. P Auburtin, A Mitchell. Elements of determination of a freckling criterion. Proceedings of the 1997 International Symposium on Liquid Metal Processing and Casting, Santa Fe, New Mexico, Feb. 16-19, 1997, A Mitchell, P Auburtin, eds. Vacuum Metallurgy Division, American Vacuum Society, 1997, pp. 18–34.
63. P Auburtin, SL Cockcroft, A Mitchell. Liquid density inversions during the solidification of superalloys and their relationships to freckle formation in castings. Superalloys 1996, Kissinger, et al., eds., Warrendale, PA: TMS, 1996, pp. 443–450.
64. KO Yu, MJ Beffel, M Robinson, DD Goettsch, BG Thomas, RG Carlson. Solidification modeling of single crystal investment casting. AFS Transactions, 1990, pp. 417–428,.
65. P Auburtin, C Edie, B Foster, I Mackenzie, A Mitchell, A Schmalz. Solidification and segregation properties of some titanium alloys. Proceedings of the 1997 International Symposium on Liquid Metal Processing and Casting, Santa Fe, New Mexico, Feb. 16-19, 1997, A Mitchell, P Auburtin, eds, Vacuum Metallurgy Division, American Vacuum Society, 1997, pp. 60-77.
66. J Szekely, AS Jassal. Experimental and analytical study of the solidification of a binary dendritic system. Metall Trans 9B:389–398, 1978.
67. DR Poirier, JC Heinrich, SD Felicelli. Simulation of transport phenomena in directionally solidified castings. Proceedings of the Julian Szekely Memorial Symposium on Materials Processing, HY Sohn, JW Evans, D Apelian, eds., Warrendale, PA: TMS, 1997, pp. 393–410.
68. SD Felicelli, JC Heinrich, DR Poirier. Numerical model for dendritic solidification of binary alloys. Numer Heat Transfer Part B, 23:461–481, 1993.
69. SD Felicelli. PhD dissertation, The University of Arizona, Tucson, AZ, 1991.
70. SD Felicelli, DR Poirier, JC Heinrich. Macrosegregation patterns in multicomponent Ni-base alloys. J Crystal Growth 177:145–161, 1997.

71. SD Felicelli, JC Heinrich, DR Poirier. Three-dimensional simulations of freckles in binary alloys. J Crystal Growth 191:879–888, 1998.

72. SD Felicelli, DR Poirier, JC Heinrich. Modeling freckle formation in three dimensions during solidification of multicomponent alloys. Metall and Matls Trans 29B:847–855, 1998.

73. MC Schneider, JP Gu, C Beckermann, WJ Boettinger, UR Kattner. Modeling of micro- and macrosegregation and freckle formation in single-crystal nickel-base superalloy directional solidification. Met Trans 28A:1517–1531, July 1997.

74. JP Bellot, E Hess, D Ablitzer, A Mitchell. Dissolution of hard-alpha defects dragged in a bath of liquid titanium. Proceedings of the 1994 International Symposium on Liquid Metal Processing and Casting, Santa Fe, NM, Sept. 11–14, A Mitchell, Fernihough, eds., Vacuum Metallurgy Division, American Vacuum Society, 1994, pp. 155–166.

18

Electron Beam Melting and Plasma Arc Melting

Yuan Pang
Concurrent Technologies Corporation, Pittsburgh, Pennsylvania

Shesh Srivatsa
GE Aircraft Engines, Cincinnati, Ohio

Kuang-O (Oscar) Yu
RMI Titanium Company, Niles, Ohio

I. INTRODUCTION

Electron beam melting (EBM) and plasma arc melting (PAM) are two relatively new melting processes, which are primarily used for the production of titanium (Ti) and Ti alloys. The original objective of using these two processes was to reduce melt-related defects such as high-density inclusions (HDIs) and type I hard-alpha defects in premium quality Ti materials for jet engine rotating components [1–7]. Currently, about 20% of the jet engine rotor-grade Ti alloys are required to use either EBM or PAM as one of the melting steps [7]. In addition to reducing detrimental inclusions, EBM and PAM also provide the advantage of increased flexibility of using lower-cost revert materials as feedstock. Currently, EBM is widely used to produce commercially pure (CP) Ti and both EBM and PAM are used to recycle Ti alloy revert materials.

The EBM and PAM processes are different from the conventional vacuum arc remelting (VAR) process in that the melting and casting steps are not inherently related and can be independently controlled. This is the most important characteristic that allows EBM and PAM to remove the detrimental inclusions as well as have the flexibility to use various forms of input materials. This chapter discusses the nature of heat transfer, fluid flow, electromagnetic field, inclusion removal, and solidification associated with the EBM and PAM processes. The approaches to model these physical phenomena are described.

II. PROCESS DESCRIPTION

Figures 1 and 2 show the schematics of the EBM and PAM processes [1, 3]. In general, there are three separate regions (melting, refining, and casting) in these processes. The solid raw materials are melted in a water-cooled copper retort or pot. The molten metal then flows into a water-cooled copper refining hearth and exits into a continuous casting ingot mold or crucible. In certain practical applications, the melting (retort) and refining (hearth) regions can be colocated at the same hearth. However, two refining hearths (Fig. 2) can be used in tandem to increase the efficiency of removing detrimental inclusions. The refining hearth is the most important component of the EBM and PAM furnace hardware for inclusion removal. As a result, EBM and PAM are commonly referred as hearth melting (HM) processes.

The forms of the input materials for EBM and PAM are quite flexible, including sponge compacts, turning compacts, loose solid scraps, and premelted or fabricated bars. This flexibility of using various forms of raw materi-

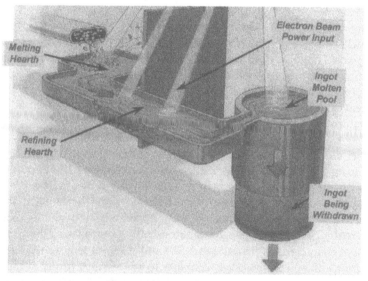

Figure 1   Schematic of an EBM furnace. (From Refs. 1, 2.)

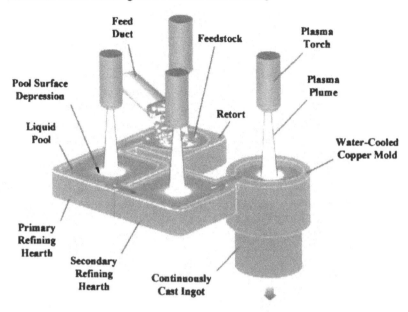

Figure 2 Schematic of a PAM furnace. (From Ref. 3.)

als is the key factor for the wide application of EBM for producing CP Ti. Both EBM and PAM are also used to recycle Ti alloy revert materials and turnings.

Several electromagnetically guided electron beam guns (EBM) or moving plasma torches (PAM) are used to heat the top surfaces of the melting pot, refining hearth, and casting mold. Heat is lost primarily by radiation from the top surface of the molten pool to the ambient and through the side and bottom surfaces of the solidified skull to the water-cooled container. For EBM, the fluid flow in the molten pool is strongly influenced by the buoyancy (natural convection) and Marangoni (surface tension) forces. The Marangoni force arises at the top surface of the molten pool due to the variation of the surface tension with temperature, which is caused by the localized heating by the electron beam. In PAM, the localized heating by a plasma plume also induces Marangoni forces. In addition, the impingement of the plasma plume on the molten pool surface results in not only a surface depression (dimple) but also a strong effect of forced convection.

The removal of detrimental inclusions happens mostly in the refining hearth with three recognized mechanisms of dissolution, mushy zone entrapment, and density separation. The motion and dissolution of the inclusion particles that enter the hearth is influenced by the flow and temperature of the molten metal. In a well-designed hearth, the inclusion particles that may enter the refining hearth are expected to dissolve completely, get trapped in the mushy zone formed at the boundary of the molten pool, or sink to the bottom of the hearth.

The outgoing molten metal from the refining hearth solidifies in a continuous casting mold or crucible to form cylindrical ingots (EBM and PAM) or rectangular slabs (EBM). Cylindrical ingots are mostly Ti alloys and typically used as electrodes for subsequent VAR processing to produce jet engine rotor-grade materials. Rectangular slabs are primarily used as input stock to produce CP Ti flat-rolled products. EBM is operated in a vacuum environment, which causes significant evaporation loss of high vapor pressure elements such as Al and Cr and can sometimes cause difficulty in controlling the ingot chemistry. Because it is operated under an inert gas (He or Ar) atmosphere, PAM ingots typically do not have this problem. However, PAM ingots sometimes can have porosities due to entrapped He and/or Ar gas [2].

III. APPLICATION

Ti and its alloys are widely used for high-performance aerospace, marine, energy, and commercial applications because of lightweight, high strength, high temperature capability, and strong resistance to corrosion. For aerospace applications, Ti alloys are used for both airframe and jet engine components. Their high reactivity with other materials requires that Ti be melted, refined, and cast into ingots or slabs in water-cooled copper crucibles under either vacuum or inert gas atmospheres. The most widely used melting process to produce Ti is VAR.

The major quality issue for Ti alloys is that of melt-related defects: type I hard-alpha defects, HDI, and chemical segregation (aluminum segregation and beta flecks). Pictures and formation mechanisms of these defects have already been presented in Chapter 17. Careful control of input materials can practically eliminate HDI defects. Understanding the mechanisms for the formation of segregation-type defects has allowed the development of improved melt practices to minimize their occurrence. Progress has also been made toward minimizing the incidence rate of hard-alpha inclusions. However, the practical inability to totally eliminate hard-alpha inclusions from the input materials and by VAR processing, along with the added difficulty in consistently finding them via nondestructive inspection, has caused hard-alpha particles to evolve as a major issue for the jet engine industry [1, 2].

Failures in gas turbine engine rotating components of Ti alloys have been few, but several of them have been attributed to hard-alpha inclusions since the early 1970s. Figure 3 shows an example of such an in-service component failure [1, 2]. Because of these failures, one of the major jet engine manufacturers, GE

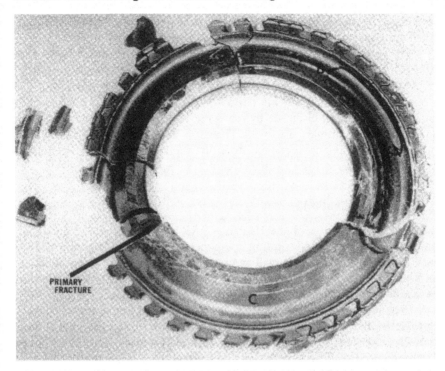

Figure 3 Example of an in-service component failure. (From Refs. 1, 2.)

Aircraft Engines, has taken major initiatives to develop HM technology (both EBM and PAM) to eliminate hard-alpha inclusions in Ti materials for jet engine rotating component applications [1, 2, 4, 5, 7]. In the meantime, Ti material suppliers realized that HM processes have an inherent flexibility to use various forms of lower-cost input materials, providing the opportunity to reduce the production cost. As a result, HM processes are also used to recycle Ti revert materials and turnings [8, 9].

A. Current Applications

EBM is currently being used extensively for melting and recycling CP Ti. Both cylindrical ingots and rectangular slabs are routinely produced. Because the melting rate and, hence, the productivity of EBM is higher than PAM and because there is no alloy element loss problem for EBM to melt CP Ti, EBM is more suitable than PAM for CP Ti applications. Currently in the USA, almost all CP products are produced by EBM. EBM is also being used to melt high

cleanliness superalloy remelt stock for investment casting of directionally solidified columnar grains and single crystal turbine airfoils.

For rotor-grade Ti alloys, the current industry standard for both PAM
and EBM ingots is the so-called Hearth + VAR process [4]. In this case, EBM
or PAM ingots are used as electrodes for the subsequent VAR processing. The
purpose of using hearth melting (EBM or PAM) before the VAR step is to
obtain a cleaner VAR electrode by removing HDI and hard-alpha particles in
the HM step. The primary use for the Hearth + VAR processed materials is for
jet engine Ti rotating components. Recently, the Hearth + VAR materials are
also being considered for airframe structural component applications.

B. Future Applications

The current application of the Hearth + VAR process is focused on quality
enhancement of jet engine rotating components. Future applications of PAM
and EBM products, however, will emphasize the cost reduction for Ti materials. The potential cost benefits for using PAM and EBM materials arise from
the flexible use of raw materials and near net shape castings.

In conventional VAR processing, many electrodes are fabricated by welding together Ti sponge compacts and large size bulk solids. The utilization of
relatively lower-cost small size scraps and turnings is limited. Due to the nature
of raw material feeding system and the capability of removing the harmful
inclusions, PAM and EBM are more flexible than VAR to use small size scraps
and turnings. Consequently, PAM and EBM can potentially have a lower raw
material cost than that of VAR [8, 9].

In current production (2 × VAR, 3 × VAR, and Hearth + VAR) of bars,
coils, plates, and sheets, the cylindrical VAR ingot has to be first hot-worked
into smaller diameter billets or rectangular slabs which are then rolled into
round bars or flat products (plates and sheets). The forming operations include
open-die forging, blooming, and rotary forging by GFM machines which are
not only costly but also can result in significant material yield losses. PAM and
EBM can potentially cast near net shape rectangular slabs and small diameter
(100–125 mm) ingots which can be rolled directly without going through those
expensive forming operations. This single-step operation prior to rolling is
called Hearth Only process because no subsequent VAR processing is required.
For jet engine rotor-grade materials, the Hearth Only process can also potentially be used to replace the currently used Hearth + VAR process. In this case,
cylindrical EBM or PAM ingots can be forged directly into billets which will be
used as the input stock for the subsequent close-die forging processes to make
jet engine rotating components.

Before the Hearth Only process can be successfully applied in production,
however, certain technical difficulties associated with the PAM and EBM pro-

cesses need to be overcome. The primary difficulty for EBM is chemistry control. In the case of PAM, there are two technical issues that need to be addressed. The first issue is the cold shuts formed on the ingot surface which reduces the ingot surface quality and material yield. The second one is helium or argon gas porosities which have a detrimental effect on material mechanical properties. The last issue that needs to be investigated for both PAM and EBM materials is the effects of ingot grain structure and macrosegregation on the microstructures and mechanical properties of the final product. Extensive development efforts on the Hearth Only processes are being undertaken by a number of Ti melters and more details will be discussed in Sec. VI.

IV. HEATING SOURCE

A. Electron Beam

The heating source for EBM is one or more electron beam guns [10]. The beam size is relatively small (usually about 5 cm diameter), resulting in a very high energy density in the beam focused area. As a result, the temperature in the beam direct heating area is very high and causes the evaporation of high vapor pressure elements such as Al and Cr. The traverse time of the electron beam is very short (typically in milliseconds). Therefore, the unsteadiness in the heating from the electron beam is limited to a very thin surface layer of the liquid pool. The energy input to the molten pool from the electron beam can be approximated as a distribution of the time-average heat flux on the top surface of the molten pool [11].

B. Plasma Arc

The heating source for PAM is a plasma torch. Figure 4 shows a schematic of the plasma torch [12, 13]. The copper wall in the torch acts as the anode, the liquid Ti in the copper container serves as the cathode. The pressurized helium or argon gas enters the torch through a number of ports uniformly distributed along an annular copper ring. Since the gas temperature at the torch center is typically between 20,000 and 30,000 K, the helium or argon gas is fully ionized. The bombardment of the positive ions on the Ti surface is the main heat source to maintain the Ti in the molten state. In addition to providing heat, the ion bombardment also exerts a pressure to cause the formation of a surface depression (dimple) and associated ripples on the molten pool surface. As the plasma torch moves, this surface depression and associated ripples also move accordingly, resulting in a strong effect on the flow pattern of the molten pool. The spread of the plasma jet and the surface tension force on the dimple surface in the radially outward direction also generates viscous shear stresses over the

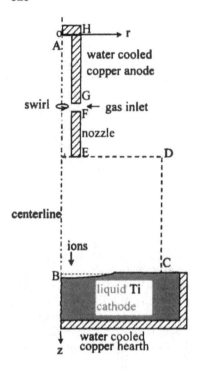

Figure 4 Schematic of a plasma torch. (From Refs. 12, 13.)

molten pool surface which have an effect on the fluid flow behavior in the molten pool.

The necessity for developing a plasma torch model is twofold. First, the plasma torch model can provide the refining hearth and ingot solidification models with the required boundary conditions (heat flux, current density, pressure, and shear stress) at the gas-molten metal interface. Second, the model can be used to understand the effects of the torch operating parameters such as gas flow rate, applied current, torch standoff, and furnace pressure on the efficiency of melting, refining, and casting.

Li and Pang [12, 13] developed a comprehensive two-dimensional axisymmetric model to simulate the pilot plant scale as well as the industrial production scale plasma torches using helium gas. In their model, the standard mass, momentum, and energy conservation equations, as well as the electrical charge conservation equation together with Ohm's law and Ampere's law, are required to account for the electromagnetic forces (Lorentz forces) and Joule heating. These source terms of momentum and energy are important to capture the main arc characteristics. Moreover, the azimuthal momentum equation is also required to take into account the swirling effect due to the strong tangential component of the helium gas velocity at the inlet port. The electromagnetic forces and the nonzero azimuthal velocity component are used as the source terms in the radial and axial momentum equations. The Joule heating, radiation loss, and energy loss due to electron drift (Thompson effect) are considered as the three additional source terms in the thermal energy equation. Because of the high viscosity at the very high temperatures, the plasma gas flow is considered as laminar. The governing equations are:

Mass conservation:
$$\frac{1}{r}\frac{\partial(\rho r u_r)}{\partial r} + \frac{\partial(\rho u_z)}{\partial z} = 0 \tag{1}$$

Radial momentum conservation:
$$\frac{1}{r}\frac{\partial(\rho r u_r u_r)}{\partial r} + \frac{\partial(\rho u_r u_z)}{\partial z} = -\frac{\partial P}{\partial r} + \frac{2}{r}\frac{\partial}{\partial r}$$
$$\left[r\mu_l\left(\frac{\partial u_r}{\partial r}\right)\right] + \frac{\partial}{\partial z}\left[\mu_l\left(\frac{\partial u_r}{\partial z} + \frac{\partial u_z}{\partial r}\right)\right] - \frac{2\mu_l u_r}{r^2} + \frac{\rho u_\theta^2}{r} - J_z B_\theta \tag{2}$$

Axial momentum conservation:
$$\frac{1}{r}\frac{\partial(\rho r u_r u_z)}{\partial r} + \frac{\partial(\rho u_z u_z)}{\partial z} = -\frac{\partial P}{\partial z} + \frac{1}{r}\frac{\partial}{\partial r}$$
$$\left[r\mu_l\left(\frac{\partial u_z}{\partial r} + \frac{\partial u_r}{\partial z}\right)\right] + 2\frac{\partial}{\partial z}\left[\mu_l\left(\frac{\partial u_z}{\partial z}\right)\right] + J_r B_\theta \tag{3}$$

Azimuthal momentum conservation:
$$\frac{1}{r}\frac{\partial(\rho r^2 u_r u_\theta)}{\partial r} + \frac{\partial(\rho r u_\theta u_z)}{\partial z} = \frac{1}{r}\frac{\partial}{\partial r}$$
$$\left[r\mu_l\left(\frac{\partial r u_\theta}{\partial r}\right)\right] + \frac{\partial}{\partial z}\left[\mu_l\left(\frac{\partial r u_\theta}{\partial z}\right)\right] - \frac{2}{r}\frac{\partial\mu_l r u_\theta}{\partial r} \tag{4}$$

Thermal energy conservation:
$$\frac{1}{r}\frac{\partial(\rho r u_r h)}{\partial r} + \frac{\partial(\rho u_z h)}{\partial z} = \frac{1}{r}\frac{\partial}{\partial r}\left[\frac{r\lambda_l}{C_p}\left(\frac{\partial h}{\partial r}\right)\right]$$
$$+ \frac{\partial}{\partial z}\left[\frac{\lambda_l}{C_p}\left(\frac{\partial h}{\partial z}\right)\right] + \frac{1}{\sigma_e}(J_r^2 + J_z^2) - S_R \tag{5}$$
$$+ \frac{5}{2}\frac{k_b}{e}\left[\frac{J_r}{C_p}\left(\frac{\partial h}{\partial r}\right) + \frac{J_z}{C_p}\left(\frac{\partial h}{\partial z}\right)\right]$$

Electrical charge conservation:
$$\frac{1}{r}\frac{\partial}{\partial r}\left[r\sigma_e\left(\frac{\partial\phi}{\partial r}\right)\right] + \frac{\partial}{\partial z}\left[\sigma_e\left(\frac{\partial\phi}{\partial z}\right)\right] = 0 \tag{6}$$

Ohm's law: $\mathbf{J} = -\sigma_e \nabla\phi$ $\tag{7}$

Ampere's law: $\mathbf{B} = \frac{\mu_0}{r}\int_0^r \mathbf{J}\cdot\xi\,d\xi$ $\tag{8}$

where r, z, and θ are the radial, axial, and azimuthal coordinates in the cylindrical coordinates system. u_r, u_z, u_θ are the corresponding velocity components. P, h, ϕ, $\mathbf{J}$, and $\mathbf{B}$ represent pressure, enthalpy, electric potential, electric current density, and magnetic field, respectively. Also included in the equations are the gas thermodynamic and transport properties, such as density ρ, specific heat C_p, molecular dynamic viscosity μ_l, molecular thermal conductivity λ_l, and electrical conductivity σ_e, as well as some universal constants, such as electronic charge e, Boltzmann constant k_b, and magnetic permeability of free space μ_0. S_k is the volumetric radiation loss and ξ is a dummy variable of integration.

Figures 5 to 7 show some of their simulation results [12, 13]. Figure 5 shows the predicted temperature distribution for a subscale plasma torch oper-

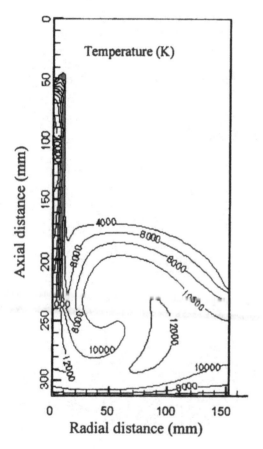

Figure 5 Calculated plasma gas temperature distribution for a subscale PAM torch. (From Refs. 12, 13)

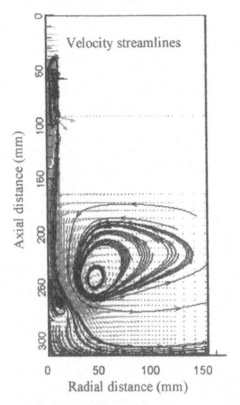

Figure 6 Calculated streamlines of the helium gas flow field for a subscale PAM torch. (From Refs. 12, 13.)

ated at a current level of 750 A and an electrical voltage of 300 V. The helium temperature reaches a maximum of 21,550 K at the upper center of the copper chamber and it decreases rapidly in both radial and axial directions. The temperature is almost uniform and lower at around 10,000 K in most of the area outside the nozzle because there is a large recirculation zone in this region. Near the liquid pool surface, which acts as the cathode, the temperature drops sharply from about 10,000 to 2000 K, due to the existence of the cathode sheath. Figure 6 presents the velocity streamlines near the cathode. The gas flows into the copper chamber with components in both radial and tangential directions. The small diameter of the chamber enforces the strong axial flow near the top of the fluid stream. Note that there is a large recirculation zone outside the torch. The large tangential inlet velocity of nearly 900 m/s has a strong swirling effect on the behavior of the helium gas. It causes a secondary

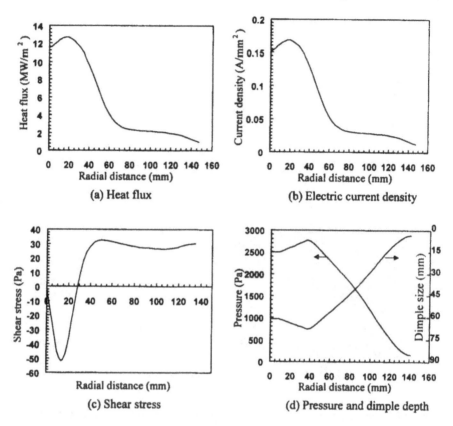

Figure 7 Calculated distributions of heat flux, electric current density, shear stress, gas pressure and molten metal dimple depth for a production scale PAM torch. (From Refs. 12, 13.)

small recirculation zone and a low-pressure area near the symmetry axis against the liquid pool surface. Distributions of heat flux, electric current density, pressure and dimple depth, and shear stress along the radial direction of the liquid pool surface are shown in Fig. 7. The peaks of heat flux and electric current density occur at about 36 mm from the centerline, the peak of pressure at about 18 mm, and the peak of shear stress at about 60 mm because its dependency on the velocity gradient. These results at the gas helium and liquid Ti interface provide the necessary input boundary conditions to the hearth and ingot models.

The thermodynamic and transport properties of helium gas have strong effects on plasma temperature, velocity, and current density distributions.

These properties are usually highly temperature dependent and can be influenced by the presence of Ti alloy vapor and furnace pressure. No experimental data of these properties are available in open literature. One approach to obtain some information regarding the effect of operating pressure and alloy evaporation on helium properties is by theoretical calculation [14]. Figure 8 shows the calculated density, specific heat, enthalpy, viscosity, thermal conductivity, and electrical conductivity. The figure also shows the properties of pure helium in a temperature range from 200 to 30,000 K for the specific pressures of 0.5 and 1.1 atm. According to these figures, the alloy effect on properties in the available temperature range is small. Thermodynamic calculation is also a viable approach to estimate thermophysical properties of molten Ti [15–17].

V. REFINING HEARTH

Refining hearths are the most important components for removing inclusions in the EBM and PAM processes [1–6, 11, 18–21]. Inclusion removal is strongly dependent on the molten metal flow pattern which, in turn, depends on the power input, i.e., electron beam and plasma torch moving patterns. In terms of modeling the heat transfer, fluid flow, and electromagnetic forces in the molten metal, PAM is significantly more complex than EBM.

A. EBM

A good EBM refining hearth model should have the following features:

- Three-dimensional geometry
- Transient and steady state flows
- Laminar and turbulent flows
- Combined radiation, conduction, and natural and forced convections
- Heat flux defined from gun patterns
- Marangoni convection
- Temperature dependent thermophysical properties
- Melting and solidification
- Inclusion trajectory, dissolution, and residence time
- Validation with production scale data.

Kelkar et al. [11, 18] developed a comprehensive EBM refining hearth model. The following paragraphs describe their model and simulation results.

Melt flow and heat transfer in the hearth is governed by the three-dimensional incompressible mass, momentum, and energy conservation equations. The flow and heat transfer are assumed to be quasi-steady since the traverse

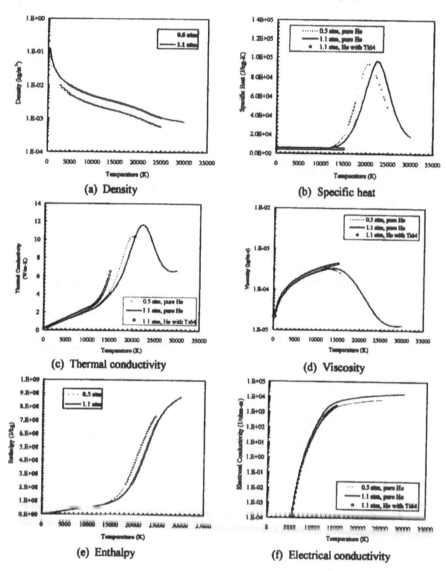

Figure 8 Calculated thermodynamic and transport properties of helium at 0.5 and 1.1 atm. (From Ref. 14.)

time of the EB guns is very short relative to the time scale of fluid flow and heat transfer. The hearth geometry is approximated as a rectangular shape (Fig. 9). The mathematical model includes all the relevant physics for predicting inclusion behavior. The flow of the molten alloy occurs in a shallow pool that is stably stratified due to heating on the top surface. This tends to suppress turbulence effects within the molten pool. The effect of turbulence was considered using a buoyancy-corrected k-ε turbulence model. Results with this turbulence model were almost the same as those obtained by assuming the flow to be laminar. Therefore, the flow was assumed to be laminar for computational simplicity. Electromagnetic body forces and Joule heating are negligibly small and so are not included in the model.

Temperature variation is governed by the energy equation, which includes heat transfer by conduction and convection within the pool and by conduction within the solidified metal. The enthalpy-porosity technique of Brent et al. [22] was employed to predict phase change processes. This technique includes the latent heat source terms in the energy equation and utilizes a porous medium based resistance in the mushy region. The enthalpy-porosity technique allows for the calculation of the phase change process in a fixed grid by decomposing the total enthalpy at a point into its sensible enthalpy and the latent heat content.

Melt velocities and temperature at the inlet are specified based on the casting rate and the inlet superheat. These quantities may either be specified as

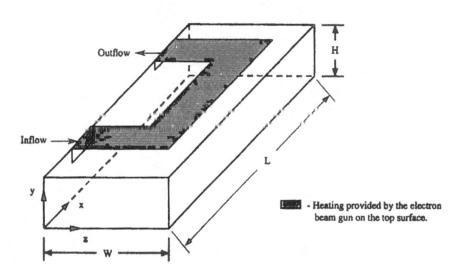

Figure 9 Geometry of the refining hearth for an EBM model. (From Ref. 18.)

constant over the inlet or as a user-defined profile. For example, the temperature of the alloy at the inlet can be assumed to vary linearly over the inflow
depth from the liquidus temperature at the bottom to the prescribed surface
temperature at the top. For the outlet of the computational domain, the standard fluid dynamics practice of zero normal gradients for all variables is
employed. This implies that flow is not reentering the hearth through the outlet
pour notch and that downstream effects do not propagate upstream at the
outlet notch. The velocity over the open portion of the outlet notch is prescribed so as to satisfy the overall mass balance.

The surface tension of Ti alloys varies with temperature. This combined
with the nonuniform temperature present on the free (top) surface creates a
spatially varying shear stress. At the melt free surface, the viscous shear force
of the liquid metal balances the temperature dependent surface tension. The
velocity components in the direction tangential to the top surface (x and z
directions) are subject to the corresponding components of the shear stress.
The vertical velocity (in the y direction) is zero at the top surface and the top
surface is assumed to be flat and horizontal.

The heat transfer boundary condition on the melt free surface is obtained
by subtracting the radiation heat loss to the surroundings from the heat flux
provided by the EB guns and the energy conducted away by the molten metal.
The steady heat flux on the top surface is determined by time-averaging the
heat flux supplied by a moving beam over one traverse. This heat flux is
nonuniform and its variation is prescribed from the gun pattern.

The velocity is specified to be zero on the side and bottom walls of the
hearth. It should be noted that the enthalpy-porosity technique ensures that the
velocity in the solidified alloy is zero. The model accounts for radiation and
contact heat transfer between the skull surface and the water-cooled copper
hearth wall. Due to high temperatures, heat loss from all surfaces of the crucible occurs predominantly by radiation.

Figures 10 to 12 show some of the simulation results [18]. Temperature
contours on the top surface and on transverse and longitudinal cross sections
are shown in Figs. 10 and 11, respectively. The profile of the liquid pool
isotherms indicates that the flow is stably stratified. There are large temperature gradients on the free surface. Therefore, the Marangoni force exerts a
strong influence on the flow field. The overall flow is from the inlet to outlet
with Marangoni forces driving the flow in transverse planes. Figure 12 shows
the pool shape in various cross sections through the hearth. The pool is shallow
with a depth of about 20 mm. The pool width to depth and length to depth
ratios are large (> 10). The depth of the pool is more or less uniform over most
of the refining hearth except near its edges. Figure 13 shows the comparison of
calculated pool top surface temperatures with experimental results. There is a

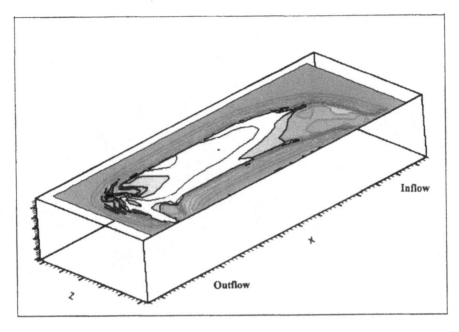

Figure 10 Top surface temperature contours for the EBM hearth shown in Fig. 9. (From Ref. 18.)

good agreement between the predicted and measured temperatures. In general, surface temperatures are underpredicted by about 20–50 K.

B. PAM

Figure 14 schematically illustrates a cut-away view of a refining hearth in the PAM furnace [23 26]. In this region, a number of complex physical phenomena are occurring simultaneously, which critically affect and control the motion and removal of inclusions. One of the most significant aspects of this region is the nature of the heat source, the plasma torch. The plasma torch provides a highly concentrated heat flux to the molten pool. This produces large temperature gradients on the pool surface, which in turn cause a strong surface tension driven flow in the same direction as the flow driven by strong shear forces from the high velocity plasma jet. Another complication to the surface flow is related to gas pressure, produced by the impinging plasma jet, which creates a surface depression (dimple) as shown in the figure. In addition, the plasma torch uses a transferred arc with reversed polarities as compared to the standard welding arcs with straight polarities. This arc, which is normally

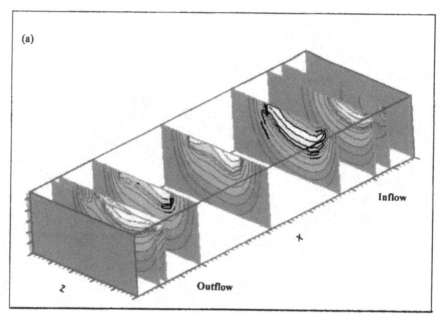

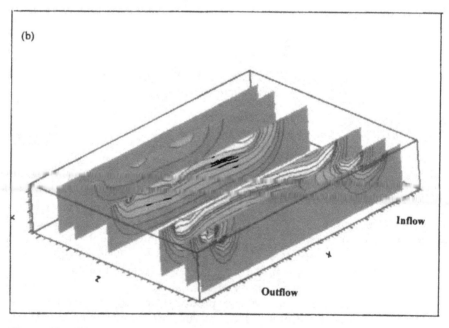

Figure 11 Calculated temperature contours of the EBM hearth shown in Fig. 9. (a) Transverse pool profile and (b) longitudinal pool profile. (From Ref. 18.)

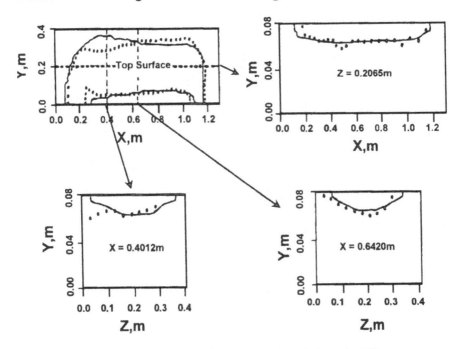

Figure 12 Calculated and measured molten metal pool shape in different cross sections of the EBM hearth shown in Fig. 9. (From Ref. 18.)

generated by an applied current in the neighborhood of 3000 amps, gives rise to electromagnetic effects that cannot be neglected. Finally, the large disturbance from the plasma jet transitions the molten pool fluid flow from the laminar to the turbulent regime. All these phenomena affect the inclusion trajectories in the molten pool and must be properly addressed in the model.

Huang and his coworkers [23–26] developed a comprehensive refining hearth model. In their model, the following conservation equations of mass, momentum, and thermal energy for three-dimensional, transient, incompressible flow are used to simulate the melt flow and heat transfer.

Mass conservation: $\dfrac{\partial}{\partial x_i}(\rho u_i) = 0$ \hfill (9)

Momentum conservation: $\dfrac{\partial}{\partial t}(\rho u_i) + \dfrac{\partial}{\partial x_j}(\rho u_j u_i)$

$$= -\frac{\partial p}{\partial x_i} + \frac{\partial}{\partial x_i}\left[\mu_{\text{eff}}\left(\frac{\partial u_i}{\partial x_j} + \frac{\partial u_j}{\partial x_i}\right)\right] + (\mathbf{J} \times \mathbf{B})_i + \rho g_i + A u_i$$

\hfill (10)

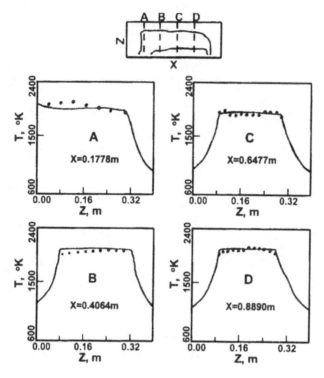

Figure 13 Calculated and measured pool top surface temperatures of the EBM hearth shown in Fig. 9. (From Ref. 18.)

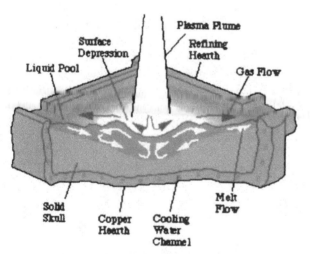

Figure 14 Schematic diagram of a cut-away PAM refining hearth. (From Refs. 23–26.)

Thermal energy conservation: $\dfrac{\partial}{\partial t}(\rho C_p T + \rho f_l \, \Delta H_f) + \dfrac{\partial}{\partial x_i}$

$$(\rho u_i C_p T + \rho u_i f_l \, \Delta H_f) = \frac{\partial}{\partial x_i}\left[\lambda_{\text{eff}}\left(\frac{\partial T}{\partial x_i}\right)\right] \tag{11}$$

Most of these symbols have been defined in Sec. IV.B. In the momentum equation, $(\mathbf{J} \times \mathbf{B})_i$ represents the Lorentz forces, ρg_i is the gravity, and $A u_i$ is the dendritic drag force. μ_{eff} is the total dynamic viscosity, which is the sum of the molecular μ_l and turbulent μ_t contributions. $\rho C_p T$ and $\rho f_l \, \Delta H_f$ in the energy equation are the volumetric sensible heat and the latent heat of the alloy, respectively. f_l is the fraction of liquid and λ_{eff} is the total thermal conductivity, which consists of both molecular λ_l and turbulent λ_t contributions.

The electromagnetic forces, namely, Lorentz forces, are calculated using a two-dimensional axisymmetric model and the results are mapped to the three-dimensional hearth. A two-dimensional electric charge conservation equation and the corresponding boundary conditions are first solved. Ohm's law and Ampere's law are then applied to obtain the current density $\mathbf{J}$ and the induced magnetic flux, $\mathbf{B}$. The Lorentz forces are finally calculated from the cross product of $\mathbf{J}$ and $\mathbf{B}$.

The impingement of the plasma jet on the molten pool surface contributes to the turbulence in the PAM hearth and crucible/mold. The standard k-ε turbulence closure model [27–29] is used to capture the effect of turbulence on the momentum and energy transport and on the inclusion behavior in the molten pool. The governing equations for the k-ε model are given below.

Turbulent kinetic energy k: $\dfrac{\partial}{\partial t}(\rho k) + \dfrac{\partial}{\partial x_i}(\rho u_i k) = \dfrac{\partial}{\partial x_i}\left[\dfrac{\mu_t}{\sigma_k}\left(\dfrac{\partial k}{\partial x_i}\right)\right] + \rho G_k - \rho \varepsilon$

$$\tag{12}$$

Dissipation rate ε of turbulent kinetic energy: $\dfrac{\partial}{\partial t}\rho\varepsilon + \dfrac{\partial}{\partial x_i}(\rho u_i \varepsilon) =$

$$\tag{13}$$

$$\frac{\partial}{\partial x_i}\left[\frac{\mu_t}{\sigma_\varepsilon}\left(\frac{\partial \varepsilon}{\partial x_i}\right)\right] + c_1 \frac{\varepsilon}{k}\rho G_k - c_2 \frac{\varepsilon}{k}\rho\varepsilon$$

$$\mu_{\text{eff}} = \mu_l + \mu_t \quad \text{and} \quad \mu_t = c_\mu \rho \frac{k^2}{\varepsilon} \tag{14}$$

$$\lambda_{\text{eff}} = \lambda_l + \lambda_t \quad \text{and} \quad \lambda_t = \frac{C_p \mu_t}{\text{Pr}_t} \tag{15}$$

$$G_k = \frac{\mu_t}{\rho}\frac{\partial u_j}{\partial x_i}\left(\frac{\partial u_i}{\partial x_j} + \frac{\partial u_j}{\partial x_i}\right) \tag{16}$$

$$\sigma_\varepsilon = 1.3 \qquad c_1 = 1.44 \qquad c_2 = 1.92 \qquad c_\mu = 0.09 \tag{17}$$

where G_k is the rate or turbulence generation and σ_k, σ_ε, c_l, c_2, and c_μ are empirical constants.

The treatment of inlet, outlet, and skull-copper interface boundary conditions is similar to that of EBM. At the molten metal free surface, the viscous shear force of the liquid metal balances the plasma gas shear force and the temperature dependent surface tension. The expression for this boundary condition is given below.

$$\mu_{eff} \frac{\partial u_i}{\partial n} = \tau_{iplasma} - \gamma_l \frac{\partial T}{\partial x_i} \tag{18}$$

where n is the normal to the free surface and γ_l is the surface tension coefficient. In addition to viscous shear force, the pressure distribution from plasma jet impingement is also required to calculate the surface depression (dimple). For heat transfer boundary condition at the molten metal surface, both the heat flux from plasma jet as well as radiation heat loss at the molten metal surface need to be included. The plasma jet pressure, gas shear stress, heat flux, and current flux contributed by the plasma gas at the liquid metal surface are calculated from the torch model discussed in Sec. IV.B.

Figures 15 to 17 show some of the calculated results [26]. Figure 15 schematically illustrates a typical torch pattern operated in a production PAM furnace. The time-averaged heat flux and gas shear stress used for the calculation are shown in Fig. 16. The predicted molten pool isotherms, liquid-

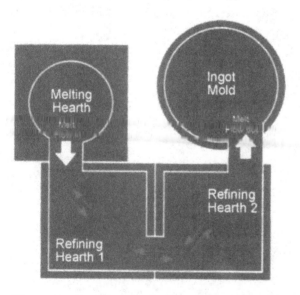

Figure 15 Typical torch moving pattern in a PAM furnace. (From Ref. 26.)

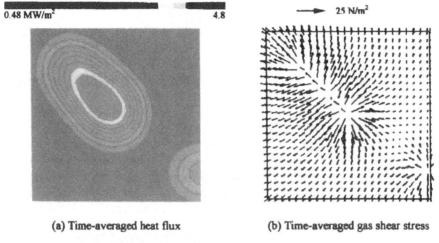

| 0.48 MW/m^2 | 4.8 | $\longrightarrow$ 25 N/m^2 |

(a) Time-averaged heat flux   (b) Time-averaged gas shear stress

Figure 16 Calculated time-averaged heat flux and gas shear stress at the molten pool surface of the Refining Hearth 1 shown in Fig. 14. (From Ref. 26.)

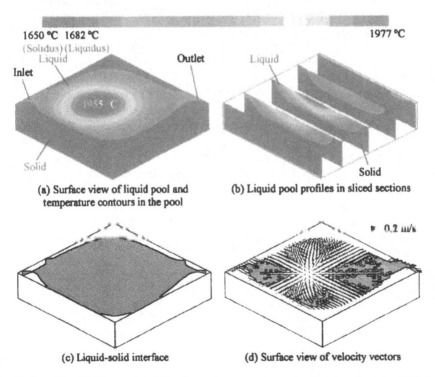

1650 °C 1682 °C   1977 °C
(Solidus)(Liquidus)

(a) Surface view of liquid pool and   (b) Liquid pool profiles in sliced sections
temperature contours in the pool

(c) Liquid-solid interface   (d) Surface view of velocity vectors

Figure 17 Model predicted liquid pool, temperature distribution, liquid-solid interface, and velocity vectors of the Refining Hearth 1 shown in Fig. 14. (From Ref. 26.)

solid interface, and velocity vectors are shown in Fig. 17a–17d. It is clear that the liquid pool occupies most of the top surface while some "necking" effect is present near both inlet and outlet (a). The liquid pool has a fairly uniform depth in the middle section of the hearth (b and c) with a maximum depth near the center where 60% of the metal is liquid. The molten metal flows predominantly outward from the torch path because of strong plasma gas shear and Marangoni forces, which are augmenting each other in the same direction (d). In the present simulation, two torches are moving in a synchronized manner (Fig. 15) to influence the primary hearth being studied. Direct heating from the primary torch would cause a high-temperature region between the middle of the hearth and the inlet. The highest temperature on the pool surface is about 1955°C close to the hearth center. The molten metal near the outlet gets hotter because of the heat from the secondary torch.

C. Inclusion Transport Model

The success of the EBM and PAM processes is critically dependent on their ability to eliminate the detrimental inclusions from the melt. There are three classes of inclusions based on their densities relative to the molten Ti alloy. The inclusions with densities higher than the Ti alloy will sink to the solidified skull. The inclusions with densities lower than the Ti alloy will float to the molten pool surface and be subsequently dissolved by the high-temperature metal under the heat source. The neutrally buoyant inclusions represent the greatest threat to the quality of the ingot since they neither sink nor float, but follow the molten metal flow and have the potential of entering the ingot. The refining step is considered to be effective if all inclusions can be removed from the EBM and PAM processes. These processes can be optimized to enhance the refining efficiency if the relationship between the inclusion trajectories and the process parameters is known.

It should be noted that the density of the inclusions can change with time. HDIs are usually solid particles, thus their densities can be treated as a constant. The situation for the hard-alpha particles is more complex. The most common chemistry of the hard-alpha particles is TiN which has a fully solid density that is higher than that of the molten Ti. Consequently, the solid hard-alpha particles would sink to the bottom of the refining hearth during melting. In reality, however, the hard-alpha particles such as those formed in the sponge manufacturing process are sometimes hollow [1] and have internal voids (Fig. 18). Depending on the actual volume fraction of the voids, hard-alpha particles can have an apparent density that is higher than, equal to, or lower than the density of the molten Ti.

Since the inclusion population density in the liquid pool is quite low, it can be assumed that any changes in the momentum and enthalpy of the inclu-

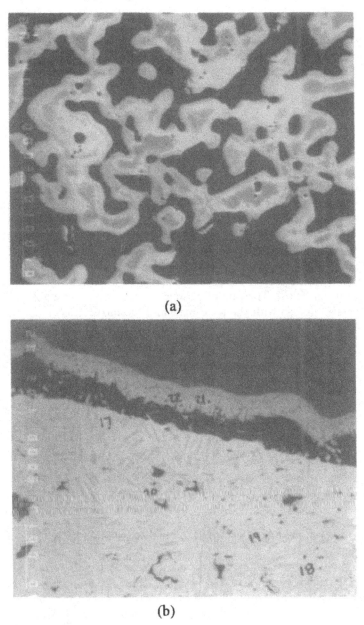

(a)

(b)

Figure 18 Photomicrographs of a (a) low N content sponge seed and (b) "natural" burned sponge seed; black regions are voids. (From Ref. 1.)

sions would not affect the flow and thermal fields of the molten metal. The likely trajectories of an inclusion can be calculated based on a stochastic Lagrangian approach [30] to account for the momentum exchange with the fixed flow field. The equation of motion for an inclusion can be written as

$$\frac{d\mathbf{u}_p}{dt} = \frac{1}{2}\frac{A_p\rho}{V_p\rho_p}C_D|\mathbf{u} - \mathbf{u}_p|(\mathbf{u} - \mathbf{u}_p) + \frac{\mathbf{g}(\rho - \rho_p)}{\rho_p} + \dot{m}_p\mathbf{u}_p \tag{19}$$

For the inclusion, V_p is the volume, ρ_p is the density, A_p is the effective drag area, $\mathbf{u}_p$ is the velocity vector, and $\dot{m}_p$ is the rate of change in mass controlled by dissolution. C_D is the drag coefficient between the inclusion and the molten metal. g is the gravitational acceleration. For the molten metal, ρ is the density and $\mathbf{u}$ is the velocity vector. $(\mathbf{u} - \mathbf{u}_p)$ is the relative velocity of the inclusion to the molten metal and t is the time. The terms on the right-hand side represent sequentially the drag force exerted by the fluid, the buoyancy force resulting from the density difference, and the rate of momentum change caused by dissolution.

The inclusion temperature T_p can be calculated from the energy conservation equation given below. The rate of change in inclusion enthalpy is governed by the heat exchange with the local melt and the heat release from the dissolved mass.

$$m_p c_p \frac{dT_p}{dt} = A_{ph}h(T - T_p) - \dot{m}_p Q_d \tag{20}$$

where m_p is the mass of the inclusion, c_p is the specific heat of the inclusion, A_{ph} is the effective area for heat exchange, h is the heat transfer coefficient at the inclusion-metal interface, T is the molten metal temperature, and Q_d is the heat of dissolution. The rate of dissolution $\dot{m}_p$ is correlated with the inclusion temperature and the parameters of dissolution kinetics through an Arrhenius-type relation [31].

$$\dot{m}_p = A_r m_p \exp\left(-\frac{E}{RT_p}\right) \tag{21}$$

where A_r is the dissolution kinetics parameter, E is the activation energy, and R is the universal gas constant.

Figures 19 and 20 show some of the calculation results [18, 26]. Figure 19 shows the particle trajectories in an EBM hearth. Detailed examination of the predicted particle trajectories shows that the residence time of the particles that escape is about 70 s. Most particles get trapped in the mushy region. There are three classes of inclusions based on their densities relative to the Ti melt. The heavy particles dip immediately, while the light particles rise to the surface and are dragged by the fluid flow at the pool surface. Therefore, both the light and the heavy particles travel only a short distance before getting trapped in the

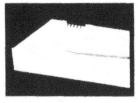

INCLUSIONS HEAVIER   INCLUSIONS LIGHTER   INCLUSIONS NEUTRALLY
THAN LIQUID   THAN LIQUID   BUOYANT

Figure 19   Calculated particle trajectories for the EBM refining hearth shown in Fig. 9. (From Ref. 18.)

mushy region. The neutrally buoyant particles (for which particle and fluid densities are within ±2%) travel the farthest and have the largest probability of survival. In general, particle trajectories in an EBM hearth follow the bulk fluid flow in the molten pool, moving smoothly from the inlet to the outlet. This is not the case for a PAM hearth (Fig. 20). From Fig. 20, it can be seen that in the PAM hearth, particles can move in a relatively random direction, do not go straight from the inlet to the outlet.

VI.   INGOT SOLIDIFICATION

A semicontinuous casting process similar to direct chill (DC) casting, VAR, and electroslag remelting (ESR) is used to cast PAM and EBM ingots. Because of the differences in the ingot top heat input conditions, the internal structure and surface quality of PAM and EBM ingots are quite different from those of DC casting, VAR, and ESR processed ingots. EBM can cast cylindrical ingots

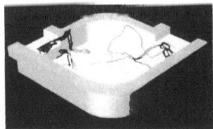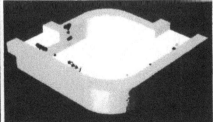

(a) Inclusion trajectories   (b) Inclusion final "destinations"

Figure 20   Calculated inclusion trajectories and final "destinations" in the PAM Refining Hearth 1 shown in Fig. 14. (From Ref. 26.)

as well as rectangular slabs, whereas PAM presently can produce only cylind-rical ingots.

A. Characteristics of Heat Input

In other semicontinuous casting processes (DC casting, VAR, and ESR), the input of the molten metal is either located at the center of (DC casting) or distributed uniformly over (VAR and ESR) the ingot top surface, resulting in a truly axisymmetric ingot structure. For PAM and EBM ingots, however, mol-ten metal flows in at one side of the ingot (Figs. 1 and 2). This situation can result in a nonaxisymmetric molten pool, grain structure, and surface quality for the resultant ingots. Whether the nonaxisymmetric molten pool will lean toward the pour-lip side or the side opposite to the pour lip (Fig. 21) depends on the temperature of the metal poured into the crucible. If the temperature of the molten metal entering the ingot crucible is lower than that of the pool surface near the pour lip, the pool depth and surface quality on lip side of the ingot will be shallower and worse than the rest of the ingot. On the other hand, if the feeding metal temperature is higher than the pool surface tempera-ture, then the ingot side opposite to the pour lip will have a shallower pool and worse surface quality than the rest of the ingot. In general, the pour-lip side of

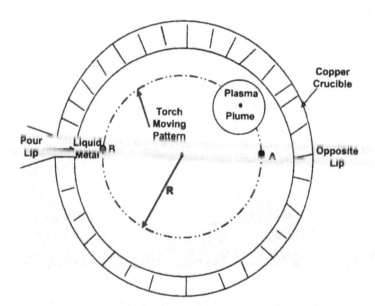

Figure 21 Schematic shows torch moving pattern for PAM ingot crucible/mold.

the ingot has a worse surface quality than the opposite side of the ingot (Fig. 22) [1]. Figure 23 is a nonaxisymmetric molten pool profile of an EBM ingot showing the pool depth on the pour-lip side is shallower than that on the opposite side [1]. Figure 24 shows a PAM ingot with a center island in the molten pool. From the figure, it can be seen that the pour-lip side has a bigger molten pool and shrinkage cavity [1]. By adjusting the electron beam or plasma torch moving pattern, the degree of nonaxisymmetry in the molten pool and grain structure can be reduced. Figure 25 shows an EBM ingot with a fairly symmetric pool profile and grain structure [1].

The heating source (electron beam for EBM and plasma arc for PAM) at the top of the ingot moves according to prespecified patterns. For cylindrical ingots, the plasma torch or electron beam is typically moving with a circular pattern to provide an axisymmetric heat distribution. In the case of EBM, the speed of the electron beam is very fast and can complete a full circle within a fraction of a second. Thus, the ingot macro heat input condition is very close to axisymmetric. On the other hand, the speed of the plasma torch is typically about 15–30 cm/s and will take about 10–20 s to complete a full circle for a typical production size (630–915 mm diameter) ingot. As a result, a strongly localized heating effect exists. The area that is directly underneath the plasma plume has a very high input heat flux, whereas the area that is outside the heating range of the plasma plume will experience a strong heat loss through radiation and helium convection. Consequently, the temperature difference between these two areas can be quite significant. During PAM processing, the ingot top surface will have a periodic temperature distribution along its circumferential direction. Consider point A on the ingot top surface (Fig. 21). When the center of the plasma plume is located at point A, the area in the vicinity of point A will have the highest temperature of the ingot top surface at that instant of time. As the torch moves along the circumferential direction and away from point A, the temperature of point A starts to drop. When the torch moves half a circle (i.e., point B) away from point A, the temperature at point A will continue to decrease. As the torch moves into its second half of the circle, when the torch gets close to point A, the temperature at point A will start to increase again and reaches the maximum temperature when the center of the plasma plume returns to point A. The temperature variation and the time period during each cycle depend on the torch pattern. The slower the torch moving speed, the larger the temperature variation and the longer the cycle time.

The radius (R in Fig. 21) of the electron beam or plasma torch moving circle has a strong effect on ingot internal structure as well as the surface quality. Because no graphite flux (DC casting) or slag (ESR) is used during PAM and EBM processing, the quality of the PAM and EBM ingot surface is almost solely controlled by the ingot surface temperature. The higher the ingot surface temperature, the better the ingot surface quality. For EBM, the size of

(a)

(b)

Figure 22   Surface of a PAM ingot (a) opposite lip side (b) lip side. (From Ref. 1.)

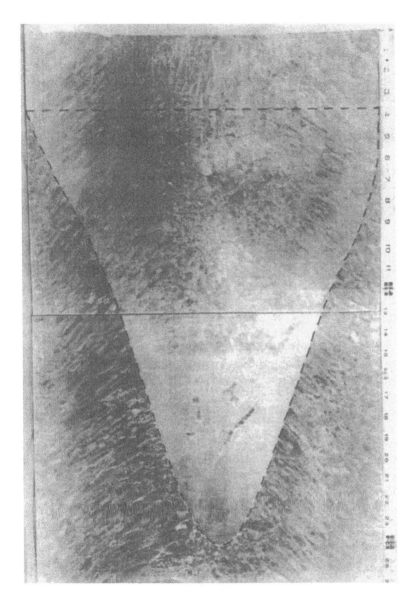

Figure 23 Nonsymmetrical molten pool profile of an EBM ingot; pour entry on left. (From Ref. 1.)

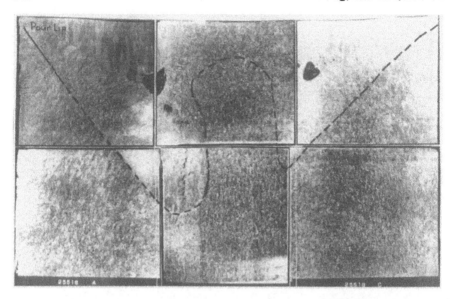

Figure 24 Center island and nonsymmetrical molten pool profile of a PAM ingot. (From Ref. 1.)

the electron beam is small and highly focused and, hence, the beam can be located very close to the ingot outside edge (i.e., R approaches to the radius of the copper crucible) to maintain a high ingot surface temperature, resulting in a fairly good ingot surface quality. On the other hand, the plasma plume is diffused and has a core diameter about 5 cm (pilot scale) to 25 cm (production scale), depending on the torch internal configuration, power rating, and furnace atmosphere pressure. The rim of the plasma plume spreads to an even bigger diameter. As a result, the center of the plasma plume is usually located quite a distance away from the ingot outside edge in order to avoid the damage of the copper crucible by direct heating from the plasma plume. This situation results in a low surface temperature and relatively poor surface quality for PAM ingots. Figure 26 shows the deep cold shuts on the surface of a PAM ingot [2]. Optimizing the torch moving pattern can improve the ingot heat transfer condition and enhance the ingot surface quality to a certain extent but will not completely eliminate the cold shuts (Fig. 27).

B. Chemistry Control

Chemistry control for PAM and EBM ingots is more complicated than that of DC casting, VAR, and ESR. For these processes, the chemistry of the

Figure 25 EBM ingot with a fairly symmetrical molten pool profile; pour entry on right. (From Ref. 1.)

input molten metal for the solidifying ingot is generally homogeneous and in good control. Thus, any nonuniform chemistry distribution in the resultant ingot is primarily due to the macrosegregation that is formed during ingot solidification. For PAM and EBM processes, however, the chemistry of the molten metal that flows out of the hearth and into the mold/crucible can vary with time.

There are three potential sources for chemistry variation in PAM and EBM ingots. First, the raw materials are mixtures of Ti sponges, alloy elements, and reverts (turnings and solid scraps). How to blend and mix all these materials to ensure a uniform and correct chemistry for input materials

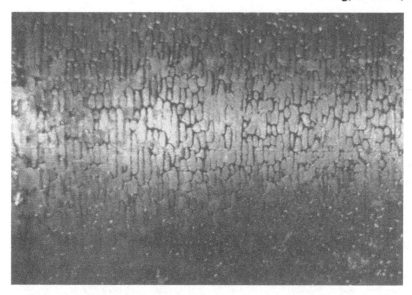

Figure 26   PAM ingot surface with deep cold shuts. (From Ref. 2.)

Figure 27   PAM ingot surface with shallow cold shuts. (From Ref. 2.)

throughout the entire melting cycle is critical to obtaining the right chemistry for the cast ingot. Second, Ti sponge and added alloy elements such as Al, V, Fe, and Mo, etc., have quite different melting points. During melting, the low melting point elements melt faster than the high melting point elements. This phenomenon can result in a nonuniform chemistry distribution in the molten pool of the refining hearth. If the molten pool is well stirred and mixed such as the PAM hearth, the chemistry variation in the molten pool will be homogenized. Otherwise, molten metals with a nonuniform chemical composition may flow into the copper mold/crucible and result in a nonuniform macro chemistry distribution in the resultant ingot. The third source for the ingot chemistry variation is due to alloy element loss during melting. Because it is operated under an inert gas atmosphere, PAM ingots typically do not have this problem. For EBM, due to the small electron beam size and high energy density in the beam direct heating area, high vapor pressure elements such as Al and Cr tend to evaporate relatively fast and result in the depletion of Al and Cr in the molten metal [32]. If the melting and casting rates can be controlled to stay in a steady state condition all the time, the rate of Al and Cr loss can be estimated accurately and compensated appropriately by adjusting the input material chemistry. However, during transient conditions such as start-up, melt interruption, and hot top, the control for an appropriate compensation of Al and Cr can be difficult. As a result, EBM ingots tend to have a large chemistry variation in the ingot start-up and hot top regions [1, 2].

C. Ingot Structure

Compared to the conventional VAR ingots, three characteristics of EBM and PAM ingots need to be addressed. The first two, structure symmetry and surface quality, are related to both EBM and PAM ingots. The third one, helium porosities, applies to PAM ingots only.

The nonsymmetric molten pool profile and grain structure of the EBM and PAM ingots may have some effects on the final product microstructures and mechanical properties. Since EBM ingots and slabs are routinely forged and/or rolled to make CP Ti mill products with satisfactory microstructures and mechanical properties, it is believed that the effect of the nonsymmetric molten pool profile and ingot grain structure on the microstructures and mechanical properties of the resultant mill product is relatively minor. For Ti alloys, however, macrosegregation can form in the solidifying ingot and may have significant effects on the microstructures and mechanical properties of the final product. At present, the effects of nonsymmetric molten pool profile and grain structure on ingot macrosegregation formation tendency and microstructures and mechanical properties of the final product are not well understood.

Another characteristic applies to both EBM and PAM ingots is the ingot surface quality. In general, the quality of both EBM and PAM ingots is not as good as that of VAR and ESR ingots. Because of the ability to flexibly control the electron beam locations, EBM ingot surface quality is better than that of PAM ingots. However, EBM slabs still need to be machined to remove the rough surface before the slabs can be used as the feedstock for the rolling machine. For Hearth Only product, the ingot surface quality has a significant impact on the ingot forgeability and the material yield of the final product.

During PAM processing, helium or argon gas bubbles can be entrapped in the solidifying ingot. The subsequent forming processes (forging and/or rolling) will flatten the gas bubbles but will not eliminate them. In the heat treatment and high temperature service conditions, these flattened bubbles can expand and form gas porosities (Fig. 28), which have a detrimental effect on material/component mechanical properties. It has been demonstrated by Shamblen [2] that operating at a reduced chamber pressure (e.g., 0.5 atm vs. normally 1.1 atm) can reduce the tendency to form helium porosities in PAM ingots. However, more work needs to be done to prove the validity of this approach for routine production environments.

D. Modeling Approach

From the theoretical point of view, except for the withdrawal process to make continuous casting ingots, the modeling procedure for the ingot casting is similar to that of the refining hearth, including heat transfer, fluid flow, electromagnetic forces, inclusion trajectory, and solidification. From a practical application point of view, however, the modeling of ingot solidification is quite different from that of the refining hearth.

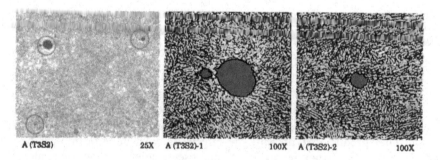

A (T3S2)   25X   A (T3S2)-1   100X   A (T3S2)-2   100X

Figure 28 Photomicrographs of clean voids (helium porosities) found in bar stock from a PAM Only ingot. (From Ref. 2.)

The most important function of the refining hearth is to remove HDI and hard-alpha inclusions. Because inclusion trajectories are strongly dependent on the hearth molten pool flow pattern, the accurate calculation of the effects of heat transfer, fluid flow, and electromagnetic forces on inclusion trajectories is crucial to establish a robust refining hearth model. On the other hand, the most important feature for ingot solidification is to obtain an ingot with a good surface quality to result in a high material yield and a grain structure that is free of macrosegregation such as beta flecks and Al segregation and can result in satisfactory material microstructures and mechanical properties. The calculation of inclusion trajectories in ingot molten pool is not needed because the molten metal that flows into the mold/crucible is considered to be inclusion free.

In general, the solidification pattern of EBM and PAM ingots is quite similar to that of VAR and ESR ingots. Thus, the discussions on VAR and ESR ingots in Chapter 17 regarding the effects of pool profile on the ingot grain structure and macrosegregation formation tendency as well as the effects of start-up and hot-top procedures on the final material yield are all applicable to EBM and PAM ingots [33, 34]. Figure 29 shows an experimentally obtained and simulated PAM ingot grain structure by Nastac et al. [33]. However, there are several other factors, which are unique to EBM and PAM ingots, need to be further discussed.

Three-dimensional models are required to simulate the nonsymmetric molten pool profile and the resultant grain structure. The most convenient way to model the ingot surface quality is by establishing an empirical relationship between the quality and temperature of the ingot surface. In general, the higher the surface temperature the better the ingot surface quality. For helium porosities, the molten pool temperature can be used as an indicator for the porosity formation tendency. A lower molten metal temperature results in a higher molten metal viscosity and, hence, a lower probability for entrapped gas to escape and a higher possibility to form helium porosities.

VII. SUMMARY

Although EBM and PAM are relatively new, the development of an integrated modeling tool for these two processes has progressed at a fairly rapid pace. Previous work has emphasized on the inclusion removal capability in the refining hearth. Future work will pay more attention to ingot solidification which is crucial for Hearth Only applications. Ongoing research work is directed towards improving the predictive capability of the model. Some of the issues being addressed are:

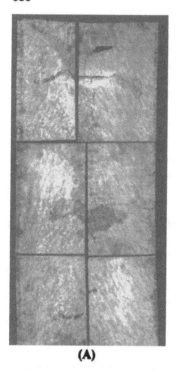

(A)   (B)

Figure 29 Comparison of an experimental and simulated PAM ingot. (A) Experiment; (B) simulation. (From Ref. 34.)

- More model validation
- Improved heat flux boundary conditions
- Realistic boundary conditions at skull/crucible interface
- Improved material properties
- Exact hearth geometry (vs. rectangular box)
- Sensitivity analysis with respect to material properties and process parameters
- Particle dissolution mechanisms
- Transient calculations
- Inclusion of evaporation mass/heat loss with realistic inputs to predict alloy composition
- Ingot pool profile and grain structure
- Microsegregation and macrosegregation
- Ingot surface quality

REFERENCES

1. CE Shamblen, GB Hunter, WH Buttrill, EL Raymond. Manufacturing technology for premium quality titanium alloys for gas turbine engine rotating components. Phase I Final Report, Vol. 1: Technical Program, WL-TR-92-8078, GE Aircraft Engines, Cincinnati, OH, 1993.
2. CE Shamblen. Manufacturing technology for premium quality titanium alloys for gas turbine engine rotating components. Phase II Final Report, WL-TR-95-8025, GE Aircraft Engines, Cincinnati, OH, 1995.
3. Y Pang, C Wang. Titanium alloy hearth melting process technology enhancement NCEMT Final Report TR No. 00-50 (Contract No. N00140-92-C-BC49), Concurrent Technologies Corporation, Johnstown, PA, 2000.
4. DJ Tilly, CE Shamblen, WH Buttrill. Premium quality Ti alloy production: HM+VAR status. Proceedings of the 1997 International Symposium on Liquid Metal Processing and Casting. A Mitchell, P Auburtin, eds., Vacuum Metallurgy Division, American Vacuum Society, Santa Fe, NM, 1997, pp. 85–96.
5. CE Shamblen, DJ Tilly. Inclusion free titanium material efforts. Proceedings of the Electron Beam Melting and Refining Conference—State of the Art 1997. R Bakish, ed., Bakish Materials Corporation, Reno, NV, 1997, pp. 39–45.
6. WR Chinnis. Present status of PAM development. Proceedings of the Electron Beam Melting and Refining Conference—State of the Art 1997. R Bakish, ed., Bakish Materials Corporation, Reno, NV, 1997, pp. 277–282.
7. CE Shamblen. Second Titanium Hard Alpha Workshop. GE Aircraft Engines, Evendale, OH, Jan. 31, 2000.
8. KO Yu. Plasma arc melting for titanium alloys. Proceedings of the Technical Program from the 1998 International Conference, International Titanium Association, Monte Carlo, Monaco, 1998, pp. 371–385.
9. KO Yu. PAM processing titanium alloys. Proceedings of Manufacturing Technology for Aerospace Materials: A Technology Demonstration and Information Exchange. Concurrent Technologies Corporation, Arlington, VA, 1999, pp. 58–74.
10. S Schiller, U Heisig, S Panzer. Electron beam technology. New York: John Wiley & Sons, 1982, 508 p.
11. KM Kelkar, SV Patankar, SK Srivatsa. Mathematical modeling of the electron beam cold hearth refining of titanium alloys. Proceedings of the Electron Beam Melting and Refining Conference—State of the Art 1997. R Bakish, ed., Bakish Materials Corporation, Englewood, NJ, 1997, pp. 238–251.
12. G Li, Y Pang. Plasma torch model development, validation, and sensitivity study. NCEMT Report TR No. 98-89 (Contract No. N00140-92-C-BC49), Concurrent Technologies Corporation, Johnstown, PA, 1999.
13. Y Pang, G Li, RTC Choo, KO Yu. Modeling of an industrial plasma torch for titanium alloy processing. 33rd Annual Meeting of the Society of Engineering Science, Tempe, AZ, 1996.

14. Y Pang. Thermodynamic and transport properties of helium plasmas. NCEMT Report TR No. 00-25 (Contract No. N00140-92-C-BC49), Concurrent Technologies Corporation, Johnstown, PA, 2000.

15. F Zhang, YA Chang, JS Chou. A thermodynamic approach to estimate titanium thermophysical properties. Proceedings of the 1997 International Symposium on Liquid Metal Processing and Casting. A Mitchell, P Auburtin, eds., Vacuum Metallurgy Division, American Vacuum Society, Santa Fe, NM, 1997, pp. 35–59.

16. JS Chou, L Nastac, CA Papesch, JJ Valencia, Y Pang. Thermophysical and solidification properties of titanium alloys. NCEMT Report TR No. 98-87 (Contract No. N00140-92-C-BC49), Concurrent Technologies Corporation, Johnstown, PA, 1999.

17. YA Chang. Numerical calculation of solidification properties for titanium alloys. A report to Concurrent Technologies Corporation, 1996.

18. S Srivatsa, K Kelkar. Mathematical modeling of the electron beam cold hearth refining of titanium alloys. 2000 TMS Annual Meeting, Nashville, TN, March 13–16, 2000.

19. JP Bellot, E Hess, D Ablitzer, A Mitchell. Dissolution of hard-alpha defects dragged in a bath of liquid titanium. Proceedings of the 1994 International Symposium on Liquid Metal Processing and Casting. A Mitchell, J Fremihough, eds., Vacuum Metallurgy Division, American Vacuum Society, Santa Fe, NM, 1994, pp. 155–166.

20. JP Bellot, E Hess, S Hans, D Ablitzer. A comprehensive simulation of electron beam cold hearth refining of titanium alloys. Proceedings of the 1997 International Symposium on Liquid Metal Processing and Casting. A Mitchell, P Auburtin, eds., Vacuum Metallurgy Division, American Vacuum Society, Santa Fe, NM, 1997, pp. 166–178.

21. JP Bellot, A Jardy, D Ablitzer. Mathematical modeling of the refining and solidification of titanium alloys during EBM. Proceedings of the Electron Beam Melting and Refining Conference State of the Art 1997. R Bakish, ed., Bakish Materials Corporation, Englewood, NJ, 1997, pp. 223–237.

22. AD Brent, VR Voller, KJ Reid. Enthalpy-porosity technique for modeling convection-diffusion phase change. Numerical Heat Transfer 13:297–318, 1988.

23. X Huang. Hearth model software. NCEMT Report TR No. 96-040 (Contract No. N00140-92-C-BC49), Concurrent Technologies Corporation, Johnstown, PA, 1996.

24. X Huang, JS Chou, KO Yu, DJ Tilly. Computer simulation of the refining hearth in a plasma arc melting process. Proceedings of the 1997 International Symposium on Liquid Metal Processing and Casting. A Mitchell, P Auburtin, eds., Vacuum Metallurgy Division, American Vacuum Society, Santa Fe, NM, 1997, pp. 179–203.

25. X Huang, JS Chou, KO Yu, DJ Tilly, V Suri. Physical modeling of the refining hearth in the plasma arc melting process. Proceedings of the 7th International Symposium on Physical Simulation of Casting, Hot Rolling, and Welding. HG Suzuki, T Sakai, F Matsuda, eds., ISIJ, Japan, 1997, pp. 489–502.

26. Y Pang. Hearth model development, validation, and sensitivity study. NCEMT Report TR No. 00-55 (Contract No. N00140-92-C-BC49), Concurrent Technologies Corporation, Johnstown, PA, 2000.
27. W Shyy, Y Pang, GB Hunter, DY Wei, MH Chen. Effect of turbulent heat transfer on continuous ingot solidification. J Engineering Materials and Technology Trans ASME 115:8–16, 1993.
28. W Shyy, Y Pang, GB Hunter, DY Wei, MH Chen. Modeling of turbulent transport and solidification during continuous ingot casting. International J Heat and Mass Transfer 35(5):1229–1245, 1992.
29. W Shyy, Y Pang, DY Wei, MH Chen. Effect of surface tension and buoyancy on continuous ingot solidification. AIAA 28th Aerospace Meeting, Paper No. 91–0506, 1991.
30. RC Sussman, M Burn, X Huang, BG Thomas. Inclusion particle behavior in a continuous slab casting mold. Iron and Steelmaker 20(2):14–16, 1993.
31. CE Shamblen, GB Hunter. Titanium base alloys clean melt process development. Proceedings of the 1989 Vacuum Metallurgy Conference on the Melting and Processing of Specialty Materials, Iron and Steel Society, Inc., Warrendale, PA, 1989, pp. 3–11.
32. KW Westerberg, TC Meier, MA McClelland, DG Braun, LV Berzins, TM Anklam. Analysis of the E-beam evaporation of titanium and Ti-6Al-4V. Proceedings of the Electron Beam Melting and Refining Conference State of the Art 1997. R Bakish, ed., Bakish Materials Corporation, Englewood, NJ, 1997, pp. 208–221.
33. L Nastac, S Sundarraj, KO Yu, Y Pang. Stochastic modeling of grain structure formation during solidification of superalloy and Ti alloy remelt ingots. Proceedings of the 1997 International Symposium on Liquid Metal Processing and Casting. A Mitchell, P Auburtin, eds., Vacuum Metallurgy Division, American Vacuum Society, Santa Fe, NM, 1997, pp. 145–165.
34. S Sundarraj, L Nastac, Y Pang, KO Yu. Numerical modeling of macrosegregation during ingot casting in the plasma arc melting process. Proceedings of the 8th International Conference on Modeling of Casting, Welding and Advanced Solidification Processes. BG Thomas, C Beckermann, eds., The Engineering Foundation, San Diego, CA, 1998, pp. 297–304.

19
Spray Forming

Huimin Liu
Ford Motor Company, Dearborn, Michigan

I. INTRODUCTION

Spray forming generally refers to a near net-shape material synthesis approach that combines melting, atomization, deposition, and consolidation into a single-step process to produce coherent, near fully dense preforms. Due to extremely low tooling costs and fewer secondary forming steps than conventional ingot/casting or powder metallurgy (PM) processes, spray forming is emerging as a cost-effective means for processing high performance materials. Consequently, this technology is being evaluated in the United States, Europe, and Japan as an alternative processing route for a variety of applications in aerospace, marine, defense, and automotive industries [1].

The earliest vestiges of modern spray forming technology can be recognized in the metal spraying process invented by M. U. Schoop in Switzerland in 1910 [2]. In this process, hot metallic droplets are projected onto a base material to apply a surface coating. More than four decades ago, J. B. Brennan [3] in the United States proposed spraying methods to produce semifinished metal products. The methods involve the disintegration of a metal melt into a spray of droplets and the deposition of the droplets onto a moving substrate to form metal strip. In the late 1960s, A. R. E. Singer [4] in the United Kingdom developed a spray rolling process for producing thin metal strip. In the spray rolling process, a metal melt is atomized by an inert gas and deposited onto a rotating drum to form thin strip. This process is conceptually a simplification of Brennan's approach. Subsequently, Singer developed and patented other spray processing concepts and designs, such as centrifugal spray deposition, simultaneous spray peening, spray coinjection for synthesizing metal matrix composites (MMCs), and arc spray forming for processing intermetallics [5–10].

Spray forming in the Osprey mode was conceived in the early 1970s by R. G. Brooks, A. G. Leatham, and J. S. Coombs [1, 11] in the United Kingdom. In the original design, discrete shaped preforms could be produced by rotating and/or translating a collector under the spray, followed by forging to final shape. By 1978, the ability to make thick section preforms in a variety of shapes and from diverse alloy compositions had been demonstrated [12, 13]. Since then, extensive research and development (R&D) on the spray forming process have been performed by corporations, government laboratories, research institutes, and universities in the major industrialized countries over the world [12], and have led to significant strides in the spray forming technology.

By making a systematic review of the latest developments in the spray forming process, the objective of this chapter is to familiarize the reader with the scientific and engineering aspects of the spray forming process, and to provide experienced researchers, scientists, and engineers in academic and industry community with in-depth information on the technology. Starting with a brief description of the spray forming process, products, and applications, the fundamental principles and phenomena associated with the spray forming process will be reviewed. It is followed by an extensive discussion of the models and computer simulations of the spray forming process, including metal and gas delivery, atomization, spray, deposition, preform cooling, and consolidation stages, as well as the basic concept of integral modeling. The emphasis will be placed on the application of numerical modeling to solving production problems.

II. SPRAY FORMING PROCESS AND PRODUCTS

A. Process and Recent Developments

Spray forming is a near net shape manufacturing route that can be used to produce near fully dense preforms of various shapes and refined microstructures at high production rates [1]. Conceptually, a spray forming process involves the energetic disintegration of a molten material into a dispersion of micro-sized droplets, followed by immediate deposition of the mixture of solid, semisolid, and liquid droplets on a substrate surface. A metallic material can be molten by, for example, induction heating or arc melting. The energetic disintegration of a molten material can be achieved by gas atomization, centrifugal atomization, or ultrasonic atomization [14]. Correspondingly, different process designs have been developed, such as atomization spray deposition [1, 12, 13], centrifugal spray deposition [15], and arc spray forming [9, 10], differing from each other mainly in the way how a material is molten and/or how a molten material is disintegrated into a spray. This chapter focuses on the atomization spray deposition process due to its demonstrated viability and compatibility with a wide range of applications.

Typically, in an atomization spray deposition process, a metal/alloy charge is induction melted and heated to a desired temperature in a crucible or tundish located on the top of a spray chamber, as schematically depicted in Fig. 1. The difference in the actual temperature and the melting temperature is referred to as melt superheat, usually 0.005–0.19 times the melting or liquidus temperature [16, 17]. The melt then issues into the chamber through a delivery tube at the bottom of the crucible. During melting, the chamber is purged with inert gas and a slight over pressure of gas (referred to as overpressure) is fed into the sealed crucible to prevent oxidation of the melt. The superheat and overpressure also serve the purpose of preventing the delivery tube from freezing up during subsequent atomization [16, 17]. In the atomizing zone below the crucible, the molten metal stream is disintegrated into a dispersion of micro-sized droplets (usually 10–200 μm in diameter [18]) by high speed inert gas jets issuing from a ring atomizer below the crucible, typically at a pressure of 0.7–1.0 MPa in the Osprey process [19], and 0.9–10 MPa in laboratory experiments. The droplets are rapidly cooled (at a maximum rate of 10^5–10^7°C/s [20, 21]) by the atomization gas through forced convective heat exchange between the two phases, and accelerated toward a deposition substrate. The post-recalescence solidification takes place at a cooling rate of 10^3–10^6°C/s, resulting in a secondary dendrite arm spacing (SDAS) of a few micrometers in magnitude [20, 21]. On the substrate, the droplets with different solid fractions (fully liquid, semisolid, fully solid) impinge and consolidate into a thick, near fully dense (>98% of theoretical density [13]) preform of the desired shape. During atomization and spray, the superheat and 60–80% of the latent heat of fusion are removed from the droplets within several milliseconds [20]. The remaining 20–40% of the latent heat is extracted on the substrate after deposition while the preform solidifies within 10–300 s at a cooling rate of 1–10°C/s [19]. Various shapes, such as disk, tube, strip, and sheet, can be produced by manipulating the substrate (substrate geometry and shape, rotation speed, withdrawal speed, and tilt angle) and/or changing the atomizer parameters (atomizer geometry, shape, and configuration, spray scanning frequency and angle). Preforms typically exhibit a uniform distribution of fine equiaxed grains, and no prior particle boundaries or discernible macroscopic segregation (a scale of segregation less than 50 μm for alloys [22] and a primary phase distribution of ~1 μm [13]). Mechanical properties are normally isotropic and equivalent or superior to those of counterpart ingot-processed alloys [13].

The spray forming process exhibits several attractive features. First, the highly efficient heat extraction during atomization ensures the maintenance of

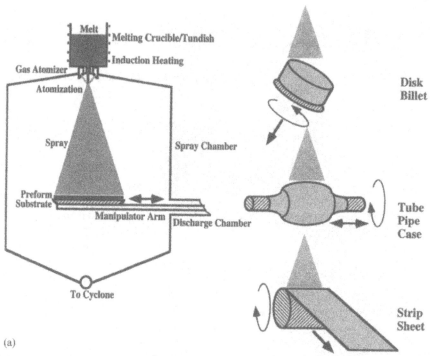

Figure 1 (a) Schematic showing a spray forming process; (b) Spray forming of a superalloy billet. (Courtesy of Dr. William T. Carter, General Electric Corporation R&D, Schenectady, NY.)

relatively low processing temperatures, which limits large scale segregation, clustering, coarsening, and surface/interface reactions. Hence, spray deposited materials exhibit some characteristics associated with rapid solidification, such as fine equiaxed grain microstructure, increased solid solubility, reduced segregation, non-equilibrium phases and small sized precipitates. Second, the inert conditions required in spray forming for melting, atomization, and deposition minimize surface oxidation and other deleterious surface reactions, reducing the sources of defect formation in sprayed preforms. On the other hand, by blending reactive gas(es) in the atomization gas and controlling surface reaction(s) during spray forming, nanometer-sized particles of nonmetallic phases (oxides, nitrides, carbides, borides, or silicides) may be synthesized in situ on droplet surfaces. The violent impact of fully liquid and partially solidified (semisolid) droplets during deposition may lead to a relatively homogeneous distribution of the nonmetallic phases (either these have existed before spray

(b)

Figure 1 (*continued*)

forming or formed in situ during spray forming) and the formation of dispersion-strengthened materials if the nonmetallic phases cannot be completely removed during preprocessing or if these phases are desired for the dispersion-strengthening effect. Therefore, spray forming has the potential of producing clean and defect-free materials with great structural uniformity, and extending alloy compositions beyond those that can be processed using conventional techniques. Third, spray forming is a near net-shape manufacturing method and involves less secondary forming steps than conventional ingot/ casting or PM processes. Hence, it can save a substantial amount of energy and reduce production costs below these processes. The fewer processing steps, operations, and secondary forming steps as well as the extremely low tooling costs make spray forming attractive from the standpoint of rapid prototyping. Spray forming is expected to offer maximum metallurgical and cost advantages compared to conventional ingot and PM processes, particularly with materials such as specialty steels, superalloys, aluminum alloys, and copper alloys. In addition, spray forming may potentially be used for near net-shape manufacturing of difficult-to-form materials, such as intermetallics and discontinuously reinforced MMCs. The fourth major attractive feature of spray forming is its high throughput rate, typically in excess of 0.3–2.5 kg/s per nozzle [13, 23].

Over the past two decades, the spray forming process has attracted considerable attention as a viable processing alternative for structural materials [11–13]. The process has been applied to highly reactive alloys (such as those based on Mg [24] and Al [25]), high temperature and high performance materials (such as those based on Fe [19, 23, 26–29], Cu [30, 31], Ti, Ni, and other alloys [32–34]), intermetallic compounds [20, 35–38], and particulate reinforced MMCs [8, 39]. Recently, significant progress has been made, such as [40]:

1. Spray forming of clad products, including those used in hot and cold strip mills
2. Elimination of porosity from the internal bore of sprayed tubes
3. Installation of a 5.4-ton spray-formed tube plant in the United States
4. Manufacture of bimetallic tubing in Sweden
5. Demonstration of low-cost manufacture of high-strength superalloy ring/casing components
6. Construction of a large pilot plant for spray forming turbine disks of ultraclean superalloys
7. Spray forming of sound aluminum-silicon alloy extrusion billets with excellent dimensional tolerances
8. Development of new copper alloys and construction of pilot plants for spray forming copper alloy billets

Over the past two years, several breakthroughs have been achieved in the spray forming technology, such as:

1. A twin-atomizing spray method to spray-form special steel billets up to 400 mm diameter with deposition yields in excess of 90% for better control of microstructures and faster production rates
2. Substrate preheating for a perfect metallurgical bond with low interfacial dilation
3. Recognition of the process by ASTM with issuance of two standards for pipe and tube
4. Trials for manufacture of 14″ pipes in a 1-ton pilot plant in Sweden

The research and development efforts have led to the maximization of plant throughputs via larger melt and spray-formed product sizes and higher deposition rates. Process efficiency has been improved through lower gas consumption, higher deposition yields, and tightened dimensional control. The technology is now matured to a manufacturing level in Europe. Moreover, reactive atomization deposition [20, 36–38] and spray coinjection deposition

[8, 39] have evolved for the synthesis of dispersion-strengthened materials and discontinuously reinforced MMCs.

B. Principal Products and Applications

Spray forming has been demonstrated capable of scaling up to production size and viable for tubular and round billets. The alloy systems focused on in commercial production are specialty steels, superalloys, aluminum alloys, and copper alloys. The current and potential applications of the spray forming technology [1, 13, 40] include the production of tool steel end-mills, superalloy tubes, aluminum automotive bodies, valve components, seawater piping components, airframe components, aircraft carrier catapult pistons, automotive engine components, aerospace engine turbine components (disks, rings and cases), shaft sleeves, clad structural steels, clad rolls, nuclear waste containers, pump stators, bimetallic gun barrels, and armor.

Tubes of stainless steels and selected Ni-base alloys (for example, 625) have been fabricated by spray forming in a single step directly from the melt [40]. The tubes are up to 8 m in length, up to 400 mm in diameter (outside), with a wall thickness of 25–50 mm. After removal of the mandrel on which droplets deposited and light machining of the inner and outer surfaces, the tubes can be used directly without the need for hot or cold working. The mechanical properties of the as-sprayed tubes are similar to those of wrought tubes. The as-sprayed tubes can also be cold-pilgered directly if the dimension or properties of the as-sprayed tubular preforms need to be changed for certain applications. The resultant mechanical properties are equivalent to those of ingot-processed tubes after similar cold working.

Spray forming technology for the production of aluminum alloy extrusion billets has been developed in Europe and Japan [40]. The spray-processed aluminum alloys include Al-Si alloys for automotive applications, the 2000 and 7000 series alloys and their modifications for automotive and aerospace applications, and Al-Li alloys for aerospace and marine applications. The spray-formed aluminum alloy billets exhibit low oxygen and hydrogen levels, and refined and uniform microstructures. The microstructure refinement leads to a significant improvement in ductility, toughness, and fatigue resistance. Billets of 400 mm in diameter and 1.3 m in length have been produced at a rate comparable to conventional single strand DC casting [40]. The billets can be extruded or isostatically pressed directly without any surface machining and further treatment.

Spray casting of copper alloy (Cu-Ni-Sn and Cu-Cr-Zr) extrusion billets of 300 mm in diameter and 2 m in length has been demonstrated [40]. The billets are converted to narrow strips or wires for applications requiring high strength and high conductivity such as connectors and welding electrodes. Cu-

Fe, Cu-Cr, and Cu-Zr alloys have also been spray-formed to strips with enhanced stress relaxation properties and tensile properties superior to those of ingot-processed materials.

High-Cr cast iron and high-C, high speed steel mill rolls have been produced by spray forming for finishing or prefinishing lines in wire, bar, flat-bar, and shaped steel mills on a commercial scale in Japan [40]. The rolls are up to 800 mm in diameter and 500 mm in length. The as-sprayed rolls exhibit fine microstructures with fine, uniformly distributed carbides throughout the roll thickness, getting rid of the eutectic and coarse carbides characteristic of the cast microstructure. Compared to the conventionally cast rolls, the mechanical properties, particularly wear resistance, of the as-sprayed rolls are improved and the roll life is substantially increased due to the refined microstructures.

Metal cleanliness and structural uniformity are two key factors governing the performance of superalloy components. It is difficult and expensive to fabricate superalloys using conventional processing techniques to satisfy these criteria. In particular, very high strength superalloys cannot be physically melted using conventional processes (VIM-VAR-forging) due to segregation and ingot cracking. The critical need for high strength, high performance, and cost-effective materials has stimulated the development of novel processing techniques. Among these, spray forming is being evaluated in the United States, Europe, and Japan as an alternative processing technology for the production of superalloy tubes, turbine rings, and turbine disks [1]. Spray forming has the potential of producing clean and defect-free superalloys with great structural uniformity, and is capable of extending the alloy compositions beyond those that can be processed using conventional techniques. Mechanical properties of as-sprayed superalloy billets are similar to those achieved via conventional PM processes. Due to fewer processing steps and operations, spray forming offers a significant cost saving for processing high performance superalloys.

Ni-base superalloy ring blanks of 300–800 mm in diameter and up to 500 mm in length have been spray-formed [40]. Superalloy billets of 300 mm in diameter and 500 kg in weight have also been spray-formed. The advantages are becoming evident, such as lower manufacturing costs, higher yields of material, and ability to process alloys that cannot be processed by ingot casting. The alloys of primary interest are IN718, Waspaloy, and Rene88. Two processing routes have been developed for shaped ring manufacturing, i.e., spray forming to near-net shape followed by hot isostatic pressing, and spray forming plus ring rolling. Mechanical properties of as-processed rings meet the ASM 5707G specification for wrought products.

One of the primary issues associated with the application of spray forming to superalloy turbine components is the defect contents in spray-formed materials. In the past, various ceramic crucible liners and transfer tubes have been employed in the melting system of the spray forming process. It has been identified that the defects in spray-formed materials originate from the ceramic particles in the melting system and serve as the initiation sites for fatigue failure. Hence, for the commercial-scale production of critical rotating components such as engine turbine disks by spray forming, refractory-free clean melting is essential. A melting process has been developed [40] in which the melt pool from an electroslag refining system (ESR) is used as the source and the liquid alloy is transferred directly from the ESR pool to the atomization system via a ceramic-free copper funnel, called cold-walled induction guide (CIG). The process relies on chemical refining in the slag to achieve the desired level of cleanliness for the rotating components. Ni-base alloys (Rene95, Rene88 and IN718) have been spray-formed into solid preforms.

Recently, a spray forming facility for the production of asymmetric, near net-shape components has been built by the U.S. Navy. Its capability to produce high-quality, complex components, such as tooling dies, has been demonstrated.

In summary, the principal products, current and potential applications, major manufacturers, and R&D participants are listed in Table 1 for an overview. A more detailed review of recent developments in the spray forming technology can be found in Ref. 40.

III. FUNDAMENTAL PHENOMENA AND PRINCIPLES IN SPRAY FORMING PROCESS

A. Atomization Mechanisms

Atomization is one of the complex physical phenomena involved in the spray forming process. During atomization, a liquid (molten) metal is impacted by a high energy gas to form a fine dispersion of droplets. Atomization of liquid metals is different from that of normal liquids because cooling of the liquid metals occurs simultaneously, leading to an increase in liquid viscosity and surface tension and eventual solidification of atomized droplets. A fundamental understanding of atomization mechanisms is needed in order to enhance atomization efficiency and reduce gas consumption. Atomization also has an important influence on the spray enthalpy and spray mass distribution.

Atomization of liquid metals may occur in several different modes, such as (1) liquid jet breakup, (2) liquid sheet breakup, (3) liquid film breakup, and (4) liquid ligament breakup [14]. Multimode breakup, i.e., simultaneous occurrence of two or more above modes, is possible, depending on the physical properties of liquid and gas, and atomizer geometry and configuration. Experimental and modeling studies [41–50] have shown that the atomization of liquid metals by gas jets in the spray forming and PM processes may take

Table 1 Spray Forming Products, Applications, Manufacturers and R&D Participants

Materials	Products	Manufacturers	R&D
Al Alloys	Castings, extrusions	ALCAN, UK Alusuisse, Switzerland Shell Billiton, Holland Pechiney, France	
		PEAK, Germany Sumitomo Light Metals, Japan	PSU, US
	Billets: 400 mm D × 1.4 m L		UCI, US
	Strips	ALCOA, US	
Cu Alloys	Castings, extrusions	Olin, US	NSWC-CD, US
	Billets: 300 mm D × 2.2 m L	Boillat, Switzerland Wieland-Werke, Germany	Univ. of Bremen, Germany
Special Steels	Castings, forgings, plates	MDH, Germany Rautaruukki OY, Finland Chaparral Steel, US	Drexel University, US NSWC-CD, US
	Billets: 400 mm D × 1.2 m L 1.3 ton	Osprey Metals, UK	Special Melted Products, UK Danish Steel, Denmark
	Tubes: 130–400 mm OD × 8 m L 20–50 mm wall thickness	Sandvik, Sweden	MDH, Germany University of Bremen,
	High Cr cast iron mill rolls: 583 mm D × 900 mm L	SHIFF, Japan	Germany
	High speed steel end mills		
	Clad rolls: 400 mm D × 1 m L Up to 800 mm D × 2 m L	Forged Rolls, UK British Rollmakers, UK	Sheffield University & Osprey Metals, UK
	Gun barrels: 25 mm D × 2 m L 120 mm D × 600 mm L	Babcock & Wilcox, US	NSWC-CD, US

	Pipes, shaft sleeves, clad structural steels, Clad rolls, nuclear waste containers, Pump stators, bimetallic armor	Babcock & Wilcox, US	
Superalloys	Billets, forgings 500 kg Engine ring blanks: 800 mm D × 500 mm L	GE, US Teledyne-Allvac, US Howmet Corporation, US	ALD, Germany Osprey Metals, UK NSWC-CD, US US Naval Academy, US
Ti Alloys	Tubes Billets	Sandvik Special Metal, US	NSWC-CD, US
MMCs			
Steel-SiC	Billets, strips, clads	ALCAN, UK	
Al Alloys-SiC	Castings, extrusions		UCI, US

place in two primary modes, i.e., liquid jet-ligament breakup and liquid film-sheet breakup. Details of the atomization mechanisms of liquid metals have been described in Ref. 14.

B. Rapid Solidification

In the spray forming process, solidification occurs in two distinct stages: (1) in spray and (2) in deposit, as schematically depicted in Fig. 2. Droplet consolidation on the deposition substrate connects these two stages and affects the microstructure of deposited preforms.

In the *spray*, droplets of different sizes are rapidly cooled, typically at a maximum cooling rate of 10^5–10^7°C/s [20, 21], by an atomization gas through forced convective heat exchange between the two phases. Such high cooling rates lead to large degrees of undercooling and rapid attainment of nucleation temperatures within short flight distances. Following nucleation, the latent heat of fusion is released at a rate that is substantially higher than the convective heat extraction rates at droplet surfaces due to high solid/liquid interface velocities driven by the large degrees of undercooling, leading to the phenomenon known as recalescence. During recalescence, the solid fraction

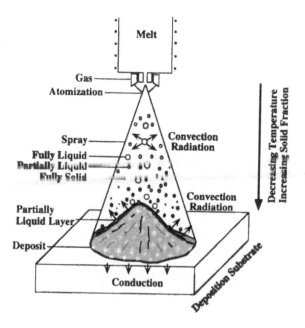

Figure 2 Cooling and solidification in spray and in deposit.

increases rapidly and the interface temperature rises up to nearly the equilibrium melting temperature within a flight time of tens of microseconds. The post-recalescence solidification takes place typically at a cooling rate of 10^3–10^6°C/s, resulting in a SDAS of a few micrometers in magnitude [20, 21]. Small droplets cool more rapidly and are fully solidified upon arrival at the deposition substrate, while large droplets remain fully liquid, and intermediately sized droplets are still semisolid. The droplets of different solid fractions impinge on the substrate and consolidate into a deposit of the desired shape and thickness.

In the *deposit*, heat extraction continues by means of convective heat exchange at the top surface with the atomization gas and via conduction at the bottom surface to the substrate. Solidification proceeds slowly, typically at a cooling rate of 1–10°C/s [19]. As with the spray, radiation heat exchange plays a little role in the cooling of the deposit.

The extremely high heat extraction rates in the spray give rise to large degrees of undercooling, causing large deviations from equilibrium, and thus offer the advantages associated with rapid solidification, i.e., fine grain sizes, small sized precipitates, reduced segregation (both in number and size), increased solid solubility (often by orders of magnitude), and new nonequilibrium phases. In general, the cooling rate during solidification has the most important influence on the resultant microstructure. However, the degree of undercooling before the onset of solidification and the post-solidification cooling rate also affect the resultant microstructure through their influences on the initial nucleation density, solid state diffusion, and coarsening processes.

C. Microstructure Evolution

Solidification in the spray is typified by the formation of a fine dendritic microstructure in atomized droplets. There exist significant microstructural variations among droplets, both in scale and morphology, owing to the complex thermal histories experienced by different droplets (undercooling, nucleation, recalescence, and post-recalescence solidification), as shown in Fig. 3. Small particles may exhibit a featureless or amorphous microstructure due to their extremely high cooling rates, whereas large particles show dendritic microstructure with various SDAS values.

However, independent of processing and materials, the as-sprayed deposits consistently exhibit a fine grained, equiaxed microstructure (Fig. 4) with grain sizes smaller than those in conventionally solidified alloys and a scale of segregation less than $50\,\mu m$ [22]. This microstructure characteristic suggests that solidification in the deposit involves a high density of nucleation sites. The equiaxed grains form either through the growth and coalescence of the deformed or fractured dendrite arms after droplets impact on the deposition

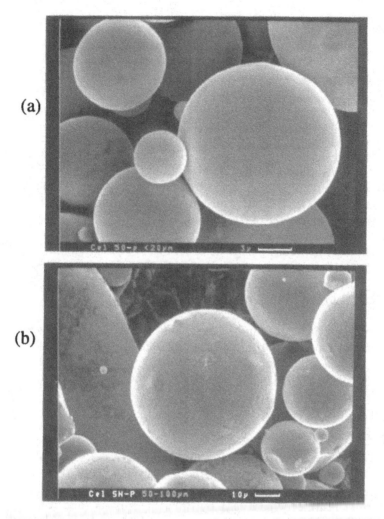

(a)

(b)

Figure 3 Microstructure of solidified steel droplets (particles) in a spray forming process. (a) Featureless (amorphous) microstructure in small particles; (b) Dendritic microstructure in large particles.

surface, or through the homogenization of the dendrites that do not deform extensively during impingement. This is a key feature of the spray forming process. This feature, along with the feature of net or near net-shape manufacture, obviates the necessity of postprocessing and/or secondary forming.

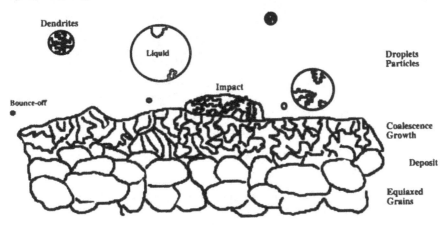

Figure 4 Schematic diagram showing microstructure evolution and mechanisms of the formation of equiaxed grain microstructure in an as-sprayed deposit: dendrite formation in spray; droplet impact on deposition surface; homogenization of dendrites that do not deform extensively during impingement; growth and coalescence of deformed or fractured dendrite arms; formation of equiaxed grains; and bounce-off of small droplets/particles.

D. Impact Deformation and Porosity Formation

In spray forming, the transient behavior associated with the impact, spreading, and consolidation of sprayed droplets on the deposition substrate is of critical importance. The physical phenomena during droplet impingement and consolidation involve fluid flow, heat transfer, and solidification. These phenomena take place simultaneously on a microscopic scale while droplets with different solid fractions at different velocities impact a microscopically nonflat, liquid, semisolid, or solid surface layer on the deposition substrate. At the same time, droplets undergoing deformation may break up, or bounce off, the deposition surface. Droplet deformation on the deposition surface causes not only the deformation and/or fracture of the dendrites, but also the formation of porosity. Thus, the droplet/substrate and droplet/droplet interactions critically determine the morphological and microstructural characteristics, and hence significantly affect the mechanical properties of as-sprayed materials.

The degree of droplet deformation and porosity content depend on many factors, such as droplet impact velocity, droplet size, droplet temperature, and substrate temperature. Detailed studies on the deposition and consolidation stages in spray deposition processes [51–60] have correlated the deformation behavior to process parameters, and identified several mechanisms responsible for pore formation in sprayed deposits. Generally, the porosity in sprayed materials may be classified into three major categories in terms of the underlying fundamental nature [61]:

1. *Chemical porosity*: Produced by a foaming agent or gas dissolution
2. *Physical porosity*: Generated due to solidification shrinkage
3. *Dynamic porosity*: Induced by droplet-gas, droplet-droplet, droplet-particle (solid or semisolid), and droplet-substrate interactions, and/or by substrate/spray motions

In terms of the locations where pores appear, three distinct porosity regions may be defined:

1. *Surface porosity*: In the upper surface region and periphery region of a deposit
2. *Internal porosity*: In the central region of a deposit
3. *Substrate porosity*: In the lower region of a deposit close to deposition substrate

The surface porosity and substrate porosity are mostly open pores with a relatively high porosity percentage ($\sim$10 vol.%) compared to the internal porosity (<2–3 vol.%).

The dynamic porosity can be further divided into the following four types according to the detailed formation mechanisms (Fig. 5):

1. *Interstitial porosity—macroporosity.* Macropores are formed primarily at interparticle boundaries, i.e., at the locations around solidified particles in a sprayed deposit. These solid particles do not flatten or flatten only to a small extent. The presence of such particles hinders the spreading and/or flow of successively incoming droplets. Therefore, some of the voids in the vicinity of solid particles or the interstices between solid particles and/or semisolid droplets cannot be completely filled by liquid due to flow stagnation and simultaneous solidification, leading to the formation of the macropores. The size scale of the macropores is comparable to droplet size. This type of porosity is frequently observed in plasma sprayed deposits and in spray processes with high solid fractions upon the impingement of droplets on the deposition surface, mostly in the periphery region and near-substrate region.

2. *Transgranular porosity—microporosity.* Transgranular pores are formed as a result of the separation of liquid from the solid/liquid interface or the breakup (ejection or rebounding) of liquid. When the detached liquid subsequently falls down to the surface, some voids may be entrapped, eventually developing into pores as solidification proceeds. Transgranular pores have smaller size scale rela-

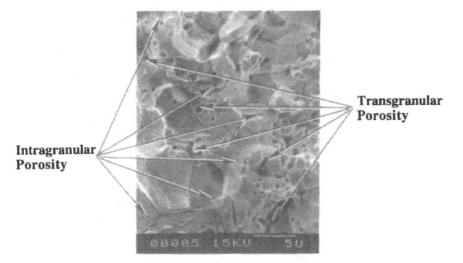

Intragranular Porosity

Transgranular Porosity

Figure 5 Microstructure of low-pressure plasma sprayed tungsten coating at a substrate temperature of 1500°C. Intragranular porosity: the very small circular cavities which exhibit a smooth circumference (periphery); transgranular porosity: the slightly larger cavities of irregular morphologies. (Courtesy of Electro Plasma Inc., Irvine, CA.)

tive to droplet size. This type of porosity may be present in the periphery region, the near-substrate region, and, to a less extent, in the central region.

3. *Intragranular/transgranular porosity—microporosity.* This type of porosity results from droplet/droplet or droplet/liquid-surface-layer interactions. The interactions lead to the formation of voids at the locations of vortices. The defects preexisting in droplets, such as extraneous inclusions and dispersoids (carbides, nitrides, or oxides), may also serve as the sites for voids to accumulate. The voids may eventually form pores when solidification is sufficiently fast and/or subsequent solidification shrinkage takes place. The defects preexisting within droplets may lead to the formation of cavities or gaps during solidification. The size scale of this type of porosity is generally smaller than droplet size.

4. *Intragranular porosity—microporosity.* Intragranular porosity has smaller size scale relative to droplets and results from cavities preexisting in droplets such as entrapped gas bubbles which may be generated during atomization and do not vanish during flight in spray chamber as a result of short flight time at high velocities.

These cavities may develop into micropores depending on the relative magnitude of deformation rates and solidification rates.

The size of the macropores should be in the order of a few micrometers to tens of micrometers. The dimensions of the micropores of the types 2, 3, and 4 might be plausibly several micrometers, several hundred nanometers, and a few nanometers, respectively. The micropores of types 3 and 4 are the internal porosity. The ultimate geometry and size of the micropores will depend on changes in molar volume during solidification, and on differences in wetting characteristics and other physical properties.

In addition to these mechanisms, porosity may also be induced by substrate motion and/or spray motion, as demonstrated by the high porosity level observed in the initial transition region and the leading edge of spray-formed tubes.

E. Thermophysical Properties and Process Parameters

The thermophysical properties of both liquid metal and atomization gas affect the droplet characteristics and subsequently the microstructure and mechanical properties of sprayed deposits. To date, the most widely used atomization gas in spray forming is nitrogen, although argon, helium, and air may be used as alternative atomization gases for different metal/alloy systems and for different end-product requirements. For example, to achieve the required cleanliness and structural uniformity of superalloys for engine turbine components, it is necessary to use an inert gas as a protective atomization gas. The use of nitrogen as an atomization gas may give rise to the formation of nitrides, which, as a type of defects, can lead to the reduction in performance of superalloy components. In this case, argon may be a preferred choice. In contrast, air or oxygen-blended nitrogen may be employed in a reactive atomization deposition process for in situ synthesis of oxide dispersion strengthened materials [20, 36 38].

The thermophysical properties of atomization gases and liquid metals/alloys are a function of temperature. Some convenient formulations and data for the thermophysical properties of commonly used and potential atomization gases and liquid metals/alloys are summarized in Ref. 14 and may be used in modeling spray forming.

The preform microstructure characteristics of critical concern are grain size and porosity. The currently achievable net or near-net preform shapes are mainly cylindrical, such as billet, ring, tube, and pipe. The geometry parameters specifying the preform dimension include thickness, diameter, and length. A number of process parameters need to be understood and controlled in spray forming in order to achieve consistency in microstructure, yield, shape and dimension of sprayed preforms. Some parameters can be selected prior to, not during, the operation of a process (referred to as preset parameters), such as (1) atomizer and nozzle designs, (2) atomization gas type, and (3) substrate geometry and configuration. Some independent parameters can be either preset or varied during the operation (referred to as on-line parameters). Important on-line parameters include (1) melt superheat and flow rate, (2) gas pressure and flow rate, (3) spray and substrate motions, and (4) atomizer-substrate distance. A more detailed list of the important preset geometry and process parameters and important on-line parameters which affect microstructure and yield is given in Table 2 for an overview.

There are some parameters which describe the intermediate states of individual stages in a spray forming process. These intermediate parameters are loosely defined and used for process analysis and modeling, as summarized in Table 3.

IV. MODELING OF SPRAY FORMING PROCESS

As in many other fields, modeling of the spray forming process provides tremendous opportunities to improve our understanding of the fundamental phenomena and the effects of process parameters on atomization efficiency, and the microstructure and mechanical properties of sprayed preforms. It enables us to establish a correlation between the process parameters and the quality of sprayed preforms. It also provides basic guidelines for on-line control over mass distribution, deposition yield, and microstructural evolution in spray forming.

The fundamental issues to be addressed in the process modeling are spray enthalpy, gas consumption, spray mass distribution, and microstructure and shape of sprayed preforms. The effects of atomization gas chemistry, alloy composition, and process parameters on the properties of the resultant preforms are also to be investigated in the process modeling. To facilitate the process modeling, a spray forming process can be divided into the following stages: (1) metal and gas delivery, (2) atomization, (3) spray, (4) deposition, and (5) preform cooling and consolidation. Solidification and microstructure evolution may start as early as in the atomization stage and continue until the end of the consolidation stage.

The primary objectives of the process modeling are as follows:

1. Control spray enthalpy
2. Increase atomization efficiency and reduce gas consumption
3. Control spray mass distribution to a sufficiently narrow pattern to significantly reduce material losses in machining and tails

Table 2   Preset and On-line Parameters

Gas atomizer	Diameter (m)
	Orifice number, diameter (m), angle (°)
	Configuration: Close-coupled, free-fall, twin
	Type of Nozzle: Laval tube, straight tube, shock wave (ultrasonic)
	Scanning angle (°), frequency (/s)
Atomization gas	Composition
	Pressure (Pa)
	Temperature (K)
	Mass flow rate (kg/s)
Metal	Composition
	Melting temperature (K)
	Superheat (K)
	Overpressure (Pa)
	Mass flow rate (kg/s)
Melting crucible (Tundish)	Material
	Diameter (m), height (m)
	Heating/melting method: Induction, resistance, electroslag remelting
Delivery nozzle	Material
	Diameter (m), length (m)
	Configuration
Spray chamber	Pressure (Pa)
	Geometry
	Atomizer-substrate distance (m)
Substrate	Material
	Cooling/heating/surface temperature (K)
	Shape and dimension
	Rotating speed (rpm)
	Withdrawal speed (m/s)
	Tilt angle (°)

4. Reduce average porosity and meet desired criterion of grain size
5. Control preform shape for cost-effective, net, or near-net shape manufacturing

This section describes in detail the process models, computational methods, and modeling results for each stage of the spray forming process.

Table 3 Intermediate Parameters

Droplet-gas interactions in spray	Droplet-surface interactions on substrate
Droplet size	State of droplets:
Droplet size distribution	temperature, solid fraction, velocity,
Spatial distribution	microstructure characteristics (SDAS)
Droplet velocity	State of deposition surface:
Droplet cooling/solidification history	liquid fraction, temperature,
Solid fraction of droplets	surface morphology,
Spatial distribution	thickness of mushy layer
Gas velocity	Droplet spreading, splashing,
Gas temperature	Solidification:
Spatial distribution	bounce-off, ejection

A. Stage I: Metal and Gas Delivery

The distribution of velocity and temperature of the atomization gas in the near-nozzle region determines the impact kinetic energy and cooling effect on the molten metal stream, and hence significantly influences the generation, initial size, shape, radial distribution, and flight direction of droplets. Melt flow and heat transfer during delivery and atomization, on the other hand, determine its temperature, viscosity and surface tension, which in turn affect the mass median size and the size distribution of droplets, as well as the continuous operation of the delivery nozzle. Therefore, the flow and heat transfer phenomena during the metal and gas delivery and in the vicinity of the atomizer are critical to the atomization efficiency and gas consumption, and have a significant impact on the droplet size, spray enthalpy, and mass distribution.

Recently, there has been an increasing number of numerical studies on the metal and gas delivery stage [16, 17, 48–50, 62, 63], and some experimental measurements in the near-nozzle region [46, 47, 64–70]. The modeling of flow and temperature fields of melt and atomization gas in the near-nozzle region predicts the necessary process parameters to prevent premature solidification of liquid metal/alloy, and determines the optimum atomizer parameters. The modeling results of the metal and gas delivery stage also provide initial conditions for the modeling of the ensuing atomization stage.

1. Modeling of Flow and Heat Transfer of Liquid Metal in Near-Nozzle Region

In spray forming, a liquid metal is normally transported from a melting crucible (tundish) to the atomizer through a delivery tube. During the flow in the delivery tube, the temperature of the melt decreases gradually because of the heat exchange with the tube wall. Premature solidification of the melt in the delivery tube, referred to as freeze-up, may occur when the melt temperature decreases down to or below the melting temperature. The freeze-up may be caused by: (1) inadequate melt superheat, (2) excessive residence time of the melt in the delivery tube, and/or (3) recirculation of undercooled droplets at the tube exit. The selection of appropriate melt superheat requires that it can offset the heat loss of the melt during the flow in the delivery tube. The heat loss and the residence time of the melt are closely related to its flow and heat transfer behavior in the tube.

In the numerical studies by Liu et al. [16, 17], a model and computer code were developed to simulate the flow and heat transfer of molten metals in the delivery tube. The model is based on the boundary layer theory and the modified van Driest and Cebeci mixing length turbulence model. Numerical calculations were conducted for liquid Al, Cu, Mg, Ni, Ti, W, In, Sn, Bi, Pb, Zn and Sb to investigate the influence of process parameters and material properties on the minimum melt superheat that is necessary to prevent the freeze-up during delivery of the molten metal prior to atomization. The modeling results showed that for the materials studied, the minimum melt superheat ranges from 0.005 T_m to $0.19 T_m$, depending on process parameters and material properties. The dependence was expressed using a correlation derived from a regression analysis of the numerical results. Processing maps were also developed on the basis of the modeling results. The correlation and processing maps can be used to determine the process conditions that are necessary to avoid the freeze-up.

The modeling results revealed that materials with high thermal conductivity, high thermal capacity and/or large density allow a small minimum melt superheat, whereas materials with high melting temperature and/or high viscosity require a large minimum melt superheat. The minimum melt superheat can be decreased by reducing the tube-length/diameter ratio, by selecting a smooth delivery tube with low thermal conductivity and/or thick tube wall, by imposing a high overpressure, and/or by enhancing atomization gas temperature. Within the range of low overpressure values (for example, $\leq 60\,kPa$), increasing the overpressure can effectively decrease the minimum melt superheat, especially for a large tube-length/diameter ratio (for example, 17) and for materials of low densities (for example, Al and Mg). This effect diminishes with further increase in overpressure. The overpressure, which is necessary to avoid the freeze-up (referred to as minimum overpressure), decreases with decreasing tube-length/diameter ratio or decreasing melting-temperature/gas-temperature ratio.

In addition to the above process parameters, atomizer geometry and configuration also influence the minimum melt superheat by altering gas flow pattern and distribution. The freeze-up may occur even for the melt superheats that are higher than the predicted minimum value, if recirculation gas flow forms in the exit region of the delivery tube. The recirculation gas flow in this region may drag relatively cool or undercooled droplets upwards, and deposit them on the tip of the delivery tube, where they eventually solidify. The solidified metal may continue to increase in thickness during atomization, eventually choking the melt flow, a phenomenon that has been described previously as freeze-up. Therefore, the minimum melt superheat calculated with the correlation is only a necessary condition, rather than a sufficient condition. A sufficient condition to avoid the freeze-up must be determined by examining the flow field below the delivery tube. If the geometry and configuration of the delivery tube and the atomization gas nozzles are arranged in such a manner that the recirculation vortex velocity is not large enough to drag cool droplets up to the delivery tube tip, then employing the melt superheat predicted by the correlation may avoid the freeze-up. Accordingly, for free-fall atomizers, in order to minimize the probability of the freeze-up, it is necessary to design the atomizer geometry and select the position of the delivery tube in such a way as to minimize the formation of the recirculation gas flow in the region below the delivery tube. However, for close-coupled atomizers, the recirculation gas flow is utilized to generate a thin liquid film and sheet prior to atomization [14].

The temperature of a liquid metal stream discharged from the delivery tube prior to primary breakup can be calculated by integrating the energy equation in time. The cooling rate can be estimated from a cylinder cooling relation for the liquid jet-ligament breakup mechanism (with free-fall atomizers), or from a laminar flat plate boundary layer relation for the liquid film-sheet breakup mechanism (with close-coupled atomizers).

2. Modeling of Gas Flow in Near-Nozzle Region

For the delivery of an atomization gas, different types of nozzles have been employed, such as straight, converging, and converging-diverging nozzles. Two major types of atomizer configurations, i.e., close-coupled and free-fall, have been used in atomization, in which gas flows may be subsonic or supersonic, depending on process parameters and gas nozzle designs. The close-coupled atomizer configuration has many variations, such as the atomizers for high pressure gas atomization (HPGA) [66], Ünal atomizers [46], truncated plug atomizers [68], NANOVAL atomizers [71], ultrasonic gas atomizers [20, 36], and the novel internal mixing atomizer proposed by Sheikhaliev and Dunkley [72]. The free-fall atomizer may also have different geometry arrangements, although the Osprey type has been the most widely used one in spray forming.

While the gas flow in a nozzle can be calculated on the basis of isentropic expansion of stagnation properties, the analysis of the gas flow in the near-nozzle region can be performed by means of computational fluid dynamics (CFD) codes. Such codes can generally treat both incompressible and compressible flows, and the k-ε two-equation model is frequently used as turbulence model.

Recently, Liu [14, 50] performed numerical modeling of gas flows in the near-nozzle region for both the free-fall and close-coupled atomizer configurations. In view of the complex nature of the gas flows, such as subsonic and supersonic flows, turbulence, shock wave, free shear flow, and recirculation flow, the full Reynolds-averaged Navier-Stokes equations were solved along with turbulence transport equations. The effect of turbulence was modeled using the standard Boussinesq approximation. Two turbulence models, i.e., combined Thomas/Baldwin-Lomax model and k-ε two-equation model, were used, switching from the former to the latter after some iterations. The calculated gas velocity distributions suggested that for a constant gas pressure, the recirculation gas flow can be minimized by arranging the geometry of the free-fall atomizer. On the other hand, the strong recirculation flow and the adjacent free shear flow are responsible for the liquid film-sheet breakup mechanism with the close-coupled atomizer. Thus, the gas flow in the recirculation region critically affects the resultant droplet properties, as will be discussed below.

B. Stage II: Atomization

Over the past decade, a number of numerical studies have been conducted to model the liquid metal delivery, spray, deposition, and consolidation stages in spray deposition processes [16, 17, 20, 21, 26, 29, 37, 51–60, 63, 73, 74]. Analytical [75, 76] and experimental studies [41–45, 66, 77] on atomization have also been performed in an effort to elucidate atomization mechanisms and control droplet size. However, only limited numerical modeling studies have been conducted for the atomization stage [48–50].

In Liu's numerical study on liquid metal atomization with a close-coupled atomizer [50], an atomization model was conceived. A computer code and computational methods were developed to model liquid metal breakup [78–80], droplet dynamics [81, 82], and heat transfer during gas atomization. Empirical and experimental correlations [78 80, 83] were used along with basic conservation equations to predict the droplet size distribution produced by atomization. The modeling results suggested that the atomization of a liquid metal may occur in the following sequence. Initially, the liquid metal flows downward through the delivery tube. At the exit of the tube, it changes its flow direction to radially outward along the tube base plane, forming a thin liquid film. The formation of the liquid film is caused by the interaction between the liquid metal and the gas flow in the recirculation region. It is also this interaction that leads to the turning of the liquid film at the edge of the tube base plane into a liquid sheet which extends downwards following gas stream lines. This is followed by the primary breakup of the sheet into droplets.

The droplets accelerate, deform, and cool in the gas flow field. The trajectories of small droplets also follow the gas stream lines closely. Secondary breakup of the droplets may occur, depending on their deformation and cooling rates.

Regarding the effects of process parameters on atomization, numerical and experimental studies [49, 64, 77] revealed that the mass median droplet diameter decreases with increasing atomization gas pressure and/or gas to metal mass flow rate ratio. The standard deviation decreases with increasing gas to metal mass flow rate ratio. As the melt superheat increases, both the mass median droplet diameter and standard deviation decrease.

C. Stage III: Spray

Numerical modeling of droplet/gas interactions in spray has been reported by a number of investigators [19, 20, 34, 63, 74, 84, 85]. In limited studies [27, 28, 34], two-dimensional experimental measurements of atomization gas and droplet velocities have been conducted. A few complex, two-dimensional Eulerian models [29, 84] have been used to calculate the droplet/gas interactions in spray. Most of these studies, however, are confined to one-dimensional, and only the droplet trajectory and thermal history along the spray centerline are addressed.

Employing the Lagrangian approach, Liu et al. developed both one-dimensional [20] and two-dimensional [21] flow and heat transfer models and rapid solidification kinetics models, including a reaction kinetics model for reactive atomization deposition, to describe droplet/gas interactions in the spray stage. The formulations are suitable for both Newtonian and non-Newtonian conditions. The models and the related numerical methods as well as the computer codes developed [20, 21] have been tested and validated for Ni_3Al intermetallic compound and Ta-W alloy droplets sprayed using nitrogen-oxygen mixture gases, argon and nitrogen, respectively. These models and codes can be used to calculate the momentum, heat and mass transfer in the spray stage and to predict the droplet velocity, temperature, cooling rate, solid fraction, spray enthalpy, and their spatial distributions prior to deposition. Some microstructural characteristics in semisolid or fully solidified droplets/particles (for example, SDAS) can be determined on the basis of the thermal history predicted by the codes.

The modeling results showed that the axial gas velocity decays exponentially along the spray axis, while the gas flow diameter increases in the axial direction. The radial profiles of axial gas velocities exhibit a shape akin to a Gaussian distribution. The axial and radial droplet velocities increase initially along the axial direction and attain rapidly their maximum values. With increasing axial distance, the radial profiles of the axial droplet velocities become wider and approach the profiles of gas velocities. At any axial distance, the droplet velocity, temperature, cooling rate, and solidification rate all exhibit a maximum at the spray axis, and decrease to a minimum at the periphery of the spray cone, except for the radial locations where solidification occurs. For a given droplet diameter, the achievable undercooling is smaller, the flight time required to reach a given axial distance is longer, and the SDAS formed during post-recalescence solidification is larger in the periphery region of the spray cone than elsewhere in the radial direction. Hence, the droplets in the periphery region of the spray cone solidify within a shorter flight distance relative to those at the spray axis due to the longer flight time in the periphery. The modeling results of the spray stage provide input data and initial conditions for the modeling of droplet/substrate interactions in the ensuing deposition stage.

D. Stage IV: Deposition

In spray forming, droplet deposition occurs when a mixture of solid, semisolid and liquid droplets from the spray impact at various velocities onto the deposition surface. During the deposition, only a fraction of droplets are directly sprayed onto the preform without being lost to the outside of the target, and only a fraction of the targeted droplets actually stick to the surface and contribute to the growth of the deposit. The former fraction is referred to as overall target efficiency and the latter fraction is defined as sticking efficiency. The ratio of the mass deposited on the preform to the mass sprayed over a specified period of time is termed deposition yield. The deposition yield is a product of the sticking efficiency and the overall target efficiency. Therefore, the overall target efficiency represents the ideal deposition yield when sticking is perfect, i.e., the sticking efficiency is 1, whereas the sticking efficiency depicts the maximum deposition yield achievable when all the sprayed droplets impinge onto the target surface, i.e., the overall target efficiency is 1. To enhance the deposition yield, both the sticking efficiency and the overall target efficiency must be increased.

The overspray can be classified into three categories in terms of its origin:

1. *Fluid-dynamic overspray (sticking effect)*: Droplets bounced off or ejected from the deposition surface due to the splashing and fragmentation of large liquid or semisolid droplets upon impact
2. *Aerodynamic overspray (target effect)*: Small presolidified droplets swept away by the atomization gas
3. *Geometric overspray (target effect)*: Droplets that do not hit the deposition surface due to geometric missing and/or shadow effects

Experiments have shown that more than half the oversprayed powder is made up of droplets bounced off and ejected from the deposition surface [13].

Therefore, the behavior of droplets during deposition critically influences the shape, yield, microstructure, adhesion and porosity of the spray-formed materials.

For stationary spray and substrate, i.e., the spray is not scanned and the substrate is not rotated and/or withdrawn continuously or inclined to the spray, a Gaussian-shaped deposition surface contour typically develops with an annular atomizer. The radial mass distribution can be formulated in terms of the mass flux in the spray or the deposition rate [86]. The deposition rate seems to scale with metal flow rate. Gas pressure tends to narrow the spray whereas melt superheat tends to flatten the spray. By manipulating the configuration and/or motion of the substrate and/or the spray, the deposit shape can be controlled. For example, cylindrical disks and billets can be made by scanning the spray across a rotating disk substrate that is inclined to the spray axis; strips and sheets can be produced by scanning the spray across the width of a horizontal belt, and any nonuniformity of thickness can be eliminated by exposing every location to an identical spray cycle; and tubes and pipes can be fabricated by scanning the spray in the longitudinal direction onto a rotating mandrel that is simultaneously withdrawn away from the spray. In addition, a linear atomizer can be employed for spray forming flat products such as strips, sheets or plates with a uniform thickness that requires a uniform distribution of the spray mass flux. The mass flux distribution in the spray formed using a linear atomizer has been found to follow the elliptical form of the Gaussian distribution [87].

Under the steady state conditions in spray forming billets, tubes or sheets, the billet diameter, the tube wall thickness, and the sheet thickness are all proportional to the square root of the metal flow rate and yield. The billet diameter and the tube wall thickness are inversely proportional to the square root of the retraction rate, and the sheet thickness is inversely proportional to the retraction rate.

The deposition yield is related to the deposit shape and substrate motion. Leatham et al. [13] indicated that the optimal strategy to maximize the yield is to raise the fraction of liquid on the preform surface close to 30% as soon as possible in the operation and then maintain it at this value throughout the process. The sticking efficiency is governed by the thermal and geometrical conditions of the deposition surface and the thermal and dynamic conditions of the spray. A sticking efficiency of nearly 1.0 has been reported for depositing tubes, whereas it ranged from 0.77 to 0.88 for depositing disk preforms as a result of the highly tilted deposition surface in disk preforming [88]. A maximum target efficiency of 0.95 and 0.99 has been achieved in spray forming disks and tubes, respectively [88].

Uhlenwinkel et al. [89] conducted extensive experimental studies on the sticking efficiency. In their experiments, the maximum of the sticking efficiency was observed at the top of the deposit as a result of the large population of larger droplets and higher spray enthalpy in the center relative to those in the periphery of the spray. The sticking efficiency may exceed 0.9 in the center. Among the parameters studied (such as droplet size, impact velocity, impact angle, spray enthalpy, and deposition surface temperature), the deposition surface temperature was found to be the most critical parameter influencing the sticking efficiency.

Cai's study [88] showed that the sticking efficiency is primarily a function of the volume fraction of liquid in the spray, and influenced by substrate motion. With increasing fraction of liquid in the spray, the sticking efficiency increases first and then turns to decrease. A maximum sticking efficiency was achieved at ~30% liquid for Cu-6wt%Ti disks and at ~53% liquid for IN625 tubes. As the rotation rate of the substrate decreases, the sticking efficiency increases. Overall, these experimental observations and qualitative analyses of the sticking conditions suggested that there exists an optimum combination of the thermal conditions of the deposition surface and the spray, as depicted in Fig. 6.

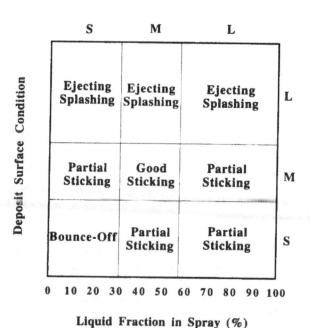

Figure 6 Sticking conditions as a function of the combination of the thermal conditions of deposition surface and spray.

In the experimental studies on the sticking efficiency using a spray forming facility (such as Ref. 89), the effects of the dynamic condition of the spray and the geometry condition of the deposition surface were tangled together with those of the thermal conditions due to the interrelated dependence. To single out the effects of the spray dynamic and surface geometry conditions, the droplet/surface interactions during deposition should be investigated at constant thermal conditions. Berg and Ulrich [90] carried out experimental studies on the impact of droplets on dry, solid surfaces under a variety of dynamic and geometry conditions. The analysis of their experimental results suggested that the amount of splashing mass increases with increasing impact velocities and increasing surface roughness of the target substrate. A large surface roughness corresponds to an earlier and more violent breakup of impact droplets and bounce-off of secondary droplets.

Several research groups have conducted numerical analyses and simulations of the transient flow of a single liquid droplet impinging onto a flat surface, into a shallow or deep pool [91], including heat transfer [92] and solidification [73, 93–95]. Recently, important progress in the modeling of droplet impingement processes has been accomplished by Liu et al. [51–60]. In their work, the interactions between multiple flattening droplets on both flat and non-flat surfaces were numerically investigated by solving the full Navier-Stokes equations coupled with the volume of fluid (VOF) function. A two-domain method was employed to treat the thermal field and solidification problem within the flattening droplets and to track the moving solid/liquid interface. The Stefan solution of solidification problem was incorporated into a two-phase flow continuum model to simulate the flow with a growing solid layer. The numerical results were used to determine the exact motion, interaction and solidification of droplets on deposition surface including arbitrary free surfaces and solid/liquid interfaces. The microporosity was quantitatively calculated from the VOF function data based on an algorithm developed by Liu et al. [54]. These studies not only provided detailed information of interactions of multiple droplets, but also improved our understanding of the effects of important process parameters on the spreading and interactions of droplets. In addition, the numerical results provided insight into the formation of vortices and the ejection of liquid from deposition surface during impingement of multiple droplets. Therefore, these studies formed a useful basis for exploring the mechanisms that govern the micropore evolution and for correlating the sticking efficiency to thermal, dynamic and geometry conditions during spray forming.

The modeling studies [51–54, 58–60] revealed that a fully liquid droplet impinging onto a solid particle on a flat substrate leads to the formation of macropores between the particle boundary and the substrate; multiple fully liquid droplets striking onto a solid particle on a flat substrate substantially eliminate the macropores, but simultaneously produce vortices within the liquid, as well as ejection and breakup of the liquid; multiple fully liquid droplets striking onto multiple solid particles on a flat substrate not only lead to vortices and ejection of the liquid, but also to the formation of the macropores due to the large solid fraction. Therefore, increasing the liquid fraction in the spray may reduce the macroporosity significantly. However, an excessive liquid fraction in the spray may decrease the sticking efficiency as a result of the increased ejection of the liquid from the deposition surface. A large droplet diameter or a high liquid density leads to a large final splat diameter, while a high impact velocity, a large roughness height, a small roughness spacing, or a small liquid viscosity (high liquid temperature) results in an early occurrence of a violent breakup of the liquid, reducing the sticking efficiency. Decreasing the roughness height and increasing the roughness spacing of the deposition surface may improve the extent of droplet flattening, reduce the breakup of the liquid and the microporosity, and enhance the sticking efficiency.

To minimize the porosity in sprayed preforms, it is necessary to maintain a certain level of liquid on the deposition surface [88]. This can be done by controlling the deposition rate, the surface temperature, and/or the liquid fraction in the spray. The latter is an important factor critically influencing the porosity. The average porosity can be minimized (below 2–3 vol.%) when the liquid fraction in the spray is increased beyond a certain value which produces a maximum sticking efficiency. A further increase in the liquid fraction can further reduce the porosity slightly at the expense of sticking efficiency. The optimum liquid fraction was found to be 31–32% and 54–55% for Cu-6wt%Ti disks and IN625 tubes, respectively.

E. Stage V: Preform Cooling and Consolidation

The thermal history experienced by droplets before and after impact onto the deposition surface is distinctly different. In the spray, the extremely rapid extraction of thermal energy by an atomization gas promotes the formation of highly refined microstructure, typified by the formation of a fine dendritic microstructure in solidified droplets. In contrast, the solidification on the deposition surface proceeds at relatively low cooling rates. The microstructure of as-sprayed materials is consistently observed to exhibit fine equiaxed grains, regardless of alloy compositions.

The microstructure across the deposit thickness is governed by two competing factors, i.e., deposition rate and solidification rate. Accordingly, solidification may occur in three distinct modes:

1. *Splat solidification*: When the solidification rate is higher than the deposition rate, successive droplets impinge onto a solidified surface and a rapid solidified microstructure is obtained throughout the deposit. The cooling and solidification are via convective heat transfer from the top surface to the atomization gas and conductive heat transfer at the bottom surface through the deposition substrate. As a result of the high heat extraction rate, splat solidification proceeds at a high cooling rate.

2. *Incremental solidification*: When the solidification rate is slightly lower than the deposition rate, a thin liquid layer may form at the deposition surface. While successive droplets impact the thin liquid layer, the thermal energy of the incoming droplets is extracted via the solidified material beneath the liquid layer, so that the solid/liquid interface moves from the substrate toward the deposition surface as the deposit grows.

3. *Growth and coarsening*: In this solidification mode, the solidification rate is much lower than the deposition rate. A thick partially liquid layer builds up due to the low heat extraction rate at the top and bottom surfaces of the deposit with a typical cooling rate of 5–10°C/s. The equiaxed grains in the final microstructure are formed through growth and coarsening of the deformed or fractured dendrite arms in the partially liquid layer after droplets impact on the deposition surface or through the homogenization of the dendrites that do not deform extensively during impingement. When a balance in the heat influx and heat extraction is reached, the thickness of the partially liquid layer will remain constant. Under such a condition, the solidification mode may switch to incremental solidification.

The *splat solidification* mode may be the operating solidification mechanism for spray forming a thin section or for the initial stage of depositing a thick section where the deposit undergoes high cooling rates due to chilling by the target substrate. Over an extended period of deposition time, the heat influx increases with increasing deposit thickness. As a thin liquid layer or partially liquid layer builds up, the *incremental solidification* and *growth and coarsening* modes may become dominant solidification mechanisms. For a moving deposition substrate with a given withdrawal rate, a minimum deposition time is needed to attain a steady state; otherwise, a minimum withdrawal rate is required for a given deposition time. Less time is required to reach a steady state with a faster retraction of the substrate [88].

To address the preform cooling and consolidation stage, one- and two-dimensional heat transfer and solidification models have been developed and implemented by Mathur et al. [19], Annavarapu et al. [23], and Cai [88], respectively. For the analysis of the consolidation of droplets/particles on a target substrate, two approaches have been utilized [23]: (1) discrete event (microscopic) approach and (2) continuum (macroscopic) approach, differing from each other mainly in the scale of resolution.

In the *discrete event approach*, the spray is not viewed as a continuous source of heat and mass flux. Instead, the consolidation on a substrate is deemed to occur by splatting of individual droplets and is modeled as a summation of discrete events. An addition of a splat at the top surface of a deposit is considered as one discrete event. The deposit is "built" up by the summation of these events, i.e., by stacking the splats one on top of another. The discrete event approach utilizes the average droplet enthalpy calculated in the spray stage and conducts an energy balance to calculate the temperature and cooling across the deposit thickness.

In the *continuum approach*, the spray is assumed as a continuous source of heat and mass flux. This approach utilizes the heat flux determined in the spray stage as an input. A mass balance is used to calculate the deposit geometry and an energy balance to predict the thermal and solidification histories at any location within the deposit.

It has been demonstrated [23] that before a liquid layer forms, the discrete event model predicts deposit solidification more accurately than the continuum model, and the continuum model underestimates the cooling rates in the chill zone. However, after an initial interaction of discrete droplets with the substrate, both models predict the same solidification rate due to the fact that consolidation proceeds via successive impingement of droplets onto previously consolidated droplets. Since the predictions of both models converge, the continuum model is more effective because it is less computation intensive and its input parameters are well defined.

The modeling studies [23] showed that the heat flux across the deposit-substrate interface is $\sim 10^7$ W/m^2 in order of magnitude during the initial stage of deposition. The heat flux falls sharply to a steady state value of $\sim 10^6$ W/m^2 over a period of 10–15 s. The peak value of the heat transfer coefficient at the deposit-gas interface is estimated as $\sim 7 \times 10^4$ W/m^{2}K, and it subsequently decreases to a steady state value of 5×10^3 W/m^2 K. The cooling rate during the solidification of the bulk deposit is less than $100°$C/s, whereas it ranges from 10^5 to 10^8 °C/s in the chill zone. In case that a sprayed preform is removed/separated from the deposition substrate, further cooling of the preform in solid state occurs at a rate of 5–10°C/s. A partially liquid layer forms on the deposit surface after a short deposition time. The fraction of solid in the layer varies from 0.95 to 1.0, and the thickness of the layer depends on the substrate velocity and temperature. Preforms typically exhibit a gradient microstructure as a result of the thermal gradient during cooling and solidification. In the absence of second-phase precipitation, grain coarsening occurs during slow cooling in bulk deposit. Grain sizes are typically in the range of 50–100 μm.

The model predictions [23] also showed that the use of a heated substrate is insufficient to eliminate the chill zone although it may be possible to seal off the surface-connected pores. As the fraction of liquid in the spray increases, both the top surface temperature of a preform and the local solidification time in the preform increase. A hotter spray will decrease the deposit cooling rate and hence yield a coarser microstructure. Increasing the metal mass flow rate influences the spray temperature and thus the deposit cooling rate. To scale up the flow rate, a possibly feasible approach is to utilize multiple nozzles with an appropriate separation from each other.

Regarding the quantitative assessment of porosity formation during preform cooling and consolidation, only a few preliminary studies have been performed in recent experimental and numerical efforts [54–57, 61]. On the basis of these studies, it is now possible to identify different porosity formation mechanisms [57] and to quantitatively calculate microporosity [54] in spray-formed materials. Further studies are needed since these previous models are more suited to the splat solidification on a flat solid surface. Therefore, the models need to be modified and extended to treat the incremental solidification in spray forming. The partially liquid surface layer, surface geometry and thermal conditions should be considered as special boundary conditions in future modeling studies. In addition, new methods for quantitative calculations of macroporosity and various microporosity should also be developed in the future.

V. SUMMARY

Over the past two decades, extensive modeling studies have been conducted in an effort to obtain a fundamental understanding of and control over the spray forming process. However, further modeling efforts are still needed in order to take full advantages of the spray forming process, such as its flexibility and potential to cost-effectively fabricate various shapes and novel materials that cannot be processed through conventional casting, forging or PM routes. It has become increasingly evident that further technological developments in spray forming, particularly in such challenging areas as computer-aided process design and control for high yield and low porosity with a variety of complex shapes and desired microstructural characteristics, necessitate a comprehensive understanding of the fundamental phenomena associated with each process stage, especially the deposition and consolidation stages. Some basic issues remain to be addressed, such as detailed mechanisms governing pore formation

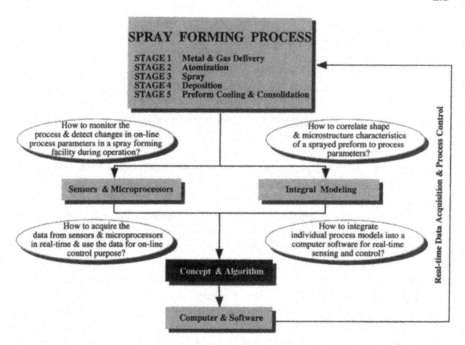

Figure 7 Basic concept of integral modeling and computer-aided control of spray forming.

and microstructural evolution in sprayed preforms, effects of different gases on atomization efficiency, and effective control over preform shape and microstructure.

By combining the individual models for each stage of spray forming, an integral model is to be developed to address the complete sequence of spray forming, i.e., metal and gas delivery, atomization, spray, deposition, preform cooling and consolidation, and concomitant microstructure evolution, as illustrated in Fig. 7. The implementation of the integral model is expected to allow us to ascertain the differences in the heat, mass, and momentum transfer characteristics between different atomization gases to determine the inherent dependence of microstructure of sprayed preforms on process parameters and to quantitatively calculate microstructural characteristics, such as grain size and porosity. Ultimately, the integral modeling results will be used to provide basic guidelines for the selection of optimum process parameters in order to satisfy the average porosity and grain size criteria required by a variety of applications and for the on-line control over mass distribution, deposition yield, and microstructural evolution in spray forming.

REFERENCES

1. AG Leatham, RG Brooks, JS Coombs, AGW Ogilvy. The past, present and future developments of the Osprey preform process. Proceedings 1st International Conference on Spray Forming, Osprey Metals, Ltd., Neath, UK, 1991.
2. AG Leatham, Y Kawashima. The past, present and future developments of the Osprey preform process. Proceedings 2nd Japan International SAMPE Symposium, Chiba, Japan, December 11–14, 1991, pp. 369–377.
3. JB Brennan. Formation of metal strip under controlled pressure. USA Patent No. 2,639,490 (1958); Apparatus and method for producing metal strip. USA Patent No. 2,864,137 (1958).
4. ARE Singer. The principles of spray rolling of metals. Metals and Materials 4:246–257, 1970.
5. ARE Singer. The challenge of spray forming. Powder Metall 25(4):195–200, 1982.
6. ARE Singer, RW Evans. Incremental solidification and forming. Metals Technol 10(2):61–68, 1983.
7. ARE Singer. Simultaneous spray deposition and peening of metals. Metals Technol 11:99–104, 1984.
8. ARE Singer. Metal matrix composites made by spray forming. Mater Sci Eng A135:13–17, 1991.
9. AP Newbery, B Cantor, RM Jordan, ARE Singer. Arc spray forming of nickel aluminides. Scripta Metall et Mater 27(7):915–918, 1992.
10. AP Newbery, RM Jordan, ARE Singer, B Cantor. Electric arc spray forming of an Ni_3Al based alloy. Scripta Metall et Mater 35(1):47–51, 1996.
11. RE Lewis, A Lawley. Spray forming of metallic materials: an overview. In: Powder Metallurgy in Aerospace and Defense Technologies, FH Froes, ed., Princeton, NJ: MPIF, 1991, pp. 173–184.
12. AG Leatham, AGW Ogilvy, L Elias. The Osprey process: current status and future possibilities. In: Powder Metallurgy in Aerospace, Defense and Demanding Applications, FH Froes, ed., Princeton, NJ: MPIF, 1993, pp. 165–175.
13. AG Leatham, A Lawley. The Osprey process: principles and applications. Intern J Powder Metall 29(4):321–329, 1993.
14. H Liu. Science and Engineering of Droplets: Fundamentals and Applications. Norwich, NY: William Andrew Publishing, 2000.
15. DRG Davies, ARE Singer. Spray forming by centrifugal spray deposition. In: Advances in Powder Metallurgy and Particulate Materials, vol. 1, JM Capus, RM German, eds., Princeton, NJ: APMI International, 1992, pp. 301–317.
16. H Liu, EJ Lavernia, RH Rangel. An analysis of freeze-up phenomena during gas atomization of metals. Intern J Heat Mass Transfer 38(12):2183–2193, 1995.
17. H Liu, DS Dandy. Modeling of liquid metal flow and heat transfer in delivery tube during gas atomization. Mater Sci Eng A197:199–208, 1995.

18. RD Payne, MA Matteson, AL Moran. Application of neural networks in spray forming technology. Intern J Powder Metall 29(4):345–351, 1993.

19. P Mathur, D Apelian, A Lawley. Analysis of the spray deposition process. Acta Metall Mater 37(2):429–443, 1989.

20. H Liu, RH Rangel, EJ Lavernia. Modeling of reactive atomization and deposition processing of Ni₃Al. Acta Metall Mater 42(10):3277–3289, 1994.

21. H Liu, RH Rangel, EJ Lavernia. Modeling of droplet-gas interactions in spray atomization of Ta-2.5W alloy. Mater Sci Eng A191(1–2):171–184, 1995.

22. S Annavarapu, RD Doherty. Evolution of microstructure in spray casting. Intern J Powder Metall 29(4):331–343, 1993.

23. S Annavarapu, D Apelian, A Lawley. Spray casting of steel strip-process analysis. Metall Trans 21A:3237–3256, 1990.

24. SJ Savage, FH Froes. Production of rapidly solidified metals and alloys. J Met 36(4):20–33, 1984.

25. J Baram. Structure and properties of a rapidly solidified Al-Li-Mn-Zr alloy for high-temperature applications, 2. Spray atomization and deposition processing. Metall Trans 22A(10):2515–2522, 1991.

26. H Liu, B Seuren. Spray atomization and deposition of steel. Second National Symposium, Society for the Advancement of Material and Process Engineering, eV, Wuppetal, Germany, Nov. 4, 1988.

27. K Bauckhage, H Liu, B Seuren, V Uhlenwinkel. Spray forming of liquid steel, local size and velocity distribution of particle in the spray cone and their reference to varying process parameters. Proceedings Powder Metallurgy Conference, PM'90, London, July 2–6, 1990, Institute of Metals, London, pp. 207–215.

28. H Liu, B Seuren, V Uhlenwinkel. On-line and in-line measurements in the spray compacting of liquid steel. Proceedings of 5th International Conference on Liquid Atomization and Spray Systems, K Bauckhage, ed., Bremen, Germany, July 1–4, 1989.

29. H Liu. Numerical modeling of the temperature and velocity fields of gas/droplets in a spray atomization deposition facility for steel. Ph.D. thesis, University of Bremen, Bremen, Germany, February 1990.

30. P Mathur, S Annavarapu, D Apelian, A Lawley. Spray casting—an integral model for process understanding and control. Mater Sci Eng A142(2):261–276, 1991.

31. RP Singh, A Lawley, B Friedman, YV Murty. Microstructure and properties of spray cast Cu-Zr alloys. Mater Sci Eng A145(?):243–255, 1991.

32. RH Bricknell. The structure and properties of a nickel-base superalloy produced by Osprey atomization-deposition. Metall Trans 17A:583–591, 1986.

33. T Harada, T Ando, RC Ohandley, NJ Grant. A microstructural study of a Nd₁₅Fe₇₇B₈ magnetic alloy produced by liquid dynamic compaction (LDC). Mater Sci Eng A133:780–784, 1991.

34. BP Bewlay, B Cantor. Gas velocity measurements from a close-coupled spray deposition atomizer. Mater Sci Eng A118:207–222, 1989.

35. DG Morris, MA Morris. Rapid solidification of Ni₃Al by Osprey deposition. J Mater Res 6(2):361–365, 1991.

36. X Zeng, H Liu, M Chu, EJ Lavernia. An experimental investigation of reactive atomization and deposition processing of Ni_3Al/Y_2O_3 using N_2-O_2 atomization. Metall Trans 23A:3394–3399, 1992.
37. H Liu, X Zeng, EJ Lavernia. Processing map for reactive atomization and deposition processing. Scripta Metall et Mater 29:1341–1344, 1993.
38. X Zeng, H Liu, EJ Lavernia. Reactive spray processing of metal matrix composites. Proceedings of 9th International Conference on Composite Materials, A Miravete, ed., Madrid, Spain, July 12–16, 1993, pp. 731–738.
39. A Lawley, D Apelian. Spray forming of metal matrix composites. Powder Metall 37(2):123–128, 1994.
40. A Leatham. Spray forming technology. Adv Mater & Processes 150(2):31–34, 1996.
41. JB See, JC Runkle, TB King. The disintegration of liquid lead streams by nitrogen jets. Metall Trans 4:2669–2673, 1973.
42. JB See, GH Johnston. Interactions between nitrogen jets and liquid lead and tin streams. Powder Technol 21:119–133, 1978.
43. SP Mehrotra. Mathematical modeling of gas atomization process for metal powder production, part 1. Powder Metall Intern 13(2):80–84, 1981.
44. SP Mehrotra. Mathematical modeling of gas atomization process for metal powder production, part 2. Powder Metall Intern 13(3):132–135, 1981.
45. H Lubanska. Correlation of spray ring data for gas atomization of liquid metals. J Metals 22:45–49, 1970.
46. A Ünal. Flow separation and liquid rundown in a gas-atomization process. Metall Trans 20B:613–622, 1989.
47. IE Anderson, RS Figliola. Observations of gas atomization process dynamics. In: Modern Developments in Powder Metallurgy 20, PU Gummeson, DA Gustafson, eds., Princeton, NJ: APMI International, 1988, pp. 205–223.
48. J Mi, RS Figliola, IE Anderson. A numerical simulation of gas flow field effects on high pressure gas atomization due to operating pressure variation. Mater Sci Eng A208:20–29, 1996.
49. DW Kuntz, JL Payne. Simulation of powder metal fabrication with high pressure gas atomization. In: Advances in Powder Metallurgy and Particulate Materials, Part I, M Phillips, J Porter, eds., Princeton, NJ: APMI International, 1995, pp. 63–77.
50. H Liu. Numerical simulation of gas atomization in spray forming process. 1997 TMS Annual Meeting & Exhibition, Orlando, FL, Feb. 9–13, 1997.
51. H Liu, EJ Lavernia, RH Rangel. Numerical simulation of substrate impact and freezing of droplets in plasma spray processes. J Phys D: Appl Phys 26:1900–1908, 1993.
52. H Liu, EJ Lavernia, RH Rangel. Numerical simulation of impingement of molten Ti, Ni and W droplets on a flat substrate in plasma spray processes. J Thermal Spray Technology 2(4):369–378, 1993.
53. H Liu, A Sickinger, E Mühlberger, EJ Lavernia, RH Rangel. Deformation and interaction behavior of molten droplets impinging on a flat substrate in plasma spray process. Conference Proceedings Thermal Spray Coatings, CC Berndt, TF

Bernecki, eds., ASM International, Anaheim, CA, June 7–11, 1993, pp. 457–462.

54. H Liu, EJ Lavernia, RH Rangel. Numerical investigation of micro-pore formation during substrate impact of molten droplets in plasma spray processes. Atomization and Sprays 4(4):369–384, 1994.

55. W Cai, H Liu, A Sickinger, E Mühlberger, D Bailey, EJ Lavernia. Low-pressure plasma deposition of tungsten. J Thermal Spray Technology 3(2):135–141, 1994.

56. H Liu, A Sickinger, E Mühlberger, EJ Lavernia, RH Rangel. Numerical investigation of micro-pore formation during substrate impact of molten droplets in spraying processes. Conference Proceedings Thermal Spray Industrial Applications, CC Berndt, S Sampath, eds., ASM International, Boston, MA, June 20–24, 1994, pp. 375–380.

57. H Liu, W Cai, RH Rangel, EJ Lavernia. Numerical and experimental study of porosity evolution during plasma spray deposition of W. NATO Workshop on Science and Technology of Rapid Solidification and Processing, MA Otooni, ed., West Point, NY, June 21–24, 1994, pp. 73–107.

58. H Liu, EJ Lavernia, RH Rangel. Modeling of molten droplet impingement on a non-flat surface. Acta Metall Mater 43(5):2053–2072, 1995.

59. H Liu, EJ Lavernia, RH Rangel. Modeling of molten droplet impingement on a non-flat surface. ASME Winter Annual Meeting, Symposium on Multiphase Flow and Heat Transfer in Materials Processing, M Chen, ed., Chicago, IL, November 6–12, 1994, pp. 1–5.

60. H Liu. Influence of droplet/substrate interactions in thermal spraying on microstructure of thermal coatings. 95' Beijing International Conference for Surf. Sci. and Eng., R Zhu, ed., International Academic Publishers, Beijing, China, May 15–19, 1995, pp. 130–134.

61. JP Delplanque, EJ Lavernia, RH Rangel. Numerical investigation of multi-phase flow induced porosity formation in spray deposited materials. 1997 TMS Annual Meeting & Exhibition, Orlando, FL, Feb. 9–13, 1997.

62. U Fritsching, V Uhlenwinkel, K Bauckhage, U Urlau. Gas und Partikel strömungen im Düsennahbereich einer Zweistoffdüse—Modelluntersuchungen zur Zerstäubung von Metallschmelzen. Chemie Ingenieur Technik 62(2):146–147, 1990.

63. U Fritsching, H Liu, K Bauckhage. Numerical modeling in the metal spray compacting process. Proceedings of 7th International Conference on Liquid Atomization and Spray Systems, HG Semerjian, ed., Gaithersburg, MD, July 15–18, 1991, pp. 491–498.

64. SD Ridder, SA Osella, PI Espina, FS Biancaniello. Intelligent control of particle size distribution during gas atomization. Intern J Powder Metall 28(2):133–147, 1992.

65. V Uhlenwinkel, U Fritsching, K Bauckhage, U Urlau. Strömungsuntersuchungen im Düsennahbereich einer Zweistoffdüse—Modelluntersuchungen für die Zerstäubung von Metallschmelzen. Chemie Ingenieur Technik 62(3):228–229, 1990.

66. J Ting, R Terpstra, IE Anderson, RS Figliola, J Mi. A novel high pressure gas atomizing nozzle for liquid metal atomization. In: Advances in Powder Metallurgy & Particulate Materials, vol. 1, part 1, TM Cadle, KS Narasimhan, eds., Princeton, NJ: MPIF and APMI International, 1996, pp. 97–108.

67. SA Miller, RS Miller. Real time visualization of close-coupled gas atomization. In: Advances in Powder Metallurgy & Particulate Materials, vol. 1, JM Capus, RM German, eds., Princeton, NJ: APMI International, 1992, pp. 113–125.

68. SP Mates, GS Settles. A flow visualization study of the gas dynamics of liquid metal atomization nozzles. In: Advances in Powder Metallurgy & Particulate Materials, part I, M Phillips, J Porter, eds., Princeton, NJ: APMI International, 1995, pp. 15–29.

69. BP Bewlay, B Cantor. Modeling of spray deposition: measurements of particle size, gas velocity, particle velocity, and spray temperature in gas-atomized sprays. Metall Trans 21B:899–912, 1990.

70. BP Bewlay, B Cantor. The relationship between thermal history and microstructure in spray-deposited tin-lead alloys. J Mater Res 6(7):1433–1454, 1991.

71. G Schulz. Laminar sonic and supersonic gas flow atomization — the NANOVAL process. In: Advances in Powder Metallurgy & Particulate Materials, vol. 1, part 1, TM Cadle, KS Narasimhan, eds., Princeton, NJ: MPIF and APMI International, 1996, pp. 43–54.

72. SM Sheikhaliev, JJ Dunkley. A novel internal mixing gas atomiser for fine powder production. In: Advances in Powder Metallurgy & Particulate Materials, vol. 1, part 1, TM Cadle, KS Narasimhan, eds., Princeton, NJ: MPIF and APMI International, 1996, pp. 161–170.

73. C San Marchi, H Liu, A Sickinger, E Mühlberger, EJ Lavernia, RH Rangel. Numerical analysis of the solidification of a single droplet by spray processing. J Mat Sci 28:3313–3321, 1993.

74. K Bauckhage, H Liu, U Fritsching. Models for the transport phenomena in a new spray compacting process. Proceedings of 4th International Conference on Liquid Atomization and Spray Systems, Tohoku Univ. Sendai, Japan, August 22–24, 1988, The Fuel Society of Japan, pp. 425–430.

75. C Tornberg. Particle size prediction in an atomization system. In: Advances in Powder Metallurgy & Particulate Materials, vol. 1, JM Capus, RM German, eds., Princeton, NJ: APMI International, 1992, pp. 137–150.

76. KP Rao, SP Mehrotra. Effect of process variables on atomization of metals and alloys. In: Modern Developments in Powder Metallurgy, Principles and Processes, vol. 12. Princeton, NJ: MPIF and APMI International, 1981, pp. 113–130.

77. FS Biancaniello, JJ Conway, PI Espina, GE Mattingly, SD Ridder. Particle size measurement of inert-gas-atomized powder. Mater Sci Eng A124:9–14, 1990.

78. H Eroglu, N Chigier, Z Farago. Coaxial atomizer liquid intact lengths. Phys Fluids A 3(2):303–308, 1991.

79. KE Knoll, PE Sojka. Flat-sheet twin-fluid atomization of high-viscosity fluids. Part I: Newtonian liquids. Atomization and Sprays 2:17–36, 1992.

80. PK Wu, GA Ruff, GM Faeth. Primary breakup in liquid-gas mixing layers. Atomization and Sprays 1:421–440, 1991.
81. PJ Thomas. On the influence of the basset history force on the motion of a particle through a fluid. Phys Fluids A 4(9):2090–2093, 1992.
82. CB Henderson. Drag coefficients of spheres in continuum and farefied flows. AIAA J 14(6):707–708, 1976.
83. LP Hsiang, GM Faeth. Near-limit drop deformation and secondary breakup. Intern J Multiphase Flow 18(5):635–652, 1992.
84. GM Trapaga. Gas-particle-deposit interactions during plasma spraying and spray forming processes. Sc.D. thesis, MIT, Cambridge, MA, September 1990.
85. PS Grant, B Cantor, L Katgerman. Modeling of droplet dynamic and thermal histories during spray forming–I. Individual droplet behavior; II. Effect of process parameters. Acta Metall Mater 41(11):3097–3118, 1993.
86. V Uhlenwinkel, K Bauckhage. Mass flux profile and local particle size in the spray cone during spray forming of steel, copper and tin. Proceedings of 2nd International Conference on Spray Forming, Osprey Metals Ltd., Swansea, UK, 1993, pp. 25–34.
87. CYA Tsao, NJ Grant. Modeling of the liquid dynamic compaction spray process. Intern J Powder Metall 30(3):323–333, 1994.
88. C Cai. A modeling study for design and control of spray forming. PhD thesis, Drexel University, Philadelphia, PA, June 1995.
89. V Uhlenwinkel, C Kramer, K Bauckhage, J Ulrich. Experimental investigation of the sticking efficiency during spray forming. In: Advances in Powder Metallurgy & Particulate Materials, vol. 1, part 9, TM Cadle, KS Narasimhan, eds., Princeton, NJ: MPIF and APMI International, 1996, pp. 29–40.
90. M Berg, J Ulrich. Experimental based detection of the splash limits for the normal and oblique impact of molten metal particles on different surfaces. 1997 TMS Annual Meeting & Exhibition, Orlando, FL, Feb. 9–13, 1997.
91. FH Harlow, JP Shannon. The splash of a liquid drop. J Appl Phys 38(10):3855–3866, 1967.
92. M Kitaura, M Yao, J Senda, H Fujimoto. Numerical calculation on deforming behavior of a single droplet impinging upon a flat surface. Proceedings of 13th Japanese Conference on Atomization of Liquids, 1984, pp. 73–78.
93. J Madejski. Solidification of droplets on a solid surface. Intern J Heat Mass Transfer 19:1009–1011, 1976.
94. G Trapaga, J Szekely. Mathematical modeling of the isothermal impingement of liquid droplets in spraying processes. Metall Trans 22B:901–914, 1991.
95. G Trapaga, EF Matthys, JJ Valencia, J Szekely. Fluid flow, heat transfer, and solidification of molten metal droplets impinging on substrates—comparison of numerical and experimental results. Metall Trans 23B:701–718, 1992.

Index

Advanced powder metallurgy, 8
Air gap, 62, 383, 503, 577
Alpha case, 106, 364, 366

Beta flecks, 596–597
Boundary conditions, 24–27, 45,
    314–315, 552–553
  at metal/mold interface, 25–26
  at metal surface, 26
  at mold exterior surface, 24–25

Casting modeling procedures, 42–52
  a. building geometry and meshing, 43
  b. assigning material properties, 44
  c. specifying initial conditions, 44
  d. specifying boundary conditions, 45
  e. selecting control parameters, 46–48
  f. selecting computational model,
     48–49
  g. running the simulation, 49–50
  h. performing post processing, 50–52
Cast products, 3
Channel segregates, 592, 601–602
Chvorinov's rule, 111, 247
CINDAS, 191
Coefficient of thermal expansion,
    203–205
Coherency point, 69, 156

Columnar-to-equiaxed transition, 127,
    138–141
Continuous casting, 5, 499–539
  basic phenomena of, 500–503
    air gap, 503
    argon gas injection, 501
    breakout, 501
    mold powder, 502
    oscillation marks, 503
    periodic oscillation, 502–503
    spray zones, 503
    submerged entry nozzle, 500
  composition variation during grade
     changes, 519–521
  consequences of mold fluid flow,
     510–519
    inclusions and gas bubbles,
      516–519
    particle trajectories, 517
    superheat dissipation, 510–512
    top-surface powder/flux layer
      behavior, 515–516
    top-surface shape and level
      fluctuation, 512–515
  crack prediction, 531–534
    CONZD, 532
    flow strain, 534
    hot tears, 532

[Continuous casting]
flow through submerged entry nozzle,
 504–506
fluid flow in mold, 506–510
 effect of argon gas injection,
  507–509
 effect of electromagnetic forces,
  509
 transient flow behavior, 409–510
longitudinal surface depressions,
 529–531
mold thermal mechanical behavior,
 524–526
process description of, 499–500
shell thermal mechanical behavior,
 524–526
solidification and mold heat transfer,
 521–524
Convection, 36–38
Conventional ingot casting, 5
Conventional powder metallurgy, 8
Crack susceptibility coefficient, 63
Criteria functions, 126–127
Critical solid fraction, 108

D'Arcy's law, 31, 114, 474
Data exchange, 270–280
 binary and character formats,
  279–280
 causes of translation problems,
  275–277
 incompatibilities, 277–278
 native file formats, 270–271
 neutral formats, 272–274
 point-to-point translation, 272–274
 translators, 271
Defect formation, 95–122, 304–311,
  348–367, 544–545, 557–559,
  590–605
Defect map, 350–351
Defect types, 96–106
 micro/grain structure related, 103–104
  dendrite arm spacing, 104
  microstructure morphology, 104
  thermal parameter, 104
 mold/core related, 104–106

[Defect types]
 alpha case, 106
 mold/filling related, 96–99
  entrapped gas, 97–98
  no-fill, 97
  weld line, 98–99
 segregation related, 101
  dendrite arm spacing, 101
  exogenous, 101
  indigenous, 101
  macrosegregation, 101
  microsegregation, 101
 shrinkage and porosity, 99–100
  gas porosity, 100
  macroshrinkage, 99
  microporosity, 99
 stress related, 102–103
  cold cracking, 103
  distortion, 102
  hot tearing, 103
  recrystallized grains, 103
Dendrite arm spacing (DAS), 101, 104,
  115, 127, 352–355
Dendrite growth, 150–158
 dendrite growth velocity models, 151
 top velocity models, 151–153
  CALPHAD, 152
  columnar growth, 152
  equiaxed dendrite, 152
 volume averaged dendrite models,
  153–158
  coherency, 156
  solid fraction, 155
Density, 205–208
 of elements, 207
Deterministic modeling, 127–130,
  159–180, 350–352
 aluminum-silicon alloys, 176–178
 cast iron, 162–175
 steel, 159–162
 superalloys, 178–180
  eutectic reaction, 180
  Laves phases, 178–179
Die casting, 4, 391–416
 commercial simulation software for,
  392–396

[Die casting]
    ABAQUS, 393–396
    ANSYS, 393–396
    capabilities and limitations of,
      396–397
    2D-BEM, 393–396
    DieCas, 393–396
    EKK, 393–396
    issues in application of, 397–400
    MAGMAsoft, 393–396
    Meltflow, 393–396
    ProCast, 393–396
    die deflection in, 400–405
      clamping forms, 405
      injection pressure, 405
      lubricant spray, 403
      thermal analysis, 402–403
    problems of, 393
Diffusion, 226–232
    Arrhenius equation, 228
    Fich's first law of, 227
    interstitial, 228
    self diffusivity of pure metals, 230, 231
    substitutional, 228
Direct chill casting, 6, 541–563
    coupled thermal, fluid and stress
      simulation, 548–559
    ALSIM, 551
    ALSPEN, 551
    modeling concepts, 549–550
    thermal boundary conditions,
      552–553
    defects, 544–545
      butt curl, 545, 557–559
      butt swell, 545
      center cracks, 544
      pull-in effect, 545
    heat transfer, 546–548
      to water film, 546–547
      between ingot and mold, 548
      to starting block, 548
    melt distribution system, 545–546
Distortion (*see* Stress analysis), 59

EBM and PAM, 613–653
    application of, 616–619

[EBM and PAM]
    Hearth Only, 618–619
    Hearth + VAR, 618
    helium/argon porosities, 619
    heating source of, 619–625
    dimple depth, 624
    electron beam, 619
    helium transport properties,
      624–625
    plasma arc, 619–625
    ingot solidification, 639–649
    characteristics of heat input,
      640–644
    chemistry control, 644–647
    element evaporation, 647
    grain structure, 647–648
    helium/argon properties, 619, 648
    modeling approach, 648–649
    pool profile, 640–644, 647–648
    pour lip, 640
    surface quality, 641–644, 647–648
    process description of, 614–616
    hearth melting (HM), 614
    inclusion removal, 615
    Marangoni forces, 615
    skull, 615
    refining hearth of, 625–639
    EBM hearth, 625–629
    electromagnetic (Lorentz) forces,
      633
    inclusion transport, 636–639
    k-ε model, 633
    PAM hearth, 629–636
    plasma jet impingingment, 629
    surface depression (dimple), 629
Electrical conductivity, 214–217
    electrical resistivities of metals, 217
Electromagnetic field, 570, 581–582, 623
Electron beam melting (EBM), 7,
    613–653
Electronic data interchange, 263–290
    CAD/CAM, 263–265
    definition, 264
    algorithm, 264
    application, 264
    data structure, 264

[Electronic data interchange]
  entity, 264
  product data, 264
Electronic representation of geometry,
  265–270
  boundary representation, 267–269
  classes of representation, 267
  computer representation of numbers,
   266–267
  constructive solid geometry, 269
  hybrid representations, 269–270
  parametric representations, 269
Electroslag remelting (ESR), 7, 565–612
Emissivity (*see* Radiation heat loss),
  220–226
  black body radiator, 223
  of alloys and refractory materials,
   226
  of prestine and oxidized metals, 225
  Plank's constant, 223
  Stefan-Boltzman equation, 220, 223
Endogenous defects, 590–602
Eutectic growth, 158–159
Exogenous defects, 602–605

Finish machining, 78
Fluid dynamics, 27–32, 344–345,
  373–378, 582–585
  back pressure, 30–31
  free surface, 30
  inflow, 32
  mold filling, 30–32, 50
Freckles, 361–363, 592–596, 600–602

Heat capacity, 191
Heat transfer, 18–21, 312–315, 341–344,
  546–548, 576–581
  by conduction, 18–19
  by convection, 19–20, 36–38
  by radiation, 20–21, 220–226
Heat transfer coefficient, 383, 385,
  579–580
High density inclusions (HDI), 604–605
Hot tears (*see* Stress analysis), 59, 63,
  308–309
Hot top, 570

Initial conditions, 24, 44
Interdendritic fluid flow, 592, 597–600
Investment casting, 3, 333–372
  application of, 338–340
   columnar grains, 338
   directionally solidified (DS), 338
   equiaxed grains (EQ), 338–339
   single crystal (SX), 338
   structural castings, 338–340
   turbine airfoils, 338
  inverse modeling of, 367–368
  microstructure related defects in,
   348–367
   alpha case, 364, 366
   as-cast gamma prime size, 352
   bigrains, 358
   CA-FE/FD, 355
   chill grains, 355–357
   defect map, 350–351
   dendrite arm spacing, 352–355
   deterministic modeling, 350–352
   freckles, 361–363
   grain morphology, 355–356
   high angle boundaries, 358
   low angle boundaries, 358
   metal ceramic reaction, 363–364
   misoriented dendrites, 359–360
   multigrains, 358
   qualification specification, 348
   recrystallized grains, 366–368
   slivers, 363–364, 366
   stochastic modeling, 350–352
   zebra grains, 360–361
  modeling of, 340–347
   fluid flow, 344–345
    no-fill, 306–308, 344
   heat transfer, 341–344
    circular symmetry, 342
    kaowool wrap, 343–344
    radiation, 341–344
    view factor, 342
   mold generation and meshing,
    340–341
    mold extrusion, 341
   thermal stress, 345–347
  process, 334–338

[Investment casting]
  centrifuge casting, 337
  copper chill, 334–338
  dynamic mode, 337
  grain selector, 337
  graphite susceptor, 337
  static mode, 337
  thermally controlled solidification,
    337–338
  withdrawal mode, 337
  wax injection in, 369–370

Latent heat of fusion/solidification,
  199
Laves phase, 178–179, 588–590
Length scales, 123–124
  macroscale, 123
  microscale, 123
  nanoscale, 124
Level rule, 147–149
Liquid film, 63
Liquidus temperature, 202
Lorentz force, 570, 581–582, 633
Lost foam casting, 3, 317–331
  advantages and difficulties of,
    317–318
  modeling of, 320–330
    contact-noncontact transition,
      325–326
    contact situation, 324–325
    interface of liquid metal/EPS
      pattern, 321–326
    noncontact situation, 326
    physical phenomena involved in,
      319–320
  process, 317
    evaporative pattern, 317
    expanded polystyrene, 317
    polymethyl methacrylate, 317

Macrosegregation, 101, 310, 590–605
Macroshrinkage prediction, 99–100,
  106–110
  by temperance field, 106–107
  for pipe location and volume, 107–110
Mechanical constrain, 59

Microporisity prediction, 99–100,
  110–120
  by comprehensive analysis, 114–117
    D'Arcy's law, 114
    gas caused porosity, 114
    permeability, 114–115
    primary dendrite arm spacing,
      115
    Sievert's law, 115
  by criteria functions, 117–120
    feeding resistance number, 119
    LCC, 118–120
    Niyama criterion, 117
  by minimum temperature gradient,
    112–113
  by total solidification time, 111–112
    Chvorinov's rule, 111
    modulus, 111
Microsegregation, 101, 146–150, 310
  level rule, 147–149
  Scheil equation, 147–150
  solute redistribution, 146–150
Microstructure evolution, 123–187
Microstructure modeling, 126–127,
  138–141
  by criteria functions, 126–127
    columnar-to-equiaxed transition,
      127, 138–141
    dendrite arm spacing, 127
    gray-to-white transition, 127
    solid-liquid interface stability, 127
  by deterministic modeling, 127–130
    nucleation, 129, 141–146
    one velocity models, 128–129
    solid fraction evolution, 130
    transformation kinetics, 127–130
    two velocity models, 129–130
  by probalistic modeling, 130–141
    cellular automation, 132–133,
      136–141
    cellular-to-equiaxed transition,
      137–138
    Monte Carlo, 130–132, 133–136
    stochastic model, 136–141
Modeling software, 40–42
  FDM, 41

[Modeling software]
  FEM, 41
  FVM, 41
  quick analysis, 42
Modulus (see Quick analysis), 111
Mold filling, 30–32, 50, 258–260,
    311–312
Mold removal, 57–60, 77–78
Mold rigidity, 85–87

Neutral formats, 280–289
  IGES, 282–283
  STEP, 283–289
    application protocol for casting,
      288–289
  STL, 281–282
Nucleation, 141–146
  continuous, 141–146
  dynamic nucleation model, 145–146
  heterogeneous, 141–145
  homogeneous, 141
  instantaneous, 142

Patternmakers' allowance, 79
Permanent mold casting, 4, 373–390
  flow analysis of, 373–378
    low pressure casting, 376–378
    permanent/semi permanent mold
      casting, 374
    tilt pour casting, 374–376
  solidification analysis of, 378–390
    artificial gaps, 383
    cooling channels, 383
    cyclic analysis, 383–386
    heat dams, 383
    hot oil lines, 383
    interfacial heat transfer coefficient,
      383, 385
    mold surface coating, 383, 385, 387
Permeability, 114–115, 474–475, 598
Plasma arc melting (PAM), 613–653
Post processing, 50–52
  cooling curves, 51
  cooling rate, 51
  isochrons, 51
  isotherms, 50–51

[Post processing]
  local solidification time, 52
  mold filling, 50
  solidification rate, 52
  temperature gradient, 52
Powder metallurgy products, 8
Probabilistic modeling, 130–141

Quick Analysis, 239–262
  geometric modulus, 247–248
    Chvorinov's rule, 247
    point modulus, 248
    section modulus, 247–248
    volume-to-surface area ratio, 247
  mold filling, 258–260
    Bernoulli approach, 258–259
    mixed approach, 259–260
  solidification times, 240–247
    in plastic injection molding, 244
    in sand casting, 241–243
    in spray casting, 243–244
    with competing heat transfer
      modes, 244–247
  thermal modulus, 249–257

Radiation heat loss, 20–21, 73–74,
    220–226, 341–344
  view factor, 74
Residual stress (see Stress analysis), 80

Sand casting, 3, 291–316
  defects of, 304–311
    gas porosity, 305–306
    hot tears and cracks, 308–309
    inclusions, 310–311
    lack of fill, 306–308, 344
    laps, 308
    macrosegregation, 309
    microsegregation, 310
    mold erosion, 310
    solidification shrinkage, 304–305
  features of, 292–304
    ceramic insert, 298
    chills, 300–301
    cores, 302
    dead weight, 303–304

[Sand casting]
  exothermic material, 298
  gating system, 294–296
  insulation, 301
  metalostatic pressure, 303–304
  pattern, 292–294
  risers, 297–300
  sand, 303
  vents, 301
  heat transfer of, 312–315
    casting / mold interface, 313–314
    exterior boundary conditions,
      314–315
    thermal properties of sand, 313
  mold filling of, 311–312
    general rules, 312
    turbulence, 311
Semi-solid metal working (SSM), 4,
  417–498
  design of robot transfer grips, 483–484
  design of shot sleeves, 484–486
  high solid fraction processing,
    473–477
    D'Arcy equation, 474
    permeability, 474–475
  induction heating of SSM billets,
    478–483
  intermediate solid fraction, 477–478
  material models, 430–473
    apparent viscosity, 433–435
    BBH model, 441–443
    Bingham fluid model, 446–450
    exponential decay model, 463–464
    general liquid suspensions, 432–433
    grain growth and coalescence,
      471–472
    internal variable viscosity model,
      450–463
    Kapranos model, 443
    Kattimis viscosity model, 440–441
    Modigell model, 467–471
    shear rate dependent model,
      436–440
    solid/liquid segregation, 464,
      472–473
    SSM slurries, 433–473

[Semi-solid metal working (SSM)]
  Xu model, 464–467
  process route, 425–430
    electromagnetic stirring, 425
    feedstock materials, 425
    rheocasting, 429–430
    stress induced melt activated
      (SIMA), 426–427
    thixocasting, 427
    thixotropic behavior, 419
Sheil equation, 147–150
Simulation control parameters, 46–48
  convergence criterion, 47–48
  data output, 48
  relaxation factor, 48
  time step size, 47
  total simulation time, 46–47
Solid fraction, 21 –22, 65–69, 108
  semiempirical model, 67–69
  solidification kinetics model, 65–67
Solidification, 21–27
  apparent specific heat, 23
  fraction of solid, 21–22
  latent heat of fusion, 22
Solidus temperature, 202
Specific heat, 191–199
  differential scanning calorimetry, 195
  Kopp-Neuman rule of mixtures,
    197–198
  of commercial alloys and mold,
    198–199
Spray forming, 9, 655–694
  modeling of, 673–687
    stage 1–metal and gas delivery,
      675–678
    stage 2–atomization, 678–679
    stage 3–spray, 679–680
    stage 4–deposition, 680–684
    stage 5–preform cooling and
      consolidation, 684–687
  phenomena and principles in, 663–673
    atomization mechanisms, 663–666
    deposit cooling rate, 667
    droplet cooling rate, 666
    impact deformation, 669
    microstructure evolution, 667–669

[Spray forming]
    porosity formation, 669–672
    process parameters, 673
    rapid solidification, 666–667
   process and product of, 656–663
    atomization spray deposition, 657
    near net-shape manufacturing, 656
Squeeze casting, 4
Stochastic modeling, 136–141,159–182
   aluminum alloys, 181
   cast iron, 181
   superalloys, 182
Strain (see Stress analysis), 63–64
   creep, 70
   mechanical, 63–64
   plastic, 70
   thermal, 63
Stress analysis, 55–93, 345–347
   basic concepts, 57–60
    after mold removal, 60
    before mold removal, 57–60
   heat loss at mold surface, 73–75
   interaction between mold and casting,
    71–73
   mushy state, 62–69
   solid state phase transformation,
    75–77
   thermal-elastic-plastic-creep material
    model, 69–71
   transient thermomechanical analysis,
    61–62
Surface tension, 208–211
   of pure metals, 212

Thermal stress (see Stress analysis),
   345–347
Thermophysical properties, 189–237,
   313
   coefficient of thermal expansion,
    203–205
   density, 205–208
   diffusion, 226–232
   electrical conductivity, 214–217
   emissivity, 220–226
   heat capacitiy, 191
   latent heat of fusion/solidification, 199

[Thermophysical properties]
   liquidus temperature, 202
   solidus temperature, 202
   specific heat, 191–199
   surface tension, 208–211
   viscosity, 211–214
Time integration, 38–39
Transforming plasticity, 76–77
Type I hard alpha, 604–605
Type II hard alpha, 597

Vacuum arc remelting (VAR), 7,
   565–612
VAR and ESR, 565–612
   applications of, 571–572
    2X VAR, 572
    3X VAR, 572
    VIM/VAR double melt, 571
    VIM/ESR/VAR triple melt, 571
   electrode melting, 574–575
   heat generation, 572–574
    energy partition, 573–574
   ingot solidification, 575
    electromagnetic field, 581–582
    fluid flow, 582–585
    gap width, 577
    grain structure, 586–587
    heat transfer, 576–581
    heat transfer coefficient, 579–580
    helium pressure, 577
    Laves phase, 588–590
    N_bC carbides, 588–590
    reverse polarity, 576
    slag skin thickness, 577
    U-shaped pool, 587–589
    V-shaped pool, 587–589
   macrosegregation defects, 590–605
    beta flecks, 596–597
    channel segregates, 592, 601–602
    endogenous defects, 590–602
    exogenous defects, 602–605
    freckles, 592–596, 600–602
    high density inclusions (HDI),
     604–605
    interdendritic fluid flow, 592,
     597–600

[VAR and ESR]
    permeability, 598
    remnant electrode, 604
    shelf, 603–604
    sonic defects, 602
    type I hard alpha, 604–605
    type II hard alpha, 597
    white spots, 602
  process description, 567–571
    consumable electrodes, 568–569
    electromagnetic body force, 570
    hot-top region, 570
    liquid metal pool, 570
    Lorentz force, 570

[VAR and ESR]
  slag, 568–569
  start-up region, 570
  steady state melting region, 570
  vacuum arc, 568–569
Viscosity, 211–214, 430–473, 633
  Andrade's relationship, 211, 213
  k-ε model, 214, 633
  of pure metals, 215
  poise, 211
Volume fraction change, 109

White spots, 602–604
Wrought products, 5